Birkhäuser

Daniel Alpay

A Complex Analysis Problem Book

Second Edition

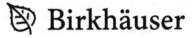 Birkhäuser

Daniel Alpay
Department of Mathematics
Ben-Gurion University of the Negev
Beer Sheva, Israel

Department of Mathematics
Chapman University
Orange, CA, USA

ISBN 978-3-319-42179-7   ISBN 978-3-319-42181-0   (eBook)
DOI 10.1007/978-3-319-42181-0

Library of Congress Control Number: 2016957416

Mathematics Subject Classification (2010): 97I80, 46C07, 46E20, 46E22, 47B32

First Edition © Springer Basel AG 2011
© Springer International Publishing AG 2016

Printed on acid-free paper

This book is published under the trade name Birkhäuser, www.birkhauser-science.com
The registered company is Springer International Publishing AG
The registered company address is: Gewerbestrasse 11, 6330 Cham, Switzerland

It is a pleasure to thank the Earl Katz family for endowing the chair (Earl Katz Family Chair in Algebraic System Theory), which supported this research. It is also a pleasure to dedicate this work to my daughter Anaëlle.

Contents

Part II   Functions of a Complex Variable

Part III   Applications and More Advanced Topics

Part IV Appendix

Prologue

Prologue

The topic of this book is the theory of complex-valued functions of a complex variable, defined on an open subset Ω of the complex plane $\mathbb{C}$ and which admit a derivative, or equivalently are $\mathbb{C}$-differentiable, at every point $z_0 \in \Omega$:

$$\forall z_0 \in \Omega, \quad \exists \lim_{z \to z_0} \frac{f(z) - f(z_0)}{z - z_0}.$$

Such functions bear various names: They are said to be *holomorphic* in Ω, or *analytic* in Ω; the terms $\mathbb{C}$-differentiable, differentiable and regular are also used. A key result in the theory is the equivalence between *holomorphicity* in an open set Ω and *analyticity* of the function there, that is, the existence of a power series expansion

$$f(z) = \sum_{n=0}^{\infty} f_n(z - z_0)^n \tag{0.0.1}$$

in a neighborhood of every point $z_0 \in \Omega$. In the first chapters we will make a distinction between the terms *holomorphic* and *analytic*. After the discussion of Cauchy's theorem we will use these terms interchangeably, and mostly use the term *analytic*. We also use mainly the term *analytic* in the discussion below.

The audience we have in mind consists of undergraduate students from mathematics and electrical engineering, with an eye on advanced students from both tracks. Analytic functions are the bread and butter of mathematicians. For engineers, analytic functions appear everywhere, in particular in the theory of linear systems, signal processing, circuit theory, sampling theorems, optimal control, to name a few. For instance, a motivation for an engineering student would be to know that transfer functions of discrete-time shift-invariant dissipative linear systems are functions analytic in the open unit disk,[1]and bounded by one in modulus there (the celebrated Schur functions); see for instance the book [6]. Unfortunately, most, if not all, electrical engineering students do not know what a transfer function is when they begin studying the theory of analytic functions. For the convenience of engineering students we give, in the second part of this prologue, a short discussion of time-invariant linear bounded systems, and their connections to the theory of analytic functions.

The book consists of four parts. The first two parts, respectively entitled *Complex numbers* and *Functions of a complex variable*, form the bulk of the book. Most of the exercises presented in these two parts have been given in the past years by the author in classes on *Introduction to the theory of functions of a complex variable* for second year electrical engineering students, and *Theory of functions of a complex variable* for mathematics students, at the department of mathematics at Ben-Gurion University. The exercises rely only on classical real analysis, but sometimes we use measure theory (mainly via the dominated convergence theorem) to avoid lengthy arguments. Study of some special Hilbert spaces of analytic functions is also scattered in the text, and requires some elementary functional analysis (these spaces are the main topic of the sequel [7] to the present book). When studying a function analytic in a domain (for instance in the open unit disk), we will usually (but not always) assume that it is analytic in a neighborhood of the closure of the domain, to avoid problems with boundary values (for instance, in the case of the open unit disk, we will assume analyticity in $|z| < 1 + \epsilon$ for some $\epsilon > 0$). The student will in particular meet in the second part of the book, in simplified forms, Bohr's inequality and the Herglotz integral representation of a function analytic in the open unit disk, and with a real positive part there. See Exercises 5.5.13, 5.5.8 and 5.5.10.

The third part, entitled *Applications and more advanced topics*, contains more advanced material, which was taught by the author to graduate students and also to undergraduate students from the double major program *mathematics and electrical engineering* at Ben-Gurion University. Topics include harmonic functions, conformal mappings, a brief introduction to the theory of linear systems, the related topic of rational functions, and a chapter on special functions and transforms.

In a book in preparation, which can be seen as a sequel to both the present book and to [7], we hope to come back to these topics, and also discuss, via exercises, various aspects function theory in the settings of several complex variables, Riemann surfaces and quaternionic analysis.

The fourth part, entitled *Advanced prerequisites*, contains material from real analysis, topology, functional analysis and measure theory, which are needed to solve the exercises (and, in fact, to fully understand a first course on complex variables). Since we mention in the text a number of Hilbert spaces of analytic

---

[1]or in a half-plane, for continuous-time systems

functions, we also have taken the liberty of mentioning the definition of a reproducing kernel Hilbert space.

For the convenience of the reader, we give in the first three parts of the book a
number of reminders of known facts from complex analysis, mostly without proofs,
in the text. The solutions of most of the exercises are presented at the end of the
chapter where they are given.

In the first weeks of a first course on complex analysis, motivations and
applications of the theory are not apparent. Moreover, some results *look like real
variable calculus.* One of the difficulties for students who take a complex variable
course is that the complex derivative obeys the same rule as the familiar derivative
from real analysis. Moreover, the familiar power series of $\sin x, \cos x, \ldots$ pop up,
and it is not clear what the novelty is. After a number of weeks into the course, the
student finally sees the proof that a function which is $\mathbb{C}$-differentiable in an open
subset of the complex plane admits derivatives of all order, and in fact, admits a
power series expansion around every point: The function is *analytic* in the given
open set. The student needs to be somewhat patient, to understand slowly the
differences between real and complex analysis.

To help the student cope with the difficulties mentioned in the previous
paragraph, one approach, sometimes taken by the author, is to skip most of the
preliminary material on complex numbers, discuss quickly the notions of continuity
and rush to the Cauchy–Riemann equations. One can then already define the
exponential function as

$$e^z = e^x(\cos y + i \sin y), \quad z = x + iy \in \mathbb{C},$$

and proceed.

In the present book, we have chosen a slower, and maybe non-standard path
for our exposition. We devote the three first chapters of the book to exercises
on complex numbers, or complex functions, but without mentioning analyticity.
There, the students already meet a variety of functions, such as Blaschke products, the Weierstrass sigma function, and the representation of $\sin z$ as an infinite
product. The definition and construction of these functions can be realized without using analyticity. Later in the book, the student will see that these are key
examples of analytic functions. Of course, such an approach delays the exercises
on analytic functions *per se*, but we think this gives time to the students to absorb
at their own pace these difficult examples.

Trying to prove the following formulas using real analysis might provide a
student motivation to study complex analysis:

$$\sum_{n=1}^{\infty} r^n \sin(n\theta) = \frac{r \sin\theta}{1 + r^2 - 2r \cos\theta}, \quad r \in [0, 1), \quad \theta \in \mathbb{R},$$

$$z^{2n} + 1 = \prod_{k=0}^{n-1} \left(z^2 - 2z \cos\left(\frac{2k+1}{2n}\pi\right) + 1 \right),$$

$$\sum_{k=1}^{m} \cot^2 \left(\frac{k\pi}{2m+1} \right) = \frac{m(2m-1)}{3}, \quad m = 1, 2, \ldots,$$

$$\int_{\mathbb{R}} \cos(t^2) dt = \sqrt{\frac{\pi}{2}},$$

$$\sum_{n=0}^{\infty} \frac{\binom{2n}{n}}{7^n} = \sqrt{\frac{7}{3}}.$$

These formulas are readily proved using complex analysis methods: The first identity is easily proved by a purely real analysis method: Multiply both sides by the denominator $1 + r^2 - 2r \cos \theta$, and use the trigonometric identity

$$\sin(n\theta) \cos \theta = \frac{\sin(n+1)\theta + \sin(n-1)\theta}{2}.$$

The second identity could also, in principle, be directly proved without resorting to complex numbers. See the discussion after the proof of Exercise 1.5.7, where one hints at such a proof using the *completing the square* argument. On the other hand, these two identities have very easy proofs using complex numbers. The third equality (see Exercise 1.3.7) can be obtained using de Moivre's formula and Newton's binomial formula. The proofs of the last two identities use the theory of analytic functions. The computation of the integral uses Cauchy's theorem (in its weak form) or, as in [192, p. 103], the theory of power series and the fundamental theorem of calculus for holomorphic functions (see Theorem 5.2.1). The proof of the formula for the sum uses the residue theorem (or, in fact, Cauchy's formula). The fourth formula, called Fresnel's integral, can also be computed by real analysis methods. We recall the references at the appropriate place in the text. Still, using complex analysis to compute this integral is, in our opinion, a striking example of the power of the methods involved.

For more information, see respectively Exercises 3.4.10, 1.5.7, 1.3.7, 5.2.7 and 7.3.13 and their solutions.

Linear time-invariant systems

This very short discussion is intended in particular for electrical engineering students, but should be of interest to mathematicians as well. We freely use notions such as measures, positive definite functions, and stochastic processes in the discussion. Some of these notions are recalled later in the book, and we send the reader to the index, to find the exact places where the definitions are given.

A *discrete time system* in engineering is often (but not always!) modeled by an input-output relation

$$(u_n)_{n \in \mathbb{N}_0} \mapsto (y_n)_{n \in \mathbb{N}_0},$$

where $u = (u_n)_{n \in \mathbb{N}_0}$ is the input sequence and $y = (y_n)_{n \in \mathbb{N}_0}$ is the output sequence. One writes this as

$$Tu = y,$$

where T is a possibly non-linear operator between spaces of sequences to be fixed depending on the context. We take in this prologue the spaces of input sequences and output sequences to be both equal to the space ℓ_2 of square summable sequences. Thus

$$\sum_{n=0}^{\infty} |u_n|^2 < \infty \quad \text{and} \quad \sum_{n=0}^{\infty} |y_n|^2 < \infty.$$

These sums can be interpreted as the (square of the) energy of the signals, and the above inequalities mean that u and y have finite energy.

We assume that the system is:

(1) Linear (that is the operator T is linear from ℓ_2 into itself).
(2) Bounded (that is, T is a bounded, or equivalently, continuous operator).
(3) Time-invariant (we also say shift-invariant): If

$$(u_n)_{n \in \mathbb{N}_0} \mapsto (y_n)_{n \in \mathbb{N}_0},$$

then

$$(u_{n-1})_{n \in \mathbb{N}_0} \mapsto (y_{n-1})_{n \in \mathbb{N}_0},$$

where we set $u_{-1} = y_{-1} = 0$. In other words, if

$$(u_0, u_1, u_2, \ldots) \mapsto (y_0, y_1, y_2, \ldots),$$

then

$$(0, u_0, u_1, u_2, \ldots) \mapsto (0, y_0, y_1, y_2, \ldots).$$

It is proved, using functional analysis tools, that such a system is defined by a convolution operator: There is a sequence of complex numbers $h_0, h_1, h_2, \ldots$ such that

$$y_n = \sum_{j=0}^{n} h_{n-j} u_j, \quad n = 0, 1, \ldots. \tag{0.0.2}$$

The z-transform of the sequence $(u_n)_{n \in \mathbb{N}_0}$ is by definition

$$u(z) = \sum_{n=0}^{\infty} u_n z^n,$$

and is convergent in the open unit disk since $\sum_{n=0}^{\infty} |u_n|^2 < \infty$. The sequence $(h_n)_{n \in \mathbb{N}_0}$ is called the *impulse response* of the system, and its z-transform $h(z) = \sum_{n=0}^{\infty} h_n z^n$ is called its *transfer function*. Taking the z-transform of (0.0.2), we obtain

$$y(z) = h(z)u(z).$$

The fact that the system is bounded (in the sense above) translates into the fact that

$$\sup_{|z|<1} |h(z)| < \infty.$$

The transfer function is a function *analytic and bounded in modulus* in the open unit disk.

This is a first example where analytic functions, blended with appropriate tools from functional analysis not detailed here, appear in electrical engineering. The theory of complex variables allows us to study various problems related to the transfer function (interpolation and approximation for instance), which in counterpart allow us to approximate, or synthesize the system.

The second example we present is related to the theory of continuous time second-order wide sense stationary processes. Such a process $(x(t))_{t\in\mathbb{R}}$ has a co-variance function

$$E(x(t)\overline{x(s)}) = r(t - s)$$

which depends only on the difference $t - s$, where we have denoted by $E(\cdot)$ the expectation. Furthermore, the function $r(t - s)$ is positive definite. Since, by the Cauchy–Schwarz inequality,

$$|r(t)| \le r(0), \tag{0.0.3}$$

the function

$$\varphi(\lambda) = \int_0^\infty e^{i\lambda t} r(t) dt \tag{0.0.4}$$

is well defined in the open upper half-plane $\mathbb{C}_+$. It is analytic and has a positive real part there, as follows from the identity (see Exercise 3.1.15 for a proof of (0.0.5))

$$\frac{\varphi(\lambda) + \overline{\varphi(w)}}{-i(\lambda - \overline{w})} = \iint_{[0,\infty)\times[0,\infty)} e^{i\lambda t} e^{-is\overline{w}} r(t - s) dt ds, \quad \lambda, w \in \mathbb{C}_+. \tag{0.0.5}$$

The fact that the function φ has a positive real part in $\mathbb{C}_+$ has a number of key consequences. In particular, various interpolation problems, which have applications to the prediction problem for the process, can be solved in an explicit way. Furthermore, the Herglotz representation theorem asserts that one can write

$$\varphi(\lambda) = \frac{1}{i} \int_{\mathbb{R}} \frac{d\mu(t)}{t - \lambda},$$

where $d\mu$ is a positive Borel measure (a more general form of this formula appears later in the book). When $d\mu$ is absolutely continuous with respect to the Lebesgue measure, its derivative is the spectral density of the process[2]. We note that discrete

---

[2] One can, equivalently, also apply Bochner's theorem directly to r to obtain these last results.

counterparts of formulas (0.0.4) and (0.0.5) are given in the sequel. See (4.4.21) and (4.4.22).

The third example we present here also pertains to the continuous time case: A continuous signal (for instance, the voice) may be modeled by an expression of the form

$$f(t) = \frac{1}{2F} \int_{[-F,F]} e^{-itu} m(u) du, \tag{0.0.6}$$

where the function m, say continuous in this prologue, denotes the spectrum. The representation (0.0.6) expresses that the signal f is built from frequencies in a limited band. The function f is analytic in the complex plane. Its special form (0.0.6) allows us to prove the *sampling theorem*

$$f(t) = \sum_{n \in \mathbb{Z}} f\left(\frac{\pi n}{F}\right) \frac{\sin(Ft - n\pi)}{Ft - n\pi},$$

where the limit is pointwise (and in a Hilbert space norm too, as explained later in the book). This formula should be a surprise to the students: How can one recover a function of a continuous argument from a discrete number of its values? This is possible because of the special properties of f as an analytic function (more precisely, as an entire function of bounded exponential type).

These three examples should at least suggest to the student that the theory of analytic functions has fruitful applications in electrical engineering and signal processing.

Last remarks

The theory of functions of a complex variable is the topic of numerous excellent books, of which we mention [5], [18], [42], [45], [144], [164], to name a few. Classics such as [192] are worth being studied in detail. Interesting sources for exercises are the book of Polya–Szegö [182], the Berkeley entrance exams book [62], and the books of exercises [75] and [176]. Giving precise references to all exercises is a Sisyphean task, and we apologize in advance for any omission.

Finally we conclude with some notation: We use $\mathbb{N} = \{1, 2, \ldots, \}$ for the positive integers, and $\mathbb{N}_0 = \mathbb{N} \cup \{0\}$. The integers are denoted by $\mathbb{Z}$, and $\mathbb{D}$ denotes the open unit disk. The unit circle is denoted by $\mathbb{T}$, and $\mathbb{C}_r$ stands for the open right half-plane. The open upper half-plane is denoted by $\mathbb{C}_+$.

Acknowledgment. It is a pleasure to thank Haim Attia for a very careful reading of various versions of this book. Comments of Sylvain Chevillard, Izchak Lewkowicz and Mamadou Mboup are also gratefully acknowledged. Finally, it is a pleasure to thank Natanael Alpay for discussions pertaining to Exercise 1.3.7 and Question 1.5.1.

Main changes in the second edition: We have added a number of exercises and comments in the text. In particular:

- We begin the section on polynomials with a discussion of C. Fefferman's very short proof of the fundamental theorem of algebra (see [79]).
- Morera's theorem has been put in a new and separate section (Section 6.2), where the former Theorem 12.5.3 has been moved.
- Section 7.5 on rational function has been moved and incorporated to a new chapter on rational functions.
- We have expanded the results on the computations of the residue. In particular we gave a formula for the residue of a function of the form $\frac{f}{g}$ where g has a zero of order $N > 1$.
- Former Sections 12.5 and 12.6 (now, Sections 14.5 and 14.6) have been slightly expanded.
- We expanded the part on conformal mapping and one-to-oneness (see, e.g., Exercise 7.3.8) and discussed Riemann's mapping theorem in greater details. In particular we wrote in details the proof that the map $z \mapsto \int_{[0,z]} \frac{ds}{\sqrt{1-s^4}}$ is conformal from the open unit disk onto a square.

Finally, we call *Question* an exercise for which no solution is provided.

Reference to the book [7] is given as [CAPB2].

Part I

Complex Numbers

Chapter 1

Complex Numbers: Algebra

This first chapter has essentially an algebraic flavor. The exercises use elementary properties of the complex numbers. A first definition of the exponential function is given, and we also meet Blaschke factors. These will appear in a number of other places in the book, and are key players in more advanced courses on complex analysis. Almost no methods from real or complex analysis are involved in the present chapter. Still, in Exercise 1.1.9 we already get a hint on difficulties which arise with respect to continuity. The argument of a complex number z is defined (modulo 2π) for $z \in \mathbb{C} \setminus \{0\}$, but not for $z = 0$, and cannot be defined in a continuous way in $\mathbb{C} \setminus \{0\}$.

1.1 First properties of the complex numbers

There are various ways to build the field of complex numbers. For the present purpose, the most appropriate seems to be the following:

Proposition 1.1.1. *The set of matrices*

$$\left\{ z = \begin{pmatrix} x & -y \\ y & x \end{pmatrix} \; ; x, y \in \mathbb{R} \right\}$$

endowed with the usual laws of addition and multiplication, is a field, which contains an isomorphic copy of the field of real numbers $\mathbb{R}$, and in which the equation $z^2 + 1 = 0$ is solvable.

The isomorphism alluded to in the proposition is the map

$$x \mapsto \tau(x) = \begin{pmatrix} x & 0 \\ 0 & x \end{pmatrix}.$$

Setting

$$i = \begin{pmatrix} 0 & -1 \\ 1 & 0 \end{pmatrix}$$

we see that

$$z = \tau(x) + iy,$$

which one writes as $z = x + iy$. Viewing a complex number as a matrix

$$z = \begin{pmatrix} x & -y \\ y & x \end{pmatrix} \tag{1.1.1}$$

$$= \rho \begin{pmatrix} \cos\theta & -\sin\theta \\ \sin\theta & \cos\theta \end{pmatrix} \tag{1.1.2}$$

gives a geometric interpretation of $z \neq 0$ as the composition of a homothety by $\rho = \sqrt{x^2 + y^2}$ and of a rotation with matrix representation

$$\begin{pmatrix} \cos\theta & -\sin\theta \\ \sin\theta & \cos\theta \end{pmatrix}, \tag{1.1.3}$$

where the angle $\theta \in [0, 2\pi)$ is defined by

$$\cos\theta = \frac{x}{\sqrt{x^2 + y^2}} \quad \text{and} \quad \sin\theta = \frac{y}{\sqrt{x^2 + y^2}}.$$

This aspect will be of importance in the sequel. See the discussion after Theorem 4.2.3 and also Remark 16.1.14.

Exercise 1.1.2. *Show that* (1.1.3) *is the matrix representation of the rotation by the angle* $\theta \in [0, 2\pi)$.

Another construction of the field of complex numbers is by considering the quotient space $\mathbb{R}[X]/(X^2 + 1)$. In this construction, i is the equivalence class of the monomial X.

We now set the notation, and recall some elementary definitions. The field of complex numbers will be denoted by $\mathbb{C}$. Let

$$z = x + iy, \tag{1.1.4}$$

be a complex number, with *real part* x and *imaginary part* y. The expression (1.1.4) is called the *cartesian form* of the complex number z. The number

$$\overline{z} = x - iy$$

is called the *conjugate* of z, and the number z is called *purely imaginary* if its real part is equal to 0. Recall that

$$x = \frac{z + \overline{z}}{2}, \quad \text{and} \quad y = \frac{z - \overline{z}}{2i}.$$

Thus:

Proposition 1.1.3. *Let $z \in \mathbb{C}$. Then:*

z is real if and only if $z = \bar{z}$.

z is purely imaginary if and only if $z = -\bar{z}$.

Proof. Indeed, $z = x + iy$ is real if and only if $y = 0$, that is, if and only if $z = \bar{z}$. Similarly, $z \in i\mathbb{R}$ if and only if $x = 0$, that is, if and only if $z = -\bar{z}$. □

The following properties of the conjugation are easily verified:

Proposition 1.1.4. *Let z, z_1 and z_2 be complex numbers. Then:*

$$\overline{z_1 + z_2} = \overline{z_1} + \overline{z_2}, \tag{1.1.5}$$

$$\overline{z_1 z_2} = \overline{z_1} \cdot \overline{z_2}, \tag{1.1.6}$$

$$\overline{z_1/z_2} = \overline{z_1}/\overline{z_2}, \quad for \quad z_2 \neq 0, \tag{1.1.7}$$

$$\overline{\overline{z}} = z, \tag{1.1.8}$$

$$\overline{z^n} = (\bar{z})^n, \quad n \in \mathbb{Z} \quad and \quad z \neq 0 \quad if \quad n < 0. \tag{1.1.9}$$

Partial proof. (1.1.5), (1.1.6) and (1.1.8) are direct computations and will be skipped. Granting (1.1.6) we have for $z \neq 0$,

$$z \cdot \frac{1}{z} = 1 = \overline{z \cdot \frac{1}{z}} = \bar{z}\overline{\frac{1}{z}},$$

where we have used (1.1.6), and so

$$\overline{1/z} = \frac{1}{\bar{z}}. \tag{1.1.10}$$

We conclude by using (1.1.10) and once more (1.1.6) to prove (1.1.7). Finally, (1.1.9) is proved by induction for $n > 0$ and using (1.1.10) for $n < 0$. □

The positive number
$$|z| = \sqrt{x^2 + y^2}$$
is called the *modulus*, or the *absolute value*, of the complex number z. Every complex number $z = x + iy$ different from 0 can be written as

$$z = |z|(\cos\theta + i\sin\theta), \tag{1.1.11}$$

where θ is uniquely determined, modulo 2π, by

$$\cos\theta = \frac{x}{|z|} \quad and \quad \sin\theta = \frac{y}{|z|}.$$

Formula (1.1.11) is a mere rewriting of (1.1.2), and is called the *polar representation* of z. One also says *polar decomposition*, and often uses the notation r or ρ rather than $|z|$ in the polar representation of z. The polar representation is convenient to compute products and quotients of complex numbers:

Proposition 1.1.5. *Let z_1 and z_2 be two complex numbers different from 0, and with polar representations*

$$z_\ell = \rho_\ell(\cos\theta_\ell + i\sin\theta_\ell), \quad \ell = 1, 2.$$

Then

$$z_1 z_2 = \rho_1\rho_2(\cos(\theta_1 + \theta_2) + i\sin(\theta_1 + \theta_2)),$$

and

$$\frac{z_1}{z_2} = \frac{\rho_1}{\rho_2}(\cos(\theta_1 - \theta_2) + i\sin(\theta_1 - \theta_2)).$$

The claims are direct consequences of the trigonometric identities, and proofs are omitted.

It follows from the preceding proposition that the nth power of a complex number $z \neq 0$ is easily computed when its polar representation

$$z = \rho(\cos\theta + i\sin\theta) \tag{1.1.12}$$

is given. We have de Moivre's formula

$$z^n = \rho^n(\cos(n\theta) + i\sin(n\theta)), \quad n \in \mathbb{Z}. \tag{1.1.13}$$

Exercise 1.1.6. *Compute $(1 + i)^n$.*

The argument θ in the polar representation is defined modulo 2π, and this is the key to the following classical result:

Theorem 1.1.7. *Let $z \neq 0$ with polar representation* (1.1.12). *Then, for every $n \in \mathbb{N}$, z has n roots of order n, i.e., there exist n different numbers $z_0, \ldots, z_{n-1}$ such that*

$$z_j^n = z, \quad j = 0, \ldots, n - 1.$$

Proof. Clearly, any root of order n of z will be different from 0 since $z \neq 0$. Let now w be a complex number different from 0, with polar decomposition

$$w = r(\cos t + i\sin t). \tag{1.1.14}$$

The condition $w^n = z$ leads to

$$r^n(\cos nt + i\sin nt) = \rho(\cos\theta + i\sin\theta).$$

It follows that $r^n = \rho$ and that $nt = \theta$ (modulo 2π). The first condition leads to $r = \rho^{1/n}$ while the second leads to

$$nt = \theta + 2k\pi, \quad k \in \mathbb{Z}.$$

Thus,

$$t = \frac{\theta}{n} + \frac{2k\pi}{n}, \quad k \in \mathbb{Z}.$$

The choices $k = 0, \ldots, n-1$ lead to n different values of t (modulo 2π) and hence the roots of order n of z are

$$z_k = \rho^{1/n} \left(\cos\left(\frac{\theta + 2k\pi}{n}\right) + i \sin\left(\frac{\theta + 2k\pi}{n}\right) \right), \quad k = 0, \ldots, n-1. \quad (1.1.15)$$

$\square$

The numbers z_j are called the *roots of order n of z*. See Exercise 2.1.1 for a geometric interpretation of the roots of order n. We note that the sums

$$\sum_{j=0}^{n-1} z_j^m, \quad m \in \mathbb{N}_0 \quad (1.1.16)$$

are computed in Exercise 1.2.3, and one has

$$\sum_{j=0}^{n-1} z_j^m = \begin{cases} nz^k, & \text{if } m = kn, \ k \in \mathbb{N}_0, \\ 0, & \text{otherwise.} \end{cases} \quad (1.1.17)$$

The preceding discussion shows that (1.1.14) is multi-valued when n is replaced by a real (or even complex, but at this stage of the book, the functions cos and sin are defined only for a real argument). When θ is restricted to a semi-open interval of length 2π (for instance $(-\pi, \pi]$ or $[0, 2\pi)$), and therefore uniquely defined, the expression (1.1.14) is also uniquely defined (and therefore defines a *function*, rather than a *multi-valued expression*) when n is allowed to be real (or even complex). For instance, one can define square root functions

$$f_1(z) = \sqrt{\rho} \left(\cos\frac{\theta}{2} + i \sin\frac{\theta}{2} \right), \quad \theta \in (-\pi, \pi],$$

and

$$f_2(z) = \sqrt{\rho} \left(\cos\frac{\theta}{2} + i \sin\frac{\theta}{2} \right), \quad \theta \in (0, 2\pi].$$

We have

$$f_1(z)^2 = f_2(z)^2 = z, \quad z \in \mathbb{C} \setminus \{0\}.$$

Note that f_1 is discontinuous along the negative axis, while f_2 is discontinuous along the positive axis. Moreover,

$$f_1(1) = 1 \quad \text{while} \quad f_2(1) = -1.$$

It is sometimes useful to have a formula for θ. We denote by $\arctan x$ the inverse of

$$\tan : (-\pi/2, \pi/2) \to \mathbb{R}.$$

Exercise 1.1.8. *Prove that*

$$\arctan u + \arctan\frac{1}{u} = \begin{cases} \dfrac{\pi}{2}, & \text{for } u > 0, \\[2mm] -\dfrac{\pi}{2}, & \text{for } u < 0. \end{cases} \quad (1.1.18)$$

Exercise 1.1.9. *Let* $z = x + iy \neq 0$. *Then:*

(a) *The function* $\theta \in (-\pi, \pi)$ *given by the formulas*

$$\theta(x,y) = \begin{cases} \arctan(y/x), & x > 0, \\ \frac{\pi}{2} - \arctan(x/y) & y > 0, \\ -\frac{\pi}{2} - \arctan(x/y), & y < 0, \end{cases} \tag{1.1.19}$$

satisfies (1.1.11).

(b) *The function* θ *is continuous in* $\mathbb{R}^2 \setminus (-\infty, 0]$ *and is discontinuous along the negative axis.*

(c) *The partial derivatives of* θ *have continuous extensions (and in fact extensions with partial derivatives of all orders) in* $\mathbb{R} \setminus \{(0,0)\}$.

In Exercise 4.1.11 it is shown that one *cannot* find a continuous function $\theta(x,y)$ defined in $\mathbb{C} \setminus \{0\}$ and satisfying (1.1.11).

We note that the functions

$$u(x,y) = \ln\sqrt{x^2 + y^2} \quad \text{and} \quad v(x,y) = \theta(x,y)$$

satisfy in $\mathbb{R}^2 \setminus (-\infty, 0]$ an important pair of partial differential equations, called the *Cauchy–Riemann equations*. See Exercise 4.2.8.

We also note the following: The negative result in (c) in the previous exercise may seem paradoxical. It stems from the following fact: In $\mathbb{R}^2$, the existence of a function with given partial derivatives in a given set depends not only on the properties of the functions (local aspect of the problem) but on the properties of the set (global aspect of the problem). See Section 5.7 for more on this question.

The following properties of the absolute value are easily verified, and proofs are omitted.

Proposition 1.1.10. *Let* z, z_1 *and* z_2 *be complex numbers. Then:*

$$|z^n| = |z|^n, \quad n \in \mathbb{Z} \quad \text{and} \quad z \neq 0 \quad \text{if} \quad n < 0, \tag{1.1.20}$$
$$|z_1 z_2| = |z_1| \cdot |z_2|, \tag{1.1.21}$$
$$|z_1/z_2| = |z_1|/|z_2|, \quad \text{for} \quad z_2 \neq 0. \tag{1.1.22}$$

The formulas

$$|z|^2 = z\bar{z}, \quad z \in \mathbb{C}, \tag{1.1.23}$$

$$\frac{1}{z} = \frac{\bar{z}}{|z|^2}, \quad z \neq 0, \tag{1.1.24}$$

are useful to compute explicitly expressions involving absolute values and inverses. See, e.g., the solution of Exercise 4.4.3 for an illustration. With $z = x + iy$ the second formula reads

$$\frac{1}{x + iy} = \frac{x - iy}{x^2 + y^2}.$$

For instance

$$\frac{1}{\cos t + i \sin t} = \frac{\cos t - i \sin t}{\cos^2 t + \sin^2 t} = \cos t - i \sin t. \tag{1.1.25}$$

Exercise 1.1.11. *Write in cartesian and polar forms the complex number*

$$z = \frac{1}{1 + \cos t + i \sin t},$$

where t is real and not an odd multiple of π.

Related to the preceding exercise, see also Exercise 3.1.1. The following exercise is related to Exercise 1.2.8.

Exercise 1.1.12. *Let x and y be real numbers. Find necessary and sufficient conditions for the complex number*

$$\cos x \cosh y - i \sin x \sinh y \tag{1.1.26}$$

to be different from 0. When these conditions hold, write in cartesian form the complex number

$$\frac{\sin x \cosh y + i \cos x \sinh y}{\cos x \cosh y - i \sin x \sinh y}. \tag{1.1.27}$$

We note that the expression (1.1.27) reduces to $\tan x$ when $y = 0$ and, when $x = 0$, to $i \tanh y = \tan iy$ (see (1.2.14) below for the latter).

The formulas

$$|z + w|^2 = |z|^2 + 2 \operatorname{Re} z\overline{w} + |w|^2, \tag{1.1.28}$$
$$= |z|^2 + 2 \operatorname{Re} \overline{z}w + |w|^2,$$
$$|z + \overline{w}|^2 = |z|^2 + 2 \operatorname{Re} zw + |w|^2,$$

prove to be quite useful in computations. They imply the *completing the square* formulas,

$$|z|^2 + 2 \operatorname{Re} zw = |z + \overline{w}|^2 - |w|^2,$$
$$|z|^2 + 2 \operatorname{Re} z\overline{w} = |z + w|^2 - |w|^2, \tag{1.1.29}$$

which are also very useful. See the solutions of Exercise 2.2.1 (and in particular equation (2.4.6)) and Exercise 2.3.16 for instance.

Exercise 1.1.13. *Prove* (1.1.28).

The following identities are classical, and play also an important role in the sequel, and in function theory in general. They appear in numerous places. See for instance [75, Exercise 1.68, p. 19].

Exercise 1.1.14. *Let $z, w \in \mathbb{C}$. Then:*

$$|z + w|^2 + |z - w|^2 = 2(|z|^2 + |w|^2), \tag{1.1.30}$$

$$|1 + z\overline{w}|^2 + |z - w|^2 = (1 + |z|^2)(1 + |w|^2), \tag{1.1.31}$$

$$|1 - z\overline{w}|^2 - |z - w|^2 = (1 - |z|^2)(1 - |w|^2), \tag{1.1.32}$$

$$|z - w|^2 - |z + \overline{w}|^2 = -4(\operatorname{Re} z)(\operatorname{Re} w). \tag{1.1.33}$$

We will focus on the geometric interpretation of the complex numbers in the following chapter, but we mention here the following: When one identifies the complex numbers and the plane $\mathbb{R}^2$, equation (1.1.30) has a nice geometric interpretation: The complex numbers z and w are the sides of a (possibly degenerate) parallelogram, with diagonals $z + w$ and $z - w$, and (1.1.30) is then the well-known relation between the lengths of the sides and of the diagonals of the parallelogram. We also send the reader to Exercise 1.4.2 for another application of (1.1.30).

It is well to recall in this first chapter that

$$\operatorname{Re} z \le |z| \quad \text{and} \quad \operatorname{Im} z \le |z|, \quad \forall z \in \mathbb{C}, \tag{1.1.34}$$

and that there is equality in the first case if and only if $z \in \mathbb{R}_+$, while there is equality in the second case if and only if $z \in i\mathbb{R}_+$. These inequalities are really of interest only for $\operatorname{Re} z > 0$ and $\operatorname{Im} z > 0$ respectively, and allow us to prove

$$\left| |z_1| - |z_2| \right| \le |z_1 - z_2| \le |z_1| + |z_2|. \tag{1.1.35}$$

For instance, the first inequality in (1.1.35) is equivalent to

$$|z_1|^2 + |z_2|^2 - 2|z_1| \cdot |z_2| \le |z_1|^2 + |z_2|^2 - 2\operatorname{Re} z_1\overline{z_2}, \tag{1.1.36}$$

that is, since $|z_1| \cdot |z_2| = |z_1\overline{z_2}|$,

$$\operatorname{Re} z_1\overline{z_2} \le |z_1\overline{z_2}|.$$

Exercise 1.1.15. *When do equalities hold in* (1.1.35)?

The first inequality in (1.1.35) is used over and over in the computations in the sequel, in particular when $|z_1| > |z_2|$. Then we have

$$|z_1| > |z_2| \quad \implies \quad |z_1| - |z_2| \le |z_1 - z_2|, \tag{1.1.37}$$

or, equivalently,

$$|z_1| > |z_2| \quad \implies \quad \frac{1}{|z_1 - z_2|} \le \frac{1}{|z_1| - |z_2|}. \tag{1.1.38}$$

This is illustrated in the next exercise, and in its continuations, namely part (c) of Exercise 1.1.23, and Exercise 3.3.1. See also (4.5.1) for another sample application. The second inequality in (1.1.35) is called the triangle inequality.

Exercise 1.1.16. *Let z be in the open unit disk $\mathbb{D}$. Show that*

$$\left| \frac{z^{2n}}{2 + z^n + z^{5n}} \right| \leq \frac{|z|^{2n}}{2(1 - |z|)}. \tag{1.1.39}$$

Using appropriately the triangle inequality one can prove:

Exercise 1.1.17 (see [60, Problem 9, p. 14]). *Show that*

$$1 + |z_1 z_2 - 1| \leq (1 + |z_1 - 1|)(1 + |z_2 - 1|), \quad z_1, z_2 \in \mathbb{C}. \tag{1.1.40}$$

The following important results follow from (1.1.32) and (1.1.33): Let $z, w \in \mathbb{C}$. Then it holds that

$$|z| < 1 \quad \text{and} \quad |w| < 1 \quad \Longrightarrow \quad \left| \frac{z - w}{1 - z\overline{w}} \right| < 1, \tag{1.1.41}$$

$$\operatorname{Re} z > 0 \quad \text{and} \quad \operatorname{Re} w > 0 \quad \Longrightarrow \quad \left| \frac{z - w}{z + \overline{w}} \right| < 1. \tag{1.1.42}$$

The proofs of (1.1.41) and (1.1.42) form the topic of Exercise 1.1.19 below. In the statements, recall that we denote by $\mathbb{C}_r$ the open right half-plane:

$$\mathbb{C}_r = \{ z \in \mathbb{C} \, ; \, \operatorname{Re} z > 0 \}. \tag{1.1.43}$$

Functions of the form

$$b_w(z) = \frac{z - w}{1 - z\overline{w}}, \quad w \in \mathbb{D}, \tag{1.1.44}$$

and

$$B_w(z) = \frac{z - w}{z + \overline{w}}, \quad w \in \mathbb{C}_r, \tag{1.1.45}$$

possibly multiplied by a constant of modulus 1, are called *Blaschke factors*, associated respectively to the open unit disk and to the open right half-plane. They are special instances of Moebius maps. See Section 2.3 for more information. They play an important role in conformal mapping and in various questions in the theory of Hilbert spaces of analytic functions; see [190]. Finite products of terms of the form (1.1.44) are called finite Blaschke products (associated to the open unit disk). One defines similarly Blaschke products associated to $\mathbb{C}_r$ as finite products of terms of the form (1.1.45). One can also define Blaschke factors associated to the open upper half-plane $\mathbb{C}_+$ as terms of the form

$$\mathscr{B}_w(z) = \frac{z - w}{z - \overline{w}}, \tag{1.1.46}$$

where $w \in \mathbb{C}_+$.

As a first consequence of (1.1.41) we have:

Exercise 1.1.18. *Let a and b be in the open unit disk. Show that*

$$c = \frac{(1-|a|^2)b + (1-|b|^2)a}{1-|ab|^2} \tag{1.1.47}$$

is also in the open unit disk.

The expression (1.1.47) has a special meaning. Applying the Schur algorithm to the degree two Blaschke product

$$\frac{z-a}{1-z\overline{a}}\frac{z-b}{1-z\overline{b}},$$

one obtains $\frac{z-c}{1-z\overline{c}}$. See Section 11.5 for more information. It also appears in the solution of the question set in Remark 2.4.2.

Exercise 1.1.19. *Prove (1.1.41) and (1.1.42).*

We note the inequality

$$\left|1 + \frac{|w|}{w}b_w(z)\right| \le \frac{2(1-|w|)}{1-|z|}$$

for w and z in the open unit disk, w being moreover different from 0. This inequality is proved in the solution of Exercise 3.7.12. The reader might want to prove it already now.

Similarly, we have the inequalities

$$|1 - B_w(z)| \le \frac{w+\overline{w}}{\operatorname{Re} z}, \tag{1.1.48}$$

for z and w in the open right half-plane, and

$$|1 - \mathscr{B}_w(z)| \le \frac{|w-\overline{w}|}{\operatorname{Im} z} \tag{1.1.49}$$

for z and w in the open upper half-plane.

These inequalities play an important role in the construction of infinite products with factors of the form $\frac{|w|}{w}b_w(z)$, $B_w(z)$ and $\mathscr{B}_w(z)$ respectively. See Exercises 3.7.12 and 3.7.13 for the first and third cases.

Exercise 1.1.20. *Prove (1.1.48).*

The function b_w makes sense also for $|w| \ge 1$. For $|w| = 1$, it is equal to a unitary constant (or more precisely, it can be continuously extended to a unitary constant) since

$$\frac{z-w}{1-z\overline{w}} = \frac{z-w}{\overline{w}(w-z)} = -\frac{1}{\overline{w}}, \quad z \ne w.$$

Formula (1.6.4) below,

$$1 - |b_w(z)|^2 = \frac{(1 - |z|^2)(1 - |w|^2)}{|1 - z\overline{w}|^2},$$

which appears in the proof of (1.1.41), can be generalized as follows:

Exercise 1.1.21. *Let $z, w, v \in \mathbb{C}$ be such that*

$$1 - z\overline{w} \neq 0, \quad 1 - z\overline{v} \neq 0 \quad and \quad 1 - v\overline{w} \neq 0. \tag{1.1.50}$$

Show that

$$\frac{1 - b_w(z)\overline{b_w(v)}}{1 - z\overline{v}} = \frac{1 - |w|^2}{(1 - z\overline{w})(1 - \overline{v}w)}. \tag{1.1.51}$$

Similarly, B_w makes sense for any complex number, and we have:

Exercise 1.1.22. *Let $z, w, v \in \mathbb{C}$ be such that*

$$z + \overline{w} \neq 0, \quad z + \overline{v} \neq 0 \quad and \quad v + \overline{w} \neq 0.$$

Show that

$$\frac{1 - B_w(z)\overline{B_w(v)}}{z + \overline{v}} = \frac{2\operatorname{Re} w}{(z + \overline{w})(\overline{v} + w)}. \tag{1.1.52}$$

We note that, similarly, for $z, w, v \in \mathbb{C}$ such that

$$z - \overline{w} \neq 0, \quad z - \overline{v} \neq 0 \quad and \quad v - \overline{w} \neq 0,$$

it holds that

$$\frac{1 - \mathscr{B}_w(z)\overline{\mathscr{B}_w(v)}}{-2i(z - \overline{v})} = \frac{\operatorname{Im} w}{(z - \overline{w})(\overline{v} - w)}. \tag{1.1.53}$$

A newcomer in complex analysis may see (1.1.51), (1.1.52) and (1.1.53) as just curiosities. While keeping in mind that we have not encountered yet at this stage the notion of analytic function, we will just mention that (1.1.52) has a far reaching generalization to function theory on a compact Riemann surface, and is basically equivalent there to an identity called *Fay's trisecant identity*. See [78] and [17] for more details.

We note that any finite product of functions of the form b_w (resp. B_w, $\mathscr{B}_w$) with $|w| \neq 1$ (resp. $\operatorname{Re} w \neq 0$, $\operatorname{Im} w \neq 0$), possibly multiplied by a constant of modulus one, is a rational function which maps the unit circle $\mathbb{T}$ (resp. the imaginary axis $i\mathbb{R}$, the real axis) onto $\mathbb{T}$. That these are the only rational functions with these properties is the topic of Exercise 6.3.4 for the first case.

Finally, we note that the formula

$$1 + z + \cdots + z^N = \frac{1 - z^{N+1}}{1 - z}, \quad z \neq 1, \tag{1.1.54}$$

for summing a finite geometric series will be of much use in the sequel, and in particular in Section 1.3. The proof is the same as for the real case.

Exercise 1.1.23.

(a) *Prove* (1.1.54).

(b) *Prove that*

$$|1 + z + \cdots + z^N| < \frac{2}{1 - |z|}, \quad for \quad z \in \mathbb{D}. \qquad (1.1.55)$$

(c) *Prove that*

$$\left| \sum_{n=0}^{N} \frac{z^{2n}}{2 + z^n + z^{5n}} \right| \leq \frac{1 - |z|^{2N+2}}{2(1 - |z|)^2}, \quad for \quad z \in \mathbb{D}.$$

Note that the right side of (1.1.55) is *independent* of N.

Other exercises which involve only complex numbers but resort to more involved methods include Exercise 3.2.5 and Exercise 3.3.3.

1.2   The exponential function

The exponential function of calculus can be defined in (at least) three equivalent ways: As a power series

$$e^x = \sum_{n=0}^{\infty} \frac{x^n}{n!}, \quad x \in \mathbb{R}, \qquad (1.2.1)$$

as a limit

$$e^x = \lim_{p \to +\infty} \left(1 + \frac{x}{p} \right)^p, \quad x \in \mathbb{R}, \qquad (1.2.2)$$

where $p \in \mathbb{N}$, and as the unique solution of the differential equation

$$f'(x) = f(x), \quad f(0) = 1 \quad (x \in \mathbb{R}).$$

Formulas (1.2.1) and (1.2.2), of an analytic nature, still make sense when x is replaced by a complex variable $z = x + iy$; see Exercise 3.4.14 in Section 3.4 for the first formula and Section 14.7 for the second one. The definition in terms of the differential equation admits also a counterpart here, but one needs first to define the complex derivative. See Exercise 4.2.4. In this section we consider another extension, more algebraic in nature: For $z = x + iy$ one defines the complex exponential function e^z as

$$e^z = e^x(\cos y + i \sin y). \qquad (1.2.3)$$

In particular, the complex exponential function coincides for real z with the exponential function of calculus. For a purely imaginary number z, $z = iy$, we have

$$e^{iy} = \cos y + i \sin y,$$

so that

$$\cos y = \frac{e^{iy} + e^{-iy}}{2} \quad \text{and} \quad \sin y = \frac{e^{iy} - e^{-iy}}{2i}, \quad y \in \mathbb{R}.$$

Thanks to the trigonometric formulas, we have

$$e^{iy_1} e^{iy_2} = e^{i(y_1 + y_2)}, \quad y_1, y_2 \in \mathbb{R},$$

and in particular

$$(e^{iy})^p = e^{iyp}, \quad p \in \mathbb{Z},$$

which is a mere rewriting of de Moivre's formula (1.1.13):

$$(\cos y + i \sin y)^n = \cos ny + i \sin ny, \quad n \in \mathbb{Z}, \, y \in \mathbb{R}.$$

In particular, it holds that

$$e^{z_1 + z_2} = e^{z_1} e^{z_2}, \quad \forall z_1, z_2 \in \mathbb{C}. \tag{1.2.4}$$

Exercise 1.2.1. *Show that*

$$|e^z| = e^{\operatorname{Re} z} \leq e^{|z|}. \tag{1.2.5}$$

Alternative, and equivalent, ways to define the function e^z are as a limit or as a power series, as in the real case. More precisely:

Theorem 1.2.2. *Let $z = x + iy \in \mathbb{C}$. Then,*

$$e^x (\cos y + i \sin y) = \sum_{n=0}^{\infty} \frac{z^n}{n!} = \lim_{\substack{p \to \infty \\ p \in \mathbb{N}}} \left(1 + \frac{z}{p}\right)^p. \tag{1.2.6}$$

As already mentioned, see Section 3.4 for the first equality and Section 14.7 for the second one.

We have chosen to define the exponential function via (1.2.3) in order to have already at hand a convenient notation, and also to compute sums as the ones appearing in the following section. Note that we will also use the notation $\exp z$ for e^z.

The reader may be interested to know that de Moivre's formula in its present form is in fact due to Euler in 1748; de Moivre himself proved another version of the formula (which we will not recall here) in 1730; see for instance [82, p. 51]. The reader might also want to know that de Moivre is the creator of the Gaussian, or normal, distribution of probability theory. See [149, p. 282] for further details on this latter point.

Exercise 1.2.3. *Let z be given by (1.1.12), and let $z_0, \ldots, z_{n-1}$ be as in (1.1.15). Compute for $m \in \mathbb{N}_0$ the sum $\sum_{j=0}^{n-1} z_j^m$.*

The formulas

$$1 - e^{ix} = e^{ix/2}(e^{-ix/2} - e^{ix/2}) = -2ie^{ix/2}\sin x/2, \tag{1.2.7}$$

$$1 + e^{ix} = e^{ix/2}(e^{-ix/2} + e^{ix/2}) = 2e^{ix/2}\cos x/2, \tag{1.2.8}$$

$$\sum_{k=0}^{n} e^{ikx} = \frac{1 - e^{i(n+1)x}}{1 - e^{ix}} = e^{inx/2}\frac{\sin\left(\frac{(n+1)x}{2}\right)}{\sin\left(\frac{x}{2}\right)}, \tag{1.2.9}$$

where x is real and, in the third formula, is not a multiple of 2π, are quite useful. They are used in particular to obtain formulas for various sums of complex numbers. Note that (1.2.9) is proved from the formula (1.1.54). It follows from (1.2.9) that

$$\sum_{k=0}^{n} \cos(kx) = \frac{\cos\left(\frac{nx}{2}\right)\sin\left(\frac{(n+1)x}{2}\right)}{\sin\left(\frac{x}{2}\right)}, \tag{1.2.10}$$

$$\sum_{k=1}^{n} \sin(kx) = \frac{\sin\left(\frac{nx}{2}\right)\sin\left(\frac{(n+1)x}{2}\right)}{\sin\left(\frac{x}{2}\right)}, \quad x \in \mathbb{R}\setminus\{2\pi m,\ m \in \mathbb{Z}\}. \tag{1.2.11}$$

See also Exercise 1.3.3. Furthermore, it also follows from (1.2.9), see also (1.1.55), that for $n \in \mathbb{N}$:

$$\left|\sum_{k=0}^{n} e^{ikx}\right| \leq \frac{1}{|\sin(\frac{x}{2})|}, \quad x \in \mathbb{R}\setminus\{2\pi m,\ m \in \mathbb{Z}\}. \tag{1.2.12}$$

Inequality (1.2.12) is very useful when applying Abel's theorem on conditionally convergent series (see Theorem 3.5.1) to compute boundary values of power series.

Formulas (1.2.7) and (1.2.8) are used in the following exercise.

Exercise 1.2.4. ([152, Exercice S9-1-3, p. 188]). *Let $a = e^{i\alpha}$ and $b = e^{i\beta}$ with α and β real numbers. Show that*

$$\frac{a+b}{a-b} = -i\cot\left(\frac{\alpha-\beta}{2}\right) \quad \text{and} \quad \frac{a+b}{1-ab} = i\frac{\cos\left(\dfrac{\alpha-\beta}{2}\right)}{\sin\left(\dfrac{\alpha+\beta}{2}\right)}.$$

When do the expressions make sense?

One defines the trigonometric functions and the hyperbolic functions for every complex number in terms of the exponential function as follows:

$$\cos z = \frac{e^{iz} + e^{-iz}}{2}, \quad \sin z = \frac{e^{iz} - e^{-iz}}{2i},$$

$$\cosh z = \frac{e^{z} + e^{-z}}{2}, \quad \sinh z = \frac{e^{z} - e^{-z}}{2}. \tag{1.2.13}$$

Note that
$$\sin(iz) = i \sinh z \quad \text{and} \quad \cos(iz) = \cosh z, \tag{1.2.14}$$

and similarly,
$$\sinh(iz) = i \sin z \quad \text{and} \quad \cosh(iz) = \cos z. \tag{1.2.15}$$

All polynomial identities involving the trigonometric functions and the hyperbolic functions proved in calculus on the real line still hold in the complex plane. One way to see this is using the notion of *analytic continuation*, see Section 6.3, which is not yet available at this stage of the book. The other way, more direct and elementary, consists in checking directly the presumed identity from the above definitions of $\cos z$ and $\sin z$ in terms of e^{iz},

$$\cos z = \frac{e^{iz} + e^{-iz}}{2}, \quad \sin z = \frac{e^{iz} - e^{-iz}}{2i},$$

and the fact that the complex exponential function is multiplicative. For instance to prove that
$$\sin(2z) = 2 \sin z \cos z, \tag{1.2.16}$$

one can do as follows:

$$2 \sin z \cos z = 2 \frac{\exp(iz) - \exp(-iz)}{2i} \cdot \frac{\exp(iz) + \exp(-iz)}{2}$$

$$= \frac{1}{2i} \{ \exp(iz)\exp(iz) + \exp(iz)\exp(-iz)$$

$$\qquad - \exp(-iz)\exp(iz) - \exp(-iz)\exp(-iz) \}$$

$$= \frac{1}{2i} \left(\exp(2iz) + 1 - 1 - \exp(-2iz) \right)$$

$$= \sin(2z).$$

See also the discussion after the proof of the next exercise for another example.

Exercise 1.2.5. *Show that*

$$\cos z = \cos x \cosh y - i \sin x \sinh y, \tag{1.2.17}$$
$$\sin z = \sin x \cosh y + i \cos x \sinh y, \tag{1.2.18}$$
$$|\cos z|^2 = \cos^2 x + \sinh^2 y,$$
$$|\sin z|^2 = \sin^2 x + \sinh^2 y. \tag{1.2.19}$$

Show directly (that is, without resorting to the maximum modulus principle) that $|\sin z|$ *has no local maximum.*

It is clear from the previous exercise that $|\cos z|$ and $|\sin z|$ are not bounded in the plane (the knowledgeable student will recognize that $\cos z$ and $\sin z$ are non-constant entire functions, and, by Liouville's theorem, cannot be bounded in modulus in the plane).

Exercise 1.2.6. *Let z_1 and z_2 be complex numbers. Show that*

$$\sin z_1 = \sin z_2 \iff \begin{cases} z_1 = z_2 + 2k\pi, & \text{for some } k \in \mathbb{Z}, \quad \text{or} \\ z_1 + z_2 = (2k+1)\pi, & \text{for some } k \in \mathbb{Z}. \end{cases}$$

Exercise 1.2.7. *Solve the equations*

(a) $\cos z = 0$,

(b) $\sin z = 5$,

(c) $\sin z = a + ib$, $\quad a, b \in \mathbb{R}$.

From (c) *follows that the range of* $\sin$ *is all of* $\mathbb{C}$. *It is clearly not one-to-one, because of the 2π-periodicity.*

(d) *Show that* $\sin z$ *is one-to-one from the strip*

$$L = \{(x, y) ; x \in (-\pi/2, \pi/2) \quad \text{and} \quad y \in \mathbb{R}\}$$

onto the set

$$\mathbb{C} \setminus \{z = x , \ x \in \mathbb{R} \quad \text{and} \quad |x| \geq 1\}.$$

The functions tan and tanh are defined as in the real case by

$$\tan z = \frac{\sin z}{\cos z} \quad \text{and} \quad \tanh z = \frac{\sinh z}{\cosh z}.$$

In view of Exercise 1.2.5 we see that $\tan z$ is given by (1.1.27),

$$\tan z = \frac{\sin x \cosh y + i \cos x \sinh y}{\cos x \cosh y - i \sin x \sinh y}.$$

Exercise 1.2.8.

(a) *Show that*

$$\tan z = \frac{\sin(2x)}{\cos(2x) + \cosh(2y)} + i\frac{\sinh(2y)}{\cos(2x) + \cosh(2y)}. \tag{1.2.20}$$

(b) *What is the image of the strip*

$$L = \{(x, y) ; x \in (-\pi/2, \pi/2) \quad \text{and} \quad y \in \mathbb{R}\} \tag{1.2.21}$$

under the function tan?

(c) *What is the image of the strip*

$$L_1 = \{(x, y) ; x \in (-\pi/4, \pi/4) \quad \text{and} \quad y \in \mathbb{R}\}$$

under the function tan?

See also Exercises 5.7.4 and 9.1.6 in connection with the previous exercise.

1.3   Computing some sums

The following exercise is taken from [103, p. 515]. It also appears as [20, §5.5, Problem 1, p. 214].

Exercise 1.3.1. *Prove that*

$$\cos\frac{\pi}{11} + \cos\frac{3\pi}{11} + \cos\frac{5\pi}{11} + \cos\frac{7\pi}{11} + \cos\frac{9\pi}{11} = \frac{1}{2}. \tag{1.3.1}$$

Hint. Let C denote the sum to be computed, and let

$$S = \sin\frac{\pi}{11} + \sin\frac{3\pi}{11} + \sin\frac{5\pi}{11} + \sin\frac{7\pi}{11} + \sin\frac{9\pi}{11}.$$

Using de Moivre's formula, compute $C + iS$.

More generally than (1.3.1), we have:

Exercise 1.3.2. *Show that, for $a, b \in \mathbb{R}$ ($b \neq 0 \bmod \pi$)*

$$\sum_{k=0}^{n-1} \cos(a + (2k+1)b) = \frac{\cos(a + bn)\sin(bn)}{\sin(b)}. \tag{1.3.2}$$

What does this formula become when $b = m\pi$ for some $m \in \mathbb{Z}$. What does this formula become when a and b are assumed to be in $\mathbb{C}$? Using (1.3.2), prove that for every real u different from 0,

$$\sum_{k=0}^{n-1} \cosh(2k+1)u = \frac{\cosh(nu)\sinh(nu)}{\sinh(u)}. \tag{1.3.3}$$

Exercise 1.3.1 corresponds to the case

$$a = 0, \quad b = \frac{\pi}{11}, \quad \text{and} \quad n = 5,$$

in (1.3.2). As a check, we see that

$$C_5 = \frac{\cos(5\pi/11)\sin(5\pi/11)}{\sin(\pi/11)} = \frac{1}{2}\frac{\sin(10\pi/11)}{\sin(\pi/11)} = \frac{1}{2},$$

as claimed in Exercise 1.3.1. In the same vein, we have the following result, taken in part from [185, p. 59]. For the second formula with $a = 0$, see also [213, p. 171].

Exercise 1.3.3. *Show that*

$$\sum_{k=0}^{n-1} \cos(a + kb) = \frac{\sin\left(\dfrac{nb}{2}\right)}{\sin\dfrac{b}{2}} \cdot \cos\left(a + (n-1)\frac{b}{2}\right), \tag{1.3.4}$$

$$\sum_{k=0}^{n} \binom{n}{k} \cos(a + kb) = 2^n \left(\cos\frac{b}{2}\right)^n \cos\left(a + \frac{nb}{2}\right), \tag{1.3.5}$$

where $a, b \in \mathbb{R}$ and where, in the first formula, $b \neq 0$ (mod 2π). What happens in the first formula when $b = 0$ (mod 2π)? What do these formulas become when a and b are assumed to be in $\mathbb{C}$?

When one sets $a = b = 0$ and $a = 0$, $b = \pi$ in formula (1.3.5) one recovers the well-known formulas

$$\sum_{k=0}^{n} \binom{n}{k} = 2^n,$$

$$\sum_{k=0}^{n} (-1)^k \binom{n}{k} = 0,$$

which also follow from Newton's binomial formula

$$(\alpha + \beta)^n = \sum_{k=0}^{n} \binom{n}{k} \alpha^k \beta^{n-k}, \quad \alpha, \beta \in \mathbb{C}, \tag{1.3.6}$$

applied to

$$(1+1)^n \quad \text{and} \quad (1-1)^n$$

respectively.

We also mention the formulas (see for instance [44, Exercice 4.7, p. 76]) where $a, b, r \in \mathbb{R}$ and n is a positive integer:

$$\sum_{k=0}^{n-1} r^k \cos(a+kb) = \frac{\cos a - r\cos(a-b) - r^n \cos(a+nb) + r^{n+1}\cos(a+(n-1)b)}{1 + r^2 - 2r\cos b}$$

$$\sum_{k=0}^{n-1} r^k \sin(a+kb) = \frac{\sin a - r\sin(a-b) - r^n \sin(a+nb) + r^{n+1}\sin(a+(n-1)b)}{1 + r^2 - 2r\cos b}$$

$$\tag{1.3.7}$$

For the next exercise, see [152, S9-1-6, p. 191].

Exercise 1.3.4. *Show that*

$$\sum_{k=0}^{[n/2]} (-1)^k \binom{n}{2k} = 2^{n/2} \cdot \cos\left(\frac{n\pi}{4}\right), \tag{1.3.8}$$

$$\sum_{k=0}^{[(n-1)/2]} (-1)^k \binom{n}{2k+1} = 2^{n/2} \cdot \sin\left(\frac{n\pi}{4}\right). \tag{1.3.9}$$

In the same vein, and a bit more difficult is the following exercise, taken from the same book ([152, p. 79]). See also [211, p. 238], and [125, pp. 203–205] for a related discussion.

Exercise 1.3.5. *Let $n \in \mathbb{N}$ and*

$$a_n = \{k \in \mathbb{N}_0 \; ; \; 0 \le 3k \le n\},$$
$$b_n = \{k \in \mathbb{N}_0 \; ; \; 0 \le (3k+1) \le n\},$$
$$c_n = \{k \in \mathbb{N}_0 \; ; \; 0 \le (3k+2) \le n\}.$$

Show that

$$\sum_{k \in a_n} \binom{n}{3k} = \frac{2^n + 2(-1)^n \cos(\frac{2n\pi}{3})}{3}, \tag{1.3.10}$$

$$\sum_{k \in b_n} \binom{n}{3k+1} = \frac{2^n + 2(-1)^n \cos(\frac{2(n+1)\pi}{3})}{3}, \tag{1.3.11}$$

$$\sum_{k \in c_n} \binom{n}{3k+2} = \frac{2^n + 2(-1)^n \cos(\frac{2(n+2)\pi}{3})}{3}. \tag{1.3.12}$$

Some sums involving trigonometric functions or their inverses are much more difficult to handle; see for instance [48]. We now present two examples. Among the places where the first one can be found, see for instance [21, Problem 2.73, p. 69] and [218, p. 60].

Exercise 1.3.6. *Show that*

$$\frac{1}{\cos \dfrac{\pi}{30}} - \frac{1}{\sin \dfrac{\pi}{15}} + \frac{1}{\sin \dfrac{2\pi}{15}} + \frac{1}{\sin \dfrac{4\pi}{15}} = 0. \tag{1.3.13}$$

The second example, which we took from [218, p. 61] (see also [186, p. 207], [185, p. 195], [198, p. 53]) has a long history (and we refer to the papers [34], [120], [177] for more general computations and historical remarks), and is conducive to an elementary way to prove that

$$\sum_{k=1}^{\infty} \frac{1}{k^2} = \frac{\pi^2}{6}, \tag{1.3.14}$$

see the remark after the proof of the exercise.

Exercise 1.3.7. *Let $m \in \mathbb{N}$. Prove that*

$$\sum_{k=1}^{m} \cot^2\left(\frac{k\pi}{2m+1}\right) = \frac{m(2m-1)}{3}. \tag{1.3.15}$$

Hint (see for instance [218, p. 61]). Compute the sum of the roots of the degree m monic polynomial p defined by

$$p(X^2) = \frac{1}{2i\binom{2m+1}{2m}} \left((X+i)^{2m+1} - (X-i)^{2m+1}\right). \tag{1.3.16}$$

Another possibility (see [177]) is to compute

$$\left(\cos \left(\frac{k\pi}{2m+1} \right) + i \sin \left(\frac{k\pi}{2m+1} \right) \right)^{2m+1}, \quad k = 1, \ldots, m,$$

in two different ways, namely using de Moivre's formula and Newton's binomial formula.

Remark 1.3.8. For $m = 1$, (1.3.15) reduces to $\cot^2 \frac{\pi}{3} = \frac{1}{3}$.

Remark 1.3.9. Some related sums can be computed using the residue theorem. See [75, pp. 276–277] and Question 7.3.15.

1.4   Confinement lemma and other bounds

Exercise 1.4.1 is taken from [88, p. 39]. The result is called a confinement lemma.

Exercise 1.4.1. *Given complex numbers* $z_1, \ldots, z_n$ *in the open unit disk, show that there exist numbers* $\epsilon_\ell = \pm 1$, $\ell = 1, \ldots, n$ *such that*

$$\left| \sum_{\ell=1}^{m} \epsilon_\ell z_\ell \right| \leq \sqrt{3}, \quad m = 1, \ldots, n. \tag{1.4.1}$$

We first give as exercises two easy results which enter in the proof of the confinement lemma.

Exercise 1.4.2. *Let* z_1 *and* z_2 *be in the closed unit disk and such that*

$$|z_1 - z_2| \geq 1.$$

Show that

$$|z_1 + z_2| \leq \sqrt{3}.$$

Exercise 1.4.3. *Let* z_1, z_2 *and* z_3 *be three pairwise different points in the closed unit disk. Show that there is a pair* $\ell, k \in \{1, 2, 3\}$ *such that* $\ell \neq k$ *and*

$$|z_\ell - z_k| \leq 1 \quad or \quad |z_\ell + z_k| \leq 1.$$

We now present another kind of bounds on complex numbers:

Exercise 1.4.4. *Given* n *complex numbers* $z_1, \ldots, z_n$, *all different from* 0, *show that there exists* $J \subset \{1, \ldots, n\}$ *such that*

$$\left| \sum_{\ell \in J} z_\ell \right| > \frac{1}{4\sqrt{2}} \sum_{\ell=1}^{n} |z_\ell|. \tag{1.4.2}$$

In connection with this section, see also Exercise 6.4.1. Another question, similar in spirit, is provided by Problem 3.3.8.

1.5   Polynomials

We begin with the fundamental theorem of algebra which states that every polynomial of degree n has n roots (counting multiplicity). In the framework of the theory
of analytic functions this key theorem is a consequence of Liouville's theorem. See
Section 6.8. It admits quite a number of other different proofs; see, e.g., [160, p.
8] for a method using differential geometry. We present an elementary proof (but
which uses notions from plane topology) in Section 15.6. We now present, as a
question, a very short proof, due to C. Fefferman [79]. From the analytic point of
view, it uses the extremum value theorem for functions of one or two real variables
continuous on a compact set.

Question 1.5.1. *Let P be a polynomial of degree $N > 0$. We assume by contradiction that $P(z) \neq 0$ for all $z \in \mathbb{C}$. The argument in [79] can be divided into two
steps.*

   Step 1: *Show that there is $z_0 \in \mathbb{C}$ such that*

$$0 < |P(z_0)| \leq |P(z)|, \quad \forall z \in \mathbb{C}. \tag{1.5.1}$$

   Step 2: *Consider the development of P in power series of $(z - z_0)$*

$$P(z) = P(z_0) + b_{n_0}(z - z_0)^{n_0} + \cdots$$

*(just write $z = z - z_0 + z_0$ in $P(z) = a_N z^N + \cdots + a_0$ to obtain it), where n_0 is the
lowest strictly positive power of $(z - z_0)$ with a non zero coefficient; thus $b_{n_0} \neq 0$.
Let w be a root of order n_0 of $-\frac{P(z_0)}{b_{n_0}}$. Compute $P(\epsilon w)$ with $\epsilon > 0$ to obtain a
contradiction with (1.5.1).*

   Even without the fundamental theorem of algebra at hand, we would still do
know the following: If $p(z) = a_n z^n + a_{n-1} z^{n-1} + \cdots + a_0$ is a polynomial of degree
n, and if $p(z_0) = 0$, then we can factor out $(z - z_0)$ from $p(z)$, that is, we can write

$$p(z) = (z - z_0)q(z), \tag{1.5.2}$$

where $q(z)$ is a polynomial of degree $n - 1$. This is called the *factor theorem*. See
[157, Theorem 6.4, p. 11]. In particular, if we know that $z_0, \ldots, z_{n-1}$ are the roots
of $p(z)$ (say, all different, for the present applications below), then

$$p(z) = a_n \prod_{k=0}^{n-1} (z - z_k), \tag{1.5.3}$$

where $a_n \neq 0$ is the coefficient of z^n in $p(z)$.

   Rewriting (1.5.3) as

$$a_n z^n + a_{n-1} z^{n-1} + \cdots + a_0 = a_n \prod_{k=0}^{n-1} (z - z_k),$$

we can relate the coefficients of the polynomial to the symmetric functions of the roots. Here we will content ourselves to note the formula

$$\sum_{k=0}^{n-1} z_k = -\frac{a_{n-1}}{a_n}. \tag{1.5.4}$$

Exercise 1.5.2. *Prove the factor theorem, that is, check (1.5.2).*

Exercise 1.5.3. *Compute the sum of the roots of the polynomial equation*

$$z^{10} + az^8 + b = 0,$$

where a and b are complex parameters.

Exercise 1.5.4. *Solve the following equations:*

$$1 - z^2 + z^4 - z^6 = 0,$$
$$1 + z + \cdots + z^7 = 0,$$
$$(1 - z)^n = (1 + z)^n,$$
$$(1 - z)^n = z^n.$$

We refer to [154, Exercise 4.5, p. 42] for the last equation in the last exercise. In the previous exercises the non-real roots of the various polynomial equations appear in conjugate pairs. If z_0 is a root so is $\overline{z_0}$. This is because the coefficients of the polynomials are real. The next result asserts that this is a general fact.

Exercise 1.5.5. (a) *Let $p(z) = a_n z^n + \cdots + a_0$ with the $a_j \in \mathbb{R}$. Then:*

$$p(z_0) = 0 \iff p(\overline{z_0}) = 0. \tag{1.5.5}$$

In particular non-real roots (if any) appear in conjugate pairs.

(b) *Check that $z = 2 + 3i$ is a solution of the equation*

$$z^4 - 5z^3 + 18z^2 - 17z + 13 = 0,$$

and find all the roots of this equation.

Exercise 1.5.6. *Let $a \in \mathbb{R}$. Show that the polynomial*

$$z^2 - 2z \cos a + 1 \tag{1.5.6}$$

divides the polynomials

$$p_n(z) = z^n \sin a - z \sin(na) + \sin((n-1)a), \quad n = 2, 3, \ldots. \tag{1.5.7}$$

What happens for complex values of a?

It will follow from the fundamental theorem of algebra that any non-constant monic polynomial with real coefficients can be factored as a product of terms of the form $(z - r)$ with $r \in \mathbb{R}$, and factors of the form

$$(z - z_0)(z - \overline{z_0}) = z^2 - 2z \operatorname{Re} z_0 + |z_0|^2, \quad z_0 \in \mathbb{C} \setminus \mathbb{R}.$$

These last factors are called *irreducible*. They cannot be factored as products of degree one polynomials with *real coefficients*. The following exercise illustrates these facts for some important polynomials. Formulas (1.5.10) and (1.5.8) are used in particular in the sequel to prove the following infinite product representations of $\sinh z$, $\cosh z$, $\sin z$ and $\cos z$. See formulas (3.7.23), (3.7.24), (3.7.25), (3.7.26), and Exercise 3.7.16:

$$\sinh z = z \prod_{k=1}^{\infty} \left(1 + \frac{z^2}{k^2 \pi^2} \right),$$

$$\cosh z = \prod_{k=0}^{\infty} \left(1 + \frac{4z^2}{(2k+1)^2 \pi^2} \right),$$

$$\sin z = z \prod_{k=1}^{\infty} \left(1 - \frac{z^2}{k^2 \pi^2} \right),$$

$$\cos z = \prod_{k=0}^{\infty} \left(1 - \frac{4z^2}{(2k+1)^2 \pi^2} \right),$$

for $z \in \mathbb{C}$.

Exercise 1.5.7. (See for instance [154, pp. 43–44].) *Prove the following classical factorizations:*

$$z^{2n} + 1 = \prod_{k=0}^{n-1} \left(z^2 - 2z \cos \left(\frac{2k+1}{2n} \pi \right) + 1 \right), \tag{1.5.8}$$

$$z^{2n+1} + 1 = (z+1) \prod_{k=1}^{n} \left(z^2 - 2z \cos \left(\frac{2k-1}{2n+1} \pi \right) + 1 \right), \tag{1.5.9}$$

$$z^{2n} - 1 = (z+1)(z-1) \underbrace{\prod_{k=1}^{n-1} \left(z^2 - 2z \cos \left(\frac{k}{n} \pi \right) + 1 \right)}_{\text{equal to 1 if } n = 1}, \tag{1.5.10}$$

$$z^{2n+1} - 1 = (z-1) \prod_{k=1}^{n} \left(z^2 - 2z \cos \left(\frac{2k}{2n+1} \pi \right) + 1 \right). \tag{1.5.11}$$

Using the third identity, decompose the polynomial $p(z) = \sum_{k=0}^{n-1} z^{2k}$ into irreducible factors and prove the identity

$$\prod_{k=1}^{n-1} \sin \left(\frac{k\pi}{2n} \right) = \frac{\sqrt{n}}{2^{n-1}}, \quad n \geq 2. \tag{1.5.12}$$

Still using the third identity, prove (see [154, p. 44]) that

$$\frac{\sin nt}{\sin t} = 2^{n-1} \prod_{k=1}^{n-1} \left(\cos t - \cos \frac{k\pi}{n} \right), \quad n \geq 2. \tag{1.5.13}$$

We note that setting $z = 1$ in (1.5.8) leads to the identity

$$2 = 2^{n-1} \prod_{k=0}^{n-1} \left(1 - \cos \left(\frac{(2k+1)\pi}{2n} \right) \right), \tag{1.5.14}$$

which will be useful in the proof of Exercise 3.7.16.

Exercise 1.5.8. *Prove that all roots of the equation*

$$z^3 + 3z + 5$$

have modulus strictly bigger than 1.

The following exercise is taken from [163, Lemma 3, p. 6].

Exercise 1.5.9. *Given complex numbers $c_1, \ldots, c_n$ not all equal to 0, show that*

$$z^n + c_1 z^{n-1} + \cdots + c_n = 0 \quad \Longrightarrow \quad |z| < 2 \max_{j=1,\ldots,n} |c_j|^{\frac{1}{j}}.$$

For other questions related to polynomials, see Exercises 3.1.10, 6.8.2 and Question 6.8.14.

1.6 Solutions

Solution of Exercise 1.1.2. Let $(u,v) \neq (0,0) \in \mathbb{R}^2$, let $r = \sqrt{u^2 + v^2}$ and let $\psi \in [0, 2\pi)$ be (uniquely) determined by

$$\cos \psi = \frac{u}{r}, \quad \text{and} \quad \sin \psi = \frac{v}{r}.$$

Then,

$$\begin{pmatrix} \cos\theta & -\sin\theta \\ \sin\theta & \cos\theta \end{pmatrix} \begin{pmatrix} u \\ v \end{pmatrix} = \begin{pmatrix} \cos\theta & -\sin\theta \\ \sin\theta & \cos\theta \end{pmatrix} \begin{pmatrix} r\cos\psi \\ r\sin\psi \end{pmatrix}$$
$$= r \begin{pmatrix} \cos\theta\cos\psi - \sin\theta\sin\psi \\ \sin\theta\cos\psi + \cos\theta\sin\psi \end{pmatrix}$$
$$= r \begin{pmatrix} \cos(\theta+\psi) \\ \sin(\theta+\psi) \end{pmatrix}. \qquad \square$$

Solution of Exercise 1.1.6. We have

$$1 + i = \sqrt{2}\left(\frac{\sqrt{2}}{2} + i\frac{\sqrt{2}}{2}\right)$$

$$= \sqrt{2}\left(\cos\frac{\pi}{4} + i\sin\frac{\pi}{4}\right),$$

and so, using (1.1.13), we obtain

$$(1 + i)^n = 2^{\frac{n}{2}}\left(\cos\frac{n\pi}{4} + i\sin\frac{n\pi}{4}\right). \qquad (1.6.1)$$

$\square$

Remark. It is of course possible to try and solve the previous exercise using Newton's binomial formula (1.3.6)

$$(\alpha + \beta)^n = \sum_{k=0}^{n}\binom{n}{k}\alpha^k\beta^{n-k}.$$

We then obtain

$$(1 + i)^n = \sum_{k=0}^{n} i^k\binom{n}{k}$$

$$= \sum_{\substack{p \\ \text{such that} \\ 2p \leq n}} (-1)^p\binom{n}{2p} + i\sum_{\substack{p \\ \text{such that} \\ 2p+1 \leq n}} (-1)^p\binom{n}{2p+1}.$$

Comparing with (1.6.1) we get the formulas

$$\sum_{\substack{p \\ \text{such that} \\ 2p \leq n}} (-1)^p\binom{n}{2p} = 2^{\frac{n}{2}}\cos\frac{n\pi}{4} \quad \text{and} \quad \sum_{\substack{p \\ \text{such that} \\ 2p+1 \leq n}} (-1)^p\binom{n}{2p+1} = 2^{\frac{n}{2}}\sin\frac{n\pi}{4},$$

and this gives in fact a proof of Exercise 1.3.4.

Solution of Exercise 1.1.8. Since

$$\frac{d\arctan u}{du} = \frac{1}{u^2 + 1} \quad \text{and} \quad \frac{d\arctan 1/u}{du} = -\frac{1}{u^2}\frac{1}{\frac{1}{u^2} + 1} = -\frac{1}{u^2 + 1},$$

we have

$$\frac{d(\arctan u + \arctan 1/u)}{du} = 0, \quad u \neq 0.$$

Thus the function $\arctan u + \arctan 1/u$ is constant on $(-\infty, 0)$ and $(0, \infty)$. Its value on each of these intervals is computed with the choices $u = \pm 1$. $\square$

Solution of Exercise 1.1.9. (a) The formula for θ follows from the definition of arctan.

(b) The continuity of θ in $\mathbb{R}^2 \setminus (-\infty, 0]$ follows from formula (1.1.18). Using the formula for θ, we have, for any given $x_0 < 0$,

$$\lim_{\substack{y \to 0 \\ y > 0}} \theta(x_0, y) = \pi \quad \text{and} \quad \lim_{\substack{y \to 0 \\ y < 0}} \theta(x_0, y) = -\pi,$$

and hence θ is discontinuous along the negative axis.

(c) Finally, in $x > 0$, $y > 0$ and in $x > 0$, $y < 0$ we have

$$\frac{\partial \theta}{\partial x} = -\frac{y}{x^2 + y^2}, \quad \text{and} \quad \frac{\partial \theta}{\partial y} = \frac{x}{x^2 + y^2},$$

and so these functions admit continuous extensions to $\mathbb{R}^2 \setminus \{(0,0)\}$. □

Solution of Exercise 1.1.11. Using (1.1.24) we have

$$\frac{1}{1 + \cos t + i \sin t} = \frac{1 + \cos t - i \sin t}{(1 + \cos t)^2 + \sin^2 t}$$
$$= \frac{1 + \cos t - i \sin t}{2(1 + \cos t)}$$
$$= \frac{1}{2} - i \frac{\sin t}{2(1 + \cos t)}.$$

Thus, for $t \neq \pi \pmod{2\pi}$ we have

$$\operatorname{Re} \frac{1}{1 + \cos t + i \sin t} = \frac{1}{2} \quad \text{and} \quad \operatorname{Im} \frac{1}{1 + \cos t + i \sin t} = -\frac{1}{2} \tan \frac{t}{2}.$$

As for the polar decomposition, recall the formulas

$$1 + \cos t = 2 \cos^2(t/2) \quad \text{and} \quad \sin t = 2 \cos(t/2) \sin(t/2).$$

Thus, using (1.1.25) and for t not an odd multiple of π we have

$$z = \frac{1}{2 \cos(t/2)} \frac{1}{\cos(t/2) + i \sin(t/2)} = \frac{1}{2 \cos(t/2)} (\cos(t/2) - i \sin(t/2)).$$

For $t \in (0, \pi) \cup (3\pi, 4\pi) \pmod{4\pi}$, we have $\cos(t/2) > 0$ and the polar representation of z is

$$z = \frac{1}{2 \cos(t/2)} (\cos(t/2) - i \sin(t/2)).$$

For $t \in (\pi, 3\pi) \pmod{4\pi}$, we have $\cos(t/2) < 0$ and the polar representation of z is

$$z = \frac{-1}{2 \cos(t/2)} (\cos((t/2) + \pi) - i \sin((t/2) + \pi)).$$ □

Solution of Exercise 1.1.12. The number (1.1.26) vanishes if and only if

$$\cos x = 0 \quad \text{and} \quad \sinh y = 0,$$

that is, if and only if

$$x = \frac{(2k+1)\pi}{2}, \quad \text{with } k \in \mathbb{Z}, \quad \text{and} \quad y = 0.$$

When this condition does not hold, the number (1.1.27) is well defined and we have

$$
\frac{\sin x \cosh y + i \cos x \sinh y}{\cos x \cos y - i \sin x \sinh y}
$$

$$
= \frac{(\sin x \cosh y + i \cos x \sinh y)(\cos x \cosh y + i \sin x \sinh y)}{(\cos x \cosh y)^2 + (\sin x \sinh y)^2}
$$

$$
= \frac{\sin x \cos x (\cosh^2 y - \sinh^2 y) + i(\cos x^2 + \sin^2 x)\sinh y \cosh y}{(\cos x \cosh y)^2 + (\sin x \sinh y)^2}
$$

$$
= \frac{\sin(2x) + i \sinh(2y)}{2\left((\cos x \cosh y)^2 + (\sin x \sinh y)^2\right)}
$$

$$
= \frac{\sin(2x) + i \sinh(2y)}{(1 + \cos(2x))\cosh^2 y + (1 - \cos(2x))\sinh^2 y}
$$

$$
= \frac{\sin(2x) + i \sinh(2y)}{\cos(2x) + \cosh(2y)}. \qquad \qquad \square
$$

Solution of Exercise 1.1.13. We have

$$
\begin{aligned}
|z + w|^2 &= (z+w)\overline{(z+w)} \\
&= (z+w)(\bar{z} + \bar{w}) \\
&= |z|^2 + z\bar{w} + w\bar{z} + |w|^2 \\
&= |z|^2 + 2\operatorname{Re}(z\bar{w}) + |w|^2.
\end{aligned}
$$

The second formula follows since

$$
\operatorname{Re}(z\bar{w}) = \frac{z\bar{w} + \bar{z}w}{2} = \operatorname{Re}(\bar{z}w).
$$

To get the third formula replace w by $\bar{w}$. $\square$

Of course one obtains other useful formulas by replacing w by $-w$ or the like. For instance

$$
|z - w|^2 = |z|^2 - 2\operatorname{Re} z\bar{w} + |w|^2, \tag{1.6.2}
$$

and

$$
|z + iw|^2 = |z|^2 + 2\operatorname{Im} z\bar{w} + |w|^2.
$$

To obtain the latter, note that

$$\mathrm{Re}(-iz\overline{w}) = \mathrm{Im}(z\overline{w}).$$

More generally, we have for any finite number of complex numbers $z_1, \ldots, z_N$,

$$\left| \sum_{\ell=1}^{N} z_\ell \right|^2 = \sum_{\ell=1}^{N} |z_\ell|^2 + 2\,\mathrm{Re}\left(\sum_{\substack{\ell,k=1 \\ \ell<k}}^{N} z_\ell \overline{z_k} \right). \tag{1.6.3}$$

Solution of Exercise 1.1.14. The first identity is proved by considering (1.1.28) for w and $-w$ and adding both identities. To prove the following two identities, one proceeds as follows: We have

$$\begin{aligned}
|1 + z\overline{w}|^2 &= (1 + z\overline{w})(1 + \overline{z}w) \\
&= 1 + z\overline{w} + \overline{z}w + |z|^2|w|^2,
\end{aligned}$$

$$\begin{aligned}
|1 - z\overline{w}|^2 &= (1 - z\overline{w})(1 - \overline{z}w) \\
&= 1 - z\overline{w} - \overline{z}w + |z|^2|w|^2,
\end{aligned}$$

and

$$\begin{aligned}
|z - w|^2 &= (z - w)(\overline{z} - \overline{w}) \\
&= |z|^2 - z\overline{w} - w\overline{z} + |w|^2.
\end{aligned}$$

Thus

$$\begin{aligned}
|1 + z\overline{w}|^2 + |z - w|^2 &= 1 + z\overline{w} + \overline{z}w + |z|^2|w|^2 + |z|^2 - z\overline{w} - w\overline{z} + |w|^2 \\
&= 1 + |z|^2|w|^2 + |z|^2 + |w|^2 \\
&= (1 + |z|^2)(1 + |w|^2),
\end{aligned}$$

and

$$\begin{aligned}
|1 - z\overline{w}|^2 - |z - w|^2 &= 1 - z\overline{w} - \overline{z}w + |z|^2|w|^2 - (|z|^2 - z\overline{w} - w\overline{z} + |w|^2) \\
&= 1 + |z|^2|w|^2 - |z|^2 - |w|^2 \\
&= (1 - |z|^2)(1 - |w|^2).
\end{aligned}$$

Finally, the fourth equality is proved as follows:

$$\begin{aligned}
|z - w|^2 - |z + \overline{w}|^2 &= |z|^2 - 2\,\mathrm{Re}(z\overline{w}) + |w|^2 - (|z|^2 + 2\,\mathrm{Re}(zw) + |w|^2) \\
&= -2\,\mathrm{Re}(z(\overline{w} + w)) \\
&= -4(\mathrm{Re}\,z)(\mathrm{Re}\,w). \qquad \square
\end{aligned}$$

Solution of Exercise 1.1.15. It follows from (1.1.36) that the first equality holds if and only if $z_1\overline{z_2} \in \mathbb{R}_+$. Similarly, the second inequality in (1.1.35) is equivalent to

$$|z_1|^2 + |z_2|^2 - 2\operatorname{Re}(z_1\overline{z_2}) \le |z_1|^2 + |z_2|^2 + 2|z_1| \cdot |z_2|,$$

and we see that there will be equality if and only if

$$-z_1\overline{z_2} = |-z_1\overline{z_2}|,$$

that is, if and only if $-z_1\overline{z_2} \in \mathbb{R}_+$.   □

Solution of Exercise 1.1.16. We have

$$|2 + z^n + z^{5n}| \ge |2 - |z^n + z^{5n}||.$$

For $|z| < 1$ we have

$$|z^n + z^{5n}| \le 2|z| < 2,$$

and so

$$|2 - |z^n + z^{5n}|| = 2 - |z^n + z^{5n}| \ge 2 - 2|z| = 2(1 - |z|),$$

and hence the result.   □

Solution of Exercise 1.1.17. It suffices to apply the triangle inequality to

$$z_1 z_2 - 1 = (z_1 - 1)(z_2 - 1) + (z_1 - 1) + (z_2 - 1),$$

and then add 1 to both sides of the obtained inequality.   □

Solution of Exercise 1.1.18. We have

$$
\begin{aligned}
|c| &\le \frac{(1 - |a|^2)|b| + (1 - |b|^2)|a|}{1 - |ab|^2} \\
&= \frac{|a| + |b| - |ab|(|a| + |b|)}{1 - |ab|^2} \\
&= \frac{(|a| + |b|)(1 - |ab|)}{(1 + |ab|)(1 - |ab|)} \\
&= \frac{|a| + |b|}{1 + |a| \cdot |b|} \\
&< 1,
\end{aligned}
$$

thanks to (1.1.41) with $z = |a|$ and $w = -|b|$.   □

Solution of Exercise 1.1.19. Let $z, w \in \mathbb{D}$. Then, $|z\overline{w}| < 1$, and thus, by (1.1.37) with $z_1 = 1$ and $z_2 = z\overline{w}$ we have

$$|1 - z\overline{w}| \ge 1 - |z\overline{w}| > 0.$$

Recall (1.1.32):

$$|1 - z\overline{w}|^2 - |z - w|^2 = (1 - |z|^2)(1 - |w|^2), \quad \forall z, w \in \mathbb{C}.$$

Dividing both sides of this equality by $|1 - z\overline{w}|^2$ we obtain

$$1 - |b_w(z)|^2 = \frac{(1 - |z|^2)(1 - |w|^2)}{|1 - z\overline{w}|^2}, \tag{1.6.4}$$

which is strictly positive since z and w are in $\mathbb{D}$, and hence (1.1.41) holds. To prove the second claim we note the following: For z and w in $\mathbb{C}_r$,

$$\mathrm{Re}(z + \overline{w}) = \mathrm{Re}(z + w) > 0, \quad \text{and thus} \quad z + \overline{w} \neq 0.$$

Dividing both sides of (1.1.33) by $|z + \overline{w}|^2$ we obtain

$$\left| \frac{z - w}{z + \overline{w}} \right|^2 - 1 = -\frac{(2\,\mathrm{Re}\,z)(\mathrm{Re}\,w)}{|z + \overline{w}|^2} < 0. \tag{1.6.5}$$

$\square$

Solution of Exercise 1.1.20. We have

$$|1 - B_w(z)| = \left| 1 - \frac{z - w}{z + \overline{w}} \right| = \frac{|w + \overline{w}|}{|z + \overline{w}|}.$$

Since w is in the open right half-plane, $\mathrm{Re}\,\overline{w} > 0$, and we have

$$\mathrm{Re}(z + \overline{w}) > \mathrm{Re}\,z,$$

and hence the result since (use the second inequality in (1.1.34) with $z + \overline{w}$ instead of z)

$$|z + \overline{w}| \geq \mathrm{Re}(z + \overline{w}) > \mathrm{Re}\,z. \qquad \square$$

Solution of Exercise 1.1.21. We have

$$\begin{aligned}
1 - b_w(z)\overline{b_w(v)} &= 1 - \frac{(z - w)(\overline{v} - \overline{w})}{(1 - z\overline{w})(1 - \overline{v}w)} \\
&= \frac{(1 - z\overline{w})(1 - \overline{v}w) - (z - w)(\overline{v} - \overline{w})}{(1 - z\overline{w})(1 - \overline{v}w)} \\
&= \frac{(1 - z\overline{v})(1 - |w|^2)}{(1 - z\overline{w})(1 - \overline{v}w)},
\end{aligned}$$

and hence we obtain the required identity. $\qquad \square$

Note that (1.1.50) will hold in particular when z, v and w belong to $\mathbb{D}$. It allows us then to show (see Definition 16.3.11 for the definition of positive definite functions):

Example 1.6.1. *Let $w \in \mathbb{D}$. The function*

$$\frac{1 - b_w(z)\overline{b_w(v)}}{1 - z\overline{v}}$$

is positive definite in $\Omega = \mathbb{D}$.

Solution of Exercise 1.1.22. We have

$$
\begin{aligned}
1 - B_w(z)\overline{B_w(v)} &= 1 - \frac{(z-w)(\overline{v}-\overline{w})}{(z+\overline{w})(\overline{v}+w)} \\
&= \frac{(z+\overline{w})(\overline{v}+w) - (z-w)(\overline{v}-\overline{w})}{(z+\overline{w})(\overline{v}+w)} \\
&= \frac{2(z+\overline{v})(\operatorname{Re} w)}{(z+\overline{w})(\overline{v}+w)},
\end{aligned}
$$

and hence

$$\frac{1 - B_w(z)\overline{B_w(v)}}{z+\overline{v}} = \frac{2\operatorname{Re} w}{(z+\overline{w})(\overline{v}+w)}. \qquad \square$$

Here too we have an example of positive definite kernel:

Example 1.6.2. *Let w be in the open right half-plane $\mathbb{C}_r$. The function*

$$\frac{1 - B_w(z)\overline{B_w(v)}}{z+\overline{v}}$$

is positive definite in $\mathbb{C}_r$.

Solution of Exercise 1.1.23. (a) It suffices to compute

$$(1 + z + \cdots + z^N)(1-z) = 1 + z + \cdots + z^N - (z + z^2 + \cdots + z^{N+1}) = 1 - z^{N+1}.$$

(b) From (1.1.38) we have

$$\frac{1}{|1-z|} \le \frac{1}{1-|z|} \quad \text{for} \quad z \in \mathbb{D}.$$

Still for z in the open unit disk,

$$|1 - z^{N+1}| < 2,$$

and inequality (1.1.55) follows.

(c) We use (1.1.39) and (1.1.54) to obtain

$$
\begin{aligned}
\left| \sum_{n=0}^{N} \frac{z^{2n}}{2 + z^n + z^{5n}} \right| &\le \sum_{n=0}^{N} \left| \frac{z^{2n}}{2 + z^n + z^{5n}} \right| \\
&\le \sum_{n=0}^{N} \frac{|z|^{2n}}{2(1-|z|)} \\
&= \frac{1 - |z|^{2N+2}}{2(1-|z|)(1-|z|^2)},
\end{aligned}
$$

and hence the result since

$$\frac{1}{1-|z|^2} \le \frac{1}{1-|z|}, \quad z \in \mathbb{D}. \qquad \square$$

Solution of Exercise 1.2.1. This follows directly from (1.2.3):

$$\begin{aligned}
|e^z| &= |e^x(\cos y + i\sin y)| \\
&= |e^x| \cdot |\cos y + i\sin y| \\
&= e^x \\
&= e^{\operatorname{Re} z} \\
&\le e^{|z|}. \qquad \square
\end{aligned}$$

Solution of Exercise 1.2.3. We write

$$z = \rho e^{i\theta} \quad \text{and} \quad z_j = \rho^{1/n} e^{i\frac{\theta+2\pi j}{n}}, \quad j = 0,\ldots,n-1.$$

Thus

$$\sum_{j=0}^{n-1} z_j^m = \rho^{m/n} \sum_{j=0}^{n-1} e^{i\frac{m\theta+2\pi mj}{n}} = \rho^{m/n} e^{i\frac{\theta m}{n}} \sum_{j=0}^{n-1} e^{i\frac{2\pi mj}{n}}.$$

First assume that m is a multiple of n: $m = kn$ for some $k \in \mathbb{N}_0$. Then,

$$e^{i\frac{2\pi mj}{n}} = e^{i2\pi jk} = 1, \quad j = 0,\ldots,n-1,$$

and the sum is equal to nz^k. On the other hand, when m is not a multiple of n, we have

$$e^{i\frac{2\pi m}{n}} \ne 1,$$

and formula (1.1.54) for the sum of a geometric series leads to

$$\sum_{j=0}^{n-1} e^{i\frac{2\pi mj}{n}} = \frac{1-e^{2i\pi m}}{1-e^{i\frac{2\pi m}{n}}} = 0,$$

and so the sum is equal to 0. $\qquad \square$

Solution of Exercise 1.2.4. The first expression will make sense if and only if $a \ne b$, i.e., if and only if $\alpha \ne \beta \pmod{2\pi}$. When this condition is in force we have

$$\frac{a+b}{a-b} = \frac{e^{i(\alpha+\beta)/2}(e^{i(\alpha-\beta)/2} + e^{i(\beta-\alpha)/2})}{e^{i(\alpha+\beta)/2}(e^{i(\alpha-\beta)/2} - e^{i(\beta-\alpha)/2})} = \frac{2\cos((\alpha-\beta)/2)}{2i\sin((\alpha-\beta)/2)},$$

and hence the result.

The second expression makes sense when $ab \ne 1$, that is, when $\alpha + \beta \ne 0 \pmod{2\pi}$. Then we have

$$\frac{a+b}{1-ab} = \frac{e^{i(\alpha+\beta)/2}(e^{i(\alpha-\beta)/2} + e^{i(\beta-\alpha)/2})}{e^{i(\alpha+\beta)/2}(e^{-i(\alpha+\beta)/2} - e^{i(\beta+\alpha)/2})} = \frac{2\cos((\alpha-\beta)/2)}{-2i\sin((\alpha+\beta)/2)},$$

and hence the result. $\qquad \square$

Solution of Exercise 1.2.5. Using the formula for the cosine of a sum (which is still valid for complex numbers) and (1.2.14), and recalling that

$$\cos(iy) = \cosh(y) \quad \text{and} \quad \sin(iy) = i\sinh(y),$$

we have

$$
\begin{aligned}
\cos z &= \cos(x + iy) \\
&= \cos x \cos(iy) - \sin x \sin(iy) \\
&= \cos x \cosh y - i \sin x \sinh y.
\end{aligned}
$$

Therefore

$$
\begin{aligned}
|\cos z|^2 &= \cos^2 x \cosh^2 y + \sin^2 x \sinh^2 y \\
&= (\cos^2 x)(1 + \sinh^2 y) + (1 - \cos^2 x)(\sinh^2 y) \\
&= \cos^2 x + \sinh^2 y.
\end{aligned}
$$

We now prove that $|\cos z|^2$ has no local maximum. Let (x, y) be a local maximum. Then

$$\frac{\partial |\cos z|^2}{\partial x} = \frac{\partial |\cos z|^2}{\partial y} = 0,$$

that is

$$\cos x \sin x = 0 \quad \text{and} \quad \sinh y = 0.$$

If $\cos x = 0$ then $\cos z = 0$ (since $\sinh y = 0$) and we have a minimum. So $\sin x = 0$ and $\cos^2 x = 1$. The Hessian at such points is equal to

$$
\begin{aligned}
H &= \begin{pmatrix} \frac{\partial^2 |\cos z|^2}{\partial x^2} & \frac{\partial^2 |\cos z|^2}{\partial x \partial y} \\ \frac{\partial^2 |\cos z|^2}{\partial x \partial y} & \frac{\partial^2 |\cos z|^2}{\partial y^2} \end{pmatrix} \\
&= \begin{pmatrix} -2(\cos^2 x - \sin^2 x) & 0 \\ 0 & 2(\sinh^2 y + \cosh^2 y) \end{pmatrix} \\
&= \begin{pmatrix} -2 & 0 \\ 0 & 2 \end{pmatrix},
\end{aligned}
$$

since $\sin x = \sinh y = 0$, and we have a saddle point. The formulas for $\sin z$ and $|\sin z|^2$ and the fact that $|\sin z|$ has no local maximum are proved in much the same way. $\qquad\square$

We note that $|\sin z|^2$ can be computed also as follows. First remark that

$$\sin(\bar{z}) = \overline{\sin z}.$$

Then, write

$$
\begin{aligned}
|\sin z|^2 &= \sin z \sin \overline{z} \\
&= \frac{1}{2}\left(\cos(z - \overline{z}) - \cos(z + \overline{z})\right) \\
&= \frac{1}{2}\left(\cos(2iy) - \cos(2x)\right) \\
&= \frac{\cosh(2y) - \cos(2x)}{2},
\end{aligned}
$$

which coincides with (1.2.19).

Solution of Exercise 1.2.6. We have

$$
\sin z_1 - \sin z_2 = 2 \cos\left(\frac{z_1 + z_2}{2}\right) \sin\left(\frac{z_1 - z_2}{2}\right)
$$

and hence

$$
\sin z_1 = \sin z_2 \iff \begin{cases} \frac{z_1 + z_2}{2} \in \frac{\pi}{2} + \pi \mathbb{Z}, & \text{or} \\ \frac{z_1 - z_2}{2} \in \pi \mathbb{Z}. \end{cases}
$$

$\square$

Solution of Exercise 1.2.7. (a) Set $z = x + iy$. By Exercise 1.2.5,

$$
\cos z = \cos x \cosh y - i \sin x \sinh y.
$$

Hence we have

$$
\cos x \cosh y = 0 \quad \text{and} \quad \sin x \sinh y = 0. \tag{1.6.6}
$$

The function $\cosh y$ has no real roots, and therefore we have $\cos x = 0$, and thus $\sin x = \pm 1$. Therefore the second equation in (1.6.6) leads to $\sinh y = 0$. Thus,

$$
x = \frac{2k + 1}{2}\pi, \quad k \in \mathbb{Z}, \quad \text{and} \quad y = 0.
$$

(b) Still by Exercise 1.2.5 we have

$$
\sin x \cosh y = 5 \quad \text{and} \quad \cos x \sinh y = 0.
$$

In the second equation, $y = 0$ would lead to $\sin x = 5$, which cannot hold for real x. Therefore, $x = \pi/2 + 2k\pi$ $(k \in \mathbb{Z})$, and $y = \text{arcosh } 5$.

(c) The best procedure is to solve the equation

$$
e^{iz} - e^{-iz} = 2iz_0, \quad \text{where we have set} \quad z_0 = a + ib.
$$

This is equivalent to

$$
(e^{iz})^2 - 2iz_0 e^{iz} - 1 = 0.
$$

Hence

$$e^{iz} = \frac{2iz_0 \pm \sqrt{-4z_0^2 + 4}}{2} = iz_0 \pm \sqrt{1 - z_0^2}.$$

We note that the number

$$iz_0 \pm \sqrt{1 - z_0^2}$$

is always different from 0, and thus one can then find z, for instance in terms of the polar representation of $iz_0 \pm \sqrt{1 - z_0^2}$.

(d) We have seen in (c) above that the range of $\sin z$ is $\mathbb{C}$. Because of the periodicity and of Exercise 1.2.6, it is enough to restrict $\sin$ to the closed strip

$$\overline{L} = \{(x, y) ; x \in [-\pi/2, \pi/2] \quad \text{and} \quad y \in \mathbb{R}\}$$

to obtain all of $\mathbb{C}$ as range. From Exercise 1.2.5 we have

$$\sin(x + iy) = \sin x \cosh y + i \cos x \sinh y,$$

and hence the image of the line $x = -\frac{\pi}{2}$ (resp. $x = \frac{\pi}{2}$) is the line $(-\infty, -1]$ (resp. $[1, \infty)$). Note that $\sin z$ is *not* one-to-one on these lines. We now show that $\sin z$ is one-to-one and onto between the asserted domains. In view of Exercise 1.2.6, the function $\sin z$ is one-to-one in the open strip

$$L = \{(x, y) ; x \in (-\pi/2, \pi/2) \quad \text{and} \quad y \in \mathbb{R}\}.$$

In view of the discussion at the beginning of the proof, the image of this open strip is exactly $\mathbb{C}$ from which the lines $(-\infty, -1]$ and $[1, \infty)$ have been removed.   □

Solution of Exercise 1.2.8. Using (1.2.14), we have

$$\begin{aligned}
\tan z &= \frac{\sin x \cos(iy) + \cos x \sin(iy)}{\cos x \cos(iy) - \sin x \sin(iy)} \\
&= \frac{\sin x \cosh y + i \cos x \sinh y}{\cos x \cosh y - i \sin x \sinh y},
\end{aligned}$$

and the computation is finished as in Exercise 1.1.12 to obtain (1.2.20):

$$\tan z = \frac{\sin(2x)}{\cos(2x) + \cosh(2y)} + i \frac{\sinh(2y)}{\cos(2x) + \cosh(2y)}.$$

We now consider the second question, and show that

$$\tan(L) = \mathbb{C} \setminus \{z = it, t \in \mathbb{R} \quad \text{and} \quad |t| \geq 1\}.$$

The function $\tan z$ is well defined in L. Furthermore, for $z \in L$,

$$\tan z \in \{z = it, t \in \mathbb{R} \quad \text{and} \quad |t| > 1\}$$

(note the inequality $>$ and not $\geq$) if and only if $x = 0$ and

$$\left|\frac{\sinh 2y}{1 + \cosh 2y}\right| > 1.$$

This inequality never holds for real y. Finally the points $\pm i$ correspond to the limit of $\tan(\pm iy)$ as $y \to \pm\infty$. This shows that the image of L under tan is included in the set $\mathbb{C} \setminus \{z = it, t \in \mathbb{R} \quad \text{and} \quad |t| \geq 1\}$. To show that equality holds, we use the fact that the function $\sin z$ has range the whole complex plane (see the previous exercise). More precisely, let $u \in \mathbb{C} \setminus \{z = it, t \in \mathbb{R} \quad \text{and} \quad |t| \geq 1\}$. Then

$$\tan z = \pm u \iff \sin^2 z = u^2(1 - \sin^2 z) \iff \sin^2 z = \frac{u^2}{1 + u^2}.$$

The number $z \in \pm\frac{\pi}{2} + i\mathbb{R}$ if and only if $\frac{u^2}{1+u^2}$ is real and of modulus greater than 1, that is, if and only if $u = it$ with $t \in \mathbb{R}$ and $|t| > 1$. Thus we can always solve $\sin^2 z = \frac{u^2}{1+u^2}$ in L.

We now turn to the third question. Let $z \in L_1$. We have

$$|\tan z| < 1 \iff \sin^2(2x) + \sinh^2(2y) < \cos^2(2x) + \cosh^2(2y) + 2\cos(2x)\cosh(2y)$$
$$\iff 1 - \cos^2(2x) + \cosh^2(2y) - 1 < \cos^2(2x) + \cosh^2(2y)$$
$$\qquad\qquad + 2\cos(2x)\cosh(2y)$$
$$\iff 0 < 2\cos(2x)(\cos(2x) + \cosh(2y)).$$

This last equality is automatically met when $x \in \left(-\frac{\pi}{4}, \frac{\pi}{4}\right)$, and therefore the image of L_1 under tan is the open unit disk. $\quad\square$

As a complement, we now compute the image of the boundary of the strip L_1. For $x = \pi/4$ we have in view of (1.2.14),

$$\tan\left(iy + \frac{\pi}{4}\right) = \frac{1 + i\sinh(2y)}{\cosh(2y)} \tag{1.6.7}$$

which is of course of modulus 1. When y goes through $\mathbb{R}$, equation (1.6.7) is a parametrization of the part of the unit circle in the right half-plane. The case $x = -\pi/4$ gives the other half of the circle. The points $\pm i$ correspond to y going to infinity.

Solution of Exercise 1.3.1. Denote by C the left side of (1.3.1), by S the analogous sum with sin instead of cos, i.e.,

$$S = \sin\frac{\pi}{11} + \sin\frac{3\pi}{11} + \sin\frac{5\pi}{11} + \sin\frac{7\pi}{11} + \sin\frac{9\pi}{11},$$

and set

$$\epsilon = \cos\frac{\pi}{11} + i\sin\frac{\pi}{11} = e^{\frac{i\pi}{11}}.$$

Then, $\epsilon^{11} = -1$. Using (1.1.54) we have

$$C + iS = \sum_{n=0}^{4} \epsilon^{2n+1} = \epsilon \sum_{n=0}^{4} (\epsilon^2)^n = \epsilon \frac{1 - (\epsilon^2)^5}{1 - \epsilon^2}$$

$$= \epsilon \frac{1 - \epsilon^{10}}{1 - \epsilon^2}$$

$$= \frac{\epsilon - \epsilon^{11}}{1 - \epsilon^2}$$

$$= \frac{\epsilon + 1}{1 - \epsilon^2}$$

$$= \frac{1}{1 - \epsilon}$$

$$= \frac{1 - \bar{\epsilon}}{(1 - \epsilon)(1 - \bar{\epsilon})}$$

$$= \frac{1 - \bar{\epsilon}}{2(1 - \operatorname{Re} \epsilon)} = \frac{\left(1 - \cos \dfrac{\pi}{11}\right) + i \sin \dfrac{\pi}{11}}{2 \left(1 - \cos \dfrac{\pi}{11}\right)}.$$

Hence

$$C = \frac{1}{2} \quad \text{and} \quad S = \frac{\sin \dfrac{\pi}{11}}{2 \left(1 - \cos \dfrac{\pi}{11}\right)} = \frac{1}{2} \cot \frac{\pi}{22}. \qquad \square$$

Remark 1.6.3. We note that $\operatorname{Re} \frac{1}{1-e^{it}} = \frac{1}{2}$ for $t \neq 0 \pmod{2\pi}$. See also Exercises 1.1.11, 3.1.1 (with $t + \pi$ instead of t there).

Solution of Exercise 1.3.2. We set C_n to be the sum to be computed and

$$S_n = \sum_{k=0}^{n-1} \sin(a + (2k+1)b).$$

Since $b \neq 0 \pmod{\pi}$ we have that $e^{2ib} \neq 1$ and we can write

$$C_n + iS_n = \sum_{k=0}^{n-1} e^{i(a+(2k+1)b)}$$

$$= e^{i(a+b)} \sum_{k=0}^{n-1} (e^{2ib})^k$$

$$= e^{i(a+b)} \frac{1 - e^{2ibn}}{1 - e^{2ib}}$$

$$= e^{i(a+b)} \frac{e^{ibn}}{e^{ib}} \frac{e^{-ibn} - e^{ibn}}{e^{-ib} - e^{ib}}$$

$$= e^{i(a+b)} \frac{e^{ibn}}{e^{ib}} \frac{\sin bn}{\sin b}$$

$$= e^{i(a+nb)} \frac{\sin bn}{\sin b}.$$

Thus taking real and imaginary parts on both sides, we get the required formula for C_n and also

$$S_n = \frac{\sin(a + bn)\sin(bn)}{\sin b}. \tag{1.6.8}$$

Assume now that $b = m\pi$ for some $m \in \mathbb{Z}$. We have

$$C_n + iS_n = \sum_{k=0}^{n-1} e^{i(a+(2k+1)b)}$$

$$= e^{i(a+b)} \sum_{k=0}^{n-1} (e^{2ib})^k$$

$$= n(-1)^m e^{ia},$$

and hence

$$C_n = (-1)^m n \cos a \quad \text{and} \quad S_n = (-1)^m n \sin a. \tag{1.6.9}$$

It is readily seen that formulas (1.3.2) and (1.6.8) reduce to the formulas in (1.6.9) when $b = 0 \pmod{\pi}$. More precisely,

$$\lim_{b \to m\pi} \frac{\cos(a + bn)\sin(bn)}{\sin(b)} = \lim_{b \to m\pi} \frac{\cos(a + bn)\frac{\sin(bn)}{b-m\pi}}{\frac{\sin(b)}{b-m\pi}}$$

$$= \frac{\cos(a + mn\pi)n \cos(nm\pi)}{\cos(m\pi)}$$

$$= (-1)^m n \cos a.$$

When a and b are assumed complex, the previous computations still make sense, and we still have

$$C_n + iS_n = e^{i(a+nb)} \frac{\sin bn}{\sin b}. \tag{1.6.10}$$

It is not true anymore that

$$\mathrm{Re}(C_n + iS_n) = C_n,$$

and one cannot take real and imaginary parts to prove the asserted formula. One obtains the result by remarking that

$$C_n(a, b) = C_n(-a, -b) \quad \text{and} \quad S_n(a, b) = -S_n(-a, -b),$$

and identify odd and even parts on both sides of (1.6.10). One could also use analytic continuation, but this is beyond the scope of the present chapter.

Taking into account the relationships (1.2.14) between the trigonometric and hyperbolic functions, the choice $a = 0$ and $b = iu$ in (1.3.2) leads to (1.3.3).   □

We note that (1.3.3) can also be proved directly as follows: Set

$$Ch_n = \sum_{k=0}^{n-1} \cosh(2k+1)u \quad \text{and} \quad Sh_n = \sum_{k=0}^{n-1} \sinh(2k+1)u.$$

Then, since $\cosh x + \sinh x = e^x$ for $x \in \mathbb{R}$ (or more generally, for a complex argument), we have

$$Ch_n + Sh_n = \sum_{k=0}^{n-1} e^{(2k+1)u}$$

$$= e^u \frac{1 - e^{2un}}{1 - e^{2u}}$$

$$= e^{un} \frac{e^{un} - e^{-un}}{e^u - e^{-u}}$$

$$= e^{un} \frac{\sinh(nu)}{\sinh(u)}.$$

One obtains formulas for Ch_n and Sh_n by taking the even and odd parts of this expression, that is

$$Ch_n = \cosh(nu)\frac{\sinh(nu)}{\sinh(u)} \quad \text{and} \quad Sh_n = \sinh(nu)\frac{\sinh(nu)}{\sinh(u)}.$$

Solution of Exercise 1.3.3. For the first formula, and assuming $b \neq 0 \pmod{2\pi}$, we compute

$$\left(\sum_{k=0}^{n-1} \cos(a+kb)\right) + i\left(\sum_{k=0}^{n-1} \sin(a+kb)\right) = \sum_{k=0}^{n-1} (\cos(a+kb) + i\sin(a+kb))$$

$$= \sum_{k=0}^{n-1} e^{i(a+kb)}$$

$$= e^{ia} \frac{1 - e^{inb}}{1 - e^{ib}}$$

$$= e^{ia} \frac{e^{i\frac{nb}{2}}}{e^{i\frac{b}{2}}} \frac{e^{-i\frac{nb}{2}} - e^{i\frac{nb}{2}}}{e^{-i\frac{b}{2}} - e^{i\frac{b}{2}}}$$

$$= e^{i(a+\frac{nb}{2}-\frac{b}{2})} \frac{\sin\frac{nb}{2}}{\sin\frac{b}{2}}.$$

By taking real parts on both sides, we obtain (1.3.4). When $b = 0 \pmod{2\pi}$, the sum $\sum_{k=0}^{n-1} \cos(a + kb)$ is clearly equal to $n \cos a$, and this is also the limit

$$\lim_{b \to 2m\pi} \frac{\sin(\frac{nb}{2})}{\sin \frac{b}{2}} \cdot \cos(a + (n-1)\frac{b}{2}) = \frac{\frac{n}{2} \cos(\frac{nb}{2})|_{b=2m\pi}}{\frac{1}{2} \cos(\frac{b}{2})|_{b=2m\pi}} \cos(a + (n-1)m\pi)$$

$$= n \frac{(-1)^{nm}}{(-1)^m} (-1)^{(n-1)m} \cos a$$

$$= n \cos a.$$

To prove the second formula, we use Newton's binomial formula (1.3.6)

$$(\alpha + \beta)^n = \sum_{k=0}^{n} \binom{n}{k} \alpha^k \beta^{n-k},$$

which for $\beta = 1$ leads to

$$(\alpha + 1)^n = \sum_{k=0}^{n} \binom{n}{k} \alpha^k.$$

To compute the sum (1.3.5) we write

$$\left(\sum_{k=0}^{n} \binom{n}{k} \cos(a + kb) \right) + i \left(\sum_{k=0}^{n} \binom{n}{k} \sin(a + kb) \right)$$

$$= \sum_{k=0}^{n} \binom{n}{k} e^{i(a+kb)}$$

$$= e^{ia} \left(\sum_{k=0}^{n} \binom{n}{k} e^{ikb} \right)$$

$$= e^{ia} (1 + e^{ib})^n$$

$$= e^{ia} e^{i\frac{bn}{2}} \left(e^{-i\frac{b}{2}} + e^{i\frac{b}{2}} \right)^n$$

$$= e^{i(a + \frac{nb}{2})} \left(2 \cos \frac{b}{2} \right)^n$$

$$= 2^n \left(\cos \frac{b}{2} \right)^n e^{i(a + \frac{nb}{2})},$$

and hence we obtain (1.3.5) by taking real parts of both sides. The case of possibly complex numbers a and b is treated as in Exercise 1.3.2.   □

We note that the proof of the preceding exercise also leads to the formulas

$$\sum_{k=0}^{n-1} \sin(a + bk) = \frac{\sin(\frac{nb}{2})}{\sin \frac{b}{2}} \sin \left(a + \frac{n-1}{2}b \right) \tag{1.6.11}$$

and

$$\sum_{k=0}^{n} \binom{n}{k} \sin(a + kb) = 2^n \cos^n \frac{b}{2} \sin\left(a + \frac{nb}{2}\right), \tag{1.6.12}$$

as is seen by taking imaginary parts rather than real parts in the arguments.

Solution of Exercise 1.3.4. On the one hand, we have $1 + i = \sqrt{2}e^{i\pi/4}$ and so

$$(1 + i)^n = 2^{n/2}e^{in\pi/4} = 2^{n/2}\left(\cos\left(\frac{n\pi}{4}\right) + \sin\left(\frac{n\pi}{4}\right)\right). \tag{1.6.13}$$

On the other hand,

$$\begin{aligned}
(1 + i)^n &= \sum_{k=0}^{n} i^k \binom{n}{k} \\
&= \sum_{k=0}^{[n/2]} i^{2k} \binom{n}{2k} + \sum_{k=0}^{[(n-1)/2]} i^{2k+1} \binom{n}{2k+1} \\
&= \left(\sum_{k=0}^{[n/2]} (-1)^k \binom{n}{2k}\right) + i\left(\sum_{k=0}^{[(n-1)/2]} (-1)^k \binom{n}{2k+1}\right).
\end{aligned}$$

The result follows by comparing real and imaginary parts of this last expression with the real and imaginary parts of $(1 + i)^n$ as given by (1.6.13). □

Solution of Exercise 1.3.5. Let us denote by A_n, B_n and C_n respectively the sums in (1.3.10)–(1.3.12). Clearly

$$A_n + B_n + C_n = (1 + 1)^n = 2^n. \tag{1.6.14}$$

Let $j = \exp(2i\pi/3)$. We have $j^3 = 1$ and $1 + j + j^2 = 0$, and so with $\ell = 3k + 1$:

$$\begin{aligned}
j^\ell &= 1 && \text{if} \quad k \in a_n, \\
j^\ell &= j && \text{if} \quad k \in b_n, \\
j^\ell &= j^2 && \text{if} \quad k \in c_n.
\end{aligned}$$

Thus

$$(1 + j)^n = \sum_{k \in a_n} \binom{n}{3k} + j \sum_{k \in b_n} \binom{n}{3k+1} + j^2 \sum_{k \in c_n} \binom{n}{3k+2} = A_n + jB_n + j^2C_n.$$

On the other hand

$$\begin{aligned}
(1 + j)^n &= (-j^2)^n \\
&= (-1)^n j^{2n} \\
&= (-1)^n \exp(4\pi in/3) = (-1)^n \exp(-2\pi in/3).
\end{aligned}$$

Similarly,

$$(1+j^2)^n = \sum_{k \in a_n} \binom{n}{3k} + j^2 \sum_{k \in b_n} \binom{n}{3k+1} + j \sum_{k \in c_n} \binom{n}{3k+2} = A_n + j^2 B_n + jC_n,$$

and on the other hand

$$\begin{aligned}
(1+j^2)^n &= (-j)^n = (-1)^n j^n \\
&= (-1)^n \exp(2\pi i n/3) \\
&= (-1)^n (\cos(2\pi n/3) + i \sin(2\pi n/3)).
\end{aligned}$$

We have thus the system of equations

$$\begin{aligned}
A_n + B_n + C_n &= 2^n, \\
A_n + jB_n + j^2 C_n &= (-1)^n \exp(-2\pi i n/3), \\
A_n + j^2 B_n + jC_n &= (-1)^n \exp(2\pi i n/3).
\end{aligned} \tag{1.6.15}$$

Note that the third equation is in fact the conjugate of the second one since $\bar{j} = j^2$. Adding the three equations together we obtain (since $1 + j + j^2 = 0$)

$$3A_n = 2^n + 2(-1)^n \cos(2\pi i n/3),$$

and hence we obtain (1.3.10). To obtain (1.3.11) first multiply the second equation in (1.6.15) by j^2 and the third one by j, and then add up the three equations. To obtain (1.3.12), first multiply the second equation in (1.6.15) by j and the third one by j^2, and then add up the three equations. □

   The reader will have remarked the following: To compute only one of the sums (1.3.8) or (1.3.9) in Exercise 1.3.4, or only one of the sums (1.3.10)–(1.3.12) in Exercise 1.3.5, and using only real analysis, seems to be a difficult task. To compute simultaneously the two sums (for Exercise 1.3.4), or the three sums (for Exercise 1.3.5), and going via the complex domain, is a much easier task. For a problem similar in spirit, see Exercise 3.3.4. The same method also allows us to compute the sums

$$\binom{n}{0} + \binom{n}{4} + \cdots = 2^{n-2} + 2^{n/2-1} \cos\left(\frac{n\pi}{4}\right),$$

$$\binom{n}{1} + \binom{n}{5} + \cdots = 2^{n-2} + 2^{n/2-1} \sin\left(\frac{n\pi}{4}\right),$$

$$\binom{n}{2} + \binom{n}{6} + \cdots = 2^{n-2} - 2^{n/2-1} \cos\left(\frac{n\pi}{4}\right),$$

$$\binom{n}{3} + \binom{n}{7} + \cdots = 2^{n-2} - 2^{n/2-1} \sin\left(\frac{n\pi}{4}\right).$$

See [213, pp. 51–52].

Solution of Exercise 1.3.6. We just outline the proof presented in [218, p. 60] and [21, Problem 2.73, p. 69]. The idea is as follows: Take $x \in \mathbb{R}$ and $k \in \mathbb{N}$. Then the quantities $\frac{1}{\cos kx}$ and $\frac{1}{\sin kx}$ are easily expressed as rational functions of $z = e^{ix}$. In the exercise at hand, we chose $x = \frac{\pi}{30}$. One then obtains for instance

$$\frac{1}{\cos \frac{\pi}{30}} = \frac{2z}{z^2 + 1} \quad \text{and} \quad \frac{1}{\sin \frac{\pi}{15}} = \frac{2iz^2}{z^4 - 1}.$$

Equation (1.3.13) is equivalent to a rational equality satisfied by z, which is verified using the fact that $z^{30} + 1 = 0$. ☐

Remark 1.6.4. The same method allows to prove that

$$\tan\left(\frac{3\pi}{11}\right) + 4\sin\left(\frac{2\pi}{11}\right) = \sqrt{11},$$

which is a question appearing in [222, Exercise 98, p. 23].

Solution of Exercise 1.3.7. The degree m polynomial $p(X)$ defined by (1.3.16) vanishes at the points

$$\cot^2\left(\frac{k\pi}{2m+1}\right), \quad k = 1, \ldots, m,$$

as is seen from de Moivre's formula. The required sum is equal to the opposite of the coefficient of X^{m-1} in the monic polynomial p defined by (1.3.16), that is, the opposite of the coefficient of the power X^{2m-2} in the degree $2m$ polynomial

$$\frac{1}{2i\binom{2m+1}{2m}} \left((X+i)^{2m+1} - (X-i)^{2m+1}\right),$$

that is,

$$\frac{2i\binom{2m+1}{2m-2}}{2i\binom{2m+1}{2m}},$$

from which the result follows. ☐

Remark 1.6.5. With the previous result at hand, the proof of (1.3.14) goes as follows (see [120], [123], [177], [224, Exercises 141.a and 145a, pp. 23–24]). One first remarks that, on $\left(0, \frac{\pi}{2}\right)$,

$$\cot^2 x \leq \frac{1}{x^2} \leq \frac{1}{\sin^2 x} = 1 + \cot^2 x.$$

Applying this inequality to $x = \frac{k\pi}{2m+1}$ for $k = 1, 2, \ldots, m$ and adding up the corresponding inequalities we obtain

$$\frac{m(2m-1)}{3} \leq \frac{(2m+1)^2}{\pi^2} \left(\sum_{k=1}^{m} \frac{1}{k^2} \right) \leq m + \frac{m(2m-1)}{3},$$

from which the result follows.

The solution of Exercise 1.4.1 is presented after the solutions of Exercises 1.4.2 and 1.4.3

Solution of Exercise 1.4.2. We use (1.1.30)

$$|z_1 + z_2|^2 + |z_1 - z_2|^2 = 2(|z_1|^2 + |z_2|^2) \leq 4,$$

and so

$$|z_1 + z_2|^2 \leq 4 - |z_1 - z_2|^2 \leq 3,$$

since $|z_1 - z_2| \geq 1$. $\square$

We note that one can put strict inequalities in the above exercise, and the proof goes in the same way.

Solution of Exercise 1.4.3. Assume by contradiction that

$$|z_\ell \pm z_k| > 1 \tag{1.6.16}$$

for all $\ell, k \in \{1, 2, 3\}$, with $\ell \neq k$. Then the numbers z_i are in particular different from 0. We claim that, furthermore,

$$\frac{z_\ell}{z_k} \notin \mathbb{R}, \quad \text{for} \quad \ell \neq k.$$

Indeed, assume by contradiction that for some pair (z_ℓ, z_k), with $\ell \neq k$ we have $z_\ell/z_k \in \mathbb{R}$. By interchanging ℓ and k we may assume that $|z_\ell/z_k| \leq 1$. Then from

$$\left| z_k \right| \cdot \left| 1 - \frac{z_\ell}{z_k} \right| = |z_k| \cdot \left(1 - \frac{z_\ell}{z_k} \right) > 1,$$

$$\left| z_k \right| \cdot \left| 1 + \frac{z_\ell}{z_k} \right| = |z_k| \cdot \left(1 + \frac{z_\ell}{z_k} \right) > 1$$

we get

$$2|z_k| > 2,$$

contradicting the fact that $|z_k| \leq 1$. We set

$$\{w_1, \ldots, w_6\} = \{\pm z_1, \pm z_2, \pm z_3\}.$$

The lines defined by the intervals $[0, w_j]$, $j = 1, \ldots, 6$, divide the plane into six angular sectors. At least one of these sectors is defined by an angle less than or equal to $\pi/3$. Let w_a and w_b be the points which define this sector, and let θ be its angle. We have $\cos\theta \geq 1/2$, and the distance between w_a and w_b is less than or equal to 1 since

$$
\begin{aligned}
|w_a - w_b|^2 &= |w_a|^2 + |w_b|^2 - 2|w_a||w_b|\cos\theta \\
&\leq |w_a|^2 + |w_b|^2 - |w_a||w_b| \\
&\leq 1.
\end{aligned}
$$

To check this last inequality, one can proceed as follows: For $u, v \in [0,1]$ the function $u^2 - uv + v^2$ is equal to $v^2 \leq 1$ for $u = 0$ and to $1 + v^2 - v \leq 1$ for $u = 1$. Furthermore, its minimum is at the point $u = v/2$ and is equal to $\frac{3v^2}{4} < 1$.

We obtain a contradiction since $|w_a - w_b| > 1$ in view of (1.6.16). □

The above argument appears in [88, p. 40]. We note that one *cannot* replace the inequalities by strict inequalities in the statement of the exercise, as is seen for instance by the choice $z_1 = 1, z_2 = 0$ and z_3 arbitrary in $\mathbb{D}$. Then,

$$
|z_1 \pm z_2| = 1.
$$

Solution of Exercise 1.4.1 (*confinement lemma*). The proof follows closely the one given in [88]. Using Exercise 1.4.2, one sees that the result is true for $n = 2$ (a different argument would lead in fact to the bound $\sqrt{2}$ rather than $\sqrt{3}$). We proceed by induction. Assume the result true for n, and let $z_1, \ldots, z_{n+1}$ be $n+1$ points in the closed unit disk. If $|z_1 + z_2| \leq 1$ or $|z_1 - z_2| \leq 1$, the induction hypothesis allows us to proceed with the points (if, say, $|z_1 + z_2| \leq 1$),

$$
z_1 + z_2 \quad \text{and} \quad z_3, \ldots, z_{n+1}.
$$

Assume now $|z_1 \pm z_2| > 1$. By Exercise 1.4.2, $|z_1 \pm z_2| \leq \sqrt{3}$, and by the preceding exercise, one of the points $z_1 \pm z_3$ and $z_2 \pm z_3$ is in the closed unit disk. The induction hypothesis allows us to proceed with the points (if, say, $|z_1 + z_3| \leq 1$),

$$
z_1 + z_3, z_2, z_4, z_5, \ldots, z_{n+1}. \qquad \Box
$$

Solution of Exercise 1.4.4. Without loss of generality we assume that the $z_\ell \neq 0$ do not lie on the lines $x \pm y = 0$. This last condition can be insured by multiplying all the z_ℓ by a common number of modulus 1. This will not change condition (1.4.2). We set

$$
\begin{aligned}
\Delta_0 &= \{z = x + iy, \text{ with } x > 0, \ y \geq 0 \text{ and } x^2 < y^2\} \\
&= \{z = \rho e^{i\theta}, \text{ with } \rho > 0, \text{ and } -\pi/2 < \theta < \pi/2\},
\end{aligned}
$$

and

$$
\Delta_j = e^{\frac{ij\pi}{2}}\Delta_0, \quad j = 1, 2, 3.
$$

We have

$$\sum_{\ell=1}^{n} |z_\ell| = \sum_{j=0}^{3} \left(\sum_{\ell \in \Delta_j} |z_\ell| \right)$$

and so there is $j_0 \in \{0, 1, 2, 3\}$ such that

$$\sum_{j \in \Delta_{j_0}} |z_j| \geq \frac{1}{4} \sum_{\ell=1}^{n} |z_\ell|. \tag{1.6.17}$$

Without loss of generality we assume that $j_0 = 0$ (if $j_0 \neq 0$, a rotation of all the z_ℓ by a multiple of $\pi/2$ will reduce the situation to this case). In Δ_0 we have

$$|z|^2 = x^2 + y^2 < 2y^2,$$

and so

$$|z| < \sqrt{2}y, \quad \text{that is,} \quad y > \frac{1}{\sqrt{2}}|z|. \tag{1.6.18}$$

Thus

$$\left| \sum_{\ell \in \Delta_0} z_\ell \right| \geq \sum_{\ell \in \Delta_0} \operatorname{Im} z_\ell \geq \frac{1}{\sqrt{2}} \sum_{\ell \in \Delta_0} |z_\ell| > \frac{1}{4\sqrt{2}} \sum_{\ell=1}^{n} |z_\ell|,$$

where we first used (1.6.18) and then (1.6.17).   □

Remark 1.6.6. The same exercise, but with the weaker requirement

$$\left| \sum_{\ell \in J} z_\ell \right| \geq \frac{1}{6} \sum_{\ell=1}^{n} |z_\ell| \tag{1.6.19}$$

appears in [189, p. 114]. It is much more difficult to prove that there exists J such that

$$\left| \sum_{\ell \in J} z_\ell \right| \geq \frac{1}{\pi} \sum_{\ell=1}^{n} |z_\ell|.$$

See [39, Exercice 1, §3, p. TGVIII.26], [88, Exercise 1.17, p. 34], [57], [58], [135] for the latter. Using a result of Reinhardt on polygons, one can in fact prove a stronger result, namely the existence of a subset J such that

$$\left| \sum_{\ell \in J} z_\ell \right| \geq \frac{\sin\left(\frac{k\pi}{2k+1}\right)}{(2k+1)\sin\left(\frac{\pi}{2k+1}\right)} \left(\sum_{\ell=1}^{n} |z_\ell| + \left| \sum_{\ell=1}^{n} z_\ell \right| \right),$$

where k is the integer part of $\frac{n+1}{2}$. See [57].

We also note that (1.6.19) is a special case of the following result from measure theory (see [135, Theorem 1, p. 672]. We refer the reader to Chapter 17 for the needed background.

Theorem 1.6.7. *Let μ be a complex-valued measure of total variation 1. Then there exists a measurable set A such that $|\mu(A)| \geq \frac{1}{\pi}$*

Solution of Exercise 1.5.2. Let $p(z) = \sum_{k=0}^{n} a_k z^k$. Since $p(z_0) = 0$ we have

$$p(z) = p(z) - p(z_0)$$

$$= \sum_{k=1}^{n} a_k (z^k - z_0^k) \quad \text{(the term with index } k = 0 \text{ vanishes)}$$

$$= \sum_{k=1}^{n} a_k (z - z_0) \left(\sum_{\ell=0}^{k-1} z^\ell z_0^{k-1-\ell} \right)$$

where we have used the equality

$$a^m - b^m = (a - b) \left(\sum_{\ell=0}^{m-1} a^\ell b^{m-1-\ell} \right) \tag{1.6.20}$$

valid for every $a, b \in \mathbb{C}$ and $m \in \mathbb{N}$. Thus we have $p(z) = (z - z_0)q(z)$ with

$$q(z) = \sum_{k=1}^{n} a_k \left(\sum_{\ell=0}^{\ell=k-1} z^\ell z_0^{k-1-\ell} \right) = \sum_{\ell=0}^{n-1} q_\ell z^\ell$$

with

$$q_\ell = \sum_{k=\ell+1}^{n} a_k z_0^{k-1-\ell}.$$

The function $q(z)$ is indeed a polynomial; its degree is $n - 1$ since $q_{n-1} = a_n \neq 0$. □

Formula (1.6.20) is trivial when $a = b$. When $a \neq b$, (1.6.20) is a mere rewriting of the formula (1.1.54)

$$1 + z + \cdots + z^{m-1} = \frac{1 - z^m}{1 - z}$$

for a geometric progression with $z = a/b$.

Solution of Exercise 1.5.3. It suffices to apply formula (1.5.4) to see that the sum is equal to 0, independently from the values of a and b. □

Solution of Exercise 1.5.4. To solve the first equation write

$$1 - z^2 + z^4 - z^6 = 0 = 1 - z^2 + z^4(1 - z^2) = (1 - z^2)(1 + z^4).$$

Thus the solutions are $z = \pm 1$ and the four roots of order 4 of -1, that is

$$z = \exp i \left\{ \frac{\pi + 2k\pi}{4} \right\}, \quad k = 0, 1, 2, 3.$$

To solve the second equation we first note that $z = 1$ is not a solution of it. Thus

$$1 + z + \cdots + z^7 = 0 \iff (1 + z + \cdots + z^7)(1 - z) = 0 \quad \text{and} \quad z \neq 1$$
$$\iff 1 - z^8 = 0 \quad \text{and} \quad z \neq 1,$$

where we have used (1.6.20). Hence, the solutions are

$$z = \exp i \left\{ \frac{2k\pi}{8} \right\}, \quad k = 1, \ldots, 7, \quad (k = 0 \quad \text{would correspond to} \quad z = 1).$$

Thus the equation has seven roots, which are

$$z = -1, \quad z = \pm i, \quad z = \pm \frac{1 + i}{\sqrt{2}} \quad \text{and} \quad z = \pm \frac{1 - i}{\sqrt{2}}.$$

We now turn to the third equation. Using the fundamental theorem of algebra we note that the cases n odd and n even will lead to a different number of solutions, since the equation is of degree $n - 1$ when n is even and of degree n when n is odd. We will recover directly this fact in the proof. The number $z = 1$ is not a solution of the equation to be solved, and so z is a solution of

$$(1 - z)^n = (1 + z)^n \tag{1.6.21}$$

if and only if

$$z \neq 1 \quad \text{and} \quad \left(\frac{1 + z}{1 - z} \right)^n = 1,$$

that is, if and only if

$$z \neq 1 \quad \text{and} \quad \frac{1 + z}{1 - z} = e^{\frac{2\pi i k}{n}}, \quad k = 0, \ldots, n - 1.$$

Thus

$$z \left(1 + e^{\frac{2\pi i k}{n}} \right) = e^{\frac{2\pi i k}{n}} - 1, \quad k = 0, \ldots, n - 1.$$

We have

$$e^{\frac{2\pi i k}{n}} + 1 = 0$$

if and only if

$$\frac{2\pi k i}{n} = \pi \quad (\text{mod } 2\pi),$$

that is, if and only if $2k = n \pmod 2$, or equivalently, if and only if n is even. Hence the roots of (1.6.21) are

$$z_k = \frac{e^{\frac{2\pi i k}{n}} - 1}{e^{\frac{2\pi i k}{n}} + 1},$$

where $k = 0, \ldots, n - 1$ for odd n while $k = 0, \ldots, \frac{n}{2} - 1, \frac{n}{2} + 1, \ldots, n - 1$ for n even.

To solve the fourth equation, we first note that $z = 0$ is not a solution of the equation, and so a number z solves

$$(1 - z)^n = z^n \tag{1.6.22}$$

if and only if $(1 - z)/z$ is a nth root of unity, that is if and only if

$$1 - z = ze^{\frac{2\pi ik}{n}}, \quad k = 0, \ldots, n - 1,$$

that is, if and only if

$$1 = z\left(1 + e^{\frac{2\pi ik}{n}}\right), \quad k = 0, \ldots, n - 1. \tag{1.6.23}$$

If n is odd, $\frac{2\pi ik}{n}$ will never be an odd multiple of π and so

$$1 + e^{\frac{2\pi ik}{n}} \neq 0, \quad k = 0, \ldots, n - 1.$$

The equation has thus n roots, namely

$$z = \frac{1}{1 + e^{\frac{2\pi ik}{n}}}, \quad k = 0, \ldots, n - 1.$$

If n is even, say $n = 2p$, the index $k = p$ corresponds to $\frac{2\pi ik}{n} = i\pi$, and thus (1.6.23) has no solution for $k = p$. Equation (1.6.22) has then $n - 1$ solutions, given by

$$z = \frac{1}{1 + e^{\frac{\pi ik}{p}}}, \quad k = 0, \ldots, p - 1, p + 1, \ldots, 2p - 1. \qquad \square$$

Solution of Exercise 1.5.5. (a) Let z_0 be a root of p. We have

$$
\begin{aligned}
p(z_0) = 0 &\iff \sum_{\ell=0}^{n} a_\ell z_0^\ell = 0 \\
&\iff \overline{\sum_{\ell=0}^{n} a_\ell z_0^\ell} = 0 \\
&\iff \sum_{\ell=0}^{n} a_\ell \overline{z_0}^\ell = 0, \quad \text{since the } a_\ell \text{ are real,} \\
&\iff p(\overline{z_0}) = 0.
\end{aligned}
$$

(b) Since the polynomial $z^4 - 5z^3 + 18z^2 - 17z + 13$ has real coefficients, it also vanishes at $z = 2 - 3i$ and hence is divisible by

$$(z - (2 + 3i))(z - (2 - 3i)) = z^2 - 4z + 13.$$

Doing the division leads to

$$z^4 - 5z^3 + 18z^2 - 17z + 13 = (z^2 - 4z + 13)(z^2 - z + 1)$$

and hence the other two zeroes of the equation are

$$z = \frac{1 \pm i\sqrt{3}}{2}.$$

□

Solution of Exercise 1.5.6. Note that $p_1(z) \equiv 0$, and so we begin with $n \geq 2$. We have

$$z^2 - 2z \cos a + 1 = (z - e^{ia})(z - e^{-ia}).$$

Since the polynomial p_n has real coefficients, it is enough to check that it vanishes for $z = e^{ia}$. We have

$$p_n(e^{ia}) = e^{ina} \sin a - e^{ia} \sin(na) + \sin((n-1)a),$$

that is

$$p_n(e^{ia}) = e^{ina} \frac{e^{ia} - e^{-ia}}{2i} - e^{ia} \frac{e^{ina} - e^{-ina}}{2i} + \frac{e^{i(n-1)a} - e^{-i(n-1)a}}{2i}.$$

But it is clear that this last expression vanishes.

The proof is almost the same for complex values of a. The fact that $p_n(e^{ia}) = 0$ does not imply automatically that $p_n(e^{-ia}) = 0$. This last fact, that is

$$p_n(e^{-ia}) = e^{-ina} \frac{e^{ia} - e^{-ia}}{2i} - e^{-ia} \frac{e^{ina} - e^{-ina}}{2i} + \frac{e^{i(n-1)a} - e^{-i(n-1)a}}{2i} = 0,$$

is readily checked. Thus, p is divisible by $(z - e^{ia})(z - e^{-ia}) = z^2 - 2z \cos a + 1$. □

Solution of Exercise 1.5.7. The idea behind the four factorizations is that the polynomials are real, and hence their non-real roots appear in pairs, which lead to second-degree real polynomials:

$$(z - z_0)(z - \overline{z_0}) = z^2 - 2(\operatorname{Re} z_0)z + |z_0|^2. \tag{1.6.24}$$

We focus on the first and third equalities, and leave to the reader the proofs of the other two.

The roots of the polynomial $z^{2n} + 1$ are $z_k = e^{i\theta_k}$, with

$$\theta_k = \frac{\pi}{2n} + \frac{k\pi}{n}, \quad k = 0, \ldots, 2n - 1.$$

The roots corresponding to $k = 0, \ldots, n-1$ are not conjugate to each other; indeed a pair of indices (k, k') corresponds to conjugate roots if

$$\frac{\pi}{2n} + \frac{k\pi}{n} = -\frac{\pi}{2n} - \frac{k'\pi}{n} \quad (\text{mod } 2\pi),$$

that is

$$\frac{1}{n} + \frac{k + k'}{n} = 0 \quad (\text{mod } 2),$$

which cannot hold if both k and k' are between 0 and $n - 1$.

Thus

$$z^{2n} + 1 = \prod_{k=0}^{2n-1} (z - z_k) = \prod_{k=0}^{n-1} (z - z_k)(z - \overline{z_k}).$$

But

$$(z - z_k)(z - \overline{z_k}) = z^2 - 2z\cos\theta_k + 1,$$

which concludes the proof of the first equality since

$$\cos\theta_k = \cos\left(\frac{(2k+1)\pi}{2n}\right).$$

We now prove the third equality, and assume that $n \geq 2$ (the case $n = 1$ is trivial). The roots of order $2n$ of unity are

$$z_k = \exp i\frac{2k\pi}{2n} = \exp i\frac{k\pi}{n}, \quad k = 0, \ldots, 2n - 1.$$

We have $z_0 = 1$ and $z_n = -1$. The other roots are not real, and appear in pairs since $p(z) = z^{2n} - 1$ has real coefficients (and thus, $p(w) = 0 \implies p(\overline{w}) = 0$; see Exercise 1.5.5). The roots from $k = 1$ to $k = n - 1$ are all different and so the roots of $p(z)$ are, besides 1 and -1,

$$z_k \quad \text{and} \quad \overline{z_k}, \quad k = 1, \ldots, n - 1.$$

Thus

$$z^{2n} - 1 = (z + 1)(z - 1) \prod_{k=1}^{n-1} (z - z_k)(z - \overline{z_k})$$

$$= (z + 1)(z - 1) \prod_{k=1}^{n-1} (z^2 - 2z \operatorname{Re} z_k + 1)$$

which concludes the proof of the third equality since $\operatorname{Re} z_k = \cos(\frac{k\pi}{n})$.

Using formula (1.1.54) for the sum of a geometric series we obtain

$$p(z) \overset{\text{def.}}{=} \sum_{k=0}^{n-1} z^{2k} = \frac{1 - z^{2n}}{1 - z^2},$$

and hence, using the previous arguments to prove the third equality and also using the third equality itself we have for $n \geq 2$

$$p(z) = \prod_{k=1}^{n-1} \left(z - \exp\frac{ik\pi}{n}\right)\left(z - \exp\frac{-ik\pi}{n}\right) = \prod_{k=1}^{n-1} (z^2 - 2z \operatorname{Re} z_k + 1).$$

We now prove (1.5.12). Setting $z = 1$ in the above equality we have

$$n = \prod_{k=1}^{n-1} \left(2 - 2\cos\left(\frac{k\pi}{n}\right) \right).$$

Recall that

$$1 - \cos\left(\frac{k\pi}{n}\right) = 2\sin^2\left(\frac{k\pi}{2n}\right).$$

Hence

$$n = \prod_{k=1}^{n-1} \left(2 - 2\cos\left(\frac{k\pi}{n}\right) \right)$$

$$= \prod_{k=1}^{n-1} 4\sin^2\left(\frac{k\pi}{2n}\right)$$

$$= 4^{n-1} \prod_{k=1}^{n-1} \sin^2\left(\frac{k\pi}{2n}\right)$$

and hence the result by taking the square root of both sides since the numbers $\sin(\frac{k\pi}{2n}) > 0$ for $k = 1, \ldots, n-1$. In view of the proof of Exercise 3.7.16 we note the formula

$$\frac{n}{2^{n-1}} = \prod_{k=1}^{n-1} \left(1 - \cos\left(\frac{k\pi}{n}\right) \right), \qquad n \geq 2, \tag{1.6.25}$$

which follows from the previous arguments. Finally, we prove (1.5.13). We set $z = e^{it}$ in (1.5.10) to obtain

$$e^{2int} - 1 = (e^{it} + 1)(e^{it} - 1)\prod_{k=1}^{n-1}\left(e^{2it} - 2e^{it}\cos\left(\frac{k\pi}{n}\right) + 1 \right).$$

Thus,

$$e^{int}(e^{int} - e^{-int}) = e^{it/2}(e^{it/2} + e^{-it/2})e^{it/2}(e^{it/2} - e^{-it/2})$$

$$\times \prod_{k=1}^{n-1} e^{it}\left(e^{it} + e^{-it} - 2\cos\left(\frac{k\pi}{n}\right) \right).$$

Dividing both sides by $2ie^{int}$ we obtain

$$\sin(nt) = 2\cos(t/2)\sin(t/2)\prod_{k=1}^{n-1}\left(2\cos t - 2\cos\left(\frac{k\pi}{n}\right) \right),$$

and hence the result (still for $n \geq 2$).   □

One can also try to prove these equalities using only real analysis. For instance, using the formula

$$a^4 + b^4 = (a^2 + b^2)^2 - 2a^2b^2 = (a^2 + b^2 + \sqrt{2}ab)(a^2 + b^2 - \sqrt{2}ab), \quad a, b \in \mathbb{C},$$

(which is basically the completing the square formula), we have

$$z^4 + 1 = (z^2 + \sqrt{2}z + 1)(z^2 - \sqrt{2}z + 1)$$

and

$$z^8 + 1 = (z^4 + \sqrt{2}z^2 + 1)(z^4 - \sqrt{2}z^2 + 1)$$
$$= \left((z^2 + 1)^2 - z^2(2 - \sqrt{2})\right)\left((z^2 + 1)^2 - z^2(2 + \sqrt{2})\right)$$
$$= \left(z^2 - z\sqrt{2 - \sqrt{2}} + 1\right)\left(z^2 + z\sqrt{2 - \sqrt{2}} + 1\right)$$
$$\times \left(z^2 - z\sqrt{2 + \sqrt{2}} + 1\right)\left(z^2 + z\sqrt{2 + \sqrt{2}} + 1\right).$$

Similarly the formula

$$a^6 + b^6 = (a^2 + b^2)(a^4 - a^2b^2 + b^4)$$
$$= (a^2 + b^2)((a^2 + b^2)^2 - 3a^2b^2)$$
$$= (a^2 + b^2)(a^2 + b^2 + \sqrt{3}ab)(a^2 + b^2 - \sqrt{3}ab), \quad a, b \in \mathbb{C},$$

will give (see [96, p. 397])

$$z^6 + 1 = (z^2 + 1)(z^2 + \sqrt{3}z + 1)(z^2 - \sqrt{3}z + 1).$$

Still, the complex variable arguments above give more insight as to the factors themselves.

Solution of Exercise 1.5.8. Assume that the equation has a solution z_0 with $|z_0| \le 1$. Then $5 = -z_0^3 - 3z_0$, and so

$$5 = |-z_0^3 - 3z_0| \le |z_0|^3 + 3|z_0| \le 4,$$

which is a contradiction.   □

Solution of Exercise 1.5.9. Let $c = \max_{j=1,\ldots,n} |c_j|^{\frac{1}{j}}$. By hypothesis $c > 0$. Let z be a root of the polynomial equation

$$z^n + c_1 z^{n-1} + \cdots + c_n = 0, \tag{1.6.26}$$

and let $u = \dfrac{z}{c}$. Dividing both sides of (1.6.26) by c^n we obtain

$$u^n + \frac{c_1}{c} u^{n-1} + \cdots + \frac{c_n}{c^n} = 0. \tag{1.6.27}$$

By definition of c we have $|c_j| \leq c^j$. Therefore, (1.6.27) leads to

$$|u|^n \leq |u|^{n-1} + \cdots + 1. \qquad (1.6.28)$$

Assume that $|u| \geq 2$. Then $1/|u| \leq 1/2$. Dividing both sides of (1.6.28) by $|u|^n$ leads to

$$1 \leq \frac{1}{|u|} + \cdots + \frac{1}{|u|^n}$$
$$\leq \frac{1}{2} + \cdots + \frac{1}{2^n}$$
$$< 1,$$

which is a contradiction. Thus $|u| < 2$, that is $|z| < 2\max_{j=1,\ldots,n} |c_j|^{\frac{1}{j}}$.   □

Chapter 2

Complex Numbers: Geometry

As is well known, the complex field can be identified with $\mathbb{R}^2$ via the map

$$z = x + iy \mapsto (x, y).$$

An important new feature with respect to real analysis is the introduction of the point at infinity, which leads to the compactification of $\mathbb{C}$. These various aspects, and some others, such as Moebius maps, are considered in this chapter.

2.1 Geometric interpretation

Exercise 2.1.1. *Describe the polygon whose vertices are defined by the roots of order n of unity.*

To have a good understanding of some forthcoming notions (for instance, limit at infinity, or the notion of pole of an analytic function), it is better to be able to leave the complex plane, and go one step further and add a point, called infinity, and denoted by the symbol ∞ (without sign, in opposition to real analysis, where you have $\pm\infty$), in such a way that the extended complex plane $\mathbb{C} \cup \{\infty\}$ is compact. The set

$$\mathbb{C} \cup \{\infty\}$$

is called the extended complex plane. See Section 15.1 for a reminder of the notion of compactness. For the topological details of the construction, see Section 15.3. In the next exercise we discuss the geometric interpretation of the point at infinity, by identifying the extended complex plane with the Riemann sphere

$$\mathbb{S}_2 = \left\{ (x_1, x_2, x_3) \in \mathbb{R}^3 \ ; x_1^2 + x_2^2 + x_3^2 = 1 \right\}.$$

Exercise 2.1.2. *For $(x_1, x_2, x_3) \in \mathbb{S}_2 \setminus \{(0,0,1)\}$, define $\varphi(x_1, x_2, x_3)$ to be the intersection of the line defined by the points $(0,0,1)$ and (x_1, x_2, x_3) with the complex plane. Show that*

$$\varphi(x_1, x_2, x_3) = \frac{x_1 + i x_2}{1 - x_3}, \tag{2.1.1}$$

and that φ is a bijection between $\mathbb{S}_2 \setminus \{(0,0,1)\}$ and $\mathbb{C}$, with inverse given by

$$\varphi^{-1}(u + iv) = \left(\frac{2u}{u^2 + v^2 + 1}, \frac{2v}{u^2 + v^2 + 1}, \frac{u^2 + v^2 - 1}{u^2 + v^2 + 1} \right). \tag{2.1.2}$$

Setting $z = u + iv$, (2.1.2) may be rewritten as

$$\varphi^{-1}(z) = \left(\frac{z + \bar{z}}{|z|^2 + 1}, \frac{z - \bar{z}}{i(|z|^2 + 1)}, \frac{|z|^2 - 1}{|z|^2 + 1} \right).$$

The map (2.1.1) is called the *stereographic projection.*

The geometrical interpretation of the point at infinity is as follows: The map φ is extended to the point $(0, 0, 1)$ by

$$\varphi(0, 0, 1) = \infty, \tag{2.1.3}$$

and going to ∞ on the complex plane means going to $(0, 0, 1)$ on the Riemann sphere. More precisely, recall that, by definition, a sequence of complex numbers $(z_n)_{n \in \mathbb{N}}$ tends to infinity if

$$\lim_{n \to \infty} |z_n| = +\infty, \tag{2.1.4}$$

that is, if and only if

$$\lim_{n \to \infty} \varphi^{-1}(z_n) = (0, 0, 1), \tag{2.1.5}$$

where this last limit can be understood in two equivalent ways: The first, and simplest, is just to say that the limit is coordinate-wise in $\mathbb{R}^3$. The second is to view $\mathbb{S}_2$ as a topological manifold, and see the limit in the corresponding topology. See also Exercise 15.1.5, where φ allows us to define a metric on the Riemann sphere, called the stereographic metric.

The intersection of $\mathbb{S}_2$ with a (non-tangent) plane is a circle. Note that the *projection* of a circle of the Riemann sphere on the plane will not be a circle in general. For instance the projection of the circle

$$x_1 = x_3,$$
$$x_1^2 + x_2^2 + x_3^2 = 1$$

onto the plane is the ellipse $2x_1^2 + x_2^2 = 1$. But we have:

Exercise 2.1.3 (see [28, Exercise 19, pp. 16–17]).

(a) *There is a one-to-one correspondence between circles on the Riemann sphere and straight lines or circles on the plane.*

(b) *Let S be a circle on the Riemann sphere. Then, $\varphi(S \setminus \{(0,0,1)\})$ is a circle on the plane if and only if $(0,0,1) \notin S$ and is a line otherwise.*

To summarize, via the map φ^{-1} the point at infinity in the extended complex plane should be seen as any other point of the complex plane. Furthermore, there is no difference between lines and circles in the extended complex plane. A line is a circle whose image under φ^{-1} goes via the point $(0,0,1)$.

The notion of a *simply-connected set* is central in complex function theory. In [19, Theorem 4.1], eleven equivalent definitions for an open connected set to be simply-connected are given. See also [42, Theorem 4.65, p. 113] for a similar result. In this book, we focus most of the time on the much simpler (but not conformally invariant) notion of star-shaped set, but we will give a number of equivalent characterizations of simply-connected sets. Recall first:

Definition 2.1.4. A set $\Omega \subset \mathbb{C}$ is called *star-shaped* if there is a point $z_0 \in \Omega$ such that, for every $z \in \Omega$, the interval

$$[z_0, z] = \{tz_0 + (1-t)z, \ t \in [0,1]\}$$

lies in Ω.

The point z_0 need not be unique. For instance, a convex set is star-shaped with respect to each of its points.

Theorem 2.1.5. *An open star-shaped subset of $\mathbb{C}$ is simply-connected.*

The first definition of a simply-connected set, which is condition (d) in [19, Theorem 4.1], is as follows:

Definition 2.1.6. A connected open subset Ω of the complex plane is *simply-connected* if the set $\mathbb{S}_2 \setminus \varphi^{-1}(A)$ is connected (in the topology of $\mathbb{S}_2$).

It is enough to check that $\mathbb{S}_2 \setminus \varphi^{-1}(A)$ is arc-connected.

Exercise 2.1.7.

(a) *Show that the punctured plane $\mathbb{C} \setminus \{0\}$ is not simply-connected.*

(b) *Show that the set*

$$\Omega = \mathbb{C} \setminus \{\{x \in \mathbb{R}; |x| \geq 1\} \cup \{iy; y \in \mathbb{R}, |y| \geq 1\}\}$$

is simply-connected.

The Riemann sphere can also be identified with the projective line. This last object is introduced in the next exercise:

Exercise 2.1.8. *In $\mathbb{C}^2 \setminus \{(0,0)\}$ define the equivalence relation:*

$$(z_1, z_2) \sim (w_1, w_2) \quad \Longleftrightarrow \quad (z_1, z_2) = c(w_1, w_2) \tag{2.1.6}$$

for some non-zero complex number c.

(1) *Show that $\sim$ indeed defines an equivalence relation. We denote by $\overset{\circ}{z}$ the equivalence class of $(z_1, z_2) \in \mathbb{C}^2$ and by $\mathbb{P}$ the set of the equivalence classes.*

(2) *Let $(\overset{\circ}{z_1, z_2}) \in \mathbb{P}$. Show that the elements in the equivalence class have all at the same time either non-zero second component or zero second component.*

We denote by $\mathbb{A}$ ($\mathbb{A}$ stands for affine) the set of equivalence classes for which the second component in any of its representative is non-zero. Show that the map

$$\psi(\overset{\circ}{z}) = \frac{z_1}{z_2}$$

is a one-to-one correspondence from $\mathbb{A}$ onto $\mathbb{C}$, and that its inverse is given by

$$\psi^{-1}(u) = (\overset{\circ}{u, 1}) \quad u \in \mathbb{C}. \tag{2.1.7}$$

The projective line $\mathbb{P}$ is the set of the equivalence classes of $\sim$.

Exercise 2.1.9. *Prove the claim made in the proof of Exercise 2.1.3 on the intersection of the plane and $\mathbb{S}_2$, that is, prove that equation (2.4.2) in Section 2.4 below is a necessary and sufficient condition for the plane and the Riemann sphere to intersect, and that the plane is tangent to the Riemann sphere if and only if equality holds in (2.4.2).*

2.2   Circles and lines and geometric sets

We recall now the formulas for equations of lines and circles in the complex plane. In the plane $\mathbb{R}^2$, a line is the set of points $M = (x, y)$ such that

$$ax + by + c = 0,$$

where $(a, b, c) \in \mathbb{R}^3$ and $(a, b) \neq (0, 0)$. Setting

$$x = \frac{z + \overline{z}}{2} \quad \text{and} \quad y = \frac{z - \overline{z}}{2i},$$

we get

$$\overline{\alpha}z + \alpha\overline{z} + \beta = 0, \tag{2.2.1}$$

with

$$\alpha = a - bi \in \mathbb{C} \setminus \{0\} \quad \text{and} \quad \beta = 2c \in \mathbb{R}.$$

Conversely, any expression (2.2.1) with $(\alpha, \beta) \in (\mathbb{C} \setminus \{0\}) \times \mathbb{R}$ is the equation of a real line.

Similarly, a circle in $\mathbb{R}^2$ is the set of points $M = (x, y)$ such that

$$x^2 + y^2 - 2ax - 2by + c = 0,$$

where a, b, c are real and such that $a^2 + b^2 - c > 0$. The center of the circle is the point (a, b) and its radius is $\sqrt{a^2 + b^2 - c}$. In the complex plane we obtain

$$|z|^2 - a(z + \overline{z}) + ib(z - \overline{z}) + c = 0,$$

that is,

$$|z|^2 - z(a - ib) - \overline{z}(a + ib) + c = 0,$$

that is, using (1.6.2),

$$|z - (a + ib)|^2 = a^2 + b^2 - c.$$

We have just seen two analytic expressions, one for lines and one for circles. There is an alternative way to write the equations of lines and circles in a unified manner as

$$|z - z_0| = \lambda |z - z_1|,$$

where $z_0 \neq z_1$ and $\lambda > 0$. This expression describes a circle when $\lambda \neq 1$, this is an Apollonius circle, and a line when $\lambda = 1$.

Exercise 2.2.1. *Show that the set of points*

$$|z - z_0| = \lambda |z - z_1| \tag{2.2.2}$$

where $\lambda > 0$, $\lambda \neq 1$ and $z_0 \neq z_1$ is the circle with center and radius

$$\frac{z_0 - \lambda^2 z_1}{1 - \lambda^2} \quad and \quad \frac{\lambda}{|1 - \lambda^2|} |z_0 - z_1|$$

respectively. Show that, conversely, a line or a circle is of the form (2.2.2) for some choice of λ and of z_0, z_1.

Remark 2.2.2. For a fixed choice of z_0, z_1 one obtains a family of coaxal circles when one lets λ vary. See [37], [179]. The notion is used in particular in [99] in a simplified proof of Riemann's mapping theorem; see Exercise 5.1.5.

Remark 2.2.3. Replacing λ by $1/\lambda$ keeps the radius inchanged, but clearly interchanges the roles of z_0 and z_1. Furthermore, the center of the Apollonius circle tends to z_0 and its radius tends to 0 as λ goes to 0. The center goes to z_1, and the radius still goes to 0, when λ goes to infinity.

Exercise 2.2.4. *Characterize and draw the sets of points in the plane $\mathbb{R}^2$ such that:*

(a) $|z - 1 + i| = 1$.

(b) $z^2 + \overline{z}^2 = 2$.

(c) $|z - i| = |z + i|$.

(d) $|z|^2 + 3z + 3\bar{z} + 10 = 0$.

(e) $|z|^2 + 3z + \bar{z} + 5 = 0$.

(f) $z^2 + 3z + 3\bar{z} + 5 = 0$.

(g) $|z| \geq 1 - \operatorname{Re} z$.

(h) $\operatorname{Re}(z(1 - i)) < \sqrt{2}$.

For a question similar to the last one, see [75, p. 13].

The following is the last exercise of the book [154].

Exercise 2.2.5. *Find the image of the unit circle under the map* $z \mapsto w(z) = z - z^n/n$ *where* $n = 2, 3, \ldots$.

2.3   Moebius maps

Recall that a Moebius map is a transformation of the form

$$\varphi(z) = \frac{az + b}{cz + d}$$

with $ad - bc \neq 0$. Such transformations are also called *linear fractional transfor-mations*, and *linear transformations* in the older literature. See for instance Ford's book [86]. We recall that the image under a Moebius map of a line or a circle in the complex plane is still a line or a circle. We have already met a special case of Moebius maps in Section 1.1, namely the Blaschke factors; see (1.1.44), (1.1.45) and (1.1.46). Finite or infinite products of Blaschke factors (of the same kind) are considered in Exercise 3.7.12.

The formula

$$\varphi(z) - \varphi(w) = \frac{(ad - bc)(z - w)}{(cz + d)(cw + d)}, \tag{2.3.1}$$

that is, for $z \neq w$,

$$\frac{\varphi(z) - \varphi(w)}{z - w} = \frac{(ad - bc)}{(cz + d)(cw + d)},$$

will prove useful in the sequel.

The first exercise expresses the fact that Moebius maps form a group isomor-phic to the group $GL(\mathbb{C}, 2)/(\mathbb{C} \setminus \{0\})$ of 2×2 invertible matrices with complex entries factored out by the invertible numbers.

Exercise 2.3.1. *Let*

$$\varphi_\ell(z) = \frac{a_\ell z + b_\ell}{c_\ell z + d_\ell}, \quad \ell = 1, 2,$$

be two Moebius transforms. Show that

$$\varphi_1(\varphi_2(z)) = \frac{az + b}{cz + d}, \tag{2.3.2}$$

where

$$\begin{pmatrix} a & b \\ c & d \end{pmatrix} = \begin{pmatrix} a_1 & b_1 \\ c_1 & d_1 \end{pmatrix} \begin{pmatrix} a_2 & b_2 \\ c_2 & d_2 \end{pmatrix}. \tag{2.3.3}$$

Sometimes it is convenient to use the following notation: Setting

$$M = \begin{pmatrix} a & b \\ c & d \end{pmatrix},$$

we define for $w \in \mathbb{C}$

$$T_M(w) = \frac{aw + b}{cw + d}. \tag{2.3.4}$$

Equation (2.3.2) can be then rewritten as

$$T_{M_1}(T_{M_2}(w)) = T_{M_1 M_2}(w). \tag{2.3.5}$$

This equation suggests that infinite products of matrices should be considered, when infinite compositions of Moebius transforms come into play. See Theorem 3.7.3 for the first issue and Section 11.5 for the second one. The matrices in these products are usually normalized. Indeed we have:

Exercise 2.3.2. *Let φ be a non-degenerate Moebius map, and let*

$$M_1 = \begin{pmatrix} a_1 & b_1 \\ c_1 & d_1 \end{pmatrix} \quad \text{and} \quad M_2 = \begin{pmatrix} a_2 & b_2 \\ c_2 & d_2 \end{pmatrix}$$

be such that

$$\varphi(z) = T_{M_1}(z) = T_{M_2}(z),$$

for every z in their common domain of definition. Show that there is a complex number $\lambda \neq 0$ such that

$$M_2 = \lambda M_1.$$

Using (2.3.3) we can study in particular the compositions of Blaschke factors of the form (1.1.44).

Exercise 2.3.3. *Let u and v be in the open unit disk $\mathbb{D}$. Show that*

$$w = \frac{u + v}{1 + u\bar{v}} \in \mathbb{D}, \tag{2.3.6}$$

and that

$$b_u(b_v(z)) = \frac{1 + u\bar{v}}{1 + v\bar{u}} b_w(z). \tag{2.3.7}$$

In Exercise 2.3.15 we compute the nth iterate of a Blaschke factor of the form (1.1.44).

Exercise 2.3.4.

(1) *Let* $w \in \mathbb{D}$ *(resp.* $w \in \mathbb{C}_r$*). Prove that* b_w *(resp.* B_w*) is a one-to-one map from* $\mathbb{D}$ *(resp.* $\mathbb{C}_r$*) onto* $\mathbb{D}$*.*

(2) *What happens if* $|w| > 1$ *in the first case and* $\mathrm{Re}\, w < 0$ *in the second case?*

Exercise 2.3.5. *Let* $z_1, z_2 \in \mathbb{D}$*. Show that there is a map of the form* cb_w *with* $c \in \mathbb{T}$ *and* $w \in \mathbb{D}$ *such that* $z_2 = cb_w(z_1)$*.*

Exercise 2.3.6 (see [184, p. 25], [75, Exercice 33.15, p. 301]). *Given two triples of complex numbers* (z_1, z_2, z_3) *and* (w_1, w_2, w_3) *such that*

$$z_\ell \neq z_j \quad and \quad w_\ell \neq w_j, \quad for \quad \ell, j = 1, 2, 3, \quad \ell \neq j,$$

show that the map $z \mapsto w$ *defined by*

$$\frac{w - w_1}{w - w_2} \cdot \frac{w_3 - w_2}{w_3 - w_1} = \frac{z - z_1}{z - z_2} \cdot \frac{z_3 - z_2}{z_3 - z_1} \tag{2.3.8}$$

is a Moebius map such that $w(z_\ell) = w_\ell$ *for* $\ell = 1, 2, 3$*.*

Suitably interpreted, the formula (2.3.8) still makes sense when one of the w_j or/and one of the z_j is equal to ∞. For instance, when $w_1 = z_2 = \infty$ we have

$$\frac{w_3 - w_2}{w - w_2} = \frac{z - z_1}{z_3 - z_1},$$

that is

$$w = w_2 + \frac{(w_3 - w_2)(z_3 - z_1)}{z - z_1}.$$

Exercise 2.3.7. *Show that four points are on the same complex circle (or on the same complex line) if and only if the number*

$$\frac{\left(\dfrac{(z_1 - z_2)}{(z_1 - z_3)} \right)}{\left(\dfrac{(z_2 - z_4)}{(z_3 - z_4)} \right)} \tag{2.3.9}$$

is real.

The number (2.3.9) is called the *cross-ratio*. See for instance [199, Theorem 2, p. 3].

Hint to the solution. Consider the case of a circle.

(1) Prove that the result is true for the unit circle.

(2) Prove that any circle can be mapped onto the unit circle by an appropriate Moebius map.

(3) Prove that (2.3.9) is invariant under Moebius transforms.

Exercise 2.3.8. *For which $k \in \mathbb{R}$ is the image of the circle $|z - 1| = k$ under the Moebius map $f(z) = \dfrac{z - 3}{1 - 2z}$ a line? Find the equation of the line.*

The next result appears in [155]; see also [127, Theorem 7, p. 67]. A natural approach would be to use the Schur algorithm. See (6.5.7). This approach does not seem to lead anywhere, and the pedestrian approach leads to a quite short proof.

Exercise 2.3.9. *Show that the non-trivial Moebius map $\varphi(z) = \dfrac{az + b}{cz + d}$ maps the open unit disk into itself if and only if*

$$|a\bar{c} - b\bar{d}| + |ad - bc| \leq |d|^2 - |c|^2. \tag{2.3.10}$$

We note the following: For the Blaschke factor (1.1.44), we have

$$a = 1, \quad b = -w, \quad c = -\overline{w}, \quad d = 1,$$

and inequality (2.3.10) reads as

$$|-1 \cdot w + w \cdot 1| + |1 - |w|^2| \leq 1 - |w|^2,$$

and thus becomes an equality. More precisely:

Exercise 2.3.10. *Using the previous result, show that a Moebius map sends the open unit disk onto itself if and only if it is of the form kb_w, where $k \in \mathbb{T}$ and b_w given by (1.1.44) with $w \in \mathbb{T}$.*

We note that equality in (2.3.10) may hold even when the image of the open unit disk is included in, but different from, the open unit disk. For instance, take the function

$$\frac{z + 1}{-z + 3},$$

which appears in the solution of Exercise 6.5.6.

We send the reader to Exercise 16.3.16 for a related exercise. As suggested by our colleague Prof. Izchak Lewkowicz, we propose:

Exercise 2.3.11. *Let $\varphi(z) = \dfrac{az + b}{cz + d}$ be a non-trivial Moebius map. Find necessary and sufficient conditions on a, b, c, d for φ to map the open left half-plane into itself.*

Exercise 2.3.12. *For which values of z_0 does the function*

$$\varphi(z) = \frac{z - 1}{z - z_0}$$

map the open unit disk into itself?

Exercise 2.3.13. *Let* $\varphi(z) = \dfrac{az+b}{cz+d}$ *be a non-trivial Moebius map, and assume that the equation* $\varphi(z) = z$ *has two distinct solutions* z_1 *and* z_2.

(a) *Prove that there is a number* $k \in \mathbb{C}$ *such that*

$$\frac{\varphi(z) - z_1}{\varphi(z) - z_2} = k\frac{z - z_1}{z - z_2}. \tag{2.3.11}$$

(b) *Give a formula for the nth iterate*

$$\underbrace{\varphi \circ \varphi \circ \cdots \varphi}_{n\text{-times}}.$$

(c) *Compute the nth iterate of*

$$\varphi(z) = \frac{1 - 3z}{z - 3}.$$

Remark 2.3.14. The number k is called the multiplier of φ (see [86, (12), p. 10]. For relations and applications to the theory of automorphic functions, see for instance [86].

We now look at the special case where $\varphi(z) = b_w(z)$, where b_w is the Blaschke factor (1.1.44).

Exercise 2.3.15.

(a) *Compute the nth iterate of the Blaschke factor (1.1.44).*
(b) *What is*

$$\lim_{n \to \infty} \underbrace{b_w \circ b_w \circ \cdots \circ b_w}_{n\text{-times}}.$$

In relation with the following exercise, see the Herglotz formula for functions holomorphic in the open upper half-plane, and with a positive real part there; see formula (5.5.21).

Exercise 2.3.16. *Let* $w \in \mathbb{C} \setminus \mathbb{R}$. *What is the image of the real line under the Moebius transform* $\frac{zw+1}{z-w}$ *?*

2.4   Solutions

Solution of Exercise 2.1.1. By formula (1.1.15) with $\rho = 1$ and $\theta = 0$ we see that the roots of order n of unity are

$$z_k = \cos\left(\frac{2k\pi}{n}\right) + i\sin\left(\frac{2k\pi}{n}\right), \quad k = 0, \ldots, n-1.$$

The points $M_k = \left(\cos(\frac{2k\pi}{n}), \sin(\frac{2k\pi}{n})\right)$, $k = 0, \ldots, n-1$, are on the unit circle, and are the vertices of a regular polygon of order n, the first vertex being the point $(1, 0)$. □

Solution of Exercise 2.1.2. The equation of the line passing through the points $(0, 0, 1)$ and (x_1, x_2, x_3) is

$$(u, v, w) = (0, 0, 1) + t(x_1, x_2, x_3 - 1) \quad t \in \mathbb{R}.$$

We want $w = 0$, and thus

$$1 + t(x_3 - 1) = 0, \quad \text{that is,} \quad t = \frac{1}{1 - x_3}.$$

The result follows. The map is clearly one-to-one. To show that it is onto, let $u + iv \in \mathbb{C}$ be given. A point $(x_1, x_2, x_3) \in \mathbb{S}_2 \setminus \{(0, 0, 1)\}$ is such that $\varphi(x_1, x_2, x_3) = u + iv$ if and only if

$$x_1 = u(1 - x_3) \quad \text{and} \quad x_2 = v(1 - x_3).$$

Thus

$$u^2 + v^2 = \frac{x_1^2 + x_2^2}{(1 - x_3)^2} = \frac{1 - x_3^2}{(1 - x_3)^2} = \frac{1 + x_3}{1 - x_3}.$$

Thus

$$x_3 = \frac{u^2 + v^2 - 1}{u^2 + v^2 + 1}.$$

So,

$$1 - x_3 = \frac{2}{u^2 + v^2 + 1},$$

and the formulas for x_1 and x_2 follow. □

Solution of Exercise 2.1.3. Let

$$ax_1 + bx_2 + cx_3 = d$$

be the equation of a plane P in $\mathbb{R}^3$. Note that $(0, 0, 1) \in P$ if and only if

$$c = d. \tag{2.4.1}$$

Using elementary analytic geometry one sees that the plane intersects the Riemann sphere if and only if

$$a^2 + b^2 + c^2 \geq d^2, \tag{2.4.2}$$

and that it is non-tangent if and only if the inequality is strict in (2.4.2):

$$a^2 + b^2 + c^2 > d^2. \tag{2.4.3}$$

These facts are proved in Exercise 2.1.9 at the end of this section. Consider now the image of $S \cap P$ under φ:

$$(u, v) \in \varphi(S \cap P) \iff \varphi^{-1}(u, v) \in S \cap P,$$

that is, if and only if it holds that

$$2au + 2bv + c(u^2 + v^2 - 1) = d(u^2 + v^2 + 1),$$

that is, if and only if

$$(c - d)(u^2 + v^2) + 2au + 2bv = c + d. \tag{2.4.4}$$

If $c = d$ we obtain the equation of a line. If $c \neq d$, we rewrite (2.4.4) as

$$\left(u + \frac{a}{c - d}\right)^2 + \left(v + \frac{b}{c - d}\right)^2 = \frac{c + d}{c - d} + \frac{a^2}{(c - d)^2} + \frac{b^2}{(c - d)^2}$$
$$= \frac{a^2 + b^2 + c^2 - d^2}{(c - d)^2},$$

which is the equation of a circle since (2.4.3) is in force.

The converse direction is done as follows: Given a line

$$au + bv = e,$$

we consider the plane

$$ax_1 + bx_2 + ex_3 = e.$$

Given a circle

$$(u - a_0)^2 + (v - b_0)^2 = R^2,$$

we may assume that $c - d = 1$ in the equation of the plane, and take the plane

$$a_0 x_1 + b_0 x_2 + cx_3 = d,$$

where c and d satisfy

$$c - d = 1 \quad \text{and} \quad R^2 - a_0^2 - b_0^2 = c + d. \qquad \Box$$

Solution of Exercise 2.1.7. (a) The set

$$\mathbb{S}_2 \setminus \varphi^{-1}\{\mathbb{C} \setminus \{0\}\}$$

consists of two points, namely $(0, 0, -1)$ and $(0, 0, 1)$, and therefore is not connected.

(b) Consider the points

$$A_1 = \varphi^{-1}(1,0) = (1,0,0),$$
$$A_2 = \varphi^{-1}(-1,0) = (-1,0,0),$$
$$A_3 = \varphi^{-1}(0,1) = (0,1,0),$$
$$A_4 = \varphi^{-1}(0,-1) = (0,-1,0)$$

in $\mathbb{S}_2$. With Ω as in the statement of the exercise, $\varphi^{-1}(\Omega)$ consists of the point $(0,0,1)$ and of four non-intersecting lines which link the points each of the point A_j to $(0,0,1)$. □

We note that the above set is in fact star-shaped with respect to the origin. In this book, we concentrate almost uniquely on star-shaped sets, and therefore have given a somewhat informal proof of the above exercise.

Solution of Exercise 2.1.8. Let $\overset{\circ}{z} \in \mathbb{P}$ and (z_1, z_2) and (w_1, w_2) be two elements in $\overset{\circ}{z}$. Since $z_1 = cw_1$ for $c \neq 0$, we see that z_1 and w_1 are simultaneously both zero or both non-zero. The map ψ is well defined; indeed, let (z_1, z_2) and (w_1, w_2) be two elements in $\overset{\circ}{z}$. If $z_1 = 0$, then $w_1 = 0$ and $\psi(\overset{\circ}{z}) = 0$. On the other hand, if $z_1 \neq 0$, then $w_1 \neq 0$ and it follows from (2.1.6) that

$$\frac{z_2}{z_1} = \frac{w_2}{w_1},$$

and so ψ is well defined. The map is one-to-one. Indeed, let $\overset{\circ}{z}$ and $\overset{\circ}{w}$ be two elements of $\mathbb{P}$, and assume that $\overset{\circ}{z} \neq \overset{\circ}{w}$. Let $(z_1, z_2) \in \overset{\circ}{z}$ and $(w_1, w_2) \in \overset{\circ}{w}$. Then, if $z_1 = 0$ we have that $w_1 \neq 0$ and so $\psi(\overset{\circ}{z}) \neq \psi(\overset{\circ}{w})$. If both z_1 and w_1 are different from 0,

$$\frac{z_2}{z_1} \neq \frac{w_2}{w_1},$$

and in this case too, $\psi(\overset{\circ}{z}) \neq \psi(\overset{\circ}{w})$. Finally, the formula (2.1.7) follows from the fact that $u \in (\overset{\circ}{u}, 1)$. □

Solution of Exercise 2.1.9. The point with coordinates

$$(x_0, y_0, z_0) = \left(\frac{da}{a^2 + b^2 + c^2}, \frac{db}{a^2 + b^2 + c^2}, \frac{dc}{a^2 + b^2 + c^2} \right)$$

$$= \frac{d}{a^2 + b^2 + c^2}(a, b, c)$$

belongs to the plane. Let u and v be a pair of unit and orthogonal vectors in $\mathbb{R}^3$, which are moreover orthogonal to (a, b, c). A point (x, y, z) is in $P \cap \mathbb{S}_2$ if and only if

$$x^2 + y^2 + z^2 = 1 \quad \text{and} \quad (x - x_0, y - y_0, z - z_0) = tu + sv$$

for some $t, s \in \mathbb{R}$.

Equivalently,

$$(x, y, z) = \frac{d}{a^2 + b^2 + c^2}(a, b, c) + tu + sv,$$

and therefore, taking norms of both sides of this equality,

$$1 = \frac{d^2}{a^2 + b^2 + c^2} + t^2 + s^2.$$

Thus

$$\frac{a^2 + b^2 + c^2 - d^2}{a^2 + b^2 + c^2} = t^2 + s^2. \tag{2.4.5}$$

Thus, the intersection will be non-empty if and only if (2.4.2) is in force, and reduced to a point if and only if

$$a^2 + b^2 + c^2 = d^2.$$

Assume now that (2.4.3) holds, and let

$$R = \sqrt{\frac{a^2 + b^2 + c^2 - d^2}{a^2 + b^2 + c^2}}.$$

It follows from (2.4.5) that there exists $\theta \in [0, 2\pi)$ such that

$$s = R\cos\theta \quad \text{and} \quad t = R\sin\theta.$$

It follows that the intersection of the plane P and of the Riemann sphere is the circle with center $\frac{d}{a^2+b^2+c^2}(a, b, c)$ and radius $R = \sqrt{\frac{a^2+b^2+c^2-d^2}{a^2+b^2+c^2}}$.   □

Solution of Exercise 2.2.1. Equality (2.2.2) is equivalent to

$$|z|^2 + |z_0|^2 - 2\operatorname{Re} z\overline{z_0} = \lambda^2(|z|^2 + |z_1|^2 - 2\operatorname{Re} z\overline{z_1}),$$

that is, since $\lambda > 0$ and $\lambda \neq 1$,

$$|z|^2 - 2\operatorname{Re} z\frac{\overline{z_0} - \lambda^2\overline{z_1}}{1 - \lambda^2} = \frac{\lambda^2|z_1|^2 - |z_0|^2}{1 - \lambda^2}.$$

Completing the square we obtain

$$\left|z - \frac{\overline{z_0} - \lambda^2\overline{z_1}}{1 - \lambda^2}\right|^2 - \left(\frac{|z_0 - \lambda^2 z_1|}{1 - \lambda^2}\right)^2 = \frac{\lambda^2|z_1|^2 - |z_0|^2}{1 - \lambda^2}, \tag{2.4.6}$$

and hence we get the circle of center $\frac{\overline{z_0}-\lambda^2\overline{z_1}}{1-\lambda^2}$ and radius R defined by

$$R^2 = \left(\frac{|z_0 - \lambda^2 z_1|}{1 - \lambda^2}\right)^2 + \frac{\lambda^2|z_1|^2 - |z_0|^2}{1 - \lambda^2} = \left(\frac{\lambda}{|1 - \lambda^2|}|z_0 - z_1|\right)^2.$$

We now study the converse, and focus only on the case of a circle. Let

$$|z - \Omega| = R$$

be the circle of center Ω and radius $R > 0$. We are looking for $\lambda > 0$ and $z_0, z_1 \in \mathbb{C}$ (with $z_0 \neq z_1$) such that

$$\Omega = \frac{z_0 - \lambda^2 z_1}{1 - \lambda^2},$$

$$R = \frac{\lambda}{|1 - \lambda^2|} |z_0 - z_1|.$$

From the first equation we get

$$z_0 = (1 - \lambda^2)\Omega + \lambda^2 z_1.$$

Plugging this expression in the second equation we obtain

$$R = \lambda |\Omega - z_1|.$$

We take

$$z_1 = \Omega + \frac{R}{\lambda}.$$

Then

$$z_0 = (1 - \lambda^2)\Omega + \lambda^2 \Omega + \lambda R = \Omega + \lambda R,$$

which ends the proof. $\square$

Remark 2.4.1. We note the equality

$$(\Omega - z_0)(\Omega - z_1) = R^2. \tag{2.4.7}$$

Solution of Exercise 2.2.4. For (a) we have the circle with center $(1, -1)$ and radius 1. Equation (b) can be rewritten as $x^2 - y^2 = 1$, and so we obtain a hyperbole. Case (c) is the line orthogonal to the interval $(0, 1)$ and $(0, -1)$ and passing by the middle of this interval, i.e., it is just the real line. More misleading are (d), (e) and (f). The equations look like the equation of a circle but this is not the case. For (d) we have the empty set. Indeed, we have

$$|z|^2 + 3z + 3\bar{z} + 10 = 0 \iff |z + 3|^2 + 10 - 9 = 0$$
$$\iff |z + 3|^2 + 1 = 0.$$

Equation (e) becomes in cartesian coordinates

$$x^2 + y^2 + 4x + y + 5 + iy = 0.$$

Equating real and imaginary parts to 0 we obtain

$$x^2 + y^2 + 4x + y + 5 = 0 \quad \text{and} \quad y = 0.$$

The equation $x^2 + 4x + 5 = 0$ has no real solution, and so (e) also corresponds to the empty set. We leave (f) and (h) to the student, and turn to (g). Condition (g) is equivalent to

$$\sqrt{x^2 + y^2} \geq 1 - x. \tag{2.4.8}$$

If $x \geq 1$, every y meets this condition. Assume now that $x < 1$. Equation (2.4.8) is then equivalent to

$$x^2 + y^2 \geq (1 - x)^2,$$

that is, to

$$y^2 \geq 1 - 2x. \tag{2.4.9}$$

We already know that $x < 1$. If $x \in [\frac{1}{2}, 1)$, every y meets this condition. If $x \leq \frac{1}{2}$, we get the points outside or on the parabola defined by (2.4.9) and for which $x \leq 1/2$. All together, the set is the complement of the points inside the parabola $y^2 = 1 - 2x$.   □

Solution of Exercise 2.2.5. Write $z(t) = e^{it}$, with $t \in [0, 2\pi]$, and $w(t) = x(t) + iy(t)$. We obtain

$$x(t) = \cos t - \frac{1}{N} \cos(Nt),$$

$$y(t) = \sin t - \frac{1}{N} \sin(Nt).$$

These are the parametric equations of an epicycloid, described by a point on a circle of radius $\frac{1}{N}$ rolling over a circle of radius $1 - \frac{1}{N}$. See also for instance [166, Exercise 7, p. 421].   □

Solution of Exercise 2.3.1. Indeed,

$$\begin{aligned}
\varphi_1(\varphi_2(z)) &= \frac{a_1 \varphi_2(z) + b_1}{c_1 \varphi_2(z) + d_1} \\
&= \frac{a_1 \dfrac{a_2 z + b_2}{c_2 z + d_2} + b_1}{c_1 \dfrac{a_2 z + b_2}{c_2 z + d_2} + d_1} \\
&= \frac{a_1(a_2 z + b_2) + b_1(c_2 z + d_2)}{c_1(a_2 z + b_2) + d_1(c_2 z + d_2)} \\
&= \frac{(a_1 a_2 + b_1 c_2)z + a_1 b_2 + b_1 d_2}{(c_1 a_2 + d_1 c_2)z + c_1 b_2 + d_1 d_2} \\
&= \frac{az + b}{cz + d},
\end{aligned}$$

where

$$\begin{pmatrix} a & b \\ c & d \end{pmatrix} = \begin{pmatrix} a_1 & b_1 \\ c_1 & d_1 \end{pmatrix} \begin{pmatrix} a_2 & b_2 \\ c_2 & d_2 \end{pmatrix}.$$ □

Solution of Exercise 2.3.2. We have for all $z \in \mathbb{C}$ where both functions are defined:

$$\varphi(z) = \frac{a_1 z + b_1}{c_1 z + d_1} = \frac{a_2 z + b_2}{c_2 z + d_2}. \tag{2.4.10}$$

Hence,

$$z^2(a_1 c_2 - a_2 c_1) + z(a_1 d_2 - a_2 d_1 + c_2 b_1 - b_2 c_1) + b_1 d_2 - b_2 d_1 \equiv 0.$$

One can proceed by remarking that the coefficients of the above polynomial are all equal to 0, and then by distinguishing various cases. We will chose another avenue to solve the problem. With

$$X(z) = \frac{c_1 z + d_1}{c_2 z + d_2},$$

and taking into account (2.4.10) we have

$$\begin{pmatrix} a_1 & b_1 \\ c_1 & d_1 \end{pmatrix} \begin{pmatrix} z \\ 1 \end{pmatrix} = (c_1 z + d_1) \begin{pmatrix} \varphi(z) \\ 1 \end{pmatrix}$$

$$= X(z)(c_2 z + d_2) \begin{pmatrix} \varphi(z) \\ 1 \end{pmatrix}$$

$$= X(z) \begin{pmatrix} a_2 & b_2 \\ c_2 & d_2 \end{pmatrix} \begin{pmatrix} z \\ 1 \end{pmatrix}.$$

Take now two points u and v, with $u \neq v$, at which all the expressions make sense. We have

$$\begin{pmatrix} a_1 & b_1 \\ c_1 & d_1 \end{pmatrix} \begin{pmatrix} u & v \\ 1 & 1 \end{pmatrix} = \begin{pmatrix} a_2 & b_2 \\ c_2 & d_2 \end{pmatrix} \begin{pmatrix} uX(u) & vX(v) \\ X(u) & X(v) \end{pmatrix}.$$

Thus

$$\begin{pmatrix} a_1 & b_1 \\ c_1 & d_1 \end{pmatrix} = \begin{pmatrix} a_2 & b_2 \\ c_2 & d_2 \end{pmatrix} \begin{pmatrix} \frac{uX(u) - vX(v)}{u - v} & \frac{uv(X(v) - X(u))}{u - v} \\ \frac{X(u) - X(v)}{u - v} & \frac{uX(v) - vX(u)}{u - v} \end{pmatrix}. \tag{2.4.11}$$

Since the map is assumed non-degenerate we can write

$$\begin{pmatrix} \frac{uX(u) - vX(v)}{u - v} & \frac{uv(X(v) - X(u))}{u - v} \\ \frac{X(u) - X(v)}{u - v} & \frac{uX(v) - vX(u)}{u - v} \end{pmatrix} = \begin{pmatrix} a_2 & b_2 \\ c_2 & d_2 \end{pmatrix}^{-1} \begin{pmatrix} a_1 & b_1 \\ c_1 & d_1 \end{pmatrix}.$$

It follows that $X(u)$ is a constant, say K, and that

$$\begin{pmatrix} a_1 & b_1 \\ c_1 & d_1 \end{pmatrix} = K \begin{pmatrix} a_2 & b_2 \\ c_2 & d_2 \end{pmatrix}. \qquad \square$$

Solution of Exercise 2.3.3. It suffices to note that

$$\begin{pmatrix} 1 & -u \\ -\overline{u} & 1 \end{pmatrix} \begin{pmatrix} 1 & -v \\ -\overline{v} & 1 \end{pmatrix} = \begin{pmatrix} 1 + u\overline{v} & -(u + v) \\ -(\overline{u} + \overline{v}) & 1 + \overline{u}v \end{pmatrix}.$$

Then, the associated transformation is

$$\frac{(1+u\bar{v})z - (u+v)}{-z(\bar{u}+\bar{v})+1+\bar{u}v} = \frac{1+u\bar{v}}{1+v\bar{u}}\, b_w(z). \qquad \square$$

Solution of Exercise 2.3.4. It follows from (1.1.51) with $z = v$ that b_w sends the open unit disk into itself. It follows from (2.3.6) and (2.3.7) that

$$b_w(b_{-w}(z)) = z,$$
$$b_{-w}(b_w(z)) = z,$$

for $z \in \mathbb{D}$. The first equation shows that b_w is onto and the second equation shows that b_w is one-to-one.

When $|w| > 1$, the map b_w is one-to-one from $\mathbb{D} \setminus \{\frac{1}{\bar{w}}\}$ onto $\{z; |z| > 1\}$. The limiting case $w = \infty$ corresponds to $b(z) = \frac{1}{z}$, which is a one-to-one map from $\mathbb{D} \setminus \{0\}$ onto $\{z; |z| > 1\}$.

The case of B_w is treated in a similar way. $\qquad \square$

Solution of Exercise 2.3.5. Let

$$u = -b_{z_1}(z_2) \quad \text{and} \quad c = \frac{1 - \overline{z_1}z_2}{1 - z_1\overline{z_2}}.$$

Then,

$$c \cdot b_u(z_1) = \frac{1 - \overline{z_1}z_2}{1 - z_1\overline{z_2}} \frac{z_1 + \dfrac{z_2 - z_1}{1 - \overline{z_1}z_2}}{1 + z_1\dfrac{\overline{z_2} - \overline{z_1}}{1 - z_1\overline{z_2}}}$$

$$= \frac{(1 - |z_1|^2)z_2}{1 - |z_1|^2}$$

$$= z_2. \qquad \square$$

Remark 2.4.2. In [8, p. 26] one asks for more: Given two different points in $\mathbb{D}$, find a map of the form $s(z) = \frac{w-z}{1-z\bar{w}}$ sending z_1 into z_2 and such that $s(s(z)) \equiv z$.

Solution of Exercise 2.3.6. Solving w in function of z in the above expression leads to

$$w(z) = \frac{w_1 - w_2\dfrac{z - z_1}{z - z_2} \cdot \dfrac{z_3 - z_2}{z_3 - z_1} \cdot \dfrac{w_3 - w_1}{w_3 - w_2}}{1 - \dfrac{z - z_1}{z - z_2} \cdot \dfrac{z_3 - z_2}{z_3 - z_1} \cdot \dfrac{w_3 - w_1}{w_3 - w_2}}$$

$$= \frac{(z - z_2)w_1 - w_2(z - z_1) \cdot \dfrac{z_3 - z_2}{z_3 - z_1} \cdot \dfrac{w_3 - w_1}{w_3 - w_2}}{(z - z_2) - (z - z_1) \cdot \dfrac{z_3 - z_2}{z_3 - z_1} \cdot \dfrac{w_3 - w_1}{w_3 - w_2}}.$$

Hence, $z \mapsto w(z)$ is indeed a Moebius map. One checks that $w(z_\ell) = z_\ell$ for $\ell = 1, 2, 3$ by direct computation:

$$w(z_1) = \frac{(z_1 - z_2)w_1 - w_2(z_1 - z_1) \cdot \dfrac{z_3 - z_2}{z_3 - z_1} \cdot \dfrac{w_3 - w_1}{w_3 - w_2}}{(z_1 - z_2) - (z_1 - z_1) \cdot \dfrac{z_3 - z_2}{z_3 - z_1} \cdot \dfrac{w_3 - w_1}{w_3 - w_2}}$$

$$= \frac{(z_1 - z_2)w_1}{(z_1 - z_2)}$$

$$= w_1.$$

$$w(z_2) = \frac{(z_2 - z_2)w_1 - w_2(z_2 - z_1) \cdot \dfrac{z_3 - z_2}{z_3 - z_1} \cdot \dfrac{w_3 - w_1}{w_3 - w_2}}{(z_2 - z_2) - (z_2 - z_1) \cdot \dfrac{z_3 - z_2}{z_3 - z_1} \cdot \dfrac{w_3 - w_1}{w_3 - w_2}}$$

$$= \frac{-w_2(z_2 - z_1) \cdot \dfrac{z_3 - z_2}{z_3 - z_1} \cdot \dfrac{w_3 - w_1}{w_3 - w_2}}{-(z_2 - z_1) \cdot \dfrac{z_3 - z_2}{z_3 - z_1} \cdot \dfrac{w_3 - w_1}{w_3 - w_2}}$$

$$= w_2.$$

$$w(z_3) = \frac{(z_3 - z_2)w_1 - w_2(z_3 - z_1) \cdot \dfrac{z_3 - z_2}{z_3 - z_1} \cdot \dfrac{w_3 - w_1}{w_3 - w_2}}{(z_3 - z_2) - (z_3 - z_1) \cdot \dfrac{z_3 - z_2}{z_3 - z_1} \cdot \dfrac{w_3 - w_1}{w_3 - w_2}}$$

$$= \frac{(z_3 - z_2)w_1 - w_2(z_3 - z_2) \cdot \dfrac{w_3 - w_1}{w_3 - w_2}}{(z_3 - z_2) - (z_3 - z_2) \cdot \dfrac{w_3 - w_1}{w_3 - w_2}}$$

$$= \frac{w_1 - w_2 \cdot \dfrac{w_3 - w_1}{w_3 - w_2}}{1 - \dfrac{w_3 - w_1}{w_3 - w_2}}$$

$$= \frac{w_1(w_3 - w_2) - w_2(w_3 - w_1)}{w_3 - w_2 - w_3 + w_1}$$

$$= \frac{w_3(w_1 - w_2)}{w_1 - w_2}$$

$$= w_3. \qquad \qquad \square$$

Solution of Exercise 2.3.7. We follow the hints given after the exercice, and focus on the case of a circle. We first assume that the four points are on a common circle. For the unit circle, we can always assume that one of the points is $z_1 = 1$, and so we have to check that for any t_2, t_3 and t_4 in $(0, 2\pi)$,

$$\frac{(1 - e^{it_2})(e^{it_3} - e^{it_4})}{(1 - e^{it_3})(e^{it_3} - e^{it_4})}$$

is real, that is, to check that

$$\frac{(1 - e^{it_2})(e^{it_3} - e^{it_4})}{(1 - e^{it_3})(e^{it_3} - e^{it_4})} = \frac{(1 - e^{-it_2})(e^{-it_3} - e^{-it_4})}{(1 - e^{-it_3})(e^{-it_3} - e^{-it_4})}.$$

This is checked by multiplying both the numerator and denominator of the right side by $e^{it_2} e^{it_3} e^{it_4}$.

Let now $|z - z_0| = R$ be the equation of another circle. The Moebius map $\varphi(z) = \frac{z - z_0}{R}$ maps this circle onto the unit circle. We now check the invariance of the cross ratio using (2.3.1). Indeed using this equation we have

$$\varphi(z_1) - \varphi(z_2) = \frac{(ad - bc)(z_1 - z_2)}{(cz_1 + d)(cz_2 + d)},$$

$$\varphi(z_1) - \varphi(z_3) = \frac{(ad - bc)(z_1 - z_3)}{(cz_1 + d)(cz_3 + d)},$$

$$\varphi(z_2) - \varphi(z_4) = \frac{(ad - bc)(z_2 - z_4)}{(cz_2 + d)(cz_4 + d)},$$

$$\varphi(z_3) - \varphi(z_4) = \frac{(ad - bc)(z_3 - z_4)}{(cz_4 + d)(cz_3 + d)},$$

and hence the result by a direct computation.

Still for the case of a circle, we consider the converse statement: Let there be therefore four pairwise points for which (2.3.9) is real. The first three points determine uniquely a circle, which we move, via a Moebius map, to be the unit circle. The corresponding quotient (2.3.9) does not change. So we are left with the following question: Given four pairwise different points for which the quotient is real, three of them being on the unit circle, show that the fourth is also on the unit circle. We set $z_1 = e^{i\theta_1}$, $z_2 = e^{i\theta_2}$, $z_3 = e^{i\theta_3}$, where $\theta_1, \theta_2, \theta_3 \in [0, 2\pi)$ are pairwise different. We have thus:

$$\frac{(e^{-i\theta_1} - e^{-i\theta_2})(e^{-i\theta_1} - e^{-i\theta_3})}{(e^{-i\theta_2} - e^{-i\theta_4})(e^{-i\theta_3} - \overline{z_4})} = \frac{(e^{i\theta_1} - e^{i\theta_2})(e^{i\theta_1} - e^{i\theta_3})}{(e^{i\theta_2} - e^{i\theta_4})(e^{i\theta_3} - z_4)}.$$

Applying the previous exercise with the Moebius map $\varphi(z) = 1/z$ we see that

$$\varphi(e^{i\theta_j}) = e^{-i\theta_j}, \quad j = 1, 2, 3,$$

and so $\overline{z_4} = \varphi(z_4)$, i.e., $|z_4| = 1$. This concludes the proof.   □

Solution of Exercise 2.3.8. Set $w = \dfrac{z - 3}{1 - 2z}$. Then, $z = \dfrac{w + 3}{1 + 2w}$. The condition $|z - 1| = k$ becomes thus $\left| \dfrac{w + 3}{1 + 2w} - 1 \right| = k$, i.e.,

$$|w + 3 - (1 + 2w)| = k|1 + 2w|,$$

which can be rewritten as

$$|2 - w| = 2k\left|w + \frac{1}{2}\right|.$$

Hence, we obtain a line if and only if $k = \frac{1}{2}$. The equation of the line is $|2 - w| = \left|w + \frac{1}{2}\right|$, i.e., $x = \frac{3}{4}$.

If one wants only the value of k but not the equation of the line, a shorter way is as follows: We know that the image of the circle is either a line or a circle. It will be a line if and only if it is not a bounded set, i.e., if and only if $z = \frac{1}{2}$ belongs to the circle $|z - 1| = k$, i.e., $\left|\frac{1}{2} - 1\right| = k$. Hence $k = \frac{1}{2}$.   □

Solution of Exercise 2.3.9. By hypothesis, $ad - bc \neq 0$ and thus the map $z \mapsto w$ is invertible, and its inverse is given by

$$z = \frac{wd - b}{a - cw}.$$

We know that $|z| < 1$ and want to find a necessary and sufficient condition for the set of images w to be in the open unit disk. We have

$$
\begin{aligned}
|z| < 1 &\iff |z|^2 < 1 \\
&\iff |wd - b|^2 < |a - wc|^2 \\
&\iff |w|^2|d|^2 + |b|^2 - 2\,\mathrm{Re}(\bar{b}dw) < |w|^2|c|^2 + |a|^2 - 2\,\mathrm{Re}(\bar{a}cw) \\
&\iff |w|^2(|d|^2 - |c|^2) - 2\,\mathrm{Re}\left\{(\bar{b}d - \bar{a}c)w\right\} + |b|^2 - |a|^2 < 0.
\end{aligned}
\tag{2.4.12}
$$

At this stage we pause and remark that, necessarily, $|c| < |d|$. Indeed, if $|d| = |c|$, the above can be rewritten as

$$-2\,\mathrm{Re}\left\{(\bar{b}d - \bar{a}c)w\right\} + |b|^2 - |a|^2 < 0, \tag{2.4.13}$$

which is an unbounded set (in fact, a half-plane). Note that, under the hypothesis $|d| = |c|$, we necessarily have

$$(\bar{b}d - \bar{a}c) \neq 0.$$

If it is equal to 0, then on the one hand (2.4.13) leads to $|b| < |a|$ and on the other hand we have

$$\bar{b}d - \bar{a}c = 0 \implies |b||d| = |a||c|$$
$$\implies |d| = |c|\frac{|a|}{|b|}$$

which together with $|d| = |c|$ leads to $|a| = |b|$.

If $|c| > |d|$, the point $-\dfrac{d}{c}$ is in $\mathbb{D}$, and φ has a pole at this point. Thus the image of $\mathbb{D}$ by φ cannot be bounded. So $|c| < |d|$. We divide both sides of (2.4.12) by $|d|^2 - |c|^2$ and obtain

$$|w|^2 - 2\,\mathrm{Re}\left\{\frac{\bar{b}d - \bar{a}c}{|d|^2 - |c|^2}w\right\} + \frac{|b|^2 - |a|^2}{|d|^2 - |c|^2} < 0.$$

Completing the square we obtain

$$\left|w - \frac{b\bar{d} - a\bar{c}}{|d|^2 - |c|^2}\right|^2 < -\frac{|b|^2 - |a|^2}{|d|^2 - |c|^2} + \left|\frac{b\bar{d} - a\bar{c}}{|d|^2 - |c|^2}\right|^2.$$

We have

$$-\frac{|b|^2 - |a|^2}{|d|^2 - |c|^2} + \left|\frac{b\bar{d} - a\bar{c}}{|d|^2 - |c|^2}\right|^2 = \frac{-(|b|^2 - |a|^2)(|d|^2 - |c|^2) + |b\bar{d} - a\bar{c}|^2}{(|d|^2 - |c|^2)^2}$$

$$= \frac{|ad - bc|^2}{(|d|^2 - |c|^2)^2}.$$

Thus the image of the open unit disk is the open disk of center

$$w_0 = \frac{b\bar{d} - a\bar{c}}{|d|^2 - |c|^2}$$

and radius

$$r_0 = \frac{|ad - bc|}{|d|^2 - |c|^2}.$$

This open disk will be included in the open unit disk if and only if

$$|w_0| + r_0 \le 1,$$

which can be rewritten as

$$\frac{|b\bar{d} - a\bar{c}|}{|d|^2 - |c|^2} + \frac{|ad - bc|}{|d|^2 - |c|^2} \le 1.$$

Multiplying both sides by the strictly positive number $|d|^2 - |c|^2$ we obtain (2.3.10).
□

Solution of Exercise 2.3.10. Necessary and sufficient conditions are that the center is $w_0 = 0$ and the radius is $r_0 = 1$. This leads to the equations

$$b\bar{d} = a\bar{c} \tag{2.4.14}$$

$$|ad - bc| = |d|^2 - |c|^2. \tag{2.4.15}$$

The second equation implies that $d \neq 0$. The first equation then gives $b = \frac{a\bar{c}}{d}$. Plugging this expression into the second equation and dividing both sides by d^2 leads to

$$\left| \frac{a}{d} - \frac{a}{d}\frac{c\bar{c}}{d\bar{d}} \right| = 1 - \left| \frac{c}{d} \right|^2.$$

Thus $k \stackrel{\text{def.}}{=} \frac{a}{d} \in \mathbb{T}$ and $\alpha \stackrel{\text{def.}}{=} -\frac{\bar{c}}{d} \in \mathbb{D}$. Equation (2.4.14) implies that $\frac{b}{d} = -k\alpha$ and the Moebius map is of the required form.   □

Remark 2.4.3. We now solve the previous question directly, without resorting to Exercise 2.3.9. Of course the computations are basically the same. We remark that $d \neq 0$ (otherwise the point 0 would go to the point at infinity), and write without loss of generality $\varphi(z) = \frac{az+b}{cz+1}$. We look at the image of the unit circle.

$$w = \frac{ae^{it} + b}{ce^{it} + 1} \quad \Longleftrightarrow \quad e^{it} = \frac{w - b}{-cw + a},$$

and so $|w - b|^2 = |a - cw|^2$, that is

$$|w|^2(1 - |c|^2) - 2\,\mathrm{Re}\,w(\bar{b} - c\bar{a}) = |a|^2 - |b|^2. \tag{2.4.16}$$

Writing that the image is $|w| = 1$ gives

$$\bar{b} = c\bar{a} \quad \text{and} \quad 1 - |c|^2 = |a|^2 - |b|^2.$$

The first equation gives

$$\varphi(z) = \frac{az + \bar{c}}{cz + 1} = a\frac{z + \bar{c}}{cz + 1}. \tag{2.4.17}$$

Plugging the first equation into the second we obtain

$$1 - |c|^2 = |a|^2 - |a|^2|c|^2 = |a|^2(1 - |c|^2).$$

We also note that $|c| < 1$ (since the image of $-\frac{1}{c}$ is the point at infinity) and so $|a| = 1$. Thus the map φ is necessarily of the form (2.4.17). The converse follows from (1.1.41).

Solution of Exercise 2.3.11. The map

$$z \mapsto \varphi_0(z) = \frac{1 - z}{1 + z}$$

is one-to-one and onto from the open unit disk onto the open left half-plane, and is equal to its inverse. Therefore the map $\varphi(z) = \frac{az + b}{cz + d}$ maps the open left half-plane into itself if and only if $\varphi_0 \circ \varphi \circ \varphi_0$ maps the open unit disk into itself. In view of (2.3.3) the coefficients of $\varphi_0 \circ \varphi \circ \varphi_0$ can be chosen to be

$$\begin{pmatrix} -1 & 1 \\ 1 & 1 \end{pmatrix} \begin{pmatrix} a & b \\ c & d \end{pmatrix} \begin{pmatrix} -1 & 1 \\ 1 & 1 \end{pmatrix} = \begin{pmatrix} -a - b - c + d & -a - b + c + d \\ -a + b - c + d & a + b + c + d \end{pmatrix}.$$

Applying (2.3.10), we obtain the condition

$$|(\bar{b}-\bar{d})(a+c)+(\bar{a}-\bar{c})(b+d)|+2|ad-bc| \le (a+c)(\bar{b}+\bar{d})+(\bar{a}+\bar{c})(b+d). \quad (2.4.18)$$

We note that, in (2.4.18), equality will hold in particular for

$$\varphi(z) = z+w \quad \text{and} \quad \varphi(z) = \frac{1}{z+w},$$

where $w + \bar{w} \ge 0$.   □

Solution of Exercise 2.3.12. Of course, from (2.3.10) with

$$a = 1, \quad b = -1, \quad c = 1, \quad \text{and} \quad d = -z_0,$$

we know that the condition is

$$2|1 - z_0| < |z_0|^2 - 1. \quad (2.4.19)$$

We reproduce this result directly as follows: Set

$$w = \frac{z-1}{z-z_0}.$$

Then

$$z = \frac{1-wz_0}{1-w},$$

and the condition $|z| < 1$ becomes

$$|1 - wz_0| < |1 - w|, \quad \text{that is,} \quad |1 - wz_0|^2 < |1 - w|^2.$$

This in turn can be rewritten as

$$|w|^2|z_0|^2 - 2\operatorname{Re}(wz_0) + 1 < |w|^2 - 2\operatorname{Re}(w) + 1,$$

i.e.,

$$|w|^2(|z_0|^2 - 1) - 2\operatorname{Re}w(z_0 - 1) < 0. \quad (2.4.20)$$

Now, if $|z_0| = 1$ the above inequality defines a half-plane, and cannot be inside the unit disk. Assume now that $|z_0| > 1$. Then (2.4.20) becomes

$$|w|^2 - 2\operatorname{Re}w\left(\frac{z_0 - 1}{|z_0|^2 - 1}\right) < 0,$$

i.e.,

$$\left|w - \frac{\bar{z_0} - 1}{|z_0|^2 - 1}\right|^2 < \left|\frac{\bar{z_0} - 1}{|z_0|^2 - 1}\right|^2.$$

Thus the image of the open unit disk is the open disk with center $C = \frac{\overline{z_0}-1}{|z_0|^2-1}$ and radius $R = |\frac{\overline{z_0}-1}{|z_0|^2-1}|$. This disk will be inside the open unit disk if and only if $|C| + R < 1$, that is

$$2\frac{|z_0-1|}{|z_0|^2-1} < 1,$$

which is (2.4.19) since $|z_0| > 1$.

Assume now $|z_0| < 1$. Then (2.4.20) becomes

$$|w|^2 - 2\operatorname{Re}\left(w\left(\frac{\overline{z_0}-1}{|z_0|^2-1}\right)\right) > 0.$$

This can be rewritten as

$$\left|w - \frac{\overline{z_0}-1}{|z_0|^2-1}\right|^2 > \left|\frac{\overline{z_0}-1}{|z_0|^2-1}\right|^2,$$

and thus defines an unbounded set, and cannot be in the open unit disk. Thus z_0 cannot be of modulus strictly less than 1, and the necessary and sufficient condition is (2.4.19). □

Solution of Exercise 2.3.13. (a) In view of formula (2.3.1) we have

$$\frac{\varphi(z) - z_1}{\varphi(z) - z_2} = \frac{\varphi(z) - \varphi(z_1)}{\varphi(z) - \varphi(z_2)}$$

$$= \frac{\dfrac{(z-z_1)(ad-bc)}{(cz+d)(cz_1+d)}}{\dfrac{(z-z_2)(ad-bc)}{(cz+d)(cz_2+d)}}$$

$$= k\frac{z-z_1}{z-z_2},$$

with

$$k = \frac{cz_2+d}{cz_1+d}. \tag{2.4.21}$$

(b) Denote by $\varphi^{\circ n}(z)$ the nth iterate of φ. Iterating (2.3.11) we have

$$\frac{\varphi^{\circ n}(z) - z_1}{\varphi^{\circ n}(z) - z_2} = k^n\frac{z-z_1}{z-z_2}, \tag{2.4.22}$$

and hence

$$\varphi^{\circ n}(z) = \frac{z_1 - z_2 k^n \dfrac{z-z_1}{z-z_2}}{1 - k^n \dfrac{z-z_1}{z-z_2}} = \frac{z_1(z-z_2) - z_2 k^n(z-z_1)}{z-z_2 - k^n(z-z_1)}.$$

Thus

$$\varphi^{\circ n}(z) = \frac{(z_1 - z_2 k^n)z - (1 - k^n)z_1 z_2}{(1 - k^n)z - (z_2 - z_1 k^n)}. \tag{2.4.23}$$

(c) When $\varphi(z) = (1 - 3z)/(z - 3)$ the equation

$$\varphi(z) = z$$

has two distinct roots, namely $z_1 = 1$ and $z_2 = -1$. We have $c = 1$, $d = -3$ and

$$k = \frac{cz_2 + d}{cz_1 + d} = 2.$$

Therefore, from (2.4.23) we have

$$\varphi^{\circ n}(z) = \frac{(1 + 2^n)z + 1 - 2^n}{(1 - 2^n)z + 1 + 2^n}. \qquad \square$$

We note the following: When $|k| > 1$, equation (2.4.22) implies that, for $z \neq z_1$,

$$\lim_{n \to \infty} \varphi^{\circ n}(z) = z_2,$$

while we have

$$\lim_{n \to \infty} \varphi^{\circ n}(z) = z_1$$

for $z \neq z_2$ when $|k| < 1$.

Solution of Exercise 2.3.15. We use the preceding exercise. For $w \neq 0$ (which is the only case of interest), the equation

$$b_w(z) = z,$$

that is,

$$z - w = z - z^2 \overline{w}$$

has two distinct solutions, say z_1 and z_2. Set $w = \rho e^{i\theta}$. Then

$$z_1 = e^{i\theta} \quad \text{and} \quad z_2 = -e^{i\theta}. \tag{2.4.24}$$

The multiplier k is given by formula (2.4.21), and hence equal to

$$k = \frac{-\overline{w}z_2 + 1}{-\overline{w}z_1 + 1} = \frac{1 + \rho}{1 - \rho}.$$

In particular, we see that k is real and belongs to $(1, +\infty)$. Formula (2.4.23) and equation (2.4.24) give

$$\underbrace{b_w \circ b_w \circ \cdots \circ b_w}_{n\text{-times}}(z) = \frac{z - w_n}{1 - z\overline{w_n}},$$

where

$$w_n = e^{i\theta} \frac{k^n - 1}{k^n + 1}.$$

(b) We have

$$\lim_{n\to\infty} w_n = e^{i\theta},$$

and thus

$$\lim_{n\to\infty} \underbrace{b_w \circ b_w \circ \cdots \circ b_w}_{n\text{-times}}(z) = e^{i\theta}.$$

In particular the limit is independent of $z \in \mathbb{D}$.   □

Solution of Exercise 2.3.16. Let $\lambda = \frac{zw+1}{z-w}$. Then $z = \frac{\lambda w+1}{\lambda - w}$. The number z is real
if and only if it holds that

$$\frac{\lambda w + 1}{\lambda - w} = \frac{\overline{\lambda} \overline{w} + 1}{\overline{\lambda} - \overline{w}},$$

which can be rewritten as

$$|\lambda|^2 - \lambda \overline{\alpha} - \overline{\lambda} \alpha + 1 = 0,$$

with

$$\alpha = \frac{|w|^2 + 1}{\overline{w} - w}.$$

Completing the square we get

$$|\lambda - \alpha|^2 = |\alpha|^2 - 1.$$

It is readily checked that

$$|\alpha|^2 - 1 = \left(\frac{|w^2 + 1|}{|w - \overline{w}|} \right)^2,$$

and we get a circle of center α and radius

$$R = \frac{|w^2 + 1|}{|w - \overline{w}|}.$$

The circle reduces to a point when $w = \pm i$.   □

Chapter 3

Complex Numbers and Analysis

We begin by discussing complex-valued functions of a real variable. This gives us more freedom for exercises in the sequel of the chapter. We discuss series and power series, and introduce the exponential function and the various trigonometric functions. Infinite products are also discussed. At the end of this chapter, we therefore will have a number of examples of functions, some highly non-trivial, which will be shown in subsequent chapters to be instances of analytic functions.

3.1   Complex-valued functions on an interval; derivatives and integrals

Let
$$f(t) = u(t) + iv(t) \tag{3.1.1}$$
where t belongs to an interval I (which can be open, closed, or half-closed), be a complex-valued function. We recall that f has a limit at the point $t_0 \in I$ if and only both u and v have limits at the point t_0. The function f is continuous at t_0 (resp. in I) if and only if both u and v are continuous at t_0 (resp. in I).

Exercise 3.1.1. *Give an example of a map $t \mapsto z(t)$ from an open interval $(a,b) \subset \mathbb{R}$ into $\mathbb{C}$ such that $\lim_{t \to b} z(t)$ does not exist, but $\lim_{t \to b} \operatorname{Re} z(t)$ exists.*

Exercise 3.1.2. *Let $a_1, \ldots, a_n$ be arbitrary complex numbers. Show that there is at least one $t \in [0,1]$ such that*

$$\left| 1 - \sum_{k=1}^{n} a_k e^{2\pi i k t} \right| \geq 1.$$

The function $1 - \sum_{k=1}^{n} a_k e^{2\pi i k t}$ in the previous exercise is periodic. The result itself can be extended to the almost-periodic case, as is illustrated in the following exercise:

Exercise 3.1.3. *Let $\alpha_1, \ldots, \alpha_n$ be real pairwise different numbers, all different from 0, and let $a_1, \ldots, a_n$ be complex numbers. Show that*

$$\forall \epsilon \in (0,1), \quad \exists t \in \mathbb{R} \quad such\ that \quad \left| 1 - \sum_{k=1}^{n} a_k e^{i\alpha_k t} \right| > \epsilon. \qquad (3.1.2)$$

Hint. Compute

$$\lim_{T \to \infty} \frac{1}{T} \int_0^T \left| 1 - \sum_{k=1}^{n} a_k e^{i\alpha_k t} \right|^2 dt. \qquad (3.1.3)$$

The function (3.1.1) has a derivative in the given interval if both u and v have derivatives there, and then

$$f'(t) = u'(t) + iv'(t).$$

The various formulas of derivation for sum, product and quotient still hold for complex-valued functions. For instance, for $a \in \mathbb{R}$,

$$(e^{iat})' = (\cos at + i \sin at)' = -a \sin at + ia \cos at = iae^{iat}.$$

Exercise 3.1.4. *Check that for any $z_0 \in \mathbb{C}$,*

$$(e^{z_0 t})' = z_0 e^{z_0 t}.$$

Differentiation allows us to obtain new and non-trivial formulas from known ones, as illustrated in the next exercises.

Exercise 3.1.5. *Compute for $n \geq 2$ in closed form the sums*

$$\sum_{k=1}^{n-1} k \cos(a + bk) \quad and \quad \sum_{k=1}^{n-1} k \sin(a + bk)$$

Exercise 3.1.6. *Compute in closed form*

$$\sum_{k=1}^{n} k e^{ikt}.$$

Integration of a complex-valued function along a path in the complex plane is one of the keystones of complex analysis. In the present section we consider a preliminary, and much easier notion, namely the integral of a (say continuous) function of a real variable, but with values in $\mathbb{C}$. We recall the definition of this integral: If $f(t) = u(t) + iv(t)$ is a continuous function from the compact interval $[a, b] \subset \mathbb{R}$ into $\mathbb{C}$, one has

$$\int_a^b f(t)dt \stackrel{\text{def.}}{=} \int_a^b u(t)dt + i \int_a^b v(t)dt.$$

A key property is that

$$\left| \int_a^b f(t) dt \right| \le \int_a^b |f(t)| dt, \quad \text{with } a \le b. \tag{3.1.4}$$

For instance, the sequence

$$\int_0^1 e^{int} e^t dt, \quad n \in \mathbb{N},$$

is bounded. In fact, an easy integration by parts shows that it goes to 0 as n goes to infinity.

The following illustration of (3.1.4) is taken from [161, pp. 476–477]:

Question 3.1.7. (*see* [161, pp. 476–477]) *Let in* (3.1.4)

$$f(t) = \frac{1}{(t + ci)^{n+1}}, \quad \text{where} \quad c > 0 \text{ and } n \in \mathbb{N}_0.$$

Show that

$$\frac{1}{n} \frac{\left| (a + ci)^n - (b + ci)^n \right|}{(a^2 + c^2)^{n/2} (b^2 + c^2)^{n/2}} \le \int_a^b \frac{dt}{(t^2 + c^2)^{\frac{n+1}{2}}},$$

and that, in particular,

$$\frac{b - a}{\sqrt{(a^2 + c^2)(b^2 + c^2)}} \le \frac{1}{c} \left(\arctan \frac{b}{c} - \arctan \frac{a}{c} \right).$$

It is also well to have in mind the Cauchy–Schwarz inequality in this setting

$$\left| \int_a^b f(t) \overline{g(t)} dt \right|^2 \le \left(\int_a^b |f(t)|^2 dt \right) \left(\int_a^b |g(t)|^2 dt \right), \tag{3.1.5}$$

where f and g are continuous on $[a, b]$. Equality holds if and only if f and g are linearly dependent, that is,

$$f(t) \equiv 0 \quad \text{or} \quad g(t) \equiv \lambda f(t) \quad \text{for some complex number } \lambda.$$

Exercise 3.1.8.

(a) *Compute* $\int_0^{2\pi} \cos^{2p} t \, dt$.
(b) *Using* (a)*, show that*

$$\lim_{p \to \infty} \frac{\binom{2p}{p}}{2^{2p}} = 0. \tag{3.1.6}$$

We note that the sequence of integrals $\int_0^{2\pi}(\cos^{2p}t)dt$ is decreasing. The associated difference sequence $\left(\int_0^{2\pi}(\cos^{2p}t)dt - \int_0^{2\pi}(\cos^{2(p+1)}t)dt\right)_{p\in\mathbb{N}_0}$ is closely related to the sequence of Catalan numbers (the sequence of moments of the semicircle law, which plays a key role in free probability and in the theory of random matrices; see [119, 158]).

Exercise 3.1.9. *Let*

$$m_p = \frac{1}{2\pi}\int_{-2}^{2}\sqrt{4-x^2}\,x^{2p}dx,\quad p=0,1,\ldots$$

denote the even moments of the semi-circle law. Compute for $p=0,1,\ldots$

$$\int_0^{2\pi}\cos^{2p}t\,dt - \int_0^{2\pi}\cos^{2(p+1)}t\,dt,$$

and show that

$$m_p = \underbrace{\frac{1}{p+1}\binom{2p+2}{p+1}}_{\text{Catalan number}}.$$

The next result would be a direct consequence of Cauchy's integral formula. At this stage, the exercise is to be proved from first principles. The exercise is taken from [52], where it is used as an intermediate tool to prove, *without integration theory*, that a complex-valued function which admits a complex derivative everywhere in an open set Ω admits a power series expansion at every point of Ω.

Exercise 3.1.10. *Let $p(z) = \sum_{\ell=0}^{n}p_\ell z^\ell$ be a polynomial bounded by 1 in modulus in the closed unit disk. Show that all $|p_\ell| \le 1$ for $\ell = 0,\ldots,n$.*

It is an interesting problem to find necessary and sufficient conditions on the coefficients p_ℓ for $|p(z)| \le 1$ to hold for all z in the closed unit disk. This question is beyond the scope of the present book, but let us briefly discuss it. Let $a(z) = \sum_{\ell=0}^{n}\alpha_\ell z^\ell$, with $\alpha_0,\ldots,\alpha_n$ arbitrary complex numbers. Writing

$$\frac{1}{2\pi}\int_0^{2\pi}|p(e^{it})a(e^{it})|^2dt \le \frac{1}{2\pi}\int_0^{2\pi}|a(e^{it})|^2dt,$$

we obtain that

$$\sum_{\ell,k=0}^{n}\alpha_\ell\overline{\alpha_k}\left(\sum_{u=\max\{\ell,k\}}^{n}p_{u-\ell}\overline{p_{u-k}}\right) \le \sum_{\ell=0}^{n}|\alpha_\ell|^2,$$

which implies the matrix inequality

$$\begin{pmatrix} p_0 & 0 & 0 & \cdots & 0 \\ p_1 & p_0 & 0 & \cdots & 0 \\ & & & & 0 \\ & & & & 0 \\ p_n & p_{n-1} & p_{n-2} & \cdots & p_0 \end{pmatrix}^{*}\begin{pmatrix} p_0 & 0 & 0 & \cdots & 0 \\ p_1 & p_0 & 0 & \cdots & 0 \\ & & & & 0 \\ & & & & 0 \\ p_n & p_{n-1} & p_{n-2} & \cdots & p_0 \end{pmatrix} \le I_{n+1},\quad (3.1.7)$$

where I_{n+1} denotes the $(n + 1) \times (n + 1)$ identity matrix, and where M^* denotes the complex conjugate of a matrix M. The above inequality means that

$$
I_{n+1} - \begin{pmatrix} p_0 & 0 & 0 & \cdots & 0 \\ p_1 & p_0 & 0 & \cdots & 0 \\ & & & & 0 \\ & & & & 0 \\ p_n & p_{n-1} & p_{n-2} & \cdots & p_0 \end{pmatrix}^* \begin{pmatrix} p_0 & 0 & 0 & \cdots & 0 \\ p_1 & p_0 & 0 & \cdots & 0 \\ & & & & 0 \\ & & & & 0 \\ p_n & p_{n-1} & p_{n-2} & \cdots & p_0 \end{pmatrix} \geq 0.
$$

See Definition 16.3.1 for the definition of a positive matrix, if need be. This inequality is not a sufficient condition for p to be bounded in modulus by 1 in the closed unit disk, but is equivalent to the existence of a power series bounded by 1 in $\mathbb{D}$ and which begins with p. The result in the preceding exercise is still true for infinite power series. This is presented in Exercise 3.4.12. Inequality (3.1.7) is still true, and expresses that a certain lower triangular operator is a contraction from ℓ_2 into itself. See Exercise 16.1.5 for the definition of the Hilbert space ℓ_2.

The integrals in the exercises above were on a bounded interval. The following exercises are related to functions defined on an unbounded interval. If $f(t) = u(t) + iv(t)$ is a continuous *complex-valued* function of the *real variable* t, we say that the integral

$$
\int_{\mathbb{R}} f(t)dt
$$

exists if both integrals

$$
\int_{\mathbb{R}} u(t)dt \quad \text{and} \quad \int_{\mathbb{R}} v(t)dt
$$

converge.

Exercise 3.1.11. *Show that*

$$
\int_{-\infty}^{0} \frac{dt}{(t - z)^2} = \frac{1}{z}
$$

for all $z \in \mathbb{C} \setminus (-\infty, 0]$.

A follow-up of the next exercise is Exercise 5.5.25.

Exercise 3.1.12. *Let m be a continuous positive function on the real line, subject to*

$$
\int_{\mathbb{R}} \frac{m(t)dt}{t^2 + 1} < \infty. \tag{3.1.8}
$$

Show that the integral

$$
f_m(z) = -i \int_{\mathbb{R}} \left\{ \frac{1}{t - z} - \frac{t}{t^2 + 1} \right\} m(t)dt, \tag{3.1.9}
$$

makes sense for z off the real line. Compute $\operatorname{Re} f_m(z)$.

In preparation to the next exercise, we mention that positive definite functions and kernels are defined in Section 16.3; see Definition 16.3.11.

Exercise 3.1.13.

(1) *Let w be in the right open half-plane $\mathbb{C}_r$ (that is, $\operatorname{Re} w > 0$). Compute*

$$\int_0^\infty e^{-tw}\,dt.$$

(2) *Show that the function*

$$\frac{1}{z + \overline{w}} \qquad\qquad (3.1.10)$$

is a positive definite kernel in $\mathbb{C}_r$.

More generally we have the following question. In the statement, we have set $z^\nu = \rho^\nu(\cos(\nu\theta) + i\sin(\nu\theta))$ where $z = \rho(\cos\theta + i\sin\theta)$ is the polar representation of z, with $\theta \in (-\pi, \pi)$. Since in the question, z and w are restricted to be in $\mathbb{C}_r$, the expressions z^ν and $(z + \overline{w})^\nu$ make sense.

Question 3.1.14. *Let $\nu > -1$, and let Γ denote Euler's Gamma function*

$$\Gamma(\nu) = \int_0^\infty e^{-t}t^{\nu-1}\,dt. \qquad\qquad (3.1.11)$$

Compute $\int_0^\infty t^\nu e^{-zt}\,dt$ for z in $\mathbb{C}_r$, and show that the function

$$\frac{\Gamma(\nu + 1)}{(z + \overline{w})^\nu} \qquad\qquad (3.1.12)$$

is positive definite in $\mathbb{C}_r$.

See (4.4.11) for the counterpart of this result in the open unit disk.

Exercise 3.1.15. *Prove* (0.0.5)

3.2 Sequences of complex numbers

Recall that a sequence $(z_n)_{n\in\mathbb{N}}$ of complex numbers converges to a number z if[3]

$$\forall \epsilon > 0 \quad \exists N, \quad n \geq N \implies |z - z_n| < \epsilon.$$

The number z is unique, and is called the *limit* of the sequence.

Exercise 3.2.1. *Compute, for $t \in \mathbb{R}$,*

$$\lim_{n\to\infty} \frac{1 + e^{it} + \cdots + e^{int}}{n}.$$

---

[3] Recall also that if, in a definition, always means *if and only if*.

With the preceding exercise at hand we will now give an indirect proof that, for any $t \neq 0 \pmod{2\pi}$, the sequence $n \mapsto e^{int}$ has no limit. The first item is a well-known result in calculus and the proof is the same in the complex case.

Exercise 3.2.2.

(a) *Assume that the sequence* $(u_n)_{n \in \mathbb{N}}$ *has a limit* $\ell \in \mathbb{C}$. *Show that the sequence*

$$v_n = \frac{u_0 + \cdots + u_n}{n}$$

tends to ℓ *too.*

(b) *Using* (a), *show that* $n \mapsto e^{int}$ *has no limit for* $t \neq 0 \pmod{2\pi}$.

Exercise 3.2.3. *Study the convergence of the sequences*

$$z_n = \frac{\sum_{k=1}^{n} k e^{ikt}}{n}$$

and

$$w_n = \frac{\sum_{k=1}^{n} k e^{ikt}}{n^2}.$$

We make now some remarks: Quite often it is difficult, or even impossible, to guess what the limit is. A more abstract criterion to study the existence of a limit is as follows: The sequence $(z_n)_{n \in \mathbb{N}}$ converges if and only if it is a Cauchy sequence, that is, if and only if

$$\forall \epsilon > 0 \quad \exists N, \quad n, m \geq N \implies |z_n - z_m| < \epsilon.$$

Most of the methods taught in calculus for *real* sequences can be seen to hold in the case of *complex* sequences, by looking separately at the real and imaginary part. For instance, if f is a complex-valued function continuous on $[0, 1]$, then

$$\lim_{n \to \infty} \frac{\sum_{k=0}^{n} f(k/n)}{n} = \int_0^1 f(t)dt. \tag{3.2.1}$$

Exercise 3.2.4. *Study the convergence of the sequence*

$$z_n = \frac{\sum_{k=0}^{n} k e^{2\pi ik/n}}{n^2}.$$

The following exercise is easily proved using the notion of radius of convergence of power series; see Exercise 5.6.4 in Section 5.6. The reader should consult [62] for a direct proof.

Exercise 3.2.5. *Let* $z_1, \ldots, z_m$ *be complex numbers different from* 0. *Show that*

$$\limsup_{n \to \infty} \left| \sum_{\ell=1}^{m} z_\ell^n \right|^{1/n} = \max_{\ell=1,\ldots,m} |z_\ell|. \tag{3.2.2}$$

In particular we have the non-trivial fact: Given any m real numbers $t_1, \ldots,$ t_m we have

$$\limsup_{n \to \infty} \left| \sum_{\ell=1}^{m} e^{int_\ell} \right|^{1/n} = 1.$$

The following exercise is used in the proof of Exercise 13.4.3 to get a formula for the Gamma function. The first claim is taken from [23, p. 119].

Exercise 3.2.6.

(a) *Show that*

$$\left(1 - \frac{t}{n}\right)^n \le e^{-t}, \quad t \in [0, n].$$

(b) *Show that, for $t > 0$ the sequence*

$$n \mapsto \left(1 - \frac{t}{n}\right)^n$$

is eventually increasing.

3.3   Series of complex numbers

We now consider series of complex numbers. See also Sections 4.4, and 5.6, where various facts on power series are given as exercises. See for instance Exercises 4.4.9, 4.4.11 and 5.6.1 there.

The easiest criterium to check convergence of a series with running term z_n is to check that the series of positive numbers $|z_n|$ converges. The series is then called *absolutely convergent*, and is in particular convergent. For example:

Exercise 3.3.1. *Show that the series*

$$\sum_{n=1}^{\infty} \frac{z^{2n}}{2 + z^n + z^{5n}}$$

converges in the open unit disk.

Exercise 3.3.2. *Show that the series*

$$\sum_{n=1}^{\infty} \left(\frac{1}{z - n} + \frac{1}{n} \right)$$

converges for every $z \notin \mathbb{N}$. Show that the convergence is uniform on any compact set which does not intersect $\mathbb{N}$.

The next exercise is taken from [154]; see Exercise 4.19, p. 49 there. In the proof, one needs the formula of the partial fraction expansion

$$\frac{nz^{n-1}}{z^n - 1} = \sum_{\ell=0}^{n-1} \frac{1}{z - z_\ell}, \tag{3.3.1}$$

where $z_0, \ldots, z_{n-1}$ are the roots of unity of order n. The proof of this simple result is postponed to Exercise 7.3.4.

Exercise 3.3.3. *Let $z_0, \ldots, z_{n-1}$ be the roots of order n of the unity. Show that*

$$\sum_{\ell=0}^{n} z_\ell^m = 0, \quad m = 1, \ldots, n - 1.$$

The next exercise is yet another example of solving a problem in real analysis by going via the complex domain. But first some preliminary discussion: To find a closed formula for the sum

$$\sum_{\ell=0}^{\infty} \frac{1}{(4\ell + 1)!},$$

or, more generally for the sums

$$A_k = \sum_{\ell=0}^{\infty} \frac{1}{(4\ell + k)!}, \quad k = 0, 1, 2, 3,$$

one needs only to know the power series expansion of the real functions $\cosh x$, $\cos x$, $\sinh x$ and $\sin x$. It is readily seen that

$$A_0 = \frac{\cosh 1 + \cos 1}{2}, \qquad A_1 = \frac{\sinh 1 + \sin 1}{2},$$
$$A_2 = \frac{\cosh 1 - \cos 1}{2} \quad \text{and} \quad A_3 = \frac{\sinh 1 - \sin 1}{2}$$

(see for instance [211, p. 238] for the computation of A_1). Such a simple approach does not seem to help for computing the sums in the next exercise:

Exercise 3.3.4. *Let $m, n \in \mathbb{N}$ be such that $0 \leq n < m$. Compute in closed form*

$$A_{m,n} = \sum_{\ell=0}^{\infty} \frac{1}{(m\ell + n)!}, \quad n = 0, 1, \ldots, m - 1.$$

The books of Polya and Szegö [182, 183] are a mine of exercises, most of them quite challenging; here are some of them (see [182, Exercises 36, 37, 38, p. 110]).

Exercise 3.3.5. *Assume that the complex numbers $z_n, n = 1, 2, \ldots$ are in the right half-plane and that both the series $\sum_{n=1}^{\infty} z_n$ and $\sum_{n=1}^{\infty} z_n^2$ converge. Show that the series $\sum_{n=1}^{\infty} |z_n|^2$ converges. Give a counterexample when the z_n are not restricted to the right half-plane.*

Exercise 3.3.6. *Find a sequence of complex numbers such that all the series $\sum_{n=1}^{\infty} z_n^k$ converge but all the series $\sum_{n=1}^{\infty} |z_n|^k$ diverge, $k = 1, 2, \ldots$.*

Exercise 3.3.7. *Let $0 < \alpha < \pi/2$ and let $z_n = \rho_n e^{i\theta_n}$, $n = 1, 2, \ldots$, be a sequence of complex numbers all different from 0 and such that*

$$-\alpha \le \theta_n \le \alpha.$$

Show that the series $\sum_{n=1}^{\infty} z_n$ and $\sum_{n=1}^{\infty} |z_n|$ converge or diverge at the same time.

We conclude with a question taken from [186, Exercise 5.1.19, p. 186]. We refer to that source for a proof.

Question 3.3.8. *Let $z_1, \ldots,$ be a sequence of complex numbers such that $|z_j - z_k| \ge 1$ for $j \ne k$. Study the convergence of the series $\sum_{n=1}^{\infty} \frac{1}{|z_n|^u}$ for $u > 0$.*

3.4   Power series and elementary functions

As in the real case, a power series is an expression of the form

$$\sum_{n=0}^{\infty} a_n (z - z_0)^n, \tag{3.4.1}$$

where now z_0 and the a_n are complex numbers. One now speaks of a disk of convergence rather than of an interval of convergence, and the proof of the following result is the same as in the real case.

Theorem 3.4.1. *The power series*

$$\sum_{n=0}^{\infty} a_n (z - z_0)^n \tag{3.4.2}$$

converges absolutely for $|z| < R$ with

$$R = \begin{cases} \frac{1}{\limsup_{n \to \infty} |a_n|^{1/n}}, & \text{if } \limsup_{n \to \infty} |a_n|^{1/n} > 0, \\ \infty, & \text{if } \limsup_{n \to \infty} |a_n|^{1/n} = 0. \end{cases} \tag{3.4.3}$$

It diverges for all $|z| > R$.

Proof. For simplicity, we set $z_0 = 0$. We assume that the series converges for some $w \ne 0$. Then, the sequence of numbers $(a_n w^n)_{n \in \mathbb{N}_0}$ goes to 0 and in particular is bounded in absolute value by a finite positive number, say M. For $|z| < |w|$ we have

$$a_n z^n = a_n w^n \left(\frac{z}{w}\right)^n,$$

and so

$$|a_n z^n| \leq M \left(\frac{|z|}{|w|} \right)^n, \tag{3.4.4}$$

and the power series (3.4.2) converges absolutely for $|z| < |w|$. Furthermore, from

$$|a_n z^n| \leq M,$$

we have for every $n \in \mathbb{N}$,

$$|a_n|^{1/n} \cdot |w| \leq M^{1/n},$$

and taking $\limsup$ on both sides we get

$$\limsup_{n \to \infty} |a_n|^{1/n} \cdot |w| \leq 1.$$

It follows that w is arbitrary if $\limsup_{n \to \infty} |a_n|^{1/n} = 0$ and

$$|w| \leq \frac{1}{\limsup_{n \to \infty} |a_n|^{1/n}}$$

otherwise. It follows that the power series converges for every z of modulus strictly less than R, where R is given by (3.4.3). Let now z be such that $|z| > R$. By definition of the $\limsup$ there exists an infinite subsequence of integers $(n_k)_{k \in \mathbb{N}}$ such that $|a_{n_k}|^{1/n_k} |z| \geq 1$. In particular $|a_{n_k} z^{n_k}| \geq 1$ and the power series (3.4.2) cannot converge. $\qquad\square$

Note that the theorem says nothing on the behaviour of the power series on the circle $|z - z_0| = R$. This is illustrated in various exercises in the section. See Exercises 3.4.4 through 3.4.6. The number R is called the *radius of convergence* of the power series. It may be equal to 0 (for instance, when $a_n = n!$), and then the power series converges only for $z = 0$ and it may be equal to ∞ (for instance, when $a_n = \frac{1}{n!}$), and then the power series converges for every complex number.

The following result is of special importance:

Corollary 3.4.2. *Let* (3.4.2) *be a power series with strictly positive radius of convergence R (possibly, $R = \infty$). Then, the power series converges uniformly and absolutely in every closed disk $|z - z_0| \leq r$ with $r \in (0, R)$.*

Often, and as in the real variable case, there are easier ways to compute the radius of convergence of a given power series, as we recall in the next proposition.

Proposition 3.4.3. *Consider the power series* (3.4.1), *and in case* (a) *assume that $a_n \neq 0$ from a certain index. Then:*

(a) *Formula using D'Alembert's (the ratio) test: Assume that*

$$\lim_{n \to \infty} \frac{|a_{n+1}|}{|a_n|} \tag{3.4.5}$$

exists. Then

$$\frac{1}{R} = \lim_{n \to \infty} \frac{|a_{n+1}|}{|a_n|}.$$

(b) *Formula using Cauchy's test: Assume that*

$$\lim_{n \to \infty} |a_n|^{1/n} \tag{3.4.6}$$

exists. Then,

$$\frac{1}{R} = \lim_{n \to \infty} |a_n|^{1/n}.$$

Moreover, we make the following remark: It follows from the proof of Theorem 3.4.1 that R is the radius of the largest *open* disk with center z_0 within which the powers series converges. This gives a *geometric* (or "picturial") way to compute R. See Exercise 5.6.4.

Exercise 3.4.4. *Find the radius of convergence of the power series*

$$\sum_{n=1}^{\infty} n^{(-1)^n} z^n. \tag{3.4.7}$$

The topic of Exercise 5.7.7 is to compute in closed form the sum (3.4.7).

Exercise 3.4.5. *Radius of convergence of the power series*

$$\sum_{n=0}^{\infty} z^{n!}. \tag{3.4.8}$$

Exercise 3.4.6. *Radius of convergence of the power series*

$$\sum_{n=0}^{\infty} z^{2^n}.$$

In the preceding two exercises, the unit circle is the natural boundary of analyticity: The functions defined by the power series cannot be extended across the unit circle. See Exercise 6.3.3.

The behaviour at the boundary of the disk of convergence is much more delicate. One can study the convergence sometimes using Abel's theorem (see Section 3.5. The behaviour at the point $z = 1$ can be sometimes studied using Raabe's test, which we now recall.

Theorem 3.4.7 (Raabe's convergence test). *Let $(a_n)_{n \in \mathbb{N}}$ be a sequence of strictly positive numbers, and assume that the limit*

$$\lim_{n \to \infty} n \left(\frac{a_n}{a_{n+1}} - 1 \right) = R \tag{3.4.9}$$

exists. Then, if $R > 1$ the series $\sum_{n=1}^{\infty} a_n$ converges while it diverges if $R < 1$. No conclusion can be given (without more information) if $R = 1$.

We recall the following result, which states that multiplication of power series corresponds to the convolution of the sequences of their coefficients. See Exercise 4.4.8 for an illustration.

Proposition 3.4.8. *Let*

$$f(z) = \sum_{n=0}^{\infty} a_n (z - z_0)^n \quad and \quad g(z) = \sum_{n=0}^{\infty} b_n (z - z_0)^n$$

be the two power series, centered at the same point z_0, and with strictly positive radiuses of convergence R_1 and R_2 respectively. Then, the product fg is a power series centered at z_0, with a strictly positive radius of convergence greater than or equal to $\min(R_1, R_2)$. Furthermore,

$$f(z)g(z) = \sum_{n=0}^{\infty} \left(\sum_{\ell=0}^{n} a_\ell b_{n-\ell} \right) (z - z_0)^n. \tag{3.4.10}$$

Proof. The result is really a particular case of Cauchy's multiplication theorem; see Section 14.2. Indeed, for z such that $|z - z_0| < \min(R_1, R_2)$, both the series

$$\sum_{n=0}^{\infty} a_n (z - z_0)^n \quad and \quad \sum_{n=0}^{\infty} b_n (z - z_0)^n$$

are absolutely convergent. Since

$$(z - z_0)^n = (z - z_0)^\ell (z - z_0)^{n-\ell},$$

Cauchy's multiplication theorem then implies that (3.4.10) holds. $\square$

Exercise 3.4.9.

(a) *Let $(t_n)_{n \in \mathbb{N}_0}$ be a sequence of non-zero numbers. Show that*

$$\liminf_{n \to \infty} t_n = \frac{1}{\limsup_{n \to \infty} \frac{1}{t_n}}.$$

(b) *Let R be the radius of convergence of the series*

$$f(z) = \sum_{n=0}^{\infty} a_n z^n,$$

and assume that $a_n \neq 0$ for all $n \in \mathbb{N}_0$. What can you say about the radius of convergence of the series

$$g(z) = \sum_{n=0}^{\infty} \frac{1}{a_n} z^n.$$

The following is a classical exercise (as most of the exercises presented in this book); see, e.g., [209, Exercice 200, p. 75]. It corresponds to set $a = 0$ and $N \to \infty$ in formulas (1.3.7).

Exercise 3.4.10. *Compute, for a real number r of absolute value strictly less than 1 and $\theta \in \mathbb{R}$, the sums*

$$S(r, \theta) = \sum_{n=1}^{\infty} r^n \sin(n\theta)$$

and

$$C(r, \theta) = \sum_{n=0}^{\infty} r^n \cos(n\theta).$$

On a similar vein, try the following ([209, Exercice 199, p. 75]):

$$\sum_{n=1}^{\infty} nr^n \sin(n\theta) = \frac{r(1 - r^2) \sin \theta}{(1 - 2r \cos \theta + r^2)^2}. \qquad (3.4.11)$$

Hint. Recall that

$$z(1 + 2z + 3z^2 + 4z^3 + \cdots) = \frac{z}{(1 - z)^2}. \qquad (3.4.12)$$

There are (at least) two ways to prove this formula. The easiest (but logically not at this place in the book) is to resort to the general theorem on differentiation of power series in their disk of convergence and to differentiate the power series of $1/(1 - z)$. The more direct one is as follows: Start from (1.1.54):

$$1 + z + \cdots + z^N = \frac{1 - z^{N+1}}{1 - z}.$$

This is an identity between two rational functions, and we can differentiate to obtain

$$1 + 2z + \cdots + Nz^{N-1} = \frac{1}{(1 - z)^2} - \left\{ \frac{(N + 1)z^N(1 - z) + z^{N+1}}{(1 - z)^2} \right\}, \qquad (3.4.13)$$

or

$$z + 2z^2 + \cdots + Nz^N = z \cdot \frac{Nz^{N+1} - (N + 1)z^N + 1}{(1 - z)^2}. \qquad (3.4.14)$$

You may also decide to prove the above formula by induction if you do not want to allow differentiation of rational functions at this stage. Then multiply both sides of (3.4.13) by z and let $N \to \infty$ to obtain (3.4.12). This last equation will be instrumental to compute (3.4.11). Exercise 4.4.3 is also connected to the previous discussion.

Yet another way to find closed formulas for sums as in (3.4.11) and in (4.4.7) below is to differentiate term by term the series defining $C(r, \theta)$ and $S(r, \theta)$ with respect to θ or r. This is legitimate because the derivatives converge uniformly on every closed disk $|z| \leq R < 1$. For instance, differentiating with respect to θ both sides of the formula giving $C(r, \theta)$ we obtain (3.4.11).

We note that equation (3.4.13) can be generalized as follows; see [203, (3.14), p. 11]:

$$\sum_{n=0}^{N-1} \binom{n+p-1}{n} z^n = \frac{1}{(1-z)^p} - \frac{z^N}{1-z} \left\{ \sum_{n=0}^{p-1} \binom{N+p-1}{N+n} \left(\frac{z}{1-z}\right)^n \right\},$$

and moreover, for $|z| < 1$,

$$\left| \frac{z^N}{1-z} \left\{ \sum_{n=0}^{p-1} \binom{N+p-1}{N+n} \left(\frac{z}{1-z}\right)^n \right\} \right| \leq \frac{(N+p-1)^{p-1}|z|^N}{(1-|z|)^p}.$$

See [203, p. 12] for the latter.

As mentioned in [125, p. 205], the formulas of Exercise 1.3.5 can be generalized as follows (see [192, Exercise 3, p. 127] for $j = 0$ and [124, p. 210], for the general case).

Exercise 3.4.11. *Let $f(z) = \sum_{k=0}^{\infty} a_k z^k$ be a convergent power series with radius of convergence R. Let $n \in \mathbb{N}$ and $j \in \{0, \dots, n-1\}$. Show that, for $|z| < R$,*

$$\sum_{k=0}^{\infty} a_{j+kn} z^{j+kn} = \frac{\sum_{u=0}^{n-1} w^{-ju} f(w^u z)}{n}, \tag{3.4.15}$$

where $w = \exp \frac{2\pi i}{n}$.

Exercise 1.3.5 corresponds to $f(z) = (1+z)^n$. See also Exercise 6.3.10 for a related question.

Exercise 3.4.12 below is [52, Lemma 4, p. 233]. See the discussion before Exercise 3.1.10 for more on this exercise. Interestingly enough, the proof uses the finite case, i.e., Exercise 3.1.10. It is of course possible to give a direct proof for all cases, for instance by computing

$$\int_0^{2\pi} |f(re^{it})|^2 dt \tag{3.4.16}$$

for $r \in (0, 1)$. See Exercise 5.6.12 in relation with (3.4.16).

Exercise 3.4.12. *Let $f(z) = \sum_{n=0}^{\infty} a_n z^n$ be a power series which converges in the open unit disk and assume moreover that*

$$|f(z)| \leq 1, \quad \forall z \in \mathbb{D}. \tag{3.4.17}$$

Show that

$$|a_n| \leq 1, \quad \forall n \in \mathbb{N}_0.$$

Exercise 3.4.13. *Let m be a continuous complex-valued function defined on the interval $[0, 1]$. Show that the expression*

$$f(z) = \int_0^1 e^{zt} m(t)\, dt$$

exists for every $z \in \mathbb{C}$ and that

$$f(z) = \sum_{n=0}^{\infty} z^n \frac{\int_0^1 t^n m(t)\, dt}{n!}, \qquad \forall z \in \mathbb{C}. \tag{3.4.18}$$

In connection with the previous exercise, see also Exercise 4.2.14.

We have defined the exponential function e^z (which we also denote by $\exp z$) in Section 1.2 by the formula (1.2.3)

$$e^z = e^x (\cos y + i \sin y).$$

As already reminded, one defines in elementary calculus the exponential function e^x in terms of a power series

$$e^x = \sum_{n=0}^{\infty} \frac{x^n}{n!}, \qquad x \in \mathbb{R}.$$

In the next exercise we show that this last formula can still be used in the complex case to define the exponential function.

Exercise 3.4.14. *Let $z = x + iy \in \mathbb{C}$. Show that*

$$e^x (\cos y + i \sin y) = \sum_{n=0}^{\infty} \frac{z^n}{n!}. \tag{3.4.19}$$

Now we therefore have (1.2.6)

$$e^x (\cos y + i \sin y) = \sum_{n=0}^{\infty} \frac{z^n}{n!} = \lim_{p \to \infty} \left(1 + \frac{z}{p}\right)^p.$$

The following exercise has a much easier proof using complex integration, see Exercise 5.2.6. At this stage, we give a proof based on the power series definition of the exponential, and using real integration of a complex-valued function.

Exercise 3.4.15. *Let z_1 and z_2 be in the left closed half-plane. Show that*

$$|e^{z_1} - e^{z_2}| \leq |z_1 - z_2|. \tag{3.4.20}$$

Hint. Show that

$$e^z - 1 = z \int_0^1 e^{tz} dt. \tag{3.4.21}$$

The next inequality plays an important role in certain infinite products; for the proof, see the solutions of Exercises 3.7.9 and 3.7.11. We repeat the proof for completeness.

Exercise 3.4.16. *Prove that, for every* $z \in \mathbb{D}$,

$$|1 - (1 - z)e^z| \leq |z|^2. \tag{3.4.22}$$

Finally, we recall the definitions of the trigonometric functions (1.2.13) in terms of power series. These definitions are known from calculus classes, when restricted to a real argument.

$$\begin{aligned}
\sin z &= \sum_{n=0}^{\infty} \frac{(-1)^n z^{2n+1}}{(2n+1)!}, \\
\cos z &= \sum_{n=0}^{\infty} \frac{(-1)^n z^{2n}}{(2n)!}, \\
\sinh z &= \sum_{n=0}^{\infty} \frac{z^{2n+1}}{(2n+1)!}, \\
\cosh z &= \sum_{n=0}^{\infty} \frac{z^{2n}}{(2n)!}.
\end{aligned} \tag{3.4.23}$$

3.5 Abel's theorem and behaviour on the boundary

The study of the convergence of a power series of radius of convergence R on the circle $|z| = R$ is a difficult problem. For instance the series

$$f(z) = \sum_{n=1}^{\infty} \frac{z^n}{n^2}$$

has radius of convergence $R = 1$ and converges for every z on the unit circle. The convergence is moreover absolute and uniform. On the other hand, the power series

$$\sum_{n=0}^{\infty} z^n$$

has the same radius of convergence, but converges for *no* z on the unit circle, since the running term z^n does not go to 0 when z is on the unit circle.

We now present Abel's summation theorem; see for instance [31, p. 127]. To ease the presentation we divide the result into two parts, namely Theorems 3.5.1 and 3.5.4.

Theorem 3.5.1. Let $(a_n)_{n\in\mathbb{N}}$ and $(b_n)_{n\in\mathbb{N}}$ be two sequences, of complex and real numbers respectively, and assume that:

(1) The $b_n \geq 0$ and the sequence $(b_n)_{n\in\mathbb{N}}$ decreases to 0.

(2) There is a number K such that all the partial sums

$$\left|\sum_{\ell=1}^{m} a_\ell\right| \leq K.$$

Then the series $\sum_{n=1}^{\infty} a_n b_n$ converges.

The proof of the above theorem is a consequence of the identity

$$\sum_{n=1}^{M} a_n b_n = \sum_{k=2}^{M} \left(\sum_{u=1}^{k-1} a_u\right)(b_{k-1} - b_k) + \left(\sum_{u=1}^{M} a_u\right) b_M, \tag{3.5.1}$$

valid for $M \geq 2$. For instance, for $M = 3$ we have

$$a_1 b_1 + a_2 b_2 + a_3 b_3 = a_1(b_1 - b_2) + (a_1 + a_2)(b_2 - b_3) + (a_1 + a_2 + a_3)b_3,$$

and

$$a_1 b_1 + a_2 b_2 + a_3 b_3 + a_4 b_4 = a_1(b_1 - b_2) + (a_1 + a_2)(b_2 - b_3)$$
$$+ (a_1 + a_2 + a_3)(b_3 - b_4) + (a_1 + a_2 + a_3 + a_4)b_4$$

for $M = 4$.

We also note that (3.5.1) implies the useful upper bound

$$\left|\sum_{n=1}^{M} a_n b_n\right| \leq K b_1, \tag{3.5.2}$$

or, more generally, for $N \leq M$,

$$\left|\sum_{n=N}^{M} a_n b_n\right| \leq K b_N, \tag{3.5.3}$$

which is obtained from (3.5.2) by a shift of index $n \mapsto n + N - 1$.

Remark 3.5.2. One can weaken the first condition in the theorem and assume that the sequence $(b_n)_{n\in\mathbb{N}}$ is of bounded variation (see Remark 3.7.10 below for the latter).

To solve Exercise 3.5.3 you need Theorem 3.5.1, and inequality (1.2.9). When $\theta = \pi$ we have the case of an alternating series.

Exercise 3.5.3. *For which real θ does the sum*

$$\sum_{n=1}^{\infty} \frac{e^{in\theta}}{\sqrt{n}} \tag{3.5.4}$$

converge?

The power series $\sum_{n=1}^{\infty} \frac{z^n}{\sqrt{n}}$ converges in $\mathbb{D}$, and the sum (3.5.4) defines "its boundary values". The following theorem makes this statement more precise.

Theorem 3.5.4. *Let $f(z) = \sum_{n=0}^{\infty} a_n z^n$ be a power series with radius of convergence R, and assume that for some $\theta \in [0, 2\pi]$ the series $\sum_{n=0}^{\infty} a_n R^n e^{in\theta}$ converges. Then,*

$$\lim_{\substack{r \to R \\ r \in (0,R)}} f(re^{i\theta}) = \sum_{n=0}^{\infty} a_n R^n e^{in\theta}. \tag{3.5.5}$$

Remark 3.5.5. Rather than a radial limit, one can also consider z tending to $Re^{i\theta}$ in a set of points z satisfying

$$|\operatorname{Re}^{i\theta} - z| \le \alpha(R - |z|) \tag{3.5.6}$$

for some $\alpha > 0$ (that is, z stays inside a Stolz angle).

The proof of Theorem 3.5.4 is a consequence of Theorem 3.5.1, and is now outlined (see also for instance [112, pp. 250–251]). Fix $\epsilon > 0$. By hypothesis there exists $N \in \mathbb{N}$ such that

$$M \ge N + 1 \implies \left| \sum_{n=N+1}^{M} a_n R^n e^{in\theta} \right| \le \frac{\epsilon}{3}. \tag{3.5.7}$$

Similarly, and with $r \in (0, R)$, applying (3.5.3) with $a_n R^n e^{in\theta}$ in place of a_n and $\frac{r^n}{R^n}$ instead of b_n leads to

$$\left| \sum_{n=N+1}^{M} a_n r^n e^{in\theta} \right| = \left| \sum_{n=N+1}^{M} a_n R^n e^{in\theta} \frac{r^n}{R^n} \right|$$

$$\le \left| \sum_{n=N+1}^{M} a_n R^n e^{in\theta} \frac{r^{N+1}}{R^{N+1}} \right| \tag{3.5.8}$$

$$\le \left| \sum_{n=N+1}^{M} a_n R^n e^{in\theta} \right| \le \frac{\epsilon}{3}.$$

It suffices now to write

$$\left| \sum_{n=0}^{\infty} a_n r^n e^{in\theta} - \sum_{n=0}^{\infty} a_n R^n e^{in\theta} \right| \le \left| \sum_{n=0}^{N} a_n (r^n - R^n) e^{in\theta} \right|$$

$$+ \left| \sum_{n=N+1}^{\infty} a_n r^n e^{in\theta} \right| + \left| \sum_{n=N+1}^{\infty} a_n R^n e^{in\theta} \right|.$$

For a given ϵ one first finds N such that each the last two terms is smaller than $\frac{\epsilon}{3}$ (see (3.5.7) and (3.5.8) with $M \to \infty$). Then one finds r_0 such that the finite sum $\left| \sum_{n=0}^{N} a_n (r^n - R^n) e^{in\theta} \right| \le \frac{\epsilon}{3}$ for $r \in (r_0, R)$.

Question 3.5.6. *Modify the proof to obtain the result when z varies in a set of the form (3.5.6).*

The following exercise will be used in Exercise 10.3.9, where a conformal map from the open unit disk onto a square is studied. The proof of the exercise itself is of interest only when the series $\sum_{n=0}^{\infty} a_n$ is not assumed to converge.

Exercise 3.5.7. *Let $(a_n)_{n \in \mathbb{N}_0}$ be a decreasing sequence of positive numbers, with limit equal to 0, and let $\theta \in [-\frac{\pi}{4}, \frac{\pi}{4}] \setminus \{0\}$, and assume that*

$$M = \sum_{n=0}^{\infty} \frac{a_n}{n+1} < \infty \tag{3.5.9}$$

(1) *Show that*

$$\int_0^1 \left(\sum_{n=0}^{\infty} a_n t^{4n} e^{i(4n+1)\theta} \right) dt = \sum_{n=0}^{\infty} a_n \frac{e^{i(4n+1)\theta}}{4n+1} \tag{3.5.10}$$

$$= M + i \int_0^{\theta} \left(\sum_{n=0}^{\infty} a_n e^{i(4n+1)u} \right) du,$$

where M is given by (3.5.9).

(2) *Show that (3.5.10) still holds for $\theta = 0$.*

3.6   Summable families

Before trying to solve the exercises in this section, the reader may want to go to Section 14.3, where the highlights on summable families are reviewed.

Exercise 3.6.1. *Show that*

$$\sum_{n=1}^{\infty} \frac{z^n}{1+z^n} = \sum_{p=1}^{\infty} \frac{(-1)^{(p-1)} z^p}{1 - z^p}, \quad |z| < 1.$$

One more exercise related to elliptic functions; see [47, p. 235].

Exercise 3.6.2. *Show that the family of functions indexed by $\mathbb{Z}^2$*

$$f_{pq}(z) = \begin{cases} \dfrac{1}{z^2}, & \text{if } (p,q) = (0,0) \\ \dfrac{1}{(z - (p+iq))^2} - \dfrac{1}{(p+iq)^2}, & \text{if } (p,q) \ne (0,0) \end{cases}$$

is summable for every $z \notin \mathbb{Z} + i\mathbb{Z}$, and that its sum $\wp(z)$ satisfies

$$\wp(z) = \wp(z+1) = \wp(z+i).$$

3.7 Infinite products

Infinite products seem to first appear in 1579 in the work of Viete, who proved

$$\frac{2}{\pi} = \sqrt{\frac{1}{2}} \sqrt{\frac{1}{2} + \frac{1}{2}\sqrt{\frac{1}{2}}} \sqrt{\frac{1}{2} + \frac{1}{2}\sqrt{\frac{1}{2} + \frac{1}{2}\sqrt{\frac{1}{2}}}} \cdots .$$

The first to have made a systematic development of infinite products is Euler. See [188, p. 3] for these and more information.

Recall that the infinite product $\prod_{k=0}^{\infty} b_k$ is said to converge if the sequence $\prod_{k=0}^{n} b_k$ converges to a number *different* from 0. In particular, all the numbers b_k are assumed different from 0. Write $b_k = 1 + a_k$. The infinite product is said to converge absolutely if the infinite product $\prod_{k=0}^{\infty}(1 + |a_k|)$ converges to a number *different* from 0. The infinite product is then convergent. Note that it may be that the product $\prod_{k=0}^{\infty} |1 + a_k|$ converges while the infinite product $\prod_{k=0}^{\infty}(1 + a_k)$ diverges, as is illustrated by the example

$$\prod_{k=1}^{\infty} \left(1 + \frac{i}{k}\right). \tag{3.7.1}$$

See [39, TG VIII.26, Exercice 4], [97, p. 354] for the latter. In this book the result is explicated in Section 4.4 since it uses the notion of logarithm. See Exercise 4.4.15. We also mention in that section an alternative condition for convergence of an infinite product in terms of logarithms. See Theorem 4.4.14.

One can find a number of approaches to infinite products in textbooks. One can discuss them after developing function theory. In particular using the properties of the function $\ln(1 - z)$, results can be proved in a quite short way. See for instance Lang's book [143, p. 372]. One can also give conditions in terms of the arguments of the elements of the products; see for instance [5, Theorem 6, p. 192]. Here, we focus on the absolute convergence, see [47] and [77, pp. 208–209], and mention the following result:

Theorem 3.7.1. *Let $(a_n)_{n \in \mathbb{N}_0}$ be a sequence of numbers all different from -1, and assume that*

$$\sum_{n=0}^{\infty} |a_n| < \infty. \tag{3.7.2}$$

Then, the infinite product $\prod_{n=0}^{\infty}(1 + a_n)$ is absolutely convergent. Let P denote the value of the infinite product. Then it holds that

$$\left| \prod_{n=0}^{N}(1 + a_n) - P \right| \le e^{2\sum_{n=0}^{\infty} |a_n|} \sum_{n=N+1}^{\infty} |a_n|. \tag{3.7.3}$$

The proof below appears for instance in [47], [77], [167, pp. 284–286]. It has the advantage of being valid in much more general settings than the complex numbers. It is important to master the details of the proof because of various upper bounds which are derived in it.

Proof of Theorem 3.7.1.
   Step 1: *It holds that*

$$\left| \prod_{j=0}^{n}(1 + a_j) - 1 \right| \le \left(\prod_{j=0}^{n}(1 + |a_j|) \right) - 1. \tag{3.7.4}$$

We proceed by induction: For $n = 0$ the inequality is trivial. Assume now that (3.7.4) holds at rank n. Then we have

$$\left| \prod_{j=0}^{n+1}(1 + a_j) - 1 \right| = \left| \left(\prod_{j=0}^{n}(1 + a_j) \right)(1 + a_{n+1}) - 1 \right|$$

$$\le \left| \left(\prod_{j=0}^{n}(1 + a_j) \right) - 1 \right| + |a_{n+1}| \left(\prod_{j=0}^{n}(1 + |a_j|) \right)$$

$$\le \left(\left(\prod_{j=0}^{n}(1 + |a_j|) \right) - 1 \right) + |a_{n+1}| \left(\prod_{j=0}^{n}(1 + |a_j|) \right)$$

$$= \left(\prod_{j=0}^{n}(1 + |a_j|) \right)(1 + |a_{n+1}|) - 1$$

$$= \left(\prod_{j=0}^{n+1}(1 + |a_j|) \right) - 1,$$

where we have used the induction hypothesis to go from the second to the third line.

Note that, replacing a_j by a_{j+n} we also get for $m > n$:

$$\left| \prod_{k=n+1}^{m}(1 + a_k) - 1 \right| \le \left(\prod_{k=n+1}^{m}(1 + |a_k|) \right) - 1. \tag{3.7.5}$$

Step 2: *We set*

$$b_n = \prod_{j=0}^{n}(1 + a_j).$$

It holds that

$$|b_n| = \prod_{k=0}^{n} |1 + a_k|$$

$$\leq \prod_{k=0}^{n} (1 + |a_k|)$$

$$\leq \prod_{k=0}^{n} e^{|a_k|} \leq e^{\sum_{k=0}^{\infty} |a_k|} = e^K < \infty,$$

with $K = \sum_{k=0}^{\infty} |a_k|$.

Step 3: $(b_n)_{n \in \mathbb{N}_0}$ *is a Cauchy sequence.*

Indeed, for $m > n$ and using (3.7.5),

$$|b_m - b_n| = \left(\prod_{k=0}^{n} |1 + a_k| \right) \cdot \left| \prod_{k=n+1}^{m} (1 + a_k) - 1 \right| \tag{3.7.6}$$

$$\leq e^K \left\{ \left(\prod_{k=n+1}^{m} (1 + |a_k|) \right) - 1 \right\}.$$

But, we have that

$$\left(\prod_{k=n+1}^{m} (1 + |a_k|) \right) - 1 \leq e^{\sum_{k=n+1}^{m} |a_k|} - 1$$

$$\leq \left(\sum_{k=n+1}^{m} |a_k| \right) e^{\sum_{k=n+1}^{m} |a_k|} \tag{3.7.7}$$

$$\leq \left(\sum_{k=n+1}^{m} |a_k| \right) e^K,$$

where we have used the inequality

$$e^x \leq 1 + xe^x, \quad x \geq 0,$$

with $x = \sum_{k=n+1}^{m} |a_k|$.

Step 4: *Let* $r_0, r_1, \ldots \in (0,1)$ *be such that* $\sum_{k=0}^{\infty} r_k < 1$. *Then it holds that*

$$\prod_{k=0}^{n} (1 - r_k) \geq 1 - \sum_{k=0}^{n} r_k, \quad n = 0, 1, \ldots. \tag{3.7.8}$$

We proceed by induction. For $n = 0$, (3.7.8) holds trivially. Assume that it holds at rank n. Then

$$\prod_{k=0}^{n+1}(1 - r_k) = \left(\prod_{k=0}^{n}(1 - r_k) \right)(1 - r_{n+1})$$

$$\geq (1 - \sum_{k=0}^{n} r_k)(1 - r_{n+1}), \quad \text{(since (3.7.8) holds at rank } n\text{)}$$

$$\geq 1 - (\sum_{k=0}^{n} r_k) - r_{n+1},$$

where we have used that (3.7.8) holds at rank 2 to go to the last line. This shows that (3.7.8) holds at rank $n + 1$, and hence for every $n \in \mathbb{N}_0$.

Step 5: $\lim_{n \to \infty} b_n \neq 0$.

Indeed, let N be such that $\sum_{k=N}^{\infty} |a_k| < 1/2$. Using the preceding step we can write for $M \geq N$,

$$\left| \prod_{k=N}^{M}(1 + a_k) \right| \geq \prod_{k=N}^{M}(1 - |a_k|) \geq 1 - \sum_{k=N}^{M} |a_k| \geq 1 - 1/2 = 1/2.$$

Hence

$$\left| \prod_{k=N}^{\infty}(1 + a_k) \right| \geq 1/2,$$

and so

$$\lim_{n \to \infty} |b_n| = \left(\prod_{k=1}^{N} |1 + a_k| \right) \left(\prod_{k=N+1}^{\infty} |1 + a_k| \right) \geq \left(\prod_{k=1}^{N} |1 + a_k| \right) \cdot \left(\frac{1}{2} \right) > 0.$$

Step 6: *The bound* (3.7.3)

$$\left| \prod_{n=0}^{N}(1 + a_n) - P \right| \leq e^{2 \sum_{n=0}^{\infty} |a_n|} \sum_{n=N+1}^{\infty} |a_n|$$

is in force.

This follows directly from (3.7.6) and (3.7.7), and letting $m \to \infty$.   □

Remarks 3.7.2.

(1) Inequality (3.7.3) is of paramount importance in the study of the convergence of the infinite product, in particular when the a_n are functions of the complex variable. It then follows from (3.7.3) that uniform convergence of the series

$\sum_{n=0}^{\infty} |a_n(z)|$ in some (usually compact) set K will imply uniform convergence of the sequence of the partial products. See [18, Proposition 4.10.1, p. 157] for a complete result related to products of functions, which uses in the proof estimates on the logarithm function, and not the above estimates.

(2) We note that (3.7.8) holds under a weaker requirement, namely $r_k \in (-\infty, 1)$ for all k. When $r_1 = r_2 = \cdots$ we then get back Bernoulli's lemma (see, e.g., [112, Exercise 2.4, p. 28]):

$$(1+x)^n \geq 1 + nx, \quad x \in (-1, \infty) \quad \text{and} \quad n \in \mathbb{N}.$$

Theorem 3.7.1 is still valid in a (possibly non-commutative) Banach algebra with identity, say $\mathcal{B}$, with norm $\| \cdot \|$ such that

$$\|ab\| \leq \|a\| \cdot \|b\|,$$

but one loses a bit the specificity of complex numbers. The proof goes in the same way, with absolute value replaced by the norm of $\mathcal{B}$. Taking $\mathcal{B} = \mathbb{C}^{N \times N}$ with the Euclidean norm (or any other norm, since all norms are equivalent in a finite-dimensional vector space) we have the following result, which will play an important role in Section 11.5.

Theorem 3.7.3. *Let $(A_n)_{n \in \mathbb{N}_0}$ be a sequence of matrices in $\mathbb{C}^{N \times N}$ such that $\det(I_N + A_n) \neq 0$ for all $n \in \mathbb{N}_0$ and*

$$\sum_{n=0}^{\infty} \|A_n\| < \infty.$$

Then the sequence of matrices

$$P_n = (I_N + A_0)(I_N + A_1) \cdots (I_N + A_n), \quad n = 0, 1, \ldots$$

converges to an invertible matrix in the norm $\| \cdot \|$.

The limit in the above theorem will be denoted by

$$\overset{\curvearrowleft}{\underset{n=0}{\overset{\infty}{\prod}}} (I_N + A_n).$$

Because of the lack of commutativity, other choices of partial products are possible. For instance one could take

$$Q_n = (I_N + A_n) \cdots (I_N + A_1)(I_N + A_0), \quad n = 0, 1, \ldots.$$

The limit is then denoted by

$$\overset{\curvearrowright}{\underset{n=0}{\overset{\infty}{\prod}}} (I_N + A_n).$$

An analog of Theorem 3.7.1 does not hold when absolute convergence is not required, as one sees from the following classical counter-example (see for instance [192, Exercise 6, p. 294], and [180, p. 43], [28, Exercise 4, p. 226] for item (a)). In that setting, see also Theorem 4.4.14.

Exercise 3.7.4.

(a) *Show that the infinite product*

$$\prod_{n=2}^{\infty}\left(1+\frac{(-1)^n}{\sqrt{n}}\right) \tag{3.7.9}$$

diverges, while the series

$$\sum_{n=2}^{\infty}\frac{(-1)^n}{\sqrt{n}} \tag{3.7.10}$$

converges.

(b) *Let*

$$a_{2n-1} = -\frac{1}{\sqrt{n}} \quad and \quad a_{2n} = \frac{1}{\sqrt{n}}+\frac{1}{n}, \quad n = 2,3,\ldots.$$

Then the series $\sum_{n=2}^{\infty} a_n$ diverges and the infinite product $\prod_{n=2}^{\infty}(1 + a_n)$ converges.

The next exercise, taken from [5, Exercise 1, p. 193], also deals with an infinite product of real numbers.

Exercise 3.7.5. *Compute, if it converges, the value of the infinite product*

$$\prod_{n=2}^{\infty}\left(1-\frac{1}{n^2}\right).$$

The following infinite products appear in the theory of fractals; see [129], [70]. It is set for a real variable t, but it can be also be chosen complex. For the case $\rho = 2$, see Exercise 3.7.15.

Exercise 3.7.6. *Let $\rho > 1$ and $t \in \mathbb{R}$. Show that the infinite product*

$$\prod_{n=1}^{\infty} \cos\left(\frac{t}{\rho^n}\right) \tag{3.7.11}$$

converges, with the exception of a countable number of values of t.

Exercise 3.7.7. *Show that the infinite product $\prod_{k=1}^{\infty}\left(1 + \frac{z^k}{k^2}\right)$ converges in the closed unit disk, with the exception of the point $z = -1$.*

It is important to express analytic functions as infinite products. An important instance is Exercise 13.1.1, which deals with elliptic functions. See [107, Exercice 604, p. 91]. We first give some other examples.

Exercise 3.7.8. *Let $0 < |q| < 1$. Show that the infinite product*

$$\prod_{\ell=1}^{\infty}(1 - q^{\ell}z)$$

converges in $\mathbb{C}$ for all z and that it vanishes only at the points $z = q^{-\ell}$, $\ell = 1, 2, \ldots$. Show that the convergence is uniform in every set of the form $|z| \leq r$, where $r \in (0, 1)$, and that the limit satisfies the functional equation

$$f(z) = (1 - qz)f(qz). \tag{3.7.12}$$

A follow-up of the previous exercise is given in Section 4.2; see Exercise 4.2.25. See Exercise 3.4.16 for a result related to the exercise below.

Exercise 3.7.9. *Let $a_0, a_1, a_2, \ldots$ be a sequence of complex numbers such that $a_0 = a_1 = 1$ and*

$$\sum_{n=0}^{\infty}|a_{n+1} - a_n| < \infty. \tag{3.7.13}$$

Show that the convergence radius of the power series $f(z) = \sum_{n=0}^{\infty} a_n z^n$ is at least 1. Assume that it is equal to infinity. Show that the infinite product

$$\prod_{n=2}^{\infty}\left(1 - \frac{z}{n}\right) f(z/n)$$

converges in $\mathbb{C}$, and that it vanishes at the points $z = 2, 3, \ldots$.

Remark 3.7.10. Sequences which satisfy (3.7.13) are called of *bounded variation*. Note that a decreasing sequence of positive numbers is always of bounded variation since

$$\sum_{n=0}^{\infty}|a_{n+1} - a_n| = \lim_{N\to\infty}\sum_{n=0}^{N}|a_{n+1} - a_n|$$

$$= \lim_{N\to\infty}\sum_{n=0}^{N}(a_n - a_{n+1})$$

$$= \lim_{N\to\infty} a_0 - a_{N+1}$$

$$\leq a_0.$$

For instance, the decreasing sequence $1, \frac{1}{2}, \frac{1}{3}, \ldots$ is of bounded variation.

The following exercise is a special case of Exercise 3.7.9 and is related to Euler products.

Exercise 3.7.11. *Show that the infinite product*

$$\prod_{n=1}^{\infty} \left(1 + \frac{z}{n}\right) e^{-z/n}$$

converges to a non-zero number for all z different from $-1, -2, \ldots$.

The solution of the preceding exercise is based on the inequality (3.4.22)

$$|1 - (1 - z)e^z| \leq |z|^2, \quad \text{for} \quad |z| \leq 1.$$

More generally, define for $p \in \mathbb{N}$,

$$E_p(z) = (1 - z)e^{\left(z + \frac{z^2}{2} + \cdots + \frac{z^p}{p}\right)}. \tag{3.7.14}$$

The function E_p is called a Weierstrass factor, and it holds that

$$|1 - E_p(z)| \leq |z|^{p+1}, \quad |z| \leq 1. \tag{3.7.15}$$

This last inequality is the key to the Weierstrass product theorem. For $p > 1$ the proof we present involves derivative, and is differed to Section 4.4. See Exercise 4.4.22.

The next exercise deals with a very important family of functions, called *Blaschke products*.

Exercise 3.7.12. *Let $z_0, z_1, \ldots$ be a sequence of numbers in the open unit disk, different from 0 and such that*

$$\sum_{n=0}^{\infty} (1 - |z_n|) < \infty. \tag{3.7.16}$$

Show that the product

$$\prod_{n=0}^{\infty} \left(\frac{|z_n|}{z_n} \frac{z_n - z}{1 - z\overline{z_n}}\right) \tag{3.7.17}$$

converges for z in the open unit disk and different from the points $z = z_n$, $n = 0, 1, \ldots$.

Exercise 3.7.13. *Let $z_0, z_1, \ldots$ be a sequence of numbers in the open upper half-plane, and such that*

$$\sum_{n=0}^{\infty} \text{Im } z_n < \infty. \tag{3.7.18}$$

Show that the product

$$\prod_{n=0}^{\infty} \frac{z - z_n}{z - \overline{z_n}} \tag{3.7.19}$$

converges for z in the open upper half-plane and different from the points $z = z_n$, $n = 0, 1, \ldots$.

Remark 3.7.14. Blaschke products play an important role in complex analysis; see for instance [69], [121], [190]. Blaschke products are examples of analytic functions; see Exercise 6.7.1. It is of interest to study the boundary behaviour of infinite Blaschke products. For instance, when (3.7.16) is strengthened to

$$\sum_{n=0}^{\infty}(1 - |z_n|)^{\alpha} < \infty, \tag{3.7.20}$$

where $\alpha \in (0, 1)$, the Blaschke product has radial limits of modulus 1 everywhere, at the possible exception of a set of α-capacity 0 (we will not recall the definition of the latter). See [51, p. 14] and [93]. We note that the case where condition (3.7.16) fails is of special importance. The product diverges, but this means that there is uniqueness in some underlying interpolation problem. In another line of research, one can find in [51, Chapter 10, p. 86] results regarding generalizations of Blaschke products, when (3.7.16) is weakened, for instance to the condition of the form

$$\sum_{n=0}^{\infty}(1 - |z_n|)^2 < \infty.$$

See [141] for this case.

You can find examples of infinite products appearing in the theory of wavelets in [90, § 4]. See also Exercise 4.2.10 below.

Exercise 3.7.15 (see for instance [203, Exercise 4, p. 66] for (3.7.21)).

(a) *Show that*

$$|1 - \cos z| \leq \frac{|z|^2}{2}e^{|z|}. \tag{3.7.21}$$

(b) *Show that the infinite product*

$$\prod_{n=0}^{\infty} \cos(z/2^n) \tag{3.7.22}$$

converges for every complex number z such that $\cos(z/2^n) \neq 0$.

The following exercise comes from [112, p. 64].

Exercise 3.7.16. *The purpose of the exercise is to show that*

$$\sinh z = z \prod_{k=1}^{\infty}\left(1 + \frac{z^2}{k^2\pi^2}\right), \tag{3.7.23}$$

$$\cosh z = \prod_{k=0}^{\infty}\left(1 + \frac{4z^2}{(2k+1)^2\pi^2}\right), \quad z \in \mathbb{C}. \tag{3.7.24}$$

(a) *Using formulas (1.5.10) and (1.5.12), show that for every $z \in \mathbb{C}$,*

$$\left(1+\frac{z}{2n}\right)^{2n} - \left(1-\frac{z}{2n}\right)^{2n} = 2z \prod_{k=1}^{n-1}\left(1+\frac{z^2}{4n^2}\frac{1+\cos\frac{k\pi}{n}}{1-\cos\frac{k\pi}{n}}\right).$$

(b) *Prove (3.7.23) using (14.7.1) (see also Theorem 14.7.2).*

(c) *Proceed in a similar way to prove (3.7.24), but now starting from (1.5.8).*

We note that replacing z by iz in (3.7.23) and (3.7.24) respectively leads to the formulas

$$\sin z = z \prod_{k=1}^{\infty}\left(1-\frac{z^2}{k^2\pi^2}\right), \tag{3.7.25}$$

$$\cos z = \prod_{k=0}^{\infty}\left(1-\frac{4z^2}{(2k+1)^2\pi^2}\right), \tag{3.7.26}$$

where $z \in \mathbb{C}$.

3.8   Multiplicable families

In Theorem 3.7.1 the infinite product will be the same for any rearrangement of the indices. More generally, one can consider an infinite product of the form

$$\prod_{w\in L}(1+a_w)$$

where L is a countable set and where

$$\sum_{w\in L}|a_w| < \infty.$$

The function appearing in the following question is called the Weierstrass sigma function (associated to the given lattice L). It plays a key role in the theory of elliptic functions. See Section 13.1.

Question 3.8.1. *Let $L = \{n+im\,;\,n,m\in\mathbb{N}_0\}$. Show that the infinite product*

$$\sigma(z) = z \prod_{\substack{w\in L \\ w\neq 0}}\left(1-\frac{z}{w}\right)e^{\left(\frac{z}{w}+\frac{z^2}{2w^2}\right)} \tag{3.8.1}$$

converges and vanishes only at the points of the lattice L.

Hint. In the proof use is made that

$$\sum_{\substack{w\in L \\ w\neq 0}}\frac{1}{|w|^3} < \infty.$$

See the proof of Exercise 3.6.2.

3.9 Solutions

Solution of Exercise 3.1.1. Consider the example of Exercise 1.1.11:

$$z(t) = \frac{1}{1 + \cos t + i \sin t},$$

where t is real and not an odd multiple of π. Then,

$$\lim_{t \to \pi} z(t)$$

does not exist, but

$$\operatorname{Re} z(t) \equiv \frac{1}{2}. \qquad \square$$

Solution of Exercise 3.1.2. Using (1.6.3), or directly, we have

$$\left| 1 - \sum_{k=1}^{n} a_k e^{2\pi i k t} \right|^2 = 1 + \sum_{k=1}^{n} |a_k|^2 - \sum_{k=1}^{n} a_k e^{2\pi i k t} - \sum_{\ell=1}^{n} \overline{a_\ell} e^{-2\pi i \ell t}$$

$$+ \sum_{\substack{k,\ell=1 \\ k \neq \ell}}^{n} a_k \overline{a_\ell} e^{2\pi i (k-\ell) t},$$

and thus

$$\frac{1}{2\pi} \int_0^1 \left| 1 - \sum_{k=1}^{n} a_k e^{2\pi i k t} \right|^2 dt = 1 + \sum_{k=1}^{n} |a_k|^2. \tag{3.9.1}$$

Note that (3.9.1) is just a particular case of Parseval's equality. From (3.9.1) it follows that

$$\frac{1}{2\pi} \int_0^1 \left| 1 - \sum_{k=1}^{n} a_k e^{2\pi i k t} \right|^2 dt \geq 1,$$

and hence that

$$\max_{t \in [0,1]} \left| 1 - \sum_{k=1}^{n} a_k e^{2\pi i k t} \right| \geq 1. \tag{3.9.2}$$

(Otherwise, the integral

$$\frac{1}{2\pi} \int_0^1 \left| 1 - \sum_{k=1}^{n} a_k e^{2\pi i k t} \right|^2 dt$$

would be strictly less than 1.) From (3.9.2) it follows that

$$\left| 1 - \sum_{k=1}^{n} a_k e^{2\pi i k t} \right| \geq 1$$

for at least one value of $t \in [0, 1]$. $\qquad \square$

Solution of Exercise 3.1.3. We first note that for any real number $\beta \neq 0$,

$$\frac{1}{T} \int_0^T e^{i\beta t} dt = \frac{e^{i\beta T} - 1}{i\beta T},$$

and thus

$$\lim_{T \to \infty} \frac{1}{T} \int_0^T e^{i\beta t} dt = 0. \tag{3.9.3}$$

A direct computation shows that

$$\frac{1}{T} \int_0^T \left| 1 - \sum_{k=1}^n a_k e^{\alpha_k it} \right|^2 dt = \frac{1}{T} \int_0^T \left\{ 1 + \sum_{k=1}^n |a_k|^2 - \sum_{k=1}^n a_k e^{\alpha_k it} \right.$$

$$\left. - \sum_{k=1}^n \overline{a_k} e^{-\alpha_k it} + \sum_{\substack{k,\ell=1 \\ k \neq \ell}}^n a_k \overline{a_\ell} e^{i(\alpha_k - \alpha_\ell)t} \right\} dt.$$

In view of (3.9.3) with $\beta = \alpha_k$ and $\beta = \alpha_k - \alpha_\ell$ $(k \neq \ell)$, we obtain that

$$\lim_{T \to \infty} \frac{1}{T} \int_0^T \left| 1 - \sum_{k=1}^n a_k e^{\alpha_k it} \right|^2 dt = 1 + \sum_{k=1}^n |a_k|^2 \geq 1. \tag{3.9.4}$$

Assume now that (3.1.2) was to fail, and let $K = \sup_{t \in \mathbb{R}} |1 - \sum_{k=1}^n a_k e^{\alpha_k it}|$. Then, $K < 1$ and we would have

$$\lim_{T \to \infty} \frac{1}{T} \int_0^T \left| 1 - \sum_{k=1}^n a_k e^{\alpha_k it} \right|^2 dt \leq \lim_{T \to \infty} \frac{1}{T} \int_0^T K^2 = K^2 < 1,$$

contradicting (3.9.4). □

Solution of Exercise 3.1.4. Write $z_0 = x_0 + iy_0$. We then have (in view of (1.2.4))

$$e^{z_0 t} = e^{x_0 t} e^{iy_0 t},$$

and the rest is plain by using the usual formulas for computing derivatives:

$$(e^{z_0 t})' = (e^{x_0 t} e^{iy_0 t})' = x_0 e^{x_0 t} e^{iy_0 t} + e^{ix_0}(iy_0 e^{iy_0 t}) = (x_0 + iy_0)e^{z_0 t}. \quad □$$

Solution of Exercise 3.1.5. We consider the second sum. Differentiating formula (1.3.4) with respect to the real variable b we obtain

$$\sum_{k=1}^{n-1} k \sin(a + bk) = \frac{-1}{\sin^2 \frac{b}{2}} \left\{ \sin\left(\frac{b}{2}\right) \left(\frac{n}{2} \cos\left(\frac{nb}{2}\right) \cos\left(a + (n-1)\frac{b}{2}\right) \right. \right.$$

$$- \frac{n-1}{2} \sin\left(\frac{nb}{2}\right) \sin\left(a + (n-1)\frac{b}{2}\right)\right)$$

$$\left. - \frac{1}{2} \cos\frac{b}{2} \sin\left(\frac{nb}{2}\right) \cos\left(a + (n-1)\frac{b}{2}\right)\right\}.$$

Writing $\frac{n}{2} = \frac{n-1}{2} + \frac{1}{2}$ in the term $\frac{n}{2}\cos\left(\frac{nb}{2}\right)\cos\left(a+(n-1)\frac{b}{2}\right)$, we get:

$$\sum_{k=1}^{n-1} k\sin(a+bk) = \frac{-1}{\sin^2\frac{b}{2}}\left\{\sin\left(\frac{b}{2}\right)\left(\frac{n-1}{2}\left(\cos\left(\frac{nb}{2}\right)\cos\left(a+(n-1)\frac{b}{2}\right)\right.\right.\right.$$

$$\left.\left.- \sin\left(\frac{nb}{2}\right)\sin\left(a+(n-1)\frac{b}{2}\right)\right)\right)$$

$$+\frac{1}{2}\left(\cos\frac{nb}{2}\sin\frac{b}{2}-\cos\frac{b}{2}\sin\frac{nb}{2}\right)\cos\left(a+(n-1)\frac{b}{2}\right)\right\}$$

$$= \frac{\sin\frac{n-1}{2}b\cos(a+\frac{n-1}{2}b)-(n-1)\sin\frac{b}{2}\cos(a+(n-\frac{1}{2})b)}{2\sin^2(\frac{b}{2})}.$$

A formula for the first sum is proved in much the same way using (1.3.5), and is left to the reader.   □

We note that the method is valid only for $n \geq 2$. As a verification of the formula, let us consider the case $n = 2$. On the left side the sum reduces to $\sin(a+b)$, while the formula gives

$$\frac{\sin\frac{b}{2}\cos(a+\frac{1}{2}b)-\sin\frac{b}{2}\cos(a+\frac{3b}{2})}{2\sin^2(\frac{b}{2})} = \frac{(\sin\frac{b}{2})2\sin(a+b)\sin\frac{b}{2}}{2\sin^2(\frac{b}{2})} = \sin(a+b),$$

where we have used the formula

$$\cos p - \cos q = 2\sin\left(\frac{p+q}{2}\right)\sin\left(\frac{q-p}{2}\right).$$

Solution of Exercise 3.1.6. We have (see (1.2.9)),

$$\sum_{k=0}^{n} e^{ikt} = \frac{1-e^{i(n+1)t}}{1-e^{it}}.$$

Differentiating with respect to t both sides of this equality, we obtain:

$$\sum_{k=1}^{n} ike^{ikt} = \frac{-i(n+1)e^{i(n+1)t}(1-e^{it})+ie^{it}(1-e^{i(n+1)t})}{(1-e^{it})^2},$$

and so

$$\sum_{k=1}^{n} ke^{ikt} = \frac{ne^{i(n+2)t}-(n+1)e^{i(n+1)t}+e^{it}}{(1-e^{it})^2}. \qquad (3.9.5)$$

                                                                                                          □

Solution of Exercise 3.1.8. (a) We have $\cos t = \dfrac{e^{it} + e^{-it}}{2}$ and so

$$\int_0^{2\pi} \cos^{2p} t\, dt = \int_0^{2\pi} \left(\frac{e^{it} + e^{-it}}{2}\right)^{2p} dt = \int_0^{2\pi} \frac{(e^{2it} + 1)^{2p}}{2^{2p} e^{2pit}} dt$$

$$= \frac{1}{2^{2p}} \sum_{\ell=0}^{2p} \binom{2p}{\ell} \int_0^{2\pi} e^{2i(\ell - p)t}\, dt = 2\pi \frac{\binom{2p}{p}}{2^{2p}},$$

(3.9.6)

since

$$\frac{1}{2\pi} \int_0^{2\pi} e^{2i(\ell - p)t}\, dt = \begin{cases} 0, & \text{if } \ell \neq p, \\ 1, & \text{if } \ell = p. \end{cases}$$

(b) The dominated convergence theorem (see Theorem 17.5.2) implies that

$$\lim_{p \to \infty} \int_0^{2\pi} \cos^{2p} t\, dt = 0,$$

and hence the result.   □

Remark. One could also prove (3.1.6) using Stirling's formula on the asymptotic behaviour of $n!$ as n goes to infinity. Indeed, Stirling's formula states in particular that

$$\lim_{n \to \infty} \frac{n!}{\sqrt{2\pi n}\, \dfrac{n^n}{e^n}} = 1.$$

(3.9.7)

See for instance [112, (10.17), p. 165] for the precise formula. Write

$$\frac{\binom{2p}{p}}{2^{2p}} = \frac{(2p)!}{(p!)!\, 4^p} = \frac{(2p)!}{\dfrac{\sqrt{4\pi p}(2p)^{(2p)}}{e^{2p}}} \cdot \frac{1}{\left(\dfrac{p!}{\dfrac{\sqrt{2\pi p}\, p^p}{e^p}}\right)^2} \cdot \frac{1}{4^p} \cdot \frac{\dfrac{\sqrt{4\pi p}(2p)^{(2p)}}{e^{2p}}}{\left(\dfrac{\sqrt{2\pi p}\, p^p}{e^p}\right)^2}$$

$$= \frac{(2p)!}{\dfrac{\sqrt{4\pi p}(2p)^{(2p)}}{e^{2p}}} \cdot \frac{1}{\left(\dfrac{p!}{\dfrac{\sqrt{2\pi p}\, p^p}{e^p}}\right)^2} \cdot \sqrt{\frac{1}{\pi p}}.$$

It then follows from (3.9.7) that

$$\lim_{p \to \infty} \frac{\binom{2p}{p}}{2^{2p}} = 0.$$

We note that the proof using Stirling's formula is more precise than the proof using the dominated convergence theorem.

Solution of Exercise 3.1.9. Making the change of variable $x = 2\cos t$ in the integral defining m_p we obtain:

$$
\begin{aligned}
m_p &= \frac{2}{2\pi} \int_0^2 \sqrt{4 - x^2} x^{2p} dx \\
&= \frac{2^{2p+3}}{2\pi} \int_0^{\frac{\pi}{2}} (\cos^{2p} t)(\sin^2 t) dt \\
&= \frac{2^{2p+3}}{2\pi} \int_0^{\frac{\pi}{2}} (\cos^{2p} t)(1 - \cos^2 t) dt \\
&= \frac{2^{2p+3}}{2\pi} \frac{1}{4} \int_0^{2\pi} (\cos^{2p} t)(1 - \cos^2 t) dt
\end{aligned}
$$

and, using (3.9.6),

$$
\begin{aligned}
&= 2 \cdot \frac{1}{4} \left\{ 4 \binom{2p}{p} - \binom{2p+2}{p+1} \right\} \\
&= \frac{1}{2} \left\{ 4 \binom{2p}{p} - \frac{(2p+2)(2p+1)}{(p+1)(p+1)} \binom{2p}{p} \right\} \\
&= \frac{1}{2} \binom{2p}{p} \left(4 - \frac{(2p+2)(2p+1)}{(p+1)(p+1)} \right) \\
&= \underbrace{\frac{\binom{2p}{p}}{p+1}}_{\text{Catalan number}}.
\end{aligned}
$$

$\square$

Solution of Exercise 3.1.10. We compute

$$
\begin{aligned}
\frac{1}{2\pi} \int_0^{2\pi} |p(e^{it})|^2 dt &= \frac{1}{2\pi} \int_0^{2\pi} \left\{ \sum_{\ell,k=0}^n p_\ell \overline{p_k} e^{i(\ell-k)t} \right\} dt \\
&= \frac{1}{2\pi} \sum_{\ell,k=0}^n p_\ell \overline{p_k} \int_0^{2\pi} e^{i(\ell-k)t} dt \\
&= \sum_{\ell=0}^n |p_\ell|^2.
\end{aligned}
$$

From the hypothesis, we have

$$
\frac{1}{2\pi} \int_0^{2\pi} |p(e^{it})|^2 dt \leq 1,
$$

and hence

$$\sum_{\ell=0}^{n} |p_\ell|^2 \le 1,$$

which implies that each of the p_ℓ is of modulus less than or equal to 1.   □

Solution of Exercise 3.1.11. For a fixed $z \in \mathbb{C}$ the complex-valued function of the real variable t,

$$f(t) = \frac{1}{t-z},$$

has derivative

$$f'(t) = -\frac{1}{(t-z)^2}.$$

Let now $z \notin (-\infty, 0)$. The function f is then continuous on the negative axis, and for every $a \in (-\infty, 0)$,

$$\int_a^0 \frac{dt}{(t-z)^2} = -\frac{1}{t-z}\Big|_a^0$$

$$= \frac{1}{z} - \frac{1}{z-a}$$

$$\longrightarrow \frac{1}{z} \quad \text{as} \quad a \longrightarrow -\infty. \qquad\qquad □$$

Solution of Exercise 3.1.12. We rewrite f_m as (see also (5.5.21))

$$f_m(z) = -i \int_{\mathbb{R}} \frac{m(t)}{t^2+1} \cdot \frac{tz+1}{t-z} dt.$$

By Exercise 2.3.16, the set

$$\frac{tz+1}{t-z}, \quad t \in \mathbb{R},$$

is a circle, and in particular is bounded, and so f_m is well defined. Furthermore,

$$\text{Re } f_m(z) = -i \int_{\mathbb{R}} \left\{ \frac{1}{t-z} - \frac{t}{t^2+1} - \frac{1}{t-\bar{z}} + \frac{t}{t^2+1} \right\} m(t) dt$$

$$= -i(z-\bar{z}) \int_{\mathbb{R}} \frac{m(t)dt}{|t-z|^2}. \qquad\qquad □$$

Solution of Exercise 3.1.13. (1) Let $w = a + ib$ with $a > 0$. Then

$$|e^{-tw}| = e^{-ta},$$

and the integral $\int_0^\infty e^{-tw} dt$ converges absolutely. Moreover,

$$\int_0^\infty e^{-tw} dt = \lim_{R\to\infty} \int_0^R e^{-tw} dt = \frac{1}{w}.$$

(2) If z and w are in $\mathbb{C}_r$ so is $z + \overline{w}$ and so

$$\int_0^\infty e^{-t(z+\overline{w})} dt = \frac{1}{z+\overline{w}}.$$

The above expression expresses $\frac{1}{z+\overline{w}}$ as an inner product, and therefore defines a positive definite kernel. More precisely, let $N \in \mathbb{N}$, $w_1, \ldots, w_N \in \mathbb{C}_r$ and $c_1, \ldots, c_N \in \mathbb{C}$. We have

$$\sum_{\ell,j=1}^N \frac{\overline{c_\ell} c_j}{w_\ell + \overline{w_j}} = \int_0^\infty \sum_{\ell,j=1}^N \overline{c_\ell} c_j e^{-t(w_\ell+\overline{w_j})} dt = \int_0^\infty \left| \sum_{\ell=1}^N \overline{c_\ell} e^{-tw_\ell} \right|^2 dt$$

$$\geq 0. \qquad \square$$

Solution of Exercise 3.1.15. In view of (0.0.3) we have for $w = c + id$ and $d > 0$ and $t \geq 0$

$$\left| \int_{-\infty}^t e^{iu\overline{w}} r(u) du \right| \leq \left| \int_{-\infty}^0 e^{iu\overline{w}} r(u) du \right| + te^{td} r(0),$$

and so

$$\lim_{t \to \infty} \frac{e^{it(\lambda-\overline{w})}}{i(\lambda-\overline{w})} \left(\int_{-\infty}^t e^{iu\overline{w}} r(u) du \right) = 0. \qquad (3.9.8)$$

We have:

$$\iint_{[0,\infty) \times [0,\infty)} e^{i\lambda t} e^{-is\overline{w}} r(t-s) dt ds = \int_0^\infty e^{i\lambda t} \left(\int_0^\infty e^{-is\overline{w}} r(t-s) ds \right) dt$$

$$= \int_0^\infty e^{i\lambda t} \left(\int_{-\infty}^t e^{-i(t-u)\overline{w}} r(u) du \right) dt$$

$$= \int_0^\infty e^{i(\lambda-\overline{w})t} \left(\int_{-\infty}^t e^{iu\overline{w}} r(u) du \right) dt.$$

We now use integration by parts and obtain, thanks (3.9.8):

$$\int_0^\infty e^{i(\lambda-\overline{w})t} \left(\int_{-\infty}^t e^{iu\overline{w}} r(u) du \right) dt$$

$$= \left[\frac{e^{it(\lambda-\overline{w})}}{i(\lambda-\overline{w})} \left(\int_{-\infty}^t e^{iu\overline{w}} r(u) du \right) \right]_0^\infty - \int_0^\infty \frac{e^{iu(\lambda-\overline{w})}}{i(\lambda-\overline{w})} e^{iu\overline{w}} r(u) du$$

$$= \frac{\int_{-\infty}^0 e^{iu\overline{w}} r(u) du + \int_0^\infty e^{iu\lambda} r(u) du}{-i(\lambda-\overline{w})}$$

$$= \frac{\varphi(\lambda) + \overline{\varphi(w)}}{-i(\lambda-\overline{w})}$$

since $r(-u) = \overline{r(u)}$. $\qquad \square$

Solution of Exercise 3.2.1. For $t = 0 \pmod{2\pi}$,

$$\frac{1 + e^{it} + \cdots + e^{int}}{n} = \frac{n+1}{n}$$

and the limit is 1. For $t \neq 0 \pmod{2\pi}$,

$$1 + e^{it} + \cdots + e^{int} = \frac{1 - e^{i(n+1)t}}{1 - e^{it}},$$

and so

$$|1 + e^{it} + \cdots + e^{int}| = \left| \frac{1 - e^{i(n+1)t}}{1 - e^{it}} \right| \leq \frac{2}{|1 - e^{it}|},$$

and so

$$\left| \frac{1 + e^{it} + \cdots + e^{int}}{n} \right| \leq \frac{2}{n|1 - e^{it}|} \to 0$$

as n goes to infinity.   □

To solve the preceding exercise, one could also have used formula (1.2.9)

$$1 + e^{it} + \cdots + e^{int} = e^{i\frac{nt}{2}} \frac{\sin \frac{(n+1)t}{2}}{\sin \frac{t}{2}}, \quad t \neq 0 \pmod{2\pi}.$$

Solution of Exercise 3.2.2. We recall the classical solution of (a). By hypothesis, for every $\epsilon > 0$ there exists an integer N such that:

$$n \geq N \quad \Longrightarrow \quad |u_n - \ell| < \epsilon/2.$$

For such an N there exists N_1, which can be assumed greater than or equal to N, such that

$$n \geq N_1 \quad \Longrightarrow \quad \left| \frac{u_0 + \cdots + u_N - N\ell}{n} \right| < \epsilon/2.$$

Let $n \geq N_1$. We have

$$|v_n - \ell| = \left| \frac{u_0 + \cdots + u_n - n\ell}{n} \right|$$

$$= \left| \frac{u_0 + \cdots + u_N - N\ell + \sum_{k=N+1}^{n}(u_j - \ell)}{n} \right|$$

$$\leq \left| \frac{u_0 + \cdots + u_N - N\ell}{n} \right| + \left| \frac{\sum_{k=N+1}^{n}(u_j - \ell)}{n} \right|$$

$$\leq \frac{\epsilon}{2} + \frac{(n-N)}{n} \frac{\epsilon}{2}$$

$$\leq \epsilon.$$

We now turn to (b). Assume that the sequence has a limit. Then, this limit has modulus 1. But the associated sequence v_n tends to 0, in view of the preceding exercise, and we have a contradiction.   □

Solution of Exercise 3.2.3. If $t = 0 \pmod{2\pi}$, then

$$z_n = \frac{n+1}{2} \quad \text{and} \quad w_n = \frac{n+1}{2n}$$

since $1 + \cdots + n = \frac{n(n+1)}{2}$. Thus

$$\lim_{n \to +\infty} z_n = +\infty \quad \text{and} \quad \lim_{n \to +\infty} w_n = \frac{1}{2}.$$

Now assume that $t \neq 0 \pmod{2\pi}$. We use formula (3.9.5) to obtain

$$z_n = \frac{e^{it} - e^{(n+1)t}}{n(1 - e^{it})^2} + e^{int} \frac{e^{it}}{e^{it} - 1}.$$

Since e^{int} has no limit as $n \to +\infty$ (see Exercise 3.2.2), it follows that the sequence z_n has no limit as $n \to \infty$. That same formula for z_n also shows that $|z_n|$ is uniformly bounded with respect to n, and hence w_n tends to 0 as $n \to +\infty$. $\quad\square$

Solution of Exercise 3.2.4. It suffices to apply (3.2.1)

$$\lim_{n \to \infty} \frac{\sum_{k=0}^{n} f(k/n)}{n} = \int_0^1 f(t)dt$$

to $f(t) = te^{2\pi it}$ to obtain that

$$\lim_{n \to \infty} z_n = \int_0^1 te^{2\pi it} dt = -\frac{i}{2\pi}. \qquad\qquad \square$$

Solution of Exercise 3.2.5. As already mentioned we send the reader to [62] for a direct proof. Another proof, using the notion of radius of convergence of power series, is presented in Exercise 5.6.4 in Section 5.6. $\quad\square$

Solution of Exercise 3.2.6. (a) For $t = 0$ or $t = n$ the claim is trivial. For $t \in (0, n)$, we have $t/n \in (0, 1)$ and we can take the logarithm on both sides. It is thus equivalent to proving

$$t + n \ln\left(1 - \frac{t}{n}\right) \leq 0, \quad t \in (0, n).$$

Using the power expansion of $\ln(1 - x)$,

$$\ln(1 - x) = -\sum_{p=1}^{\infty} \frac{x^p}{p}, \quad x \in (0, 1),$$

this inequality is in turn equivalent to

$$t + n\left(-\frac{t}{n} - \frac{t^2}{2n^2} - \cdots\right) \leq 0.$$

This in turn is obvious.

(b) Given $t > 0$, we take n's such that $0 < t/n < 1$. Claim (b) is equivalent to showing that the sequence

$$n \mapsto n \ln\left(1 - \frac{t}{n}\right)$$

is increasing. Once more using the power expansion of $\ln(1 - x)$, it is equivalent to proving that

$$-n\left(\sum_{p=1}^{\infty} \frac{t^p}{pn^p}\right) + (n+1)\left(\sum_{p=1}^{\infty} \frac{t^p}{p(n+1)^p}\right) \leq 0.$$

This in turn is equivalent to checking that

$$\sum_{p=1}^{\infty} \frac{t^p}{p}\left(\frac{1}{(n+1)^{p-1}} - \frac{1}{n^{p-1}}\right) \leq 0.$$

But this is clear.   □

Solution of Exercise 3.3.1. In view of (1.1.39)

$$\left|\frac{z^{2n}}{2 + z^n + z^{5n}}\right| \leq \frac{|z|^{2n}}{2(1 - |z|)}, \quad z \in \mathbb{D},$$

the series with running term $\frac{|z|^{2n}}{2(1-|z|)}$ converges for $|z| < 1$ and hence the series at hand converges absolutely.   □

Solution of Exercise 3.3.2. For $z = 0$ the series trivially converges since every term is then equal to 0. Write

$$\frac{1}{z - n} + \frac{1}{n} = -\frac{z}{n^2(1 - \frac{z}{n})}.$$

Given any $z \in \mathbb{C}$ there exists $n_0 \in \mathbb{N}$ such that

$$n \geq n_0 \longrightarrow |z/n| < 1/2.$$

For $n \geq n_0$ we have:

$$\left|\frac{1}{z - n} + \frac{1}{n}\right| = \left|\frac{z}{n^2(1 - \frac{z}{n})}\right| \leq \frac{|z|}{n^2(1 - |\frac{z}{n}|)} \leq \frac{2|z|}{n^2}. \qquad (3.9.9)$$

Hence the series is absolutely convergent. Let now K be a compact set which does not intersect $\mathbb{N}$. It is included in a closed set of the form

$$K_0 = \{|z| \leq R\} \setminus \cup \{|z - n_j| < \epsilon_j\},$$

where we take $R \notin \mathbb{N}$, and where the n_j are the elements of $\mathbb{N}$ (if any) such that $|n_j| < R$. The ϵ_j are small enough strictly positive numbers. We now take n_0 such that, moreover

$$R/n_0 \leq 1/2,$$

and write

$$\sum_{n=1}^{\infty}\left(\frac{1}{z-n}+\frac{1}{n}\right) = \sum_{n=1}^{N}\left(\frac{1}{z-n}+\frac{1}{n}\right) + \sum_{n=N+1}^{\infty}\left(\frac{1}{z-n}+\frac{1}{n}\right).$$

In this decomposition, we take $N > n_0$. The first sum, which is finite, is well defined since the positive integers smaller than R do not belong to K_0. In view of (3.9.9), the terms of the second sum are bounded by $\frac{2R}{n^2}$, and hence the second sum converges uniformly in K_0, and hence in K. □

Solution of Exercise 3.3.3. We have

$$z_\ell = e^{\frac{2\pi i \ell}{n}}, \quad \ell = 0, 1, \dots, n-1,$$

and (3.3.1):

$$\frac{nz^{n-1}}{z^n - 1} = \sum_{\ell=0}^{n-1}\frac{1}{z - z_\ell}.$$

But, for $|z| < 1$,

$$\frac{1}{z - z_\ell} = -\frac{1}{z_\ell(1 - zz_\ell^{-1})} = -\sum_{m=0}^{\infty}z^m z_\ell^{-1-m}.$$

Thus, the mth coefficient in the power series expansion of $\sum_{\ell=0}^{n-1}\frac{1}{z-z_\ell}$ is

$$-\sum_{\ell=0}^{n-1}z_\ell^{-1-m}.$$

The result follows by taking conjugates since

$$\frac{nz^{n-1}}{z^n - 1} = -n\sum_{k=0}^{\infty}z^{nk+n-1}.$$ □

Solution of Exercise 3.3.4. Let $\theta \in \mathbb{R}$, and set

$$A_{m,n}(\theta) = \sum_{\ell=0}^{\infty}\frac{e^{im\ell\theta}}{(m\ell + n)!}, \quad n = 0, 1, \dots, m-1. \tag{3.9.10}$$

We have

$$e^{e^{i\theta}} = \sum_{n=0}^{m-1}e^{in\theta}A_{m,n}(\theta). \tag{3.9.11}$$

Setting in (3.9.10)

$$\theta_k = \frac{2\pi k}{m}, \quad k = 0, \ldots, m-1,$$

we have $A_{m,n}(\theta_k) = A_{m,n}$. Therefore, we obtain from (3.9.11):

$$e^{e^{i\theta_k}} = \sum_{n=0}^{m-1} e^{\frac{2\pi i n k}{m}} A_{m,n}, \quad k = 0, 1, \ldots, m-1.$$

Set $w_k = e^{e^{i\theta_k}}$ and $z_k = e^{i\theta_k}$. This system of equations can be rewritten in matrix form as

$$\begin{pmatrix} w_0 \\ w_1 \\ \vdots \\ w_{m-1} \end{pmatrix} = \begin{pmatrix} 1 & 1 & \cdots & 1 \\ 1 & z_1 & \cdots & z_{m-1} \\ & & & \\ 1 & z_1^{m-1} & \cdots & z_{m-1}^{m-1} \end{pmatrix} \begin{pmatrix} A_{m,0} \\ A_{m,1} \\ \vdots \\ A_{m,m-1} \end{pmatrix}. \tag{3.9.12}$$

Since

$$1 + z_\ell + \cdots + z_\ell^{m-1} = 0, \quad \ell = 1, \ldots, m-1,$$

multiplying both sides of (3.9.12) by the row vector

$$\begin{pmatrix} 1 & 1 & \cdots & 1 \end{pmatrix},$$

we obtain (since the $A_{m,n}$, and in particular $A_{m,0}$, are real):

$$A_{m,0} = \frac{\sum_{k=0}^{m-1} e^{\cos(\frac{2\pi k}{m})} \cos(\sin(\frac{2\pi k}{m}))}{m}.$$

More generally, one computes $A_{m,n}$ by multiplying both sides of (3.9.12) by the row vector

$$\begin{pmatrix} 1 & \overline{z_n} & \cdots & \overline{z_n^{m-1}} \end{pmatrix},$$

and obtains

$$A_{m,n} = \frac{\sum_{k=0}^{m-1} \overline{z_n}^{-k} w_k}{m},$$

that is,

$$A_{m,n} = \frac{\sum_{k=0}^{m-1} e^{\cos(\frac{2\pi k}{m})} \cos(\sin(\frac{2\pi k}{m}) - \frac{2\pi k n}{m})}{m}. \qquad \square$$

The formula above seems difficult to obtain using only real analysis.

Solution of Exercise 3.3.5. Let $z_n = x_n + iy_n$. By hypothesis $x_n \geq 0$ and so

$$\sum_{n=1}^{\infty} x_n \quad \text{converges} \quad \Longrightarrow \quad \sum_{n=1}^{\infty} x_n^2 < \infty \tag{3.9.13}$$

since $x_n^2 \leq x_n \leq 1$ for n large enough. Since $\sum_{n=1}^{\infty} z_n^2$ converges we have that the series

$$\sum_{n=1}^{\infty} (x_n^2 - y_n^2) = \sum_{n=1}^{\infty} \operatorname{Re} z_n^2$$

converges, and hence so does the series

$$-\sum_{n=1}^{\infty} (x_n^2 - y_n^2) + 2\sum_{n=1}^{\infty} x_n^2 = \sum_{n=1}^{\infty} (x_n^2 + y_n^2) = \sum_{n=1}^{\infty} |z_n|^2.$$

Remark that (3.9.13) does not necessarily hold when the x_n are not in the right half-plane. Take for instance $x_n = \dfrac{(-1)^n}{\sqrt{n}}$. The series $\sum_{n=1}^{\infty} x_n$ converges (use Abel's theorem), but $\sum_{n=1}^{\infty} x_n^2$ is the harmonic series and is divergent. More generally, consider

$$z_n = \frac{e^{in\theta}}{\sqrt{n}} \quad n = 1, 2, \ldots$$

where θ is a real number such that both θ and 2θ are not multiples of 2π (take for instance $\theta = \pi/4$). Here too, by Abel's theorem, both the series $\sum_{n=1}^{\infty} z_n$ and

$$\sum_{n=1}^{\infty} z_n^2 = \sum_{n=1}^{\infty} \frac{e^{2in\theta}}{n}$$

converge. But the series with term $|z_n|^2 = \dfrac{1}{n}$ diverges.   □

Solution of Exercise 3.3.6. (see [182, p. 306]). It suffices to take

$$z_n = \frac{e^{2\pi i\theta n}}{\ln(n+1)},$$

where θ is irrational.   □

Solution of Exercise 3.3.7. (see [182, p. 305]). Let $z_n = x_n + iy_n$. By hypothesis

$$|z_n| \leq \frac{x_n}{\cos \alpha},$$

from which follows the result.   □

Solution of Exercise 3.4.4. For $|z| > 1$ the running term of the series diverges, and so the radius of convergence is at most equal to 1. It is equal to 1 since the sums

$$\sum_{p=1}^{\infty} (2p)^{(-1)^{2p}} z^{2p} = \sum_{p=1}^{\infty} 2p z^{2p} \qquad\qquad (3.9.14)$$

and

$$\sum_{p=0}^{\infty}(2p+1)^{(-1)^{2p+1}}z^{2p+1} = \sum_{p=0}^{\infty}\frac{z^{2p+1}}{2p+1}$$

both converge there.

Another, and quicker way, is of course to apply formula (3.4.3) with

$$a_n = n^{(-1)^n}.$$

Then,

$$\limsup_{n\to\infty}|a_n|^{1/n} = 1. \qquad\qquad \square$$

The sum of the series in the above exercise is computed in closed form in Exercise 5.7.7.

Solution of Exercise 3.4.5. For $|z| < 1$ we have

$$\sum_{n=0}^{\infty}|z^{n!}| = \sum_{n=0}^{\infty}|z|^{n!} \le \sum_{p=0}^{\infty}|z|^p = \frac{1}{1-|z|}.$$

Thus the series (3.4.8) converges absolutely for every z in the open unit disk. It does not converge for $|z| \ge 1$. Indeed, the running term of a converging series goes to 0, and this is not the case for the series with term

$$u_p = \begin{cases} z^p, & p = n!, \\ 0, & \text{otherwise}, \end{cases}$$

when $|z| \ge 1$. $\qquad\qquad \square$

Solution of Exercise 3.4.6. The proof is as in the previous exercise. The series converges absolutely for every z in the open unit disk. It does not converge for $|z| \ge 1$, since its running term does not go to 0 for $|z| > 1$. $\qquad \square$

As mentioned after the statement of Exercise 3.4.6, much more is true: The unit circle is the natural boundary of the two power series considered in Exercises 3.4.4 and 3.4.6. See Exercise 6.3.3.

Solution of Exercise 3.4.9. We skip the solution of (a). To prove (b), set R_{-1} be the radius of convergence of g. We have

$$R_{-1} = \frac{1}{\limsup_{n\to\infty}\frac{1}{|a_n|^{1/n}}} = \liminf_{n\to\infty}|a_n|^{1/n},$$

and this last limit is in general different from

$$\limsup_{n\to\infty}|a_n|^{1/n},$$

so that, in general,

$$R_{-1} \neq \frac{1}{R}.$$

To illustrate this, consider the sequence

$$a_n = \begin{cases} 2^n, & \text{if } n \text{ is even,} \\ 3^{-n}, & \text{if } n \text{ is odd.} \end{cases}$$

We have

$$\limsup_{n \to \infty} |a_n|^{1/n} = 2, \qquad \liminf_{n \to \infty} |a_n|^{1/n} = \frac{1}{3}$$

and

$$\limsup_{n \to \infty} |1/a_n|^{1/n} = 3, \qquad \liminf_{n \to \infty} |1/a_n|^{1/n} = \frac{1}{2}.$$

With this choice of a_n, the radius of convergence of the series $\sum_{n=0}^{\infty} a_n z^n$ is equal to $1/2$, while the radius of convergence of the series $\sum_{n=0}^{\infty} (1/a_n) z^n$ is $1/3$.

We note that, in the general case, we will have $R_{-1} = 1/R$ if and only if the sequence $|a_n|^{1/n}$ has a limit. $\qquad \square$

Solution of Exercise 3.4.10. We have

$$C(r, \theta) + iS(r, \theta) = \sum_{n=0}^{\infty} r^n e^{in\theta}$$

$$= \frac{1}{1 - re^{i\theta}}$$

$$= \frac{1 - re^{-i\theta}}{|1 - re^{i\theta}|^2}$$

$$= \frac{1 - r\cos\theta + ir\sin\theta}{1 + r^2 - 2r\cos\theta},$$

and so

$$C(r, \theta) = \frac{1 - r\cos\theta}{1 + r^2 - 2r\cos\theta} \quad \text{and} \quad S(r, \theta) = \frac{r\sin\theta}{1 + r^2 - 2r\cos\theta}. \qquad (3.9.15)$$
$\qquad \square$

We refer the reader to formula (8.5.1) for a related computation.

Solution of Exercise 3.4.11. In the notation of the statement of the exercise, we have:

$$\frac{\sum_{u=0}^{n-1} w^{-ju} f(w^u z)}{n} = \frac{\sum_{u=0}^{n-1} w^{-ju} \left(\sum_{m=0}^{\infty} a_m w^{um} z^m \right)}{n}$$

$$= \sum_{m=0}^{\infty} a_m z^m \left(\frac{\sum_{u=0}^{n-1} w^{u(m-j)}}{n} \right).$$

Recall that $w = \exp \frac{2\pi i}{n}$. Thus

$$\sum_{u=0}^{n-1} w^{u(m-j)} = \begin{cases} n, & \text{if } m-j \in n\mathbb{Z}, \\ 0, & \text{otherwise.} \end{cases}$$

Since $j \in \{0, \ldots, n-1\}$ and $m \in \mathbb{N}_0$, we see that in fact $m - j \in n\mathbb{N}$ when $m - j \in n\mathbb{Z}$. The result follows.   □

Solution of Exercise 3.4.12. We follow the proof of Lemma 4 in [52]. Take r and r_0 in $(0,1)$ such that $r < r_0$. For $z \in \mathbb{D}$ we have that $|rz| \le r$ and so (3.4.4) leads to

$$|a_n r^n z^n| \le M \left(\frac{r}{r_0}\right)^n,$$

where $M = \sup_{n \in \mathbb{N}_0} |a_n r_0^n|$. Therefore, given r and $\epsilon > 0$ there exists $N_0 \in \mathbb{N}$ such that

$$N \ge N_0 \Longrightarrow \left| \sum_{n=N+1}^{\infty} a_n r^n z^n \right| \le \epsilon,$$

and so, still for $N \ge N_0$ and in view of (3.4.17),

$$\left| \sum_{n=0}^{N} a_n r^n z^n \right| = \left| \sum_{n=0}^{\infty} a_n r^n z^n - \sum_{n=N+1}^{\infty} a_n r^n z^n \right| \le 1 + \epsilon.$$

Using Exercise 3.1.10 we get

$$\left| \frac{|a_n| r^n}{1 + \epsilon} \right| \le 1, \quad n = 0, \ldots, N.$$

For a given choice of r and ϵ we may let $N \to \infty$ and then

$$|a_n| \le (1 + \epsilon) r^n, \quad n \in \mathbb{N}_0.$$

Letting r vary in $(0,1)$ and $\epsilon \to 0$ we get that $|a_n| \le 1$ for all $n \in \mathbb{N}_0$.   □

We remark the following: Computing (3.4.16) leads to the much stronger statement

$$\sum_{n=0}^{\infty} |a_n|^2 \le 1.$$

From this last equality the following claim is quite clear: If $|a_{n_0}| = 1$ for some $n_0 \in \mathbb{N}_0$ then

$$f(z) = a_{n_0} z^{n_0}.$$

We leave it to the industrious to find a direct proof of this fact.

Solution of Exercise 3.4.13. Let $M = \max_{t \in [0,1]} |m(t)|$. For a given $z \in \mathbb{C}$, let

$$f_n(t) = \frac{t^n z^n}{n!} m(t).$$

We have

$$|f_n(t)| \leq M \frac{|z|^n}{n!}.$$

Since

$$\sum_{n=0}^{\infty} M \frac{|z|^n}{n!} < \infty,$$

Weierstrass theorem (see Theorem 14.4.1) insures that

$$\int_0^1 e^{zt} m(t) dt = \int_0^1 \left(\sum_{n=0}^{\infty} \frac{z^n t^n}{n!} \right) m(t) dt = \sum_{n=0}^{\infty} z^n \frac{\int_0^1 t^n m(t) dt}{n!}. \qquad \Box$$

Remark. The same result will hold if m is only measurable with respect to Lebesgue measure, and $\int_{[0,1]} |m(t)| dt < \infty$. One then has to resort to the dominated convergence theorem. The same remark holds if we replace $m(t) dt$ by a general signed measure $d\mu(t)$ for which $\int_{[0,1]} d|\mu|(t) < \infty$.

Solution of Exercise 3.4.14. Let us denote by $\psi(z)$ the power series on the right side of (3.4.19). Since the series

$$\sum_{n=0}^{\infty} \frac{|z|^n}{n!}$$

converges for every $z \in \mathbb{C}$, it follows that the series (3.4.19) converges absolutely in $\mathbb{C}$. We note that for real z, the function $\psi(z)$ reduces to the usual exponential function from calculus. On the other hand, for $z = iy$ and using the known power series expansions for $\sin y$ and $\cos y$ for *real* y, we have

$$\psi(iy) = \sum_{n=0}^{\infty} \frac{(iy)^n}{n!}$$

$$= \sum_{p=0}^{\infty} \frac{(iy)^{2p}}{(2p)!} + \sum_{p=0}^{\infty} \frac{(iy)^{2p+1}}{(2p+1)!} \qquad (3.9.16)$$

$$= \cos y + i \sin y.$$

We now show that for every pair of complex numbers z_1 and z_2 it holds that

$$\psi(z_1 + z_2) = \psi(z_1)\psi(z_2). \qquad (3.9.17)$$

To prove (3.9.17), we apply Cauchy's multiplication theorem (see Theorem 14.2.1 below) to the sequences $f_n = (z_1^n/n!)$ and $g_n = (z_2^n/n!)$. Then,

$$\sum_{p=0}^{n} f_p g_{n-p} = \sum_{p=0}^{n} \frac{z_1^p z_2^{n-p}}{p!(n-p)!} = \frac{1}{n!} \sum_{p=0}^{n} \frac{n!}{p!(n-p)!} z_1^p z_2^{n-p} = \frac{(z_1 + z_2)^n}{n!},$$

and we obtain

$$\psi(z_1)\psi(z_2) = \left(\sum_{n=0}^{\infty} \frac{z_1^n}{n!}\right)\left(\sum_{n=0}^{\infty} \frac{z_2^n}{n!}\right) = \sum_{n=0}^{\infty} \frac{(z_1 + z_2)^n}{n!} = \psi(z_1 + z_2).$$

(3.4.19) will then follow with the choices $z_1 = x$ and $z_2 = iy$ since

$$e^x(\cos y + i \sin y) = \psi(x)\psi(iy) = \psi(x + iy) = \sum_{n=0}^{\infty} \frac{(x + iy)^n}{n!}. \qquad \square$$

Solution of Exercise 3.4.15. Using Weierstrass' theorem we can write for $z \in \mathbb{C}$,

$$e^z - 1 = \sum_{n=0}^{\infty} \frac{z^{n+1}}{(n+1)!} = \sum_{n=0}^{\infty} z^{n+1} \int_0^1 \frac{t^n}{n!} dt = \int_0^1 \left(\sum_{n=0}^{\infty} z^{n+1} \frac{t^n}{n!}\right) dt,$$

so that

$$e^z - 1 = z \int_0^1 e^{tz} dt.$$

Thus,

$$|e^z - 1| \le |z| \int_0^1 e^{t(\operatorname{Re} z)} dt. \qquad (3.9.18)$$

Let now z_1 and z_2 be in the closed left half-plane. Since the formula we want to prove is symmetric in z_1 and z_2, we can assume without loss of generality that

$$\operatorname{Re} z_2 \le \operatorname{Re} z_1.$$

Then, using (3.9.18) with $z = z_2 - z_1$, we have

$$|e^{z_1} - e^{z_2}| = |e^{z_1}| \cdot |e^{z_2 - z_1} - 1| \le |z_2 - z_1| \int_0^1 e^{t(\operatorname{Re} z_2 - \operatorname{Re} z_1)} dt \le |z_2 - z_1|,$$

since we assumed $\operatorname{Re} z_2 \le \operatorname{Re} z_1$. $\qquad \square$

Solution of Exercise 3.4.16. We have

$$(1 - z)e^z - 1 = -1 + (1 - z)\left(1 + z + \frac{z^2}{2!} + \frac{z^3}{3!} + \cdots\right)$$

$$= -1 + 1 + z^2 \left(\frac{1}{2!} - 1\right) + z^3 \left(\frac{1}{3!} - \frac{1}{2!}\right) + \cdots.$$

Therefore, for $|z| \le 1$,

$$|(1 - z)e^z - 1| \le |z^2| \left(1 - \frac{1}{2!}\right) + |z^3| \left(\frac{1}{2!} - \frac{1}{3!}\right) + \cdots$$

$$\le |z|^2 \left(1 - \frac{1}{2!} + \frac{1}{2!} - \frac{1}{3!} + \cdots\right)$$

$$= |z|^2. \qquad \square$$

Solution of Exercise 3.5.3. For $\theta = 0 \pmod{2\pi}$ we have $e^{in\theta} = 1$ and the series diverges. Let us now assume that $\theta \neq 0 \pmod{2\pi}$. Then, we can apply (1.2.9). More precisely, $|1 - e^{i\theta}| > 0$ and we have

$$\left| \sum_{n=1}^{m} e^{in\theta} \right| = \left| \frac{1 - e^{im\theta}}{1 - e^{i\theta}} \right|$$

$$\leq \frac{2}{|1 - e^{i\theta}|} \overset{\text{def.}}{=} M.$$

Thus, condition (2) in Abel's theorem (Theorem 3.5.1) is met. The sequence $(\frac{1}{\sqrt{n}})_{n \in \mathbb{N}}$ decreases to 0, and so the first condition is also met, and the series converges. $\qquad\square$

Solution of Exercise 3.5.7. (1) For preassigned $t \in [0, 1]$ and $\theta \in [-\frac{\pi}{4}, \frac{\pi}{4}] \setminus \{0\}$ we use the bound (3.5.3) (and (1.2.9) to find K in (3.5.3)) to the sequences $a_n = e^{4in\theta}$ and $b_n = \alpha_n t^n$, to obtain

$$\left| \sum_{n=0}^{N} \alpha_n t^n e^{i(4n+1)\theta} \right| \leq \frac{2\alpha_N t^N}{|\sin 2\theta|}.$$

Furthermore, by Abel's theorem, the sequence of functions (still for fixed $\theta \in [-\frac{\pi}{4}, \frac{\pi}{4}] \setminus \{0\}$)

$$f_N(t) = \sum_{n=0}^{N} \alpha_n t^n e^{i(4n+1)\theta}$$

converges pointwise for $t \in [0, 1]$. Applying the dominated convergence theorem (see Theorem 17.5.2) to the sequence $(f_N)_{N \in \mathbb{N}_0}$ we obtain that

$$\int_0^1 (\lim_{N \to \infty} f_N(t))dt = \lim_{N \to \infty} \int_0^1 f_N(t)dt = \lim_{N \to \infty} \sum_{n=0}^{N} \alpha_n \frac{e^{i(4n+1)\theta}}{4n + 1}$$

and, using Exercise 3.5.7,

$$= \sum_{n=0}^{\infty} \alpha_n \frac{e^{i(4n+1)\theta}}{4n + 1}$$

$$= \underbrace{\sum_{n=0}^{\infty} \frac{\alpha_n}{4n + 1}}_{\text{denoted by } M} + \sum_{n=0}^{\infty} \alpha_n \frac{e^{i(4n+1)\theta} - 1}{4n + 1}$$

$$= M + \sum_{n=0}^{\infty} i \int_0^{\theta} \alpha_n e^{i(4n+1)u} du$$

$$= M + i \int_0^{\theta} \left(\sum_{n=0}^{\infty} \alpha_n e^{i(4n+1)u} \right) du,$$

where, as above, to go from the penultimate line to the last line, we have used (3.5.2) and (1.2.9) and the dominated convergence theorem, now with the functions

$$g_N(u) = \sum_{n=0}^{N} a_n e^{i(4n+1)u}, \quad N = 0, 1, \dots.$$

(2) For $\theta = 0$, we need to prove that

$$\int_0^1 \left(\sum_{n=0}^{\infty} a_n t^{4n} \right) dt = \sum_{n=0}^{\infty} \frac{a_n}{4n+1},$$

and this identity follows directly from the monotone convergence theorem. □

Solution of Exercise 3.6.1. That the series on the left is absolutely convergent for $|z| < 1$ follows from

$$\left| \frac{z^n}{1+z^n} \right| \le \frac{|z|^n}{1-|z|^n} \le \frac{|z|^n}{1-|z|}.$$

Similarly the series on the right is absolutely convergent for $|z| < 1$. Here we want to compute explicitly the sum on the left. We consider the family

$$a_{n,p} = (-1)^p z^{n(1+p)}, \quad n \in \mathbb{N} \text{ and } p \in \mathbb{N}_0.$$

For $|z| \le r < 1$ we have

$$\sum_{n=1}^{\infty} \sum_{p=0}^{\infty} |a_{n,p}| \le \sum_{n=1}^{\infty} \frac{r^n}{1-r^n} < \sum_{n=1}^{\infty} \frac{r^n}{1-r} < \infty,$$

and so the family $(a_{n,p})_{\substack{n \in \mathbb{N} \\ p \in \mathbb{N}_0}}$ is absolutely summable. We have

$$\sum_{n=1}^{\infty} \sum_{p=0}^{\infty} a_{n,p} = \sum_{n=1}^{\infty} \frac{z^n}{1+z^n},$$

and

$$\sum_{p=0}^{\infty} \sum_{n=1}^{\infty} a_{n,p} = \sum_{p=0}^{\infty} \frac{(-1)^p z^{(1+p)}}{1-z^{(1+p)}},$$

and hence the result. □

Solution of Exercise 3.6.2. We follow the argument in the book of Choquet on topology; see [46, 47]. For $z \notin \mathbb{Z} + i\mathbb{Z}$ we have

$$\left| \frac{1}{(z-(p+iq))^2} - \frac{1}{(p+iq)^2} \right| = \left| \frac{z(z-2(p+iq))}{(z-(p+iq))^2(p+iq)^2} \right|$$

$$\le \frac{|z|(|z|+2\sqrt{p^2+q^2})}{(|z|-\sqrt{p^2+q^2})^2(p^2+q^2)}.$$

Fix now $R > 0$. For (p, q) such that $\sqrt{p^2 + q^2} > 2R$ and for $|z| < R$ we have

$$|z| + 2\sqrt{p^2 + q^2} < R + 2\sqrt{p^2 + q^2} < \frac{\sqrt{p^2 + q^2}}{2} + 2\sqrt{p^2 + q^2} = \frac{5\sqrt{p^2 + q^2}}{2},$$

and

$$\left||z| - \sqrt{p^2 + q^2}\right| \geq \sqrt{p^2 + q^2} - R \geq \sqrt{p^2 + q^2} - \frac{\sqrt{p^2 + q^2}}{2} = \frac{\sqrt{p^2 + q^2}}{2}.$$

Hence,

$$\left|\frac{1}{(z - (p + iq))^2} - \frac{1}{(p + iq)^2}\right| \leq \frac{5R\sqrt{p^2 + q^2}}{2\frac{(p^2 + q^2)}{4}(p^2 + q^2)} = \frac{10R}{(\sqrt{p^2 + q^2})^3}.$$

To conclude it remains to show that the family

$$\left(\frac{1}{(\sqrt{p^2 + q^2})^3}\right)_{(p,q) \in \mathbb{Z}^2 \setminus (0,0)}$$

is summable. Since $p^2 + q^2 \geq \frac{(|p| + |q|)^2}{2}$ it is enough to show that the family $\left(\frac{1}{(|p| + |q|)^3}\right)_{(p,q) \in \mathbb{Z}^2 \setminus (0,0)}$ is summable. It is enough in turn to check that the family is summable for (p, q) both greater than or equal to 1. But, for $p \geq 1$,

$$\sum_{q=1}^{\infty} \frac{1}{(p + q)^3} \leq \int_0^{\infty} \frac{dx}{(p + x)^3} = \frac{1}{2p^2}.$$

Thus

$$\sum_{p=1}^{\infty} \left(\sum_{q=1}^{\infty} \frac{1}{(p + q)^3}\right) < \infty,$$

and the summability follows from Theorem 14.3.1 of Section 14.3. □

Solution of Exercise 3.7.4. (a) Abel's theorem on alternating series (see Theorem 3.5.1) insures that the series (3.7.10) converges. We now check that the infinite product (3.7.9) diverges. Using Taylor's formula with remainder we have

$$\ln(1 + x) = x - \frac{x^2}{2} + \frac{1}{2}\int_0^x \frac{2(x - u)^2}{(1 + u)^3}\,du, \quad x \in (-1, 1),$$

and so

$$\ln(1 + x) = x - \frac{x^2}{2} + t(x), \quad x \in (-1, 1),$$

where, for $|x| \leq 1/2$,

$$|t(x)| \leq \frac{8|x|^3}{3}.$$

Thus, for $n \geq 4$,

$$\ln\left(1 + \frac{(-1)^n}{\sqrt{n}}\right) = \frac{(-1)^n}{\sqrt{n}} - \frac{1}{2n} + t_n, \quad |t_n| \leq \frac{8}{3n^{3/2}}.$$

It follows that the series $\sum_{n=2}^{\infty} \ln\left(1 + \frac{(-1)^n}{\sqrt{n}}\right)$ diverges to $-\infty$, and so the infinite product has limit 0.

(b) The series is divergent since

$$\sum_{n=3}^{2p} a_n = \sum_{n=2}^{p} \frac{1}{n}.$$

As for the product, we have

$$a_{2n-1} a_{2n} = 1 - \frac{1}{n\sqrt{n}},$$

and so the sequence of products $P_{2p} = \prod_{n=3}^{2p}(1 + a_n)$ converges since

$$P_{2p} = \prod_{n=3}^{2p}(1 + a_n) = \prod_{n=2}^{p}\left(1 - \frac{1}{n\sqrt{n}}\right).$$

Furthermore, we have

$$P_{2p+1} = P_{2p} \cdot \left(1 - \frac{1}{\sqrt{p+1}}\right),$$

and so the sequence P_{2p+1} converges to the same limit as the sequence P_{2p}, and hence the result.   □

Solution of Exercise 3.7.5. It is readily shown by induction that

$$\prod_{n=2}^{N}\left(1 - \frac{1}{n^2}\right) = \frac{N+1}{2N}.$$

Thus the infinite product is convergent, and its value is $1/2$.   □

Solution of Exercise 3.7.6. The infinite product will in particular diverge at the points t where one of the factors vanishes, that is at those t such that

$$\cos\left(\frac{t}{\rho^n}\right) = 0$$

for some $n \in \mathbb{N}$, that is

$$t = \frac{\pi}{2}\rho^n(2k+1), \quad n \in \mathbb{N}, \quad k \in \mathbb{Z}.$$

Consider now a point different from those points. Since cos is an even function we focus on $t > 0$. Since

$$\sin u \leq u, \quad u \geq 0,$$

we have

$$1 - \cos u = \int_0^u \sin v dv \leq \int_0^u v dv = \frac{u^2}{2},$$

and therefore

$$1 - \cos\left(\frac{t}{\rho^n}\right) \leq \frac{t^2}{\rho^{2n}}.$$

Since $\rho > 1$ the series with running term $\frac{t^2}{\rho^{2n}}$ converges and so the infinite product also converges. $\qquad\square$

Solution of Exercise 3.7.7. The series

$$\sum_{k=1}^{\infty} \frac{|z|^k}{k^2}$$

converges in the closed unit disk, and only there. Furthermore, the equation

$$1 + \frac{z^k}{k^2} = 0$$

has a solution in the closed unit disk only for $k = 1$; then $z = -1$. $\qquad\square$

Solution of Exercise 3.7.8. The series with running term $q^\ell z$ converges absolutely for every $z \in \mathbb{C}$. The infinite product converges thus to a function (which is entire) that vanishes only at the point $q^{-\ell}$. Furthermore,

$$(1 - qz)f(qz) = (1 - qz)\prod_{\ell=1}^{\infty}(1 - q^{\ell+1}z) = f(z). \qquad\square$$

Solution of Exercise 3.7.9. We have for $n \geq 2$,

$$a_n = a_1 + \sum_{\ell=2}^{n}(a_\ell - a_{\ell-1})$$

and hence

$$|a_n| \leq |a_1| + \sum_{\ell=2}^{n}|a_\ell - a_{\ell-1}| \leq |a_1| + \sum_{\ell=2}^{\infty}|a_\ell - a_{\ell-1}|$$

and so the sequence $(|a_n|)_{n\in\mathbb{N}_0}$ is bounded and thus the radius of convergence of the power series $\sum_{n=0}^{\infty} a_n z^n$ is at least 1. Set $g(z) = (1 - z)f(z)$. We have, for $|z| < 1$,

$$g(z) = (1 - z)\left(\sum_{n=0}^{\infty} a_n z^n\right)$$

$$= 1 + z(-1 + a_1) + z^2(-a_1 + a_2) + \cdots + z^n(-a_{n-1} + a_n) + \cdots.$$

Since $a_1 = 1$ we have

$$sg(z) = 1 + \sum_{n=2}^{\infty} z^n(-a_{n-1} + a_n),$$

and so, for $|z| < 1$, we have:

$$|g(z) - 1| \leq \sum_{n=2}^{\infty} |z|^n | - a_{n-1} + a_n| \tag{3.9.19}$$

$$\leq |z|^2 \sum_{n=2}^{\infty} | - a_{n-1} + a_n|.$$

We now assume that the radius of convergence is equal to infinity. Let $z \in \mathbb{C}$ and let $n_0 \in \mathbb{N}$ be such that $|z/n_0| < 1$. Using (3.9.19) we have

$$|g(z/n) - 1| \leq \frac{|z|^2}{n^2} \left(\sum_{n=2}^{\infty} | - a_{n-1} + a_n| \right)$$

for $n \geq n_0$. By Theorem 3.7.1, the product converges and vanishes at the points $z = 2, 3, \ldots.$ □

Solution of Exercise 3.7.11. The sequence $1/n$ is decreasing, and in particular of bounded variation, and hence the preceding exercise applies. More precisely, (3.9.19) reads now

$$|1 - (1 - z)e^z| \leq |z|^2, \quad z \in \mathbb{D}. \tag{3.9.20}$$

Solution of Exercise 3.7.12. We write

$$\frac{|z_n|}{z_n} \frac{z_n - z}{1 - z\overline{z_n}} = 1 + a_n(z).$$

To prove the claim of the exercise it is enough to show that, for $|z| < 1$,

$$\sum_{n=0}^{\infty} |a_n(z)| < \infty. \tag{3.9.21}$$

We have

$$a_n(z) = \frac{|z_n|}{z_n} \frac{z_n - z}{1 - z\overline{z_n}} - 1$$

$$= \frac{|z_n|z_n - |z_n|z - z_n + |z_n|^2 z}{z_n(1 - z\overline{z_n})}$$

$$= \frac{(|z_n| - 1)(|z_n|z + z_n)}{z_n(1 - z\overline{z_n})}$$

$$= \frac{(|z_n| - 1) \left(\dfrac{|z_n|}{z_n} z + 1 \right)}{(1 - z\overline{z_n})}.$$

Thus, for $|z| < 1$,

$$|a_n(z)| \leq (1 - |z_n|)\frac{2}{1 - |z|}, \tag{3.9.22}$$

and (3.9.21) holds since (3.7.16) holds. □

Solution of Exercise 3.7.13. It suffices to write

$$\frac{z - z_n}{z - \overline{z_n}} = 1 + \frac{\overline{z_n} - z_n}{z - \overline{z_n}}$$

and take into account (3.7.18). □

Solution of Exercise 3.7.15. (a) Using the power expansion for $\cos z$ we have

$$|1 - \cos z| = |z|^2 \cdot \left| \sum_{n=1}^{\infty} \frac{(-1)^n z^{2n-2}}{(2n)!} \right|$$

$$\leq \frac{|z|^2}{2} \cdot \sum_{n=1}^{\infty} \frac{2|z|^{2n-2}}{(2n)!}$$

$$\leq \frac{|z|^2}{2} \sum_{n=1}^{\infty} \frac{|z|^{2n-2}}{(2n-2)!} \quad \text{since, for } n \geq 1, \ 2(2n-2)! \leq (2n)!,$$

$$\leq \frac{|z|^2}{2} e^{|z|}.$$

(b) Take $M > 0$ and consider z such that $|z| \leq M$. In view of (a), we have

$$\left| 1 - \cos\left(\frac{z}{2^n}\right) \right| \leq \frac{M^2 e^{\frac{M}{2^n}}}{2^{2n+1}} < 1$$

for n large enough (which depends on M), and hence the given infinite product converges at those points z such that $\cos(z/2^n) \neq 0$. □

Solution of Exercise 3.7.16. We begin with (3.7.23), and first replace z in (1.5.10) by z/λ to obtain the formula

$$z^{2n} - \lambda^{2n} = (z^2 - \lambda^2) \prod_{k=1}^{n-1} \left(z^2 - 2z\lambda \cos\left(\frac{k\pi}{n}\right) + \lambda^2 \right).$$

Replacing z and λ by $(1 + z/2n)$ and $(1 - z/2n)$ respectively we obtain

$$\left(1 + \frac{z}{2n} \right)^{2n} - \left(1 - \frac{z}{2n} \right)^{2n} = \frac{2z}{n} \prod_{k=1}^{n-1} \left(2 + \frac{z^2}{2n^2} - 2\left(1 - \frac{z^2}{4n^2} \right) \cos\left(\frac{k\pi}{n}\right) \right)$$

$$= \frac{2z}{n} \prod_{k=1}^{n-1} \left(2\left(1 - \cos\left(\frac{k\pi}{n}\right) \right) \right) \prod_{k=1}^{n-1} \left(1 + \frac{z^2}{4n^2} \frac{1 + \cos\frac{k\pi}{n}}{1 - \cos\frac{k\pi}{n}} \right)$$

$$= \frac{2z}{n} 2^{n-1} \prod_{k=1}^{n-1} \left(1 - \cos\left(\frac{k\pi}{n}\right)\right) \prod_{k=1}^{n-1} \left(1 + \frac{z^2}{4n^2} \frac{1 + \cos\dfrac{k\pi}{n}}{1 - \cos\dfrac{k\pi}{n}}\right)$$

$$= 2z \prod_{k=1}^{n-1} \left(1 + \frac{z^2}{4n^2} \frac{1 + \cos\dfrac{k\pi}{n}}{1 - \cos\dfrac{k\pi}{n}}\right),$$

where we have used (1.6.25) to go from the penultimate line to the last line. We conclude using Theorem 14.7.2 since

$$\lim_{n\to\infty} \frac{z^2}{4n^2} \frac{1 + \cos\dfrac{k\pi}{n}}{1 - \cos\dfrac{k\pi}{n}} = \frac{z^2}{k^2\pi^2}.$$

The proof of (3.7.24) is done in much the same way. We now start from (1.5.8) and, with the same change of variables as above, obtain

$$\left(1 + \frac{z}{2n}\right)^{2n} + \left(1 - \frac{z}{2n}\right)^{2n} = \prod_{k=0}^{n-1} \left(2 + \frac{z^2}{2n^2} - 2\left(1 - \frac{z^2}{4n^2}\right)\cos\left(\frac{(2k+1)\pi}{2n}\right)\right)$$

$$= \prod_{k=0}^{n-1} \left(2\left(1 - \cos\left(\frac{(2k+1)\pi}{2n}\right)\right)\right) \prod_{k=0}^{n-1} \left(1 + \frac{z^2}{4n^2} \frac{1 + \cos\dfrac{(2k+1)\pi}{2n}}{1 - \cos\dfrac{(2+1)k\pi}{2n}}\right)$$

$$= 2 \prod_{k=0}^{n-1} \left(1 + \frac{z^2}{4n^2} \frac{1 + \cos\dfrac{(2k+1)\pi}{2n}}{1 - \cos\dfrac{(2k+1)\pi}{2n}}\right),$$

where now we use (1.5.14) to go from the penultimate line to the last line. As above, we conclude using Theorem 14.7.2 since

$$\lim_{n\to\infty} \frac{z^2}{4n^2} \frac{1 + \cos\dfrac{(2k+1)\pi}{n}}{1 - \cos\dfrac{(2k+1)\pi}{2n}} = \frac{4z^2}{(2k+1)^2\pi^2}. \qquad \square$$

Part II

Functions of a Complex Variable

Chapter 4

Cauchy–Riemann Equations and ℂ-differentiable Functions

In this chapter we present exercises on ℂ-differentiable functions and the Cauchy-Riemann equations. We begin with exercises related to continuity in Section 4.1. We then study derivatives. Recall that ℂ-differentiability of a function $f(z) = u(x,y) + iv(x,y)$ at a given point $z_0 = x_0 + y_0$ implies the Cauchy-Riemann equations at that point. The converse in general does not hold. It will hold in particular when the real and imaginary parts of f are differentiable (as real-valued functions of two real variables) at (x_0, y_0). Section 4.2 gives a geometric interpretation of the lack of derivative at z_0, when u and v are differentiable at (x_0, y_0). In Section 4.3 we present various counterexamples that exhibit functions which are not ℂ-differentiable, but for which the Cauchy-Riemann equations hold. Exercises related to analytic functions are given in Section 4.4. A complex-valued function f continuous in an open set $\Omega \subset \mathbb{C}$ is called *holomorphic* if it has a derivative at every point of Ω. At this stage of the book, we do not know yet that a holomorphic function has a power expansion in a neighborhood of every point of its domain of definition (that is, is *analytic*) but we know that power series are examples of holomorphic functions. Section 4.4 contains exercises on analyticity and power series.

4.1 Continuous functions

> *Il est naturellement gênant de ne pas pouvoir définir dans le corps ℂ une authentique fonction continue $\sqrt{z}$ qui vérifirait $(\sqrt{z})^2 = z$.*
>
> Jean Dieudonné, [63, p. 202]

Let us first review some definitions. Let f be a complex-valued function defined in a neighborhood of a point $z_0 \in \mathbb{C}$. We will say that

$$\lim_{z \to z_0} f(z)$$

exists and is equal to $\ell \in \mathbb{C}$ if the following condition holds:

$$\forall \epsilon > 0, \quad \exists \eta > 0: \quad |z - z_0| < \eta \Longrightarrow |f(z) - \ell| < \epsilon.$$

It is an easy exercise to check that the limit, if it exists, is unique. Furthermore, we have:

Proposition 4.1.1. *Let f be a complex-valued function defined in a neighborhood of the point z_0. Then,*

$$\lim_{z \to z_0} f(z) = \ell \quad \Longleftrightarrow \quad \begin{cases} \lim_{z \to z_0} \operatorname{Re} f(z) = \operatorname{Re} \ell \\ and \\ \lim_{z \to z_0} \operatorname{Im} f(z) = \operatorname{Im} \ell. \end{cases}$$

The usual results on limits proved in calculus for functions of a real variable still hold here, and we list them without proof.

Theorem 4.1.2. *Let f and g be defined in a neighborhood of the point z_0, and admitting limits at z_0. Then the functions $af + bg$ (where a and b are arbitrary points in $\mathbb{C}$) and fg have a limit at z_0, and we have*

$$\lim_{z \to z_0} (af + bg) = a \lim_{z \to z_0} f + b \lim_{z \to z_0} g,$$

$$\lim_{z \to z_0} (fg) = \left(\lim_{z \to z_0} f \right) \left(\lim_{z \to z_0} g \right).$$

Assume moreover that $\lim_{z \to z_0} g$ is different from 0. Then $\lim_{z \to z_0} (f/g)$ exists and it holds that

$$\lim_{z \to z_0} \frac{f}{g} = \frac{\lim_{z \to z_0} f}{\lim_{z \to z_0} g}.$$

Sometimes the function will be defined only in a punctured neighborhood of z_0. Then, the condition $|z - z_0| < \eta$ is replaced by $0 < |z - z_0| < \eta$ and the definition of the limit becomes

$$\forall \epsilon > 0, \quad \exists \eta > 0: \quad 0 < |z - z_0| < \eta \Longrightarrow |f(z) - \ell| < \epsilon.$$

The limit is then sometimes denoted by $\lim_{\substack{z \to z_0 \\ z \neq z_0}} f(z)$, but we will here mostly stick to the notation $\lim_{z \to z_0} f(z)$. The uniqueness of the limit and Theorem 4.1.2 still hold in the case of punctured neighborhoods.

Theorem 4.1.3. *Let f and g be defined in a punctured neighborhood of the point z_0, and admitting limits at z_0. Then the functions $af + bg$ (where a and b are arbitrary points in $\mathbb{C}$) and fg have a limit at z_0, and we have*

$$\lim_{z \to z_0} (af + bg) = a \lim_{z \to z_0} f + b \lim_{z \to z_0} g,$$

$$\lim_{z \to z_0} (fg) = \left(\lim_{z \to z_0} f \right) \left(\lim_{z \to z_0} g \right).$$

Assume moreover that $\lim_{z \to z_0} g$ is different from 0. Then $\lim_{z \to z_0} (f/g)$ exists and it holds that

$$\lim_{z \to z_0} \frac{f}{g} = \frac{\lim_{z \to z_0} f}{\lim_{z \to z_0} g}.$$

A new feature is that we may allow ∞ both as a limit point and at the point where the limit is taken. Definitions are as follow:

Definition 4.1.4. Let f be a complex-valued function whose domain of definition contains a set of the form $|z| > R_0$ for some $R_0 \ge 0$, and let $\ell \in \mathbb{C}$. One says that

$$\lim_{z \to \infty} f(z) = \ell$$

if

$$\forall \epsilon > 0, \quad \exists R > 0, \quad |z| > R \Longrightarrow |f(z) - \ell| < \epsilon.$$

One says that

$$\lim_{z \to \infty} f(z) = \infty$$

if

$$\lim_{|z| \to \infty} |f(z)| = +\infty. \tag{4.1.1}$$

Note that condition (4.1.1) can be rewritten as

$$\forall S > 0, \quad \exists R > 0, \quad |z| > R \Longrightarrow |f(z)| > S.$$

Thus, for a complex-valued function f whose domain of definition contains a set of the form $|z| > R_0$ for some $R_0 \ge 0$, and which does not vanish for $|z| > R_0$, we have the equivalence

$$\lim_{z \to \infty} f(z) = \infty \quad \Longleftrightarrow \quad \lim_{z \to \infty} \frac{1}{f(z)} = 0.$$

Question 4.1.5. *Let p and q be two polynomials with $\deg q \ge \deg p + 2$. Show that*

$$\lim_{z \to \infty} \frac{p(z)}{q(z)} = 0.$$

Definition 4.1.6. The function f defined in a neighborhood of the point $z_0 \in \mathbb{C}$ is said to be *continuous at* z_0 if

$$\lim_{z \to z_0} f(z) = f(z_0).$$

The definitions of limit and continuity depend really on the metric space structure of $\mathbb{C}$. All the usual results on continuity of sums, products, quotient and composition still hold here, and we will not recall them. These are *local properties*. The specific structure of $\mathbb{C}$, or of its subsets, will come into play when one studies the existence of a continuous function in a given set.

The following example is taken from [115, 14.2, p. 269].

Exercise 4.1.7. *As usual, we set $z = x + iy$ to be the cartesian representation of the complex number z, and, for $z \neq 0$ we write $re^{i\theta}$, where $\theta \in [-\pi, \pi)$ and $r > 0$ its polar representation. Let f be defined by*

$$f(z) = \begin{cases} 0, & \text{if } z = 0 \text{ or } \theta = 0, \\ \dfrac{r}{\theta}, & \text{if } z \neq 0. \end{cases}$$

(a) *Show that f is continuous along every straight line passing through the origin.*
(b) *Show that f is not continuous at the origin.*

The function (4.1.2) below is a special case of the functions defined in Exercise 4.4.21, and is in fact analytic in $\mathbb{C} \setminus [0, 1]$. In Exercise 4.1.8 we focus on the continuity.

Exercise 4.1.8.

(a) *Show that the function*

$$F(z) = \int_0^1 \frac{dt}{t - z} \qquad (4.1.2)$$

is continuous in $\mathbb{C} \setminus [0, 1]$.
(b) *Show that for $s \in (0, 1)$ both the limits*

$$F_+(s) = \lim_{\epsilon \downarrow 0} F(s + i\epsilon),$$
$$F_-(s) = \lim_{\epsilon \downarrow 0} F(s - i\epsilon) \qquad (4.1.3)$$

exist and are finite.
(c) *Show that the limits*

$$\lim_{\substack{z \to 0 \\ z \notin [0,1]}} F(z) \quad \text{and} \quad \lim_{\substack{z \to 1 \\ z \notin [0,1]}} F(z) \qquad (4.1.4)$$

do not exist.

Roughly speaking, continuity creates no special problems when the function is defined in an (open) simply-connected set. The precise definition of a simply-connected set is somewhat beyond the level of a first course on complex variables. We will use here the notion of star-shaped sets. See Definition 2.1.4 above. These sets are much easier to handle. For instance, $\mathbb{C}\setminus(-\infty, 0]$ is star-shaped, but $\mathbb{C}\setminus\{0\}$ is not. Any non-zero complex number z admits a logarithm, that is, there always exists a number w such that $z = \exp w$. To see this write $z = \rho\exp(i\theta)$. It suffices to take

$$w = \ln\rho + i\theta.$$

On the other hand, it is false that there always exists a *continuous* logarithm function on a given set Ω. The same problem holds for continuous square root functions. The following two exercises discuss the non-existence of a continuous square root of the function $f(z) = z$.

Exercise 4.1.9. *Show that there is no function f continuous on $\mathbb{C}\setminus\{0\}$ such that $f(z)^2 = z$. To that purpose, let $\Omega = \mathbb{C}\setminus\mathbb{R}_-$, and for $z = \rho e^{i\theta} \in \Omega$ with $\theta \in (-\pi, \pi)$, let $f_0(z) = \sqrt{\rho}e^{i\frac{\theta}{2}}$. Assume by contradiction that f exists and consider the function $f(z)/f_0(z)$ for $z \in \Omega$.*

A pure topological proof of the previous exercise uses homotopy groups. An elementary proof (but which uses the fact that the continuous image of a connected set is connected) is as follows (see, e.g., [223, p. 106], where the case of a square root of order n is considered). Assume that a function f exists with the required properties. Then, for every $t \in \mathbb{R}$, the number $f(e^{it})$ is a square root of e^{it}. So we can write

$$f(e^{it}) = s(t)e^{\frac{it}{2}},$$

where $s(t) \in \{-1, 1\}$. The function $s(t) = e^{-\frac{it}{2}}f(e^{it})$ is continuous and with values in $\{-1, 1\}$, and so is constant. Thus we have

$$f(e^{it}) \equiv e^{\frac{it}{2}} \quad \text{or} \quad f(e^{it}) \equiv -e^{\frac{it}{2}}.$$

Setting $t = 0$ and $t = 2\pi$ we obtain a contradiction in both cases.

Another very elementary approach to the same question is illustrated by the following exercise, taken and from [91, p. 41], [92, p. 34]. See [42, Exercise 4.22, p. 93] for a more general result. We also refer to [42, Theorem 4. 18, p. 92] for the abstract monodromy theorem which characterizes when a continuous logarithm exists.

Exercise 4.1.10.

(a) *Show that there is no function from $\mathbb{C}\setminus\{0\}$ into itself such that*

$$f(zw) = f(z)f(w) \quad \text{and} \quad f(z)^2 = z \tag{4.1.5}$$

for all $z, w \in \mathbb{C}\setminus\{0\}$.

(b) *Show that there is no continuous function from $\mathbb{C} \setminus \{0\}$ into itself such that $f(z)^2 = z$.*

(c) *Show that there is no continuous logarithm on $\mathbb{C} \setminus \{0\}$.*

(d) *Show that there is a continuous logarithm on $\mathbb{C} \setminus \mathbb{R}_-$. Explain.*

Related to this exercise and the discussion preceding it, we mention that a continuous function from the closed unit disk into the unit circle always has a continuous logarithm. This can be seen from the above-mentioned monodromy theorem, or by direct arguments; see [87, Theorem 1, p. 372]. In this last work, this result is used to prove Brouwer's theorem, and as a consequence of the latter, to prove the fundamental theorem of algebra.

As a corollary of Exercise 4.1.10 we have:

Exercise 4.1.11. *Show that there is no continuous real-valued function θ defined on $\mathbb{R}^2 \setminus \{(0,0)\}$ and such that*

$$z = \sqrt{x^2 + y^2}\, e^{i\theta(x,y)}.$$

Indeed, should such a function exist, z would have a continuous square root in $\mathbb{C} \setminus \{0\}$.

We recall that in any star-shaped open set (or more generally, in any open simply-connected domain) a non-vanishing holomorphic function (that is a function which has everywhere a derivative) has a holomorphic (and in particular continuous) logarithm. We also recall that one can give necessary and sufficient conditions in terms of integrals for a holomorphic logarithm to exist when the set is not simply-connected.

Remark 4.1.12. By Exercise 4.2.18 below, any function f continuous in $\mathbb{C} \setminus \{0\}$ such that $f(z)^2 = z$ there, would be automatically holomorphic. One can then proceed as in Exercise 7.2.21 to prove that no such function exists.

The following exercise has follow-up exercises. See Exercises 6.1.9 and 10.2.7.

Exercise 4.1.13. *Let Ω be a connected open subset of $\mathbb{C}$ and let a and b be continuous functions in Ω. Assume that the sets*

$$\Omega_+ = \{z \in \Omega \,;\, |b(z)| < |a(z)|\} \quad and \quad \Omega_- = \{z \in \Omega \,;\, |b(z)| > |a(z)|\}$$

are both non-empty. Show that the set

$$\Omega_0 = \{z \in \Omega \,;\, |b(z)| = |a(z)|\}$$

is also non-empty.

4.2 Derivatives

The complex-valued function $f(z) = u(x, y) + iv(x, y)$ defined in a neighborhood of the point z_0 is said to have a derivative at z_0 (or to be differentiable at z_0, or to be $\mathbb{C}$-differentiable at z_0) if the limit

$$\lim_{\substack{z \to z_0 \\ z \neq z_0}} \frac{f(z) - f(z_0)}{z - z_0} \tag{4.2.1}$$

exists.

The various rules of derivations still hold in the present case.

Question 4.2.1. *Let f and g be defined in an open set Ω, and assume that they have derivatives of order N at the point z_0. Then, the product fg has a derivative of order N at z_0 and we have*

$$(fg)^{(N)}(z_0) = \sum_{k=0}^{N} \binom{N}{k} f^{(k)}(z_0) g^{(N-k)}(z_0). \tag{4.2.2}$$

Let f be a function $\mathbb{C}$-differentiable at a point z_0 and not vanishing there. The number

$$L(f) = \frac{f'(z_0)}{f(z_0)}$$

is called the logarithmic derivative of f at the point z_0.

Question 4.2.2. *Let $f_1, \ldots, f_n, g_1, \ldots, g_m$ be functions differentiable at the point $z_0 \in \mathbb{C}$ and assume moreover that $f_\ell(z_0) \neq 0$ for $\ell = 1, \ldots, n$ and $g_j(z_0) \neq 0$ for $j = 1, \ldots, m$. Show that*

$$\frac{\left(\dfrac{f_1 \cdots f_n}{g_1 \cdots g_m} \right)'(z_0)}{\left(\dfrac{f_1 \cdots f_n}{g_1 \cdots g_m} \right)(z_0)} = \sum_{\ell=1}^{n} \frac{f_\ell'(z_0)}{f_\ell(z_0)} - \sum_{j=1}^{m} \frac{g_j'(z_0)}{g_j(z_0)}. \tag{4.2.3}$$

Condition (4.2.1) is equivalent to the existence of a complex number ℓ_{z_0} such that

$$\lim_{z \to z_0} \frac{|f(z) - f(z_0) - \ell_{z_0}(z - z_0)|}{|z - z_0|} = 0. \tag{4.2.4}$$

In other words, f is differentiable when we view $\mathbb{C}$ as a Banach space over $\mathbb{C}$, both for the domain and the range of f (and hence the terminology $\mathbb{C}$-differentiability at the point z_0; we have $\mathcal{B}_1 = \mathcal{B}_2 = \mathbb{C}$ in Definition 16.1.13). We have the following key result (the notion of differentiability of a real-valued function of two real variables has been recalled in Section 14.1):

Theorem 4.2.3. *Let $f(z) = u(x, y) + iv(x, y)$ be defined in a neighborhood of the point $z_0 = x_0 + iy_0 \in \mathbb{C}$. Then, f is $\mathbb{C}$-differentiable at the point z_0 if and only if both the real part and the imaginary part of f are differentiable at the point (x_0, y_0) and satisfy the Cauchy–Riemann equations*

$$\frac{\partial u}{\partial x}(x_0, y_0) = \frac{\partial v}{\partial y}(x_0, y_0),$$
$$\frac{\partial u}{\partial y}(x_0, y_0) = -\frac{\partial v}{\partial x}(x_0, y_0). \tag{4.2.5}$$

Furthermore the following formula holds for the derivative at the point z_0:

$$f'(z_0) = \frac{\partial u}{\partial x}(x_0, y_0) - i\frac{\partial u}{\partial y}(x_0, y_0). \tag{4.2.6}$$

See for instance [192, (6.4), p. 59], [74, Theorem 1.1, p. 25]. As mentioned in the statement of the theorem, the equations in (4.2.5) are called the Cauchy–Riemann equations. Formula (4.2.6) for the derivative is of central importance, and will be used over and over in the sequel. Using the previous theorem we get the complex counterpart of the third definition of the function exponential discussed at the beginning of Section 1.2, (and hence the fourth definition of e^z that we have, including defining e^z via the formula $e^z = e^x(\cos y + i \sin y)$).

Exercise 4.2.4. *Show that the function e^z is the only function which admits a derivative at every point in $\mathbb{C}$ and is such that*

$$f'(z) = f(z), \quad \text{and} \quad f(0) = 1.$$

We now give a geometric interpretation of $\mathbb{C}$-differentiability at a point. One can view a complex-valued function of a complex variable as a map from an open subset of $\mathbb{R}^2$ into $\mathbb{R}^2$:

$$r : (x, y) \mapsto (u(x, y), v(x, y)).$$

For the geometric discussion below it is better to view both the domain of r and the range space as column vectors, that is

$$r : \begin{pmatrix} x \\ y \end{pmatrix} \mapsto \begin{pmatrix} u(x, y) \\ v(x, y) \end{pmatrix}.$$

When u and v are differentiable at the point (x_0, y_0), the Jacobian matrix of the map r at the point (x_0, y_0) is equal to

$$dr_{(x_0, y_0)} = \begin{pmatrix} \frac{\partial u}{\partial x}(x_0, y_0) & \frac{\partial u}{\partial y}(x_0, y_0) \\ \frac{\partial v}{\partial x}(x_0, y_0) & \frac{\partial v}{\partial y}(x_0, y_0) \end{pmatrix}. \tag{4.2.7}$$

When the Cauchy–Riemann equations hold, this Jacobian becomes

$$dr_{(x_0, y_0)} = \begin{pmatrix} \frac{\partial u}{\partial x}(x_0, y_0) & \frac{\partial u}{\partial y}(x_0, y_0) \\ -\frac{\partial u}{\partial y}(x_0, y_0) & \frac{\partial u}{\partial x}(x_0, y_0) \end{pmatrix}, \tag{4.2.8}$$

that is, in the notation (1.1.1) with $\frac{\partial u}{\partial x}(x_0, y_0)$ and $-\frac{\partial u}{\partial y}(x_0, y_0)$ in place of x and y,

$$f'(z_0) = dr_{(x_0,y_0)}.$$

$dr(x_0, y_0)$ is a rotation composed with a homothety; it is therefore a complex number. This can be seen also as follows: The map r has an expansion of the form

$$r(x_0 + h, y_0 + k) = r(x_0, y_0) + (dr_{(x_0,y_0)}) \begin{pmatrix} h \\ k \end{pmatrix} + \sqrt{h^2 + k^2} \begin{pmatrix} E(h, k) \\ F(h, k) \end{pmatrix},$$

where the functions E and F tend to 0 as h and k go to 0. When the function

$$f(z) = u(x, y) + iv(x, y)$$

is $\mathbb{C}$-differentiable at the point (x_0, y_0), the Cauchy–Riemann equations hold at that point, and the transformation

$$\begin{pmatrix} h \\ k \end{pmatrix} \mapsto \begin{pmatrix} h_1 \\ k_1 \end{pmatrix} = (dr_{(x_0,y_0)}) \begin{pmatrix} h \\ k \end{pmatrix}$$

can be rewritten as

$$h_1 + ik_1 = \left(\frac{\partial u}{\partial x}(x_0, y_0) - i \frac{\partial u}{\partial y}(x_0, y_0) \right) (h + ik).$$

The transformation keeps angles invariant. It is said to be *conformal* at the point (x_0, y_0).

Exercise 4.2.5. *At which points are the following functions* $\mathbb{C}$*-differentiable?*

(a) $f(z) = x^2 + y^2 + 2ixy$.
(b) $f(z) = z \operatorname{Re} z$.
(c) $f(z) = e^z = e^x(\cos y + i \sin y)$.
(d) *The function f defined by*

$$f(z) = \begin{cases} \dfrac{\bar{z}^2}{z}, & \text{if } z \neq 0, \\ 0, & \text{if } z = 0. \end{cases}$$

(e) *The function* $f(z) = \bar{z}$.

Compute their derivative at these points where they are $\mathbb{C}$*-differentiable.*

Exercise 4.2.6 (see also Exercise 6.3.9). *Let* $\Omega \subset \mathbb{C}$ *be symmetric with respect to the real axis (that is,* $z \in \Omega \to \bar{z} \in \Omega$*). Assume that the pair of functions* $(u(x, y), v(x, y))$ *defined in* Ω *satisfy the Cauchy–Riemann equations at the point* (x_0, y_0)*. Show that the pair of functions* $(U(x, y), V(x, y))$ *defined by*

$$U(x, y) = u(x, -y) \quad \text{and} \quad V(x, y) = -v(x, -y)$$

satisfy the Cauchy–Riemann equations at the point $(x_0, -y_0)$*. Explain the result when u and v are differentiable at the point* (x_0, y_0)*.*

Remark 4.2.7. As a corollary of the previous exercise, and under the hypothesis of the exercise, we have the formula

$$\left(\overline{f(\overline{z})}\right)' = \overline{f'(\overline{z})}. \tag{4.2.9}$$

For $(x, y) \in \mathbb{R}^2 \setminus (-\infty, 0]$, define

$$\ln z = \ln \sqrt{x^2 + y^2} + i\theta(x, y), \tag{4.2.10}$$

where $\theta(x, y)$ is defined by (1.1.19).

Exercise 4.2.8. *Let $u(x, y) = \ln \sqrt{x^2 + y^2}$. Show that for $(x, y) \in \mathbb{R}^2 \setminus (-\infty, 0]$ it holds that*

$$\begin{aligned}
\frac{\partial u}{\partial x} &= \frac{\partial \theta}{\partial y}, \\
\frac{\partial u}{\partial y} &= -\frac{\partial \theta}{\partial x}.
\end{aligned} \tag{4.2.11}$$

It follows from Theorem 4.2.3 that the function (4.2.10) is ℂ-differentiable in $\mathbb{C} \setminus (-\infty, 0]$. Formula (4.2.6) implies that

$$(\ln z)' = \frac{1}{z}.$$

In Exercise 4.2.5 we saw that the function e^z is differentiable in the whole complex plane (such functions are called entire functions) and is its own derivative. Recall that the functions, $\sin z$, $\cos z$, $\sinh z$ and $\cosh z$ have been defined in (1.2.13) in terms of e^z (their definitions in terms of power series is recalled in (3.4.23)).

Exercise 4.2.9. *Show that*

$$\begin{aligned}
(\sin z)' &= \cos z, \\
(\cos z)' &= -\sin z, \\
(\sinh z)' &= \cosh z, \\
(\cosh z)' &= \sinh z.
\end{aligned}$$

Exercise 4.2.10. *Show that*

$$\cos(z/2)\cos(z/2^2) \cdots \cos(z/2^n) = \frac{\sin z}{2^n \sin(z/2^n)}, \quad n = 1, 2, \ldots,$$

and compute

$$\lim_{n \to \infty} \cos(z/2)\cos(z/2^2) \cdots \cos(z/2^n).$$

See also Exercise 3.7.15 in connection with the previous exercise.

Exercise 4.2.11 (see [207, p. 3]). *Let $f = u + iv$ be differentiable in an open connected set Ω, and assume that the real and imaginary parts of f are related by*

$$au(x, y) + bv(x, y) + c = 0, \tag{4.2.12}$$

where the numbers a, b, c are real, the numbers a and b not being simultaneously 0. Find f.

Let $f(z) = u(x, y) + iv(x, y)$ be a complex-valued function defined in an open set Ω and such that u and v admit first-order continuous partial derivatives in Ω. We introduce

$$\frac{\partial f}{\partial z} = \frac{1}{2}\left(\frac{\partial f}{\partial x} - i\frac{\partial f}{\partial y}\right) \quad \text{and} \quad \frac{\partial f}{\partial \bar{z}} = \frac{1}{2}\left(\frac{\partial f}{\partial x} + i\frac{\partial f}{\partial y}\right). \tag{4.2.13}$$

In other words,

$$\begin{aligned}
\frac{\partial f}{\partial z} &= \frac{1}{2}\left\{\left(\frac{\partial u}{\partial x} - i\frac{\partial u}{\partial y}\right) + \left(\frac{\partial v}{\partial y} + i\frac{\partial v}{\partial x}\right)\right\}, \\
\frac{\partial f}{\partial \bar{z}} &= \frac{1}{2}\left\{\left(\frac{\partial u}{\partial x} + i\frac{\partial u}{\partial y}\right) + \left(-\frac{\partial v}{\partial y} + i\frac{\partial v}{\partial x}\right)\right\}.
\end{aligned} \tag{4.2.14}$$

These operators will also be denoted by

$$\partial_z f \quad \text{and} \quad \partial_{\bar{z}} f \quad \text{or} \quad \partial f \quad \text{and} \quad \bar{\partial} f. \tag{4.2.15}$$

respectively. Their significance is explained in the following exercise, taken from [215, pp. 8–9].

Exercise 4.2.12. *Let $u(x, y)$ and $v(x, y)$ be real-valued and differentiable at the point (x_0, y_0). Let $z_0 = x_0 + iy_0$. For $z = x + iy$ in an open neighborhood Ω of z_0, let*

$$f(z) = u(x, y) + iv(x, y).$$

Show that

$$f(z) - f(z_0) = \frac{\partial f}{\partial z}\Big|_{z=z_0}(z - z_0) + \frac{\partial f}{\partial \bar{z}}\Big|_{z=z_0}(\bar{z} - \overline{z_0}) + o(z - z_0), \quad z \in \Omega, \tag{4.2.16}$$

where $o(z - z_0)$ denotes an expression such that

$$\lim_{z \to z_0} \frac{|o(z - z_0)|}{|z - z_0|} = 0.$$

Describe the set of limits

$$\lim_{z_n \to z_0} \frac{f(z_n) - f(z_0)}{z_n - z_0},$$

as $(z_n)_{n \in \mathbb{N}}$ is a sequence which tends to z_0. When is this set reduced to a single element (that is, when is the limit independent of the given sequence)?

Exercise 4.2.13. *Assume that the function $f(z)$ admits a derivative at the point z_0. Show that*

$$\frac{\partial f}{\partial z}\Big|_{z=z_0} = f'(z_0),$$
$$\frac{\partial f}{\partial \bar{z}}\Big|_{z=z_0} = 0.$$

(4.2.17)

Recall that a function which is ℂ-differentiable in an open set Ω is called holomorphic in Ω. It is called entire if it is holomorphic in $\Omega = \mathbb{C}$.

Exercise 4.2.14. *Let $m \in C_0[0,1]$, that is m is a (possibly complex-valued) continuous function on the interval $[0,1]$. From the definition of the derivative show that the function*

$$F(z) = \int_0^1 e^{izt} m(t) dt$$

is entire and compute its derivative.

Exercise 4.2.15. *Let $m \in C_0[1,2]$. Show that the function*

$$F(z) = \int_1^2 \frac{m(t)dt}{(z-t)^2}$$

is holomorphic in $\mathbb{C} \setminus [1,2]$ and show that

$$F'(z) = -2 \int_1^2 \frac{m(t)dt}{(z-t)^3}, \quad \forall z \in \mathbb{C} \setminus [1,2].$$

The next exercise is taken from [42, Theorem 2.34, p. 48]. In the statement, $[0,1]$ can be replaced by any piecewise differentiable (and not necessarily connected) path.

Exercise 4.2.16. *Let $m \in C_0[0,1]$. Show that for every $n \geq 1$, the function*

$$F_n(z) = \int_0^1 \frac{m(t)dt}{(t-z)^n}$$

is holomorphic in $\mathbb{C} \setminus [0,1]$ and that

$$F_n'(z) = n F_{n+1}(z), \quad \forall z \in \mathbb{C} \setminus [0,1].$$

(4.2.18)

For the next exercise, see [42, Exercise 3.41, p. 79]. The aim is to show that a continuous logarithm on an open set is holomorphic.

Exercise 4.2.17. *Let Ω be an open set and let f be a function continuous on Ω and such that*

$$\exp(f(z)) = z, \quad z \in \Omega.$$

Show that f is holomorphic in Ω.

The following exercise is taken from [62, Exercise 5.7.9, p. 70]. For yet another proof, which uses much deeper machinery, but allows f to vanish in Ω, see Exercise 5.4.3.

Exercise 4.2.18. *Let f be a function continuous in an open set $\Omega \subset \mathbb{C}$ and assume that f^2 is holomorphic in Ω. Assume moreover that f does not vanish in Ω. Show that f is holomorphic in Ω.*

We conclude this section with exercises on further properties of the operators (4.2.13).

Exercise 4.2.19. *Show that, for a complex-valued function with continuous second-order derivatives,*

$$\frac{\partial^2 f}{\partial z \partial \overline{z}} \overset{\text{def.}}{=} \frac{\partial}{\partial z}\left(\frac{\partial f}{\partial \overline{z}}\right)$$

$$= \frac{\partial}{\partial \overline{z}}\left(\frac{\partial f}{\partial z}\right) \tag{4.2.19}$$

$$= \frac{1}{4}\left(\frac{\partial^2 f}{\partial x^2} + \frac{\partial^2 f}{\partial y^2}\right).$$

The operator

$$\Delta \overset{\text{def.}}{=} \frac{\partial^2}{\partial x^2} + \frac{\partial^2}{\partial y^2}$$

is called the Laplacian. Real-valued functions u such that

$$\Delta u = 0$$

in an open set Ω are called *harmonic functions*. See Chapter 9 for more on these functions.

Exercise 4.2.20.

(i) *Show that the operators $\frac{\partial}{\partial z}$ and $\frac{\partial}{\partial \overline{z}}$ satisfy the usual rule of differentiation for a product and a quotient.*

(ii) *Show that for $z_0 \in \mathbb{C}$,*

$$\frac{\partial}{\partial \overline{z}}\left(\frac{g(z)}{z_0 - z}\right) = \frac{\frac{\partial g}{\partial \overline{z}}(z)}{z_0 - z}, \tag{4.2.20}$$

where g is in the domain of definition of $\partial_{\overline{z}}$. More generally, show that, if $f(z)$ is $\mathbb{C}$-differentiable,

$$\frac{\partial}{\partial \overline{z}}(f(z)g(z)) = f(z)\frac{\partial g}{\partial \overline{z}}(z). \tag{4.2.21}$$

The function (4.2.22) which appears in the following exercise has been also studied in Exercise 4.2.5.

Exercise 4.2.21. *Compute $\dfrac{\partial f}{\partial z}$ and $\dfrac{\partial f}{\partial \bar{z}}$ for the function*

$$f(z) = \begin{cases} \dfrac{\bar{z}^2}{z}, & \text{if } z \neq 0, \\ 0, & \text{if } z = 0. \end{cases} \tag{4.2.22}$$

Exercise 4.2.22.

(a) *Compute*

$$\frac{\partial |z|^n}{\partial z} \quad and \quad \frac{\partial |z|^n}{\partial \bar{z}}, \quad n = 1, 2, \ldots.$$

(b) *Compute*

$$\frac{\partial \ln |z|}{\partial z} \quad and \quad \frac{\partial \ln |z|}{\partial \bar{z}}.$$

It is well known that $1/z$ has no primitive in $\mathbb{C} \setminus \{0\}$; see Exercise 5.2.3 below. On the other hand from the previous exercise we see that

$$2\frac{\partial \ln |z|}{\partial z} = \frac{1}{z}.$$

The result in the following exercise is used in the computation of the Chern class of a complex line bundle of a compact Riemann surface; see for instance [111, pp. 101–102].

Exercise 4.2.23. *Let Ω be an open subset of $\mathbb{C}$, and let f, g be functions taking strictly positive values in Ω, and with real and imaginary parts having continuous second derivatives there. Let furthermore h be analytic and non-vanishing in Ω, and assume that*

$$f(z) = |h(z)|^2 g(z), \quad z \in \Omega. \tag{4.2.23}$$

Then,

$$\Delta \ln f(z) = \Delta \ln g(z), \quad z \in \Omega. \tag{4.2.24}$$

The following question is adapted from [163, Proposition 3, p. 22].

Question 4.2.24. *Let $f \in C^\infty(\overline{\mathbb{D}})$ be such that f vanishes on the unit circle. Show that there is $u \in C^\infty(\overline{\mathbb{D}})$ such that*

$$\frac{\partial u}{\partial \bar{z}} = f.$$

For $q \in (0,1)$ and f defined in a neighborhood Ω of the origin, and $\mathbb{C}$-differentiable at the origin we define (see for instance [216, p. 4747]):

$$(R_q f)(z) = \begin{cases} \frac{f(z)-f(qz)}{z(1-q)}, & z \in \Omega \setminus \{0\}, \\ f'(0), & z = 0. \end{cases} \qquad (4.2.25)$$

For $q = 0$ we have the celebrated backward-shift operator

$$(R_0 f)(z) = \begin{cases} \frac{f(z)-f(0)}{z}, & z \in \Omega \setminus \{0\}, \\ f'(0), & z = 0, \end{cases} \qquad (4.2.26)$$

while $q = 1$ corresponds to $R_1 f(z) = f'(z)$.

Exercise 4.2.25. *Solve the equation*

$$R_q f = \lambda f, \qquad (4.2.27)$$

for $\lambda \in \mathbb{C}$.

4.3   Various counterexamples

The Cauchy–Riemann equations at a point do not imply differentiability at the given point. The next three exercises illustrate this phenomenon. In all three exercises, the point under consideration is $z = 0$. The first example is not very strong since the function is not even continuous at the point where the Cauchy–Riemann equations hold. The second is a bit more involved. The function is continuous, but all radial derivatives at the origin (the point where the Cauchy–Riemann equations hold)

$$\lim_{\epsilon \to 0} \frac{f(\epsilon e^{i\theta}) - f(0)}{\epsilon e^{i\theta}}, \quad \text{where } \theta \text{ is fixed,}$$

exist, but depend on the angle θ. In the last exercise, the function is continuous, all radial derivatives are the same at the origin, but the function still is not differentiable.

Without the continuity hypothesis, differentiability in an open set is not insured even if one assumes that the Cauchy–Riemann equations hold in the given open set. Still, it is well to recall the not-so-well-known Looman–Menchoff theorem (see, e.g., [164, Chapter I, §6]). If f is continuous in the open set Ω and if the Cauchy–Riemann equations hold there, then f is analytic in Ω. One can weaken the continuity hypothesis, and merely assume that f is bounded in Ω. This is a result, stated by Montel in 1913, and proved by G.P. Tolstov.

In the first example below, due to Montel, the Cauchy–Riemann equations hold in the whole plane, but the function f is not bounded, and this shows that the above-mentioned result of Montel does not hold without the hypothesis of boundedness. See the discussion in the introduction of [215] for more information. The paper [108], with the suggestive title *When is a function that satisfies the Cauchy–Riemann equations analytic*, is also recommended.

Exercise 4.3.1 (see [91, p. 66]). *Let*

$$f(z) = \begin{cases} \exp\left(-\dfrac{1}{z^4}\right), & \text{if } z \neq 0, \\ 0, & \text{if } z = 0. \end{cases}$$

(a) *Where does $f(z)$ have a derivative?*

(b) *Show that the Cauchy–Riemann equations hold in $\mathbb{C}$.*

For the next exercise, see [53, Exercise 1, p. 50].

Exercise 4.3.2. *Let*

$$f(z) = \begin{cases} \dfrac{x^3 y(y - ix)}{x^6 + y^2}, & \text{if } z \neq 0, \\ 0, & \text{if } z = 0. \end{cases}$$

(a) *Show that f is continuous at the origin.*

(b) *Show that $\lim_{z \to 0} \dfrac{f(z) - f(0)}{z}$ exists along any fixed direction, that all these limits are equal to 0, but that f is not differentiable at the origin.*

(c) *Show that the Cauchy–Riemann equations hold at the origin.*

After these exercises, the second item in the following question is quite easy (see [138, Exercise 1, p. 12]).

Exercise 4.3.3.

(a) *Can a function of a complex variable which is defined and continuous for $|z| < 1$ be such that it is only differentiable at the origin?*

(b) *Can a function which is continuous in a region have a derivative only along certain lines of that region?*

4.4 Analytic functions

Recall that a continuous function f which is defined in an open set $\Omega \subset \Omega$ is called *analytic* if it admits at every point of Ω a power series expansion: For every $z_0 \in \Omega$ there exists $R > 0$ such that

$$f(z) = \sum_{n=0}^{\infty} a_n (z - z_0)^n, \quad \forall z \in B(z_0, R), \tag{4.4.1}$$

where $B(z_0, R)$ is defined by (15.4.1). The coefficients a_n of course depend on z_0. The largest R for which (4.4.1) holds is called the radius of convergence of the power series, and is given by the formula (3.4.3):

$$R = \frac{1}{\limsup_{n \to \infty} |a_n|^{1/n}}.$$

Theorem 4.4.1. *A power series is differentiable in its open disk of convergence. The derivative of f defined by (4.4.1) is given by*

$$f'(z) = \sum_{n=1}^{\infty} n a_n (z - z_0)^{n-1}. \tag{4.4.2}$$

Hence analytic functions are holomorphic. The converse is also true. This is one of the keystones of the theory. See Chapter 5.

Formula (4.4.2) can be iterated, and we have:

$$f^{(p)}(z) = \sum_{n=p}^{\infty} n(n-1) \cdots (n-p+1) a_n (z - z_0)^{n-p}. \tag{4.4.3}$$

The simplest such power series is maybe the geometric series

$$\frac{1}{1-z} = \sum_{n=0}^{\infty} z^n, \quad |z| < 1,$$

for which $R = 1$. Its derivative is thus given by the formula

$$\frac{1}{(1-z)^2} = \sum_{n=1}^{\infty} n z^{n-1}, \quad |z| < 1. \tag{4.4.4}$$

Exercise 4.4.2. *Show that*

$$\sum_{n=1}^{\infty} \frac{n}{2^n} = 2. \tag{4.4.5}$$

Exercise 4.4.3. *Show that*

$$\sum_{n=2}^{\infty} n(n-1) z^{n-2} = \frac{2}{(1-z)^3}, \quad |z| < 1, \tag{4.4.6}$$

and find a closed-form formula for the sum

$$\sum_{n=2}^{\infty} n(n-1) r^{n-2} \cos((n-2)\theta), \quad \theta \in \mathbb{R}, \quad r \in [0, 1). \tag{4.4.7}$$

We set for $\alpha \in \mathbb{C}$

$$\binom{\alpha}{n} = \frac{\alpha(\alpha-1)\cdots(\alpha-n+1)}{n!} \tag{4.4.8}$$

For $\alpha = \pm\frac{1}{2}$, see (4.5.7) below.

Exercise 4.4.4. *Define for $\alpha \in \mathbb{C}$,*

$$f_\alpha(z) = 1 + \sum_{n=1}^{\infty} \binom{\alpha}{n} z^n. \tag{4.4.9}$$

Show that f_α is analytic in the open unit disk (and is a polynomial when α is a natural number) and that the following formulas hold:

$$f'_\alpha(z) = \frac{\alpha f_\alpha(z)}{1+z}, \tag{4.4.10}$$

$$f_\alpha(z)f_\beta(z) = f_{\alpha+\beta}(z), \quad |z| < 1.$$

Remark 4.4.5. The function $f_{1/2}$ is used in particular in Exercises 4.4.8 and 7.3.13, where it is needed to construct an analytic square root of $z^2 - 4z$ in $|z| > 4$.

We also remark the following: for $\nu > -1$ the coefficients of the power series

$$(1-z)^{-\nu-1} = 1 + \sum_{n=1}^{\infty} \frac{(1+\nu)\cdots(n+\nu)}{n!} z^n$$

are positive, and hence the function

$$\frac{1}{(1-z\overline{w})^{\nu+1}} \tag{4.4.11}$$

is positive definite in the open unit disk. One can also get to this conclusion by replacing z and w in (3.1.12) by $\frac{1-z}{1+z}$ and $\frac{1-w}{1+w}$ respectively.

Exercise 4.4.6. *Let*

$$J_0(z) = \sum_{p=0}^{\infty} \frac{(-1)^p}{(p!)^2} \left(\frac{z}{2}\right)^{2p}. \tag{4.4.12}$$

Show that J_0 is an entire function. Show that

$$J_0(z) = \frac{1}{2\pi} \int_{[0,2\pi]} e^{iz\cos u} du. \tag{4.4.13}$$

The function J_0 is the Bessel function of order 0. It is one of the solutions of the differential equation (called Bessel's equation of order 0)

$$x^2 y^{(2)}(x) + xy^{(1)}(x) + x^2 y(x) = 0.$$

We also note that the function $J_0(z)$ is positive definite (see Definition 16.3.11) since

$$J_0(z - \overline{w}) = \frac{1}{2\pi} \int_0^{2\pi} e^{iz\cos u} \overline{e^{iw\cos u}} \, du.$$

In connection with the following two exercises, see Exercise 7.2.20. The first one, Exercise 4.4.7, is just a rewriting of Proposition 3.4.8, to which we send back the reader if the proof has been skipped in a first reading.

Exercise 4.4.7. *Let* $f(z) = \sum_{n=0}^{\infty} a_n z^n$ *and* $g(z) = \sum_{n=0}^{\infty} b_n z^n$ *be two convergent power series with radiuses of convergence* R_1 *and* R_2 *respectively.*

(a) *Show that the product* fg *is a convergent power series*

$$(fg)(z) = \sum_{n=0}^{\infty} c_n z^n$$

with radius of convergence bigger or equal to $\text{Min} \, (R_1, R_2)$.

(b) *Show that*

$$c_n = \sum_{\ell=0}^{n} a_\ell b_{n-\ell}, \quad n = 0, 1, \dots. \tag{4.4.14}$$

The sequence $(c_n)_{n \in \mathbb{N}}$ above is called the convolution, or Cauchy product, of the sequences $(a_n)_{n \in \mathbb{N}}$ and $(b_n)_{n \in \mathbb{N}}$. The convolution of two sequences plays an important role in the theory of discrete signals. See Section 11.4 for more details.

The following exercise is taken from [126, pp. 18–19]. The number a_n is equal to the number of possible n-ary compositions of n elements for a non associative binary law.

Exercise 4.4.8. *Let* $a_1, a_2, \dots,$ *be a sequence of numbers satisfying* $a_1 = 1$ *and*

$$a_n = \sum_{k=1}^{n-1} a_k a_{n-1-k}. \tag{4.4.15}$$

Compute $\sum_{n=1}^{\infty} a_n z^n$ *and show that*

$$a_n = \frac{(2n-1)!}{n!(n-1)!}.$$

Exercise 4.4.9. *Given a sequence* $a_\ell, \ell = 0, 1, 2, \dots$ *of complex numbers such that* $\limsup |a_{\ell+1}|^{\frac{1}{\ell+1}} < 1$, *let* $S_n = \sum_{\ell=0}^{n} a_\ell$. *Show that*

$$\sum_{n=0}^{\infty} S_n z^n = \frac{1}{1-z} \sum_{\ell=0}^{\infty} a_\ell z^\ell, \quad |z| < 1.$$

Exercise 4.4.10. *Let* $f(z) = \sum_{n=0}^{\infty} \frac{a_n}{n!} z^n$ *and* $g(z) = \sum_{n=0}^{\infty} \frac{b_n}{n!} z^n$ *be two convergent power series with radiuses of convergence* R_1 *and* R_2 *respectively. Show that the product* fg *(which is a convergent power series by Exercise 4.4.7) can be written as*

$$(fg)(z) = \sum_{n=0}^{\infty} \frac{c_n}{n!} z^n$$

where

$$c_n = \sum_{\ell=0}^{n} \binom{n}{\ell} a_\ell b_{n-\ell}, \quad n = 0, 1, \dots.$$

Exercise 4.4.11 (Fibonacci numbers). *Let* a_n, $n = 0, 1, \dots$ *be defined by*

$$
\begin{aligned}
a_0 &= 1, \\
a_1 &= 1, \\
a_{n+2} &= a_{n+1} + a_n.
\end{aligned}
\tag{4.4.16}
$$

Show that

$$\sum_{n=0}^{\infty} a_n z^n = \frac{1}{1 - z - z^2}, \quad |z| < \frac{\sqrt{5} - 1}{2}.$$

Remark 4.4.12. The coefficients a_n in the previous exercise are called the Fibonacci numbers, and can be computed as follows. Let $z_- = \frac{-1-\sqrt{5}}{2}$ and $z_+ = \frac{-1+\sqrt{5}}{2}$ be the zeroes of the polynomial $1 - z - z^2$, and write

$$
\begin{aligned}
\frac{1}{1 - z - z^2} &= \frac{1}{z_- - z_+} \left(\frac{1}{z - z_+} - \frac{1}{z - z_-} \right) \\
&= \frac{1}{z_+ - z_-} \left(\frac{1}{z_+(1 - z/z_+)} - \frac{1}{z_-(1 - z/z_-)} \right) \\
&= \frac{1}{z_+ - z_-} \sum_{n=0}^{\infty} z^n \left(z_+^{-n-1} - z_-^{-n-1} \right) \\
&= \frac{1}{z_+ - z_-} \sum_{n=0}^{\infty} z^n z_+^{-n-1} z_-^{-n-1} \left(z_-^{n+1} - z_+^{n+1} \right) \\
&= \sum_{n=0}^{\infty} z^n (-1)^n \frac{z_+^{n+1} - z_-^{n+1}}{z_+ - z_-},
\end{aligned}
$$

since $z_- z_+ = -1$. It follows that

$$a_n = (-1)^n \frac{z_+^{n+1} - z_-^{n+1}}{z_+ - z_-}, \quad n = 0, 1, \dots.$$

Equations (4.4.16) can also be checked directly from this explicit expression.

Exercise 4.4.13. *Define* $F(z) = -\sum_{n=1}^{\infty}(-1)^n \dfrac{z^n}{n}.$

(1) *Show that F is analytic in the open unit disk and that*

$$\exp F(z) = 1 + z, \quad \text{for} \quad |z| < 1.$$

(2) *The function $F(z)$ is denoted by $\ln(1+z)$. Show that*

$$|\ln(1+z) - z| \le |z|^2, \quad |z| \le \frac{1}{2}, \tag{4.4.17}$$

and that, in particular (see [4, p. 192]),

$$\frac{|z|}{2} \le |\ln(1+z)| \le \frac{3|z|}{2}, \quad |z| \le \frac{1}{2}. \tag{4.4.18}$$

The function $F(z)$ is the analytic extension to the open unit disk of the function $\ln(1+x)$ of calculus defined on $(-1,1)$ by the same power series (and hence the notation $\ln(1+z)$). For another proof of Exercise 4.4.13 using analytic continuation, see Exercise 6.3.1.

We can now continue the discussion on infinite products began in Section 3.7, and mention the following result (see [4, p. 191]):

Theorem 4.4.14 ([4, p. 191]). *Let $\ln(1+z)$ be defined as in Exercise 4.4.13, and let $(a_n)_{n \in \mathbb{N}}$ be a sequence of complex numbers, all different from -1. The infinite product $\prod_{n=1}^{\infty}(1+a_n)$ converges if and only if the series $\sum_{n=N_0}^{\infty} \ln(1+a_n)$ converges, where N_0 is such that $|a_n| < \frac{1}{2}$ for $n \ge N_0$.*

For an illustration of this theorem for real a_n's, see Exercise 3.7.4 above. The proof of the theorem may be outlined as follows (see [4, p. 191]): Since the condition $\lim_{n \to \infty} a_n = 0$ is necessary for the infinite product to converge, we may assume that $|a_n| \le \frac{1}{2}$ for all n. This assumption will allow us to use (4.4.18). Assume first that the infinite product converges to $\ell \ne 0$, and set $\ell = e^s$ (of course, s is defined up to an additive multiple of $2\pi i$). Write $1 + a_n = e^{\ln(1+a_n)}$, where the function $\ln(1+z)$ is defined as in Exercise 4.4.13, and $z_N = -s + \sum_{n=1}^{N} \ln(1+a_n)$. We have

$$e^{z_N} = \frac{\prod_{n=1}^{N}(1+a_n)}{\ell}.$$

Hence there exists a sequence of integers $k_1, k_2, \ldots$ such that

$$\lim_{N \to \infty} |z_N - 2\pi i k_N| = 0.$$

Thus there exists M such that:

$$N \ge M \implies |z_N - 2\pi i k_N| \le \frac{1}{4}.$$

In particular,

$$|z_N - z_{N+1} - 2\pi i(k_N - k_{N+1})| \leq \frac{1}{2}, \quad N \geq M,$$

that is

$$|\ln(1 + a_{N+1}) + 2\pi i(k_N - k_{N+1})| \leq \frac{1}{2}, \quad N \geq M,$$

and so

$$||\ln(1 + a_{N+1})| - 2\pi|k_N - k_{N+1}|| \leq \frac{1}{2}, \quad N \geq M,$$

Assuming $k_N \neq k_{N+1}$ we have, using (4.4.18),

$$2\pi|k_N - k_{N+1}| - \frac{3|a_{N+1}|}{2} \leq \frac{1}{2}, \quad N \geq M.$$

Hence (since $|a_{N+1}| \leq \frac{1}{2}$)

$$2\pi|k_N - k_{N+1}| \leq \frac{3}{4} + \frac{1}{2},$$

which is impossible. Hence $k_M = k_N$ for $N \geq M$. If follows that $\lim_{N \to \infty} z_N$ exists, and this is equivalent to the convergence of the series $\sum_{n=N_0}^{\infty} \ln(1 + a_n)$.

The converse statement is trivial.

Question 4.4.15. *Prove directly (without Theorem 4.4.14) that the infinite product (3.7.1):*

$$\prod_{k=1}^{\infty}\left(1 + \frac{i}{k}\right)$$

diverges.

Exercise 4.4.16. (adapted from [202, p. 484]). *Show that*

$$\frac{\ln(1 - z)}{1 - z} = -\sum_{n=1}^{\infty}\left(\sum_{\ell=1}^{n}\frac{1}{\ell}\right)z^n, \quad |z| < 1.$$

Is there a continuous function ψ such that

$$\frac{\ln(1 - z)}{1 - z} = \int_0^{2\pi} \frac{\psi(e^{it})dt}{e^{it} - z}, \quad |z| < 1.$$

Exercise 4.4.17. *Let $F(z) = \sum_{n=0}^{\infty} a_n z^n$ be a convergent power series with convergence radius greater than or equal to 1. Show that*

$$\int_0^1 F'(tz)dt = \begin{cases} \dfrac{F(z) - F(0)}{z}, & \text{for } z \neq 0, \\ F'(0), & \text{for } z = 0. \end{cases}$$

Exercise 4.4.18. *Let* $F(z) = \sum_{n=0}^{\infty} a_n z^n$ *be a power series with radius of convergence* R *(R may be infinite). Let* $a > 0$*, and let* f *be a continuous function from* $[0, a]$ *into* $\mathbb{C}$*. Where is the function*

$$G(z) = \int_0^a F(zt)f(t)dt$$

analytic? Compute its derivative, both as power series and in closed form.

Same question for $H(z) = \int_1^a F(z/t)f(t)dt$*, where now* $a > 1$ *and* f *is continuous on* $[1, a]$*.*

Exercise 4.4.19. *Let* m *be a complex-valued continuous function in the interval* $[0, 1]$*. Show that the function*

$$f(z) = \int_0^1 e^{izt}m(t)dt \tag{4.4.19}$$

is entire by computing its power series expansion centered at the origin, and compute its derivative.

In connection with the previous exercise, see Exercises 3.4.13 and 4.2.14.

One way to solve the following exercise would be to use Theorem 6.2.3. Here, we suggest a simpler, and more effective way: Show that φ has a power series expansion centered at the origin, with radius of convergence equal to 1, and show a symmetry property satisfied by $\varphi(z)$; see (4.5.9).

Exercise 4.4.20. *Let* $t \mapsto m(t)$ *be a complex-valued continuous function on the closed interval* $[0, 2\pi]$*. Show that the function*

$$\varphi(z) = \int_0^{2\pi} \frac{e^{it} + z}{e^{it} - z} m(t)dt \tag{4.4.20}$$

is analytic in $\mathbb{C} \setminus \mathbb{T}$*.*

Food for thought. Here are some related additional questions.

(1) Show that

$$\varphi'(z) = \int_0^{2\pi} \frac{2e^{it}}{(e^{it} - z)^2} m(t)dt, \quad \text{for} \quad |z| \neq 1.$$

(2) Assume that m vanishes on some closed subinterval $I \subset [0, 2\pi]$. Then, the function φ is analytic across the interior of the corresponding arc of circle.

(3) Assume that the function m is real-valued and that moreover it takes positive values: $m(t) \geq 0$ on $[0, 2\pi]$. Show that

$$\operatorname{Re} \varphi(z) \geq 0, \quad \text{for} \quad |z| < 1.$$

See also Exercise 6.5.12 in Section 6.5.

Functions φ of the form (4.4.20) with a positive $m(t)$ are called Carathéodory functions. They play an important role in the prediction of stationary second-order processes. Indeed, let $(x_n)_{n\in\mathbb{N}_0}$ be a, say real-valued, second-order stationary process with covariance function

$$E(x_n x_m) = r(n - m), \quad n, m \in \mathbb{N}_0,$$

where E denotes expectation in the probability space. Then, by the Cauchy–Schwarz inequality,

$$|r(n - m)| \leq \sqrt{r(0)}\sqrt{r(0)} = r(0),$$

(see (0.0.3) for the analogous fact in the case of processes indexed by $\mathbb{R}$), and the function

$$\varphi(z) = r(0) + 2 \sum_{n=1}^{\infty} z^n r(n) \tag{4.4.21}$$

is analytic in the open unit disk, and has a real positive part there. To check this, we apply Theorem 14.3.1 to check that the family $(z^n \overline{w}^m r(n - m))_{n,m\in\mathbb{N}_0}$ is absolutely summable for fixed $z, w \in \mathbb{D}$. That same theorem allows us to write (see (14.3.1))

$$\sum_{n,m=0}^{\infty} z^n \overline{w}^m r(n - m) = \frac{r(0)}{1 - z\overline{w}} + \frac{r(1)z}{1 - z\overline{w}} + \cdots$$

$$+ \frac{r(-1)\overline{w}}{1 - z\overline{w}} + \frac{r(-2)\overline{w}^2}{1 - z\overline{w}} + \cdots \tag{4.4.22}$$

$$= \frac{\varphi(z) + \overline{\varphi(w)}}{2(1 - z\overline{w})},$$

where one uses that $r(-n) = r(n)$ (or, more generally, $r(-n) = \overline{r(n)}$ when the process is complex-valued). By the Herglotz representation theorem (see the discussion after Exercise 5.5.10 for the statement), φ is of the form (4.4.20) (but with a positive finite measure $d\mu$ rather than a positive function $m(t)$ in general). See [6] for a survey. Note that (4.4.21) and (4.4.22) are the discrete counterparts of (0.0.4) and (0.0.5).

Exercise 4.4.21. *Let $m(t)$ be a complex-valued continuous function defined in the closed interval $[0, 1]$. Show that the function*

$$F(z) = \int_0^1 \frac{m(t)dt}{(t - z)^2} \tag{4.4.23}$$

is analytic in $\mathbb{C} \setminus [0, 1]$ and that

$$F'(z) = \int_0^1 \frac{2m(t)dt}{(t - z)^3}.$$

The *Laurent expansion* of the function (4.4.23) in $|z| > 1$ is considered in Exercise 7.1.14.

Exercise 4.4.22. *Prove* (3.7.15).

Hint. Compute the derivative of the function $E_p(z)$.

Exercise 4.4.23. *Show that the infinite product*

$$\prod_{n=1}^{\infty} \left(1 - \frac{z}{\sqrt{n}}\right) e^{\frac{z}{\sqrt{n}} + \frac{z^2}{2n}} \tag{4.4.24}$$

converges to a function which vanishes at the points $z = \sqrt{n}$, $n = 1, 2, \ldots$, *and only at these points.*

4.5 Solutions

Solution of Exercise 4.1.7. The function f is continuous at the origin on the real axis, since

$$f(x) = \begin{cases} 0, & \text{if } x \geq 0, \\ \frac{-x}{\pi}, & \text{if } x < 0. \end{cases}$$

Let us now consider another straight line, different from the real axis, and passing through the origin. Its equation is

$$z = te^{i\theta_0}, \quad t \in \mathbb{R},$$

where $\theta_0 \in (0, \pi)$ is fixed, and different from 0 and π. For $t < 0$ we can write

$$te^{i\theta_0} = (-t)e^{i(-\pi + \theta_0)},$$

and $\pi + \theta_0 \in (-\pi, 0)$. Hence, we have

$$f(z) = \begin{cases} \frac{t}{\theta_0}, & \text{if } t > 0, \\ \frac{-t}{\theta_0 - \pi}, & \text{if } t < 0. \end{cases}$$

Hence, f is continuous at the origin on the given line.

To show that f is not continuous at the origin, it suffices to take, for $n \geq 1$,

$$z_n = \frac{e^{\frac{i}{n}}}{n}.$$

Then

$$\lim_{n \to \infty} z_n = 0 \quad \text{while} \quad f(z_n) = 1. \qquad \square$$

Solution of Exercise 4.1.8. (a) Let $z \in \mathbb{C} \setminus [0, 1]$, and let d be the distance from z to $[0, 1]$:

$$d = \min_{t \in [0,1]} |t - z|.$$

For $h \in \mathbb{C}$ such that $|h| < d/2$ we have

$$\frac{1}{|t - z - h|} \leq \frac{1}{|t - z| - |h|} < \frac{1}{d - d/2} = \frac{2}{d}. \tag{4.5.1}$$

Thus, for z and h as above, we have

$$F(z + h) - F(z) = \int_0^1 \frac{h\,dt}{(t - z)(t - z - h)}.$$

Inequalities (3.1.4) and (4.5.1) lead to

$$|F(z + h) - F(z)| \leq |h| \int_0^1 \frac{2\,dt}{d^2} = \frac{2|h|}{d^2},$$

and the continuity of F at the point z follows.

(b) Let $s \in (0, 1)$ and $\epsilon > 0$. We have

$$\operatorname{Re} F(s + i\epsilon) = \int_0^1 \frac{(t - s)\,dt}{(t - s)^2 + \epsilon^2} \quad \text{and} \quad \operatorname{Im} F(s + i\epsilon) = \int_0^1 \frac{\epsilon\,dt}{(t - s)^2 + \epsilon^2}.$$

These integrals are easily evaluated with the change of variable $u = (t - s)/\epsilon$:

$$\int_0^1 \frac{(t - s)\,dt}{(t - s)^2 + \epsilon^2} = \int_{-\frac{s}{\epsilon}}^{\frac{1-s}{\epsilon}} \frac{u\,du}{u^2 + 1} = \frac{1}{2} \ln\left(\frac{(1 - s)^2 + \epsilon^2}{s^2 + \epsilon^2}\right),$$

$$\int_0^1 \frac{\epsilon\,dt}{(t - s)^2 + \epsilon^2} = \int_{-\frac{s}{\epsilon}}^{\frac{1-s}{\epsilon}} \frac{du}{u^2 + 1} = \arctan\left(\frac{1 - s}{\epsilon}\right) + \arctan\left(\frac{s}{\epsilon}\right).$$

The same computations hold for the real and imaginary parts of $F(s - i\epsilon)$, and we obtain

$$F_+(s) = \ln\left|\frac{1 - s}{s}\right| + i\pi,$$

$$F_-(s) = \ln\left|\frac{1 - s}{s}\right| - i\pi.$$

(c) We show that the first limit does not exist. The other one is treated in the same way. First let h be real and strictly negative. We have for $z = re^{i\theta}$, where $r > 0$ and $\theta \in (0, 2\pi)$,

$$\operatorname{Re} F(z) = \sin\theta \int_0^1 \frac{r\,dt}{(t - r\cos\theta)^2 + r^2 \sin^2\theta}$$

$$= \sin\theta \int_0^{1/r} \frac{du}{u^2 - 2u\cos\theta + 1} \quad \text{(change of variable } t = ru\text{)}.$$

Thus, for fixed $\theta \in (0, 2\pi)$,

$$\lim_{r \to 0} F(re^{i\theta}) = \sin\theta \int_0^\infty \frac{du}{u^2 - 2u\cos\theta + 1}. \tag{4.5.2}$$

In Exercise 8.3.4 we will compute the integral appearing in (4.5.2) for every $\theta \in (0, 2\pi)$ using residues, and it will follow from (8.6.15) that

$$\sin\theta \int_0^\infty \frac{du}{u^2 - 2u\cos\theta + 1} = \theta, \quad \theta \in (0, 2\pi).$$

Here it suffices to note that this integral is equal to 1 if $\theta = \pi$ and to $\pi/2$ for $\theta = \pi/2$ and hence the corresponding limits in (4.5.2) are then equal to 0 and $\pi/2$. Hence the first limit in (4.1.4) does not exist. As already mentioned, the second limit is treated in much the same way. □

Solution of Exercise 4.1.9. Assume by contradiction that there is such a function f, and consider $f_0(z) = \sqrt{\rho}e^{i\frac{\theta}{2}}$ in $\mathbb{C} \setminus \mathbb{R}_-$. In $\mathbb{C} \setminus \mathbb{R}_-$ we have $(f(z)/f_0(z))^2 = 1$, so that the range of f/f_0 is included in $\{-1, +1\}$. The continuous image of a connected set is connected, see Theorem 15.1.2, and thus a continuous function which takes a discrete set of values on a connected set is constant. Thus $f(z) \equiv f_0(z)$ or $f(z) \equiv -f_0(z)$ in $\mathbb{C} \setminus \mathbb{R}_-$. So f_0 would be equal in $\mathbb{C} \setminus \mathbb{R}_-$ to a function which is continuous on $\mathbb{C} \setminus \{0\}$. This cannot be since f_0 is not continuous across the negative axis. □

Solution of Exercise 4.1.10. (see also the remark after the proof).

(a) We assume by contradiction that such a function exists. Set in (4.1.5) first $z = w = 1$ and then $z = w = -1$. We get on the one hand

$$f(1) = f(1)^2 \quad \text{and} \quad f(1)^2 = 1,$$

so that $f(1) = f(1)^2 = 1$. On the other hand

$$f(1) = f(-1)^2 \quad \text{and} \quad f(-1)^2 = -1,$$

so that $f(1) = f(-1)^2 = -1$, which leads to a contradiction.

(b) We also proceed by contradiction. Assume that f exists. Let z_0 be in $\mathbb{C} \setminus \{0\}$ and define $g(z) = \frac{f(z_0 z)}{f(z)f(z_0)}$. The function g is continuous in $\mathbb{C} \setminus \{0\}$ since f is continuous there and since f does not vanish on $\mathbb{C} \setminus \{0\}$ (recall that $f(z)^2 = z$). Since

$$g(z)^2 = \frac{f(z_0 z)^2}{f(z_0)^2 f(z)^2} = \frac{z_0 z}{z_0 z} = 1$$

we get that $g(z)$ takes only the values 1 and -1. The open set $\mathbb{C} \setminus \{0\}$ is arc-connected (see for instance Exercise 15.1.8), and therefore connected (see Lemma

15.4.7). Since g is continuous and since the image of a connected set under a continuous map is connected (see Theorem 15.1.2) we have

$$g(z) \equiv 1 \quad \text{or} \quad g(z) \equiv -1.$$

In the first case $g(z) \equiv 1$ implies that $f(z)f(z_0) = f(zz_0)$; since $f(z)^2 = z$, we obtain a contradiction with (a). In the second case it suffices to replace f by $-f$ to obtain a contradiction.

(c) Assume that there is a function $h(z)$ continuous on $\mathbb{C} \setminus \{0\}$ such that $\exp(h(z)) = z$ for all $z \in \mathbb{C} \setminus \{0\}$. Set $g(z) = \exp(h(z)/2)$. Then $h(z)$ is continuous on $\mathbb{C} \setminus \{0\}$ and $h(z)^2 = z$, a contradiction with (b).

(d) The function

$$\ln z \stackrel{\text{def.}}{=} \ln \rho + i\theta,$$

where $z = \rho \exp(i\theta)$ and where $-\pi < \theta < \pi$, is continuous (in fact differentiable, with derivative $1/z$) and satisfies $\exp(\ln z) = z$ in $\mathbb{C} \setminus \{0\}$. The difference between (c) and (d) is that the set $\mathbb{C} \setminus \{0\}$ is not simply-connected whereas $\mathbb{C} \setminus \mathbb{R}_-$ is simply-connected.   □

We have divided the previous proof into two steps to help the student, but a slightly quicker proof, based on the same arguments, goes as follows: Assume a continuous function exists in $\mathbb{C} \setminus \{0\}$ such that $f(z)^2 = z$ (it can be extended continuously to $z = 0$, but we will not use this). Then,

$$(f(z^2))^2 = z^2, \quad \forall z \in \mathbb{C} \setminus \{0\}.$$

Thus

$$\frac{f(z^2)}{z} = \pm 1, \quad \forall z \in \mathbb{C} \setminus \{0\}.$$

Using that $\frac{f(z^2)}{z}$ is continuous on the connected set $\mathbb{C} \setminus \{0\}$ we have

$$f(z^2) \equiv z \quad \text{or} \quad f(z^2) \equiv -z.$$

Taking $z = \pm 1$ leads then to a contradiction.

Solution of Exercise 4.1.13. The set $\mathcal{Z}(a)$ of points in Ω where a vanishes is closed since a is assumed continuous. The function

$$\sigma(z) = \frac{b(z)}{a(z)}$$

is continuous in the open set $\Omega \setminus \mathcal{Z}(a)$, and therefore

$$\Omega_+ = \sigma^{-1}(\mathbb{D})$$

is open in $\Omega \setminus \mathcal{Z}(a)$, and therefore in Ω since Ω is open. A similar argument shows that Ω_- is open. If Ω_0 is empty, we will have

$$\Omega = \Omega_+ \cup \Omega_-,$$

which contradicts the connectedness of Ω.   □

Solution of Exercise 4.2.4. That the function

$$e^z = e^x \cos y + ie^x \sin y$$

satisfies the asserted conditions is clear, and is also checked in item (c) of Exercise 4.2.5. We consider the converse statement, and set $f(z) = u(x, y) + iv(x, y)$ to be a function which admits a derivative at every point in $\mathbb{C}$ and such that $f'(z) = f(z)$ and $f(0) = 1$. Formula (4.2.6) for the derivative implies that

$$\frac{\partial u}{\partial x}(x, y) = u(x, y),$$

$$-\frac{\partial u}{\partial x}(x, y) = v(x, y).$$

In particular, $u(x, y) = e^x c(y)$ and (taking into account the second Cauchy–Riemann equation) $v(x, y) = e^x s(y)$, for some functions c and s of the variable y. These functions admit first-order derivatives since u and v are differentiable, and the Cauchy–Riemann equations lead to

$$c'(y) = s(y) \quad \text{and} \quad s'(y) = -c(y).$$

It follows that $c(y) = \cos y$ and $s(y) = \sin y$ and hence the result. □

Solution of Exercise 4.2.5. (a) We have $f(z) = u(x, y) + iv(x, y)$ with

$$u(x, y) = x^2 + y^2 \quad \text{and} \quad v(x, y) = 2xy.$$

The functions $u(x, y)$ and $v(x, y)$ are polynomials and in particular differentiable (in the sense of functions of two real variables) everywhere and so the Cauchy–Riemann equations give a necessary and sufficient condition (as opposed to only necessary in general) for $f(z)$ to be differentiable (in the complex variable sense). We have

$$\frac{\partial u}{\partial x}(x, y) = 2x,$$

$$\frac{\partial u}{\partial y}(x, y) = 2y,$$

$$\frac{\partial v}{\partial x}(x, y) = 2y,$$

$$\frac{\partial v}{\partial y}(x, y) = 2x.$$

Hence the Cauchy–Riemann equations are

$$2x = 2x \quad \text{and} \quad 2y = -2y.$$

Thus $y = 0$ and x is arbitrary, and the Cauchy–Riemann equations are satisfied at any real point. The function $f(z)$ is differentiable on the real line. By formula (4.2.6) the derivative is given by

$$f'(x) = 2x - i2y = 2x, \quad \text{since } y = 0.$$

(b) Here $u(x, y) = x^2$ and $v(x, y) = xy$, and as in the previous case, the existence of a derivative at a given point is equivalent to the Cauchy–Riemann equations being satisfied at that point. We have now

$$\frac{\partial u}{\partial x}(x, y) = 2x,$$

$$\frac{\partial u}{\partial y}(x, y) = 0,$$

$$\frac{\partial v}{\partial x}(x, y) = y,$$

$$\frac{\partial v}{\partial y}(x, y) = x.$$

Hence the Cauchy–Riemann equations are

$$2x = x \quad \text{and} \quad 0 = -y.$$

Thus $x = y = 0$ and the Cauchy–Riemann equations are satisfied if and only if $z = 0$. The function $f(z)$ is differentiable if and only if $z = 0$. By formula (4.2.6) we have for $z = 0$ that $f'(0) = 0$.

(c) The Cauchy–Riemann equations are satisfied at every point of $\mathbb{R}^2$, and $u(x, y) = e^x \cos y$ and $v(x, y) = e^x \sin y$ are differentiable (as real-valued functions of two real variables) in all of $\mathbb{R}^2$. Thus, e^z is $\mathbb{C}$-differentiable in $\mathbb{C}$ and, using once more formula (4.2.6) we obtain

$$(e^z)' = \frac{\partial u}{\partial x}(x, y) - i\frac{\partial u}{\partial y}(x, y)$$

$$= e^x \cos y + ie^x \sin y$$

$$= e^z.$$

(d) This is a standard counterexample. It can be found, e.g., in [49, Exercise 7, p. 48]. See also Exercise 4.2.21. We will show that:

(i) f is continuous at the origin.

(ii) The Cauchy–Riemann equations hold at the origin.

(iii) f is not differentiable at the origin.

(iv) f is not differentiable in $\mathbb{C} \setminus \{0\}$.

(i) We have for $z \neq 0$,

$$|f(z) - f(0)| = \left| \frac{\bar{z}^2}{z} - 0 \right|$$
$$= \frac{|z^2|}{|z|}$$
$$= |z|,$$

and so we get that

$$\lim_{z \to 0} |f(z) - f(0)| = \lim_{z \to 0} |z| = 0.$$

Hence f is continuous at the origin.

(ii) Write

$$f(z) = \frac{(x - iy)^2}{x + iy} = u(x, y) + iv(x, y).$$

Since

$$f(x) = x \quad \text{and} \quad f(iy) = iy$$

we obtain

$$u(x, 0) = x, \quad v(x, 0) = 0, \quad u(0, y) = 0, \quad \text{and} \quad v(0, y) = y,$$

for $x \neq 0$ and $y \neq 0$. Moreover, since $f(0) = 0$, we have $u(0, 0) = v(0, 0) = 0$. Thus

$$\lim_{x \to 0} \frac{u(x, 0) - u(0, 0)}{x} = 1 \quad \text{and} \quad \lim_{y \to 0} \frac{v(0, y) - v(0, 0)}{y} = 1.$$

Thus $\dfrac{\partial u}{\partial x}(0, 0)$ and $\dfrac{\partial v}{\partial y}(0, 0)$ exist and are equal to 1. Hence the first Cauchy–Riemann equation holds at $z = 0$. The second one is proved in the same way, and one has

$$\frac{\partial u}{\partial y}(0, 0) = -\frac{\partial v}{\partial x}(0, 0) = 0.$$

(iii) We now prove that f is not differentiable at the origin. Write $z = \epsilon e^{i\theta}$. Then,

$$\frac{f(z) - f(0)}{z} = \frac{\bar{z}^2}{z^2} = \frac{\epsilon^2 e^{-2i\theta}}{\epsilon^2 e^{2i\theta}} = e^{-4i\theta}.$$

Hence, for a fixed θ,

$$\lim_{\epsilon \to 0} \frac{f(z) - f(0)}{z} = \lim_{\epsilon \to 0} e^{-4i\theta} = e^{-4i\theta}.$$

Thus $\lim_{z \to 0} \dfrac{f(z) - f(0)}{z} = 1$ when $\theta = 0$ and $= -1$ when $\theta = \dfrac{\pi}{4}$.

So $\lim_{z \to 0} \dfrac{f(z) - f(0)}{z}$ does not exist and f is not differentiable at the origin.

(iv) One can check that the Cauchy–Riemann equations do not hold at points different from the origin. One can also note that

$$\lim_{\substack{z=\rho z_0 \\ \rho\in\mathbb{R} \\ \rho\to 1}} \frac{f(z)-f(z_0)}{z-z_0} = \frac{\overline{z_0^2}}{z_0^2},$$

while

$$\lim_{\substack{z=(1+i\epsilon)z_0 \\ \epsilon\in\mathbb{R} \\ \epsilon\to 0}} \frac{f(z)-f(z_0)}{z-z_0} = -3\frac{\overline{z_0^2}}{z_0^2}.$$

(e) The function $f(z)=\overline{z}$ is nowhere differentiable since the Cauchy–Riemann equations do not hold at any point. As pointed out in various places (see for instance [203, p. 21]), this is a very simple example of a continuous function which is nowhere $\mathbb{C}$-differentiable while the counterparts in real analysis are much more difficult to obtain. □

Solution of Exercise 4.2.6. Using the chain rule for differentiation and the fact that the Cauchy–Riemann equations hold for the pair (u,v) at the point (x_0,y_0), we obtain

$$\frac{\partial U}{\partial x}(x_0,-y_0) = \frac{\partial u}{\partial x}(x_0,y_0)$$
$$= \frac{\partial v}{\partial y}(x_0,y_0)$$
$$= \frac{\partial V}{\partial y}(x_0,-y_0)$$

and

$$\frac{\partial U}{\partial y}(x_0,-y_0) = -\frac{\partial u}{\partial y}(x_0,y_0)$$
$$= \frac{\partial v}{\partial x}(x_0,y_0)$$
$$= -\frac{\partial V}{\partial x}(x_0,-y_0).$$

Let $f(z)=u(x,y)+iv(x,y)$. When the functions u and v are differentiable at the point (x_0,y_0), the functions U and V are differentiable at the point $(x_0,-y_0)$, and the result expresses that the function $\overline{f(\overline{z})}$ is $\mathbb{C}$-differentiable at the point $z_0=x_0+iy_0$ when the function $f(z)$ is $\mathbb{C}$-differentiable at the point z_0. □

Solution of Exercise 4.2.8. Although the computation is just an application of the chain rule, we prove the first equation. We have

$$\frac{\partial \ln\sqrt{x^2+y^2}}{\partial x} = \frac{1}{2}\cdot\frac{2x}{x^2+y^2} = \frac{x}{x^2+y^2},$$

and, for $x > 0$,

$$\frac{\partial \arctan(y/x)}{\partial y} = \frac{1}{x}\frac{1}{1 + \frac{y^2}{x^2}} = \frac{x}{x^2 + y^2}.$$

Therefore the first Cauchy–Riemann equation holds when $x > 0$. The other cases are treated in the same way, and so is the second Cauchy–Riemann equation. □

Solution of Exercise 4.2.9. We will prove that the $\mathbb{C}$-derivative of $\cos z$ is $-\sin z$. Recall that

$$\cos z = \cos x \cosh y - i \sin x \sinh y.$$

(See (1.2.17).) Thus $\operatorname{Re} \cos z = \cos x \cosh y$. We use formula (4.2.6) and obtain

$$(\cos z)' = -\sin x \cosh y - i \cos x \sinh y = -\sin z,$$

where we have used (1.2.18). The other formulas are proved in the same way. □

Solution of Exercise 4.2.10. Set

$$f_n(z) = \cos(z/2) \cos(z/2^2) \cdots \cos(z/2^n).$$

If $z = 0$, then $f_n(0) = 1$ for every n and so is the limit. We now assume that $z \neq 0$. We use the identity (1.2.16) to show by induction that

$$f_n(z) = \frac{\sin z}{2^n \sin(z/2^n)}.$$

For $n = 1$ the claim is true since

$$f_1(z) = \cos(z/2) = \frac{\sin(z)}{2 \sin(z/2)}.$$

Assume the claim true at n. Then

$$\sin(z/2^{n+1}) f_{n+1}(z) = f_n(z) \sin(z/2^{n+1}) \cos(z/2^{n+1})$$
$$= f_n(z) \frac{\sin(z/2^n)}{2}$$
$$= \frac{\sin(z)}{2^n \sin(z/2^n)} \frac{\sin(z/2^n)}{2}$$
$$= \frac{\sin(z)}{2^{n+1}},$$

where we have used (1.2.16) to go from the first line to the second and the induction hypothesis to go from the second line to the third. Thus

$$f_{n+1}(z) = \frac{\sin(z)}{2^{n+1}} \frac{1}{\sin(z/2^{n+1})},$$

and the induction hypothesis is true at rank $n + 1$. Since it holds at rank $n = 1$ it holds for every positive integer.

Thus

$$\lim_{n\to\infty} f_n(z) = \lim_{n\to\infty} \frac{\sin(z)}{z} \left(\frac{\sin(z/2^{n+1})}{z/2^{n+1}} \right)^{-1}.$$

The function $\sin(z)$ is differentiable at every point z and its derivative is $\sin(z)$. Thus, by definition of the derivative,

$$\sin'(0) = \lim_{n\to\infty} \frac{\sin(z/2^{n+1})}{z/2^{n+1}},$$

and so

$$\lim_{n\to\infty} f_n(z) = \frac{\sin(z)}{z}, \quad (z \neq 0).$$

The formula also holds for $z = 0$ when one extends the function $\dfrac{\sin(z)}{z}$ to be 1 at the origin.   □

For a (more complicated) proof using wavelets and Haar systems, see [90, Exercise 5.4.3, p. 423].

Solution of Exercise 4.2.11. Since f is differentiable in Ω, the functions u and v have first-order derivatives and the Cauchy–Riemann equations hold. We can differentiate (4.2.12) with respect to x and y and obtain

$$a\frac{\partial u}{\partial x} + b\frac{\partial v}{\partial x} = 0,$$

$$a\frac{\partial u}{\partial y} + b\frac{\partial v}{\partial y} = 0.$$

These equations together with the Cauchy–Riemann equations can be written as

$$\begin{pmatrix} 1 & 0 & 0 & -1 \\ 0 & 1 & 1 & 0 \\ a & 0 & b & 0 \\ 0 & a & 0 & b \end{pmatrix} \begin{pmatrix} \dfrac{\partial u}{\partial x} \\[2mm] \dfrac{\partial u}{\partial y} \\[2mm] \dfrac{\partial v}{\partial x} \\[2mm] \dfrac{\partial v}{\partial y} \end{pmatrix} = \begin{pmatrix} 0 \\ 0 \\ 0 \\ 0 \end{pmatrix}.$$

We have

$$\det \begin{pmatrix} 1 & 0 & 0 & -1 \\ 0 & 1 & 1 & 0 \\ a & 0 & b & 0 \\ 0 & a & 0 & b \end{pmatrix} = a^2 + b^2 > 0$$

since a and b are not simultaneously equal to 0. It follows that

$$\frac{\partial u}{\partial x}(x, y) = \frac{\partial u}{\partial y}(x, y) = \frac{\partial v}{\partial x}(x, y) = \frac{\partial v}{\partial y}(x, y) = 0,$$

and u and v are constant in Ω since Ω is connected. So f is a constant function $u + iv$, with the real numbers u and v satisfying (4.2.12) (that is, the point (u, v) belongs to the straight line defined by (4.2.12)). $\qquad\square$

Solution of Exercise 4.2.12. We have, with $x - x_0 = h$ and $y - y_0 = k$,

$$u(x, y) = u(x_0, y_0) + h\frac{\partial u}{\partial x}(x_0, y_0) + k\frac{\partial u}{\partial y}(x_0, y_0) + o(x - x_0, y - y_0),$$

$$v(x, y) = v(x_0, y_0) + h\frac{\partial v}{\partial x}(x_0, y_0) + k\frac{\partial v}{\partial y}(x_0, y_0) + o(x - x_0, y - y_0),$$

where we denote by the same letter o expressions which, after division by

$$\sqrt{(x - x_0)^2 + (y - y_0)^2},$$

tend to 0 as (x, y) tends to (x_0, y_0). Since

$$h = \frac{z - z_0 + \overline{z - z_0}}{2} \quad \text{and} \quad k = \frac{z - z_0 - \overline{z - z_0}}{2i},$$

these expressions become

$$u(x, y) = u(x_0, y_0) + \frac{z - z_0 + \overline{z - z_0}}{2}\frac{\partial u}{\partial x}(x_0, y_0) + \frac{z - z_0 - \overline{z - z_0}}{2i}\frac{\partial u}{\partial y}(x_0, y_0)$$
$$+ o(x - x_0, y - y_0),$$

$$v(x, y) = v(x_0, y_0) + \frac{z - z_0 + \overline{z - z_0}}{2}\frac{\partial v}{\partial x}(x_0, y_0) + \frac{z - z_0 - \overline{z - z_0}}{2i}\frac{\partial v}{\partial y}(x_0, y_0)$$
$$+ o(x - x_0, y - y_0).$$

We obtain the result by multiplying both sides of the second equality by i and adding the first and the second equality side by side.

From (4.2.16) we obtain

$$\frac{f(z) - f(z_0)}{z - z_0} = \frac{\partial f}{\partial z}\Big|_{z=z_0} + \frac{\partial f}{\partial \bar{z}}\Big|_{z=z_0}\frac{\overline{z - z_0}}{z - z_0} + \frac{o(z - z_0)}{z - z_0},$$

and hence the set of possible limits is the circle with center $\frac{\partial f}{\partial z}\Big|_{z=z_0}$ and radius $\frac{\partial f}{\partial \bar{z}}\Big|_{z=z_0}$.

This circle reduces to a single point if and only if the function f admits a derivative at the point z_0. $\qquad\square$

Solution of Exercise 4.2.13. The result is a direct consequence of the Cauchy–Riemann equations and of (4.2.14). More precisely, assume that F admits a derivative at the point $z_0 = x_0 + iy_0$. Then,

$$\frac{\partial F}{\partial \overline{z}}\Big|_{z=z_0} = \frac{1}{2}\left(\frac{\partial}{\partial x} + i\frac{\partial}{\partial y}\right)(u + iv)\Big|_{x=x_0,y=y_0}$$

$$= \frac{1}{2}\left\{\frac{\partial u}{\partial x} + i\frac{\partial v}{\partial x} + i\frac{\partial u}{\partial y} - \frac{\partial v}{\partial y}\right\}\Big|_{x=x_0,y=y_0}$$

$$= \frac{1}{2}\left\{\frac{\partial u}{\partial x} - i\frac{\partial u}{\partial y} + i\frac{\partial u}{\partial y} - \frac{\partial u}{\partial x}\right\}\Big|_{x=x_0,y=y_0}$$

$$= 0$$

where we have used the Cauchy–Riemann equations to go from the second to the third equality. Similarly,

$$\frac{\partial F}{\partial z}\Big|_{z=z_0} = \frac{1}{2}\left(\frac{\partial}{\partial x} - i\frac{\partial}{\partial y}\right)(u + iv)\Big|_{x=x_0,y=y_0}$$

$$= \frac{1}{2}\left\{\frac{\partial u}{\partial x} + i\frac{\partial v}{\partial x} - i\frac{\partial u}{\partial y} + \frac{\partial v}{\partial y}\right\}\Big|_{x=x_0,y=y_0}$$

$$= \frac{1}{2}\left\{\frac{\partial u}{\partial x} - i\frac{\partial u}{\partial y} - i\frac{\partial u}{\partial y} + \frac{\partial u}{\partial x}\right\}\Big|_{x=x_0,y=y_0}$$

$$= \frac{\partial u}{\partial x}(x_0, y_0) - i\frac{\partial u}{\partial y}(x_0, y_0),$$

and this last expression is exactly $F'(z_0)$.   □

Solution of Exercise 4.2.14. We will show that

$$\lim_{h \to 0} \frac{F(z+h) - F(z)}{h} = \int_0^1 ite^{izt}m(t)dt.$$

We note that, for $|h| \leq 1$

$$\left|\frac{e^{ith} - 1}{h} - it\right| = \left|\frac{e^{izt} - 1 - ith}{h}\right|$$

$$= \left|\frac{\sum_2^\infty \dfrac{h^n t^n i^n}{n!}}{h}\right|$$

$$\leq |h|\sum_2^\infty \frac{|h|^{n-2}}{n!}$$

$$\leq |h|\sum_2^\infty \frac{|h|^{n-2}}{(n-2)!}$$

$$\leq |h|e^{|h|} \leq e|h|.$$

Thus, for $|h| \le 1$, and with $M = \max_{t \in [0,1]} |m(t)|$, we have

$$
\left| \frac{F(z+h) - F(z)}{h} - \int_0^1 ite^{itz} m(t)dt \right| = \left| \int_0^1 \left(\frac{e^{ith} - 1}{h} - it \right) m(t)dt \right|
$$

$$
\le \int_0^1 \left| \frac{e^{ith} - 1}{h} - it \right| |m(t)| dt
$$

$$
\le |h| eM \to 0 \quad \text{as } h \to 0. \qquad \square
$$

An alternative solution of this result uses Weierstrass' theorem on interchanging the order of summation and integration. We then have (in a way similar to (3.4.18) in Exercise 3.4.13)

$$
F(z) = \sum_{n=0}^{\infty} z^n \frac{\int_0^1 i^n t^n m(t)dt}{n!}, \qquad \forall z \in \mathbb{C}. \tag{4.5.3}
$$

A function F of the form (4.4.19) can be seen as a signal with band limited spectrum $m(t)$; see Section 11.1. An interesting question is to recover m from F. The function F is, up to normalization, equal to the Fourier transform of m, and one can of course use the inverse Fourier transform. Another interesting way, when $m(t) \ge 0$, is to view (4.5.3) as a moment problem:

$$
\int_0^1 t^n m(t)dt = i^{-n} F^{(n)}(0), \quad n = 0, 1, \ldots.
$$

Solution of Exercise 4.2.15. Recall that the distance from a complex number z_0 to $[1, 2]$ is

$$
\text{dist} \, (z_0, [1, 2]) = \text{Min}_{t \in [1,2]} |z_0 - t|.
$$

Let z be such that $d \stackrel{\text{def.}}{=} \text{dist} \, (z_0, [1, 2]) < 1$. For z such that $|z - z_0| < d$ we have:

$$
F(z) = \int_1^2 \frac{m(t)dt}{(t - z)^2}
$$

$$
= \int_1^2 \frac{m(t)dt}{(t - z_0 - (z - z_0))^2}
$$

$$
= \int_1^2 \frac{m(t)dt}{(t - z_0)^2 (1 - \frac{z - z_0}{t - z_0})^2}
$$

$$
= \int_1^2 \frac{m(t)}{(t - z_0)^2} \left(\sum_{n=1}^{\infty} n \left(\frac{z - z_0}{t - z_0} \right)^{n-1} \right) dt
$$

$$
= \sum_{n=1}^{\infty} (z - z_0)^{n-1} F_n,
$$

where we have used Weierstrass' theorem (Theorem 14.4.1) to interchange summation and integration, and where we have defined

$$F_n = n \int_1^2 \frac{m(t)dt}{(t - z_0)^{n+1}}.$$

Thus F is locally equal to a Maclaurin series around every point $z_0 \in \mathbb{C} \setminus [1, 2]$, and hence is analytic there.   □

Solution of Exercise 4.2.16. As in the previous exercise, we have

$$\frac{F_n(z) - F_n(z_0)}{z - z_0} - n \int_0^1 \frac{m(t)dt}{(t - z_0)^{n+1}}$$

$$= \int_0^1 \left\{ \frac{\frac{1}{(t - z)^n} - \frac{1}{(t - z_0)^n}}{z - z_0} - \frac{n}{(t - z_0)^{n+1}} \right\} m(t)dt$$

$$= \int_0^1 \left\{ \frac{(\sum_{k=0}^{n-1}(t - z)^k(t - z_0)^{n-1-k})(z - z_0)}{(t - z)^n(t - z_0)^n(z - z_0)} - \frac{n}{(t - z_0)^{n+1}} \right\} m(t)dt$$

$$= \int_0^1 \left\{ \frac{(\sum_{k=0}^{n-1}(t - z)^k(t - z_0)^{n-1-k})}{(t - z)^n(t - z_0)^n} - \frac{n}{(t - z_0)^{n+1}} \right\} m(t)dt.$$

We now note that, for every $t \in [0, 1]$,

$$\lim_{z \to z_0} \frac{(\sum_{k=0}^{n-1}(t - z)^k(t - z_0)^{n-1-k})}{(t - z)^n(t - z_0)^n} - \frac{n}{(t - z_0)^{n+1}} = 0.$$

With the notation as in the solution of the previous exercise, we have that, for $d(z) < d(z_0)/2$ and $|z| \le 2|z_0|$,

$$\left| \frac{(\sum_{k=0}^{n-1}(t - z)^k(t - z_0)^{n-1-k})}{(t - z)^n(t - z_0)^n} - \frac{n}{(t - z_0)^{n+1}} \right|$$

$$\le \sum_{k=0}^{n-1} \frac{2^n(1 + |z_0|)^{n-1-k}(1 + 2|z_0|)^k}{d(z_0)^{2n}} + \frac{n}{d(z_0)^{n+1}},$$

we can conclude using the dominated convergence theorem. One can also avoid this theorem and proceed in a direct way; see [42, pp. 48–49].   □

Solution of Exercise 4.2.17. Let $z \in \Omega$ and let $(z_n)_{n \in \mathbb{N}}$ be a sequence of points in Ω converging to z. Since f is continuous, we have that

$$\lim_{n \to \infty} f(z_n) = f(z).$$

We set $f(z_n) = w_n$ and $f(z) = w$. We have

$$\lim_{n \to \infty} \frac{f(z_n) - f(z)}{z_n - z} = \lim_{n \to \infty} \frac{f(z_n) - f(z)}{e^{f(z_n)} - e^{f(z)}}$$

$$= \lim_{n \to \infty} \frac{w_n - w}{e^{w_n} - e^w} = \frac{1}{\lim_{n \to \infty} \frac{e^{w_n} - e^w}{w_n - w}}$$

$$= \frac{1}{e^w} = \frac{1}{e^{f(z)}} = \frac{1}{z}. \qquad \square$$

Solution of Exercise 4.2.18. Let $z_0, z \in \Omega$. We have:

$$\frac{f(z)^2 - f(z_0)^2}{z - z_0} = (f(z) + f(z_0)) \frac{f(z) - f(z_0)}{z - z_0}.$$

By hypothesis the limit $\lim_{z \to z_0} \dfrac{f(z)^2 - f(z_0)^2}{z - z_0}$ exists and, in view of the continuity of f, the limit $\lim_{z \to z_0} f(z) + f(z_0) = 2f(z_0)$. Since $f(z_0) \neq 0$ we have that
$\lim_{z \to z_0} \dfrac{f(z) - f(z_0)}{z - z_0}$ exists, and is equal to

$$\frac{\lim_{z \to z_0} \dfrac{f(z)^2 - f(z_0)^2}{z - z_0}}{2f(z_0)}. \qquad \square$$

Solution of Exercise 4.2.19. Recall the notation (4.2.15). We have

$$\partial_z(\partial_{\bar z} f) = \frac{1}{2} \left(\frac{\partial}{\partial x}(\partial_{\bar z} f) - i \frac{\partial}{\partial y}(\partial_{\bar z} f) \right)$$

$$= \frac{1}{4} \left(\frac{\partial}{\partial x} \left(\frac{\partial f}{\partial x} + i \frac{\partial f}{\partial y} \right) - i \frac{\partial}{\partial y} \left(\frac{\partial f}{\partial x} + i \frac{\partial f}{\partial y} \right) \right)$$

$$= \frac{1}{4} \Delta f,$$

since

$$\frac{\partial^2 f}{\partial x \partial y} = \frac{\partial^2 f}{\partial y \partial x},$$

due to the smoothness of the real and imaginary parts of f. $\qquad \square$

Solution of Exercise 4.2.20. Let $F_1 = u_1 + iv_1$ and $F_2 = u_2 + iv_2$ where u_1, v_1, u_2
and v_2 have smooth partial derivatives. We want to show that

$$\frac{\partial F_1 F_2}{\partial z} = \frac{\partial F_1}{\partial z} F_2 + F_1 \frac{\partial F_2}{\partial z}$$

and similarly for $\frac{\partial}{\partial \bar z}$. We have

$$\frac{\partial F_1 F_2}{\partial x} = \frac{\partial F_1}{\partial x} F_2 + F_1 \frac{\partial F_2}{\partial x}$$

and

$$\frac{\partial F_1 F_2}{\partial y} = \frac{\partial F_1}{\partial y} F_2 + F_1 \frac{\partial F_2}{\partial y}$$

and the result follows by linearity.   □

Solution of Exercise 4.2.21. At the origin, we have

$$\frac{\partial f}{\partial z}(0,0) = \frac{\partial f}{\partial \overline{z}}(0,0) = 1$$

since

$$\frac{\partial u}{\partial x}(0,0) = 1 \quad \text{and} \quad \frac{\partial u}{\partial y}(0,0) = 0.$$

At a point $z \neq 0$ we have, using Exercise 4.2.20,

$$\frac{\partial f}{\partial z} = -\left(\frac{\overline{z}}{z}\right)^2 \quad \text{and} \quad \frac{\partial f}{\partial \overline{z}} = \frac{2\overline{z}}{z}.$$ □

    In the previous exercise, the function f is continuous in all of $\mathbb{C}$, and the real and imaginary parts of f have partial derivatives of first order in all of $\mathbb{R}^2$. The Cauchy–Riemann equations hold only at the origin. The function is not differentiable at the origin. There is no contradiction with Theorems 4.2.3 and 14.1.3, since the partial derivatives are not continuous at the origin.

Solution of Exercise 4.2.22. To prove (a), we first consider the case $p = 1$.

$$\frac{\partial |z|}{\partial \overline{z}} = \frac{1}{2}\left(\frac{\partial \sqrt{x^2 + y^2}}{\partial x} + i\frac{\partial \sqrt{x^2 + y^2}}{\partial y}\right)$$

$$= \frac{1}{2}\left(\frac{2x}{2\sqrt{x^2 + y^2}} + i\frac{2y}{2\sqrt{x^2 + y^2}}\right)$$

$$= \frac{z}{2|z|}.$$

Similarly,

$$\frac{\partial |z|}{\partial z} = \frac{\overline{z}}{2|z|}.$$ (4.5.4)

For $n = 2p$ we have

$$|z|^{2p} = z^p \overline{z}^p,$$

and formula (4.2.21) leads to

$$\frac{\partial |z|^{2p}}{\partial \overline{z}} = pz^p \overline{z}^{p-1} = pz|z|^{2p-2}.$$

For $n = 2p + 1$ we have

$$\frac{\partial |z|^{2p+1}}{\partial \bar{z}} = |z|^{2p} \frac{\partial |z|}{\partial \bar{z}} + \frac{\partial |z|^{2p}}{\partial \bar{z}} |z| = \left(\frac{z}{2} + pz \right) |z|^{2p-1} = \frac{2p+1}{2} z |z|^{2p-1}.$$

We now turn to (b).

$$\begin{aligned}
\frac{\partial \ln |z|}{\partial \bar{z}} &= \frac{1}{4} \left(\frac{\partial \ln(x^2 + y^2)}{\partial x} + i \frac{\partial \ln(x^2 + y^2)}{\partial y} \right) \\
&= \frac{1}{4} \left(\frac{2x}{x^2 + y^2} + i \frac{2y}{x^2 + y^2} \right) \\
&= \frac{2z}{4|z|^2} \\
&= \frac{1}{2\bar{z}}
\end{aligned}$$

and

$$\begin{aligned}
\frac{\partial \ln |z|}{\partial z} &= \frac{1}{4} \left(\frac{\partial \ln(x^2 + y^2)}{\partial x} - i \frac{\partial \ln(x^2 + y^2)}{\partial y} \right) \\
&= \frac{1}{4} \left(\frac{2x}{x^2 + y^2} - i \frac{2y}{x^2 + y^2} \right) \\
&= \frac{2\bar{z}}{4|z|^2} \\
&= \frac{1}{2z}.
\end{aligned}$$
$\square$

Solution of Exercise 4.2.23. Taking logarithm on both sides of (4.2.23), we obtain

$$\ln f(z) = \ln |h(z)|^2 + \ln g(z). \tag{4.5.5}$$

Now, if one knows that the logarithm of the modulus of an analytic function is harmonic (see Exercise 9.1.5), the result is immediate by applying the operator Δ on both sides of the above equality. Otherwise, applying the operator ∂ (we here use notation (4.2.15)) to both sides of (4.5.5) we have:

$$\begin{aligned}
\partial \ln f(z) &= \partial \ln |h(z)|^2 + \partial \ln g(z) \\
&= \frac{\partial |h(z)|^2}{|h(z)|^2} + \partial \ln g(z) \\
&= \frac{\overline{h(z)} \partial h(z)}{|h(z)|^2} + \partial \ln g(z) \\
&= \frac{\partial h(z)}{h(z)} + \partial \ln g(z), \qquad z \in \Omega,
\end{aligned}$$

where we have used the fact that $\partial \overline{h(z)} = 0$ (since h is analytic in Ω) and the equality

$$\partial(h(z)\overline{h(z)}) = \overline{h(z)}\partial h(z).$$

The result follows by applying $\overline{\partial}$ on both sides of the last equality since $\overline{\partial}\frac{\partial h(z)}{h(z)} = 0$. $\quad\square$

Solution of Exercise 4.2.25. By definition of R_q we assume any solution of the equation (4.2.27) is defined in some neighborhood, say Ω, of the origin (*a priori* depending on f), and which we assume invariant under the map $z \mapsto qz$. From (4.2.27) we have

$$(1 - \lambda z(1-q))f(z) = f(qz), \quad z \in \Omega.$$

Iterating this equality we obtain

$$\left(\prod_{j=0}^{n-1}(1 - \lambda q^j z(1-q))\right)f(z) = f(q^n z), \quad z \in \Omega.$$

We can let $n \to \infty$ since the infinite product $\prod_{j=0}^{\infty}(1 - \lambda q^j z(1-q))$ converges (see Exercise 3.7.8) and since f is differentiable, and hence continuous, at the origin. We obtain thus $f(z) \equiv f(0)$ if $\lambda = 0$ and

$$f(z) = \frac{f(0)}{\prod_{j=0}^{\infty}(1 - \lambda q^j z(1-q))}, \quad z \in \mathbb{C} \setminus \left\{\frac{1}{\lambda(1-q)q^j}, \ j = 0, 1, \ldots\right\}$$

if $\lambda \neq 0$. $\quad\square$

Solution of Exercise 4.3.1. The function $-\dfrac{1}{z^4}$ is a rational function (quotient of polynomials) and as such admits a derivative at all points where it is defined, that is in $\mathbb{C} \setminus \{0\}$. The function $\exp z$ admits a derivative at all points of (that is, it is an *entire* function). By composition of differentiable functions the function $f(z)$ admits a derivative in all of $\mathbb{C} \setminus \{0\}$, and in particular the Cauchy–Riemann equations hold there.

The function $f(z)$ is not continuous at the origin. Indeed, for $z = x \in \mathbb{R}$,

$$\lim_{x \to 0} f(x) = 0$$

while for $z = \rho \exp \dfrac{i\pi}{4}$, $f(z) = \exp \dfrac{1}{\rho^4}$ and so

$$\lim_{\rho \to 0} f(\rho e^{\frac{i\pi}{4}}) = \infty.$$

Since the function f is not continuous at the origin, it is in particular not differentiable there. We show that the Cauchy–Riemann equations hold at $z = 0$. Write

$$f(z) = \exp\left(-\frac{1}{(x+iy)^4}\right) = u(x,y) + iv(x,y).$$

We have

$$u(x,0) = \exp\left(-\frac{1}{x^4}\right),$$

$$v(x,0) \equiv 0,$$

$$u(0,y) = \exp\left(-\frac{1}{y^4}\right),$$

$$v(0,y) \equiv 0.$$

By hypothesis, $f(0) = 0$ and hence $u(0,0) = v(0,0) = 0$. Moreover,

$$\lim_{x\to0}\frac{u(x,0)-u(0,0)}{x} = \lim_{x\to0}\frac{1}{x}\exp\left(-\frac{1}{x^4}\right) = 0,$$

$$\lim_{y\to0}\frac{v(0,y)-v(0,0)}{y} = \lim_{y\to0}\frac{0-0}{y} = 0.$$

Thus $\frac{\partial u}{\partial x}(0,0)$ and $\frac{\partial v}{\partial y}(0,0)$ exist and are equal to 0. Hence the first Cauchy–Riemann equation holds at $z = 0$. The second one is proved in the same way. $\quad\square$

Solution of Exercise 4.3.2. Recall that for any two real numbers a and b,

$$2|ab| \le a^2 + b^2. \qquad (4.5.6)$$

We have

$$|f(z) - f(0)| = \frac{|x|^3|y|}{x^6+y^2}\sqrt{x^2+y^2} \le \frac{1}{2}\sqrt{x^2+y^2},$$

where we have used (4.5.6) with $a = x^3$ and $b = y$. Thus

$$\lim_{z\to0}|f(z)-f(0)| \le \lim_{z\to0}\frac{1}{2}\sqrt{x^2+y^2} = \lim_{z\to0}\frac{|z|}{2} = 0.$$

Thus f is continuous at the origin. This proves (a). To prove (b) we compute with θ fixed and $z = \epsilon e^{i\theta}$ (so that $x = \epsilon\cos\theta$ and $y = \epsilon\sin\theta$):

$$\frac{f(z) - f(0)}{z} = \frac{\epsilon^5(\cos^3\theta)(\sin\theta)(\sin\theta - i\cos\theta)}{(\epsilon^6(\cos\theta)^6 + \epsilon^2(\sin\theta)^2)\epsilon(\cos\theta + i\sin\theta)}.$$

Hence,

$$\left|\frac{f(z)-f(0)}{z}\right| = \frac{\epsilon^2|(\cos\theta)^3\sin\theta|}{\epsilon^4(\cos\theta)^6 + (\sin\theta)^2}.$$

If $\sin\theta = 0$, the function is identically equal to 0 and so is the limit. If $\sin\theta \ne 0$ the limit is equal to 0 since the denominator tends to $\sin^2\theta > 0$.

We now show that f is not ℂ-differentiable at the origin. Take $z = t + it^3$.
Then
$$\frac{f(z) - f(0)}{z} = \frac{t^3 t^3 (t^3 - it)}{2t^6(t + it^3)} = \frac{1}{2}\frac{t^2 - i}{1 + it^2}.$$

This expression tends to $-i/2$ when $t \to 0$ and so is different from the limit along the rays and so f is not differentiable at the origin.

Finally, since the real and imaginary parts of $f(z)$ are given by

$$u(x, y) = \frac{x^3 y^2}{x^6 + y^2} \quad \text{and} \quad v(x, y) = \frac{-x^4 y}{x^6 + y^2},$$

we have

$$u(x, 0) = v(0, y) = u(0, y) = v(x, 0) \equiv 0 \quad \text{for} \quad x \neq 0 \quad \text{and} \quad y \neq 0.$$

Hence the various partial derivatives of first order all exist at the origin and are equal to 0, that is, the Cauchy–Riemann equations hold at the origin.   □

Solution of Exercise 4.3.3. (a) We have

$$\operatorname{Re} f(z) = x^2 + y^2 \quad \text{and} \quad \operatorname{Im} f(z) \equiv 0.$$

Since the real part and imaginary part of f are smooth, a necessary and sufficient condition for f to be ℂ-differentiable is that the Cauchy–Riemann equations hold at that point. Here, these equations take the form

$$2x = 0 \quad \text{and} \quad 2y = 0.$$

Therefore, f is differentiable only at the point 0.
(b) The function in the preceding exercise answers the question.   □

Solution of Exercise 4.4.2. It suffices to look at the development (4.4.4) for $z = 1/2$:

$$\frac{1}{(1/2)^2} = \sum_{n=1}^{\infty} \frac{n}{2^{n-1}},$$

so that

$$4 = 2 \sum_{n=1}^{\infty} \frac{n}{2^n},$$

and hence the result.   □

As a matter of fact, since $\sum_{n=1}^{\infty} 1/2^n = 1$, the sum (4.4.5) corresponds to an entropy calculation; it appears in [140, p. 61].

Solution of Exercise 4.4.3. Applying formula (4.4.3) with $p = 2$ to the function $\frac{1}{1-z}$ we obtain (4.4.6):

$$\frac{2}{(1-z)^3} = \sum_{n=2}^{\infty} n(n-1)z^{n-2}, \quad z \in \mathbb{D}.$$

We put $z = re^{i\theta}$ and, taking the real part of both sides of (4.4.6), we obtain

$$\sum_{n=2}^{\infty} n(n-1)r^{n-2}\cos((n-2)\theta) = 2 \cdot \operatorname{Re} \frac{1}{(1-re^{i\theta})^3}$$

$$= 2 \cdot \frac{\operatorname{Re}(1-e^{-i\theta}r)^3}{(1-2r\cos\theta+r^2)^3}$$

$$= 2 \cdot \frac{1 - 3r\cos\theta + 3r^2\cos(2\theta) - r^3\cos(3\theta)}{(1-2r\cos\theta+r^2)^3}. \quad \square$$

Solution of Exercise 4.4.4. If α is a natural integer, then f_α is a polynomial, and $R = \infty$. If $\alpha \notin \mathbb{N}$ all the coefficients

$$a_n = \frac{\alpha(\alpha-1)\cdots(\alpha-n+1)}{n!}$$

are non-zero. Since

$$\frac{|a_{n+1}|}{|a_n|} = \frac{|\alpha-n|}{n+1} \longrightarrow 1 \quad \text{as} \quad n \longrightarrow \infty,$$

we have that $R = 1$. Using the result on the differentiation of a complex power series we have

$$f_\alpha'(z) = \sum_{n=1}^{\infty} \frac{\alpha(\alpha-1)\cdots(\alpha-n+1)}{(n-1)!}z^{n-1},$$

and

$$zf_\alpha'(z) = \sum_{n=1}^{\infty} \frac{\alpha(\alpha-1)\cdots(\alpha-n+1)}{(n-1)!}z^n,$$

and the coefficient of z^n $(n > 0)$ of $(1+z)f_\alpha'(z)$ is

$$\frac{\alpha(\alpha-1)\cdots(\alpha-n)}{n!} + \frac{\alpha(\alpha-1)\cdots(\alpha-n+1)}{(n-1)!}$$

$$= \frac{\alpha(\alpha-1)\cdots(\alpha-n+1)}{n!}(\alpha-n+n)$$

$$= \alpha a_n,$$

and hence the result. To prove the second equality, it suffices to differentiate the function $H = f_\alpha f_\beta - f_{\alpha+\beta}$. By Proposition 3.4.8 the function H is defined by a power series, which is convergent in the open unit disk. Furthermore, $H(0) = 0$. We have

$$H'(z) = f'_\alpha(z)f_\beta(z) + f_\alpha(z)f'_\beta(z) - f'_{\alpha+\beta}(z)$$

$$= \frac{\alpha f_\alpha(z)}{1+z}f_\beta(z) + f_\alpha(z)\frac{\beta f_\beta(z)}{1+z} - \frac{(\alpha+\beta)f_{\alpha+\beta}(z)}{1+z}$$

$$= (\alpha+\beta)\frac{H(z)}{1+z},$$

from which it follows by induction that

$$H'(0) = \cdots = 0.$$

When checking the induction, formula (4.2.2) will prove useful. Since H is defined by a power series, it vanishes identically.   □

We see in particular that, for every $N \in \mathbb{N}$,

$$(f_{1/N}(z))^N = 1 + z, \quad z \in \mathbb{D},$$

and special cases of α give the well-known power series expansions (see [204, p. 137], if need be)

$$(1+z)^{1/2} = 1 + \frac{1}{2}z - \frac{1}{2\cdot4}z^2 + \frac{1\cdot3}{2\cdot4\cdot6}z^3 + \cdots,$$

$$(1+z)^{-1/2} = 1 - \frac{1}{2}z + \frac{1\cdot3}{2\cdot4}z^2 - \frac{1\cdot3\cdot5}{2\cdot4\cdot6}z^3 + \cdots. \tag{4.5.7}$$

One uses the notation $f_\alpha(z) = (1+z)^\alpha$ for $\alpha \in \mathbb{C}$. One should not forget that, for $\alpha \notin \mathbb{Z}$, this is, *a priori*, just a notation for the power series (4.4.9). We will see in Exercise 6.3.2 that, for real α, the function f_α is the analytic extension to $\mathbb{D}$ of the classical function from calculus $(1+x)^\alpha = \exp(\alpha \ln(1+x))$.

Solution of Exercise 4.4.6. It follows from the very rough estimate

$$\left|\frac{z^{2p}}{2^{2p}(p!)^2}\right| \le \frac{\left(\frac{|z^2|}{4}\right)^p}{p!}$$

that the power series (4.4.12) converges for every z. We now prove (4.4.13). Using Weierstrass' theorem (Theorem 14.4.1) we have

$$\int_{[0,2\pi]} e^{iz\cos u}du = \sum_{n=0}^\infty \frac{z^n i^n}{n!}\int_{[0,2\pi]}\cos^n u\,du$$

$$= \sum_{p=0}^\infty \frac{z^{2p}(-1)^p}{(2p)!}\int_{[0,2\pi]}\cos^{2p}u\,du,$$

since

$$\int_{[0,2\pi]} (\cos u)^{2p+1} du = 0.$$

See the discussion following the proof of Exercise 3.1.8 for the latter. The integrals

$$\int_{[0,2\pi]} (\cos u)^{2p} du = 2\pi \binom{2p}{p} 2^{2p} = 2\pi \frac{(2p)!2^{2p}}{(p!)^2}$$

have been computed in Exercise 3.1.8. The result follows. □

Solution of Exercise 4.4.8. We follow [126, pp. 18–19], and assume first that the power series $f(z) = \sum_{n=1}^{\infty} a_n z^n$ has a strictly positive radius of convergence. Then equation (4.4.15) implies that

$$f(z)^2 = f(z) - z.$$

This equation has a unique solution for which $f'(0) = 1$ (see Remark 4.4.5), namely

$$f(z) = \frac{1 - (1 - 4z)^{1/2}}{2},$$

and so the power series f has in fact radius of convergence equal to 1. The formula for a_n follows from (4.4.9). □

Solution of Exercise 4.4.9. Since $1/(1-z) = 1 + z + z^2 + \cdots$, the sequence (S_n) is the convolution of the sequence identically equal to 1 with the sequence (a_n). □

Solution of Exercise 4.4.10. From Exercise 4.4.7 (or from Proposition 3.4.8) the sequence $(\frac{c_n}{n!})_{n \in \mathbb{N}_0}$ is the convolution of the sequences $(\frac{a_n}{n!})_{n \in \mathbb{N}_0}$ and $(\frac{b_n}{n!})_{n \in \mathbb{N}_0}$, that is

$$\frac{c_n}{n!} = \sum_{\ell=0}^{n} \frac{a_\ell}{\ell!} \frac{b_{n-\ell}}{(n-\ell)!} = \frac{1}{n!} \sum_{\ell=0}^{n} \binom{n}{\ell} a_\ell b_{n-\ell}.$$ □

Solution of Exercise 4.4.11. The zeroes of the polynomial $1 - z - z^2$ are

$$z_- = \frac{-1 - \sqrt{5}}{2} \quad \text{and} \quad z_+ = \frac{-1 + \sqrt{5}}{2},$$

and therefore $\frac{1}{1-z-z^2}$ is analytic in the open disk $|z| < \min\{|z_-|, |z_+|\} = |z_+|$. Hence, there is a power series expansion

$$\frac{1}{1 - z - z^2} = \sum_{n=0}^{\infty} a_n z^n, \quad |z| < |z_-|.$$

From the equality,

$$(1 - z - z^2) \left(\sum_{n=0}^{\infty} a_n z^n \right) = 1,$$

and comparing coefficients we then obtain (4.4.16). □

Solution of Exercise 4.4.13.

(1) The function F is defined by a power series with radius of convergence equal to

$$R = \frac{1}{\lim_{n\to\infty} \dfrac{|a_{n+1}|}{|a_n|}} = \frac{1}{\lim_{n\to\infty} \dfrac{n+1}{n}} = 1.$$

Thus it is analytic in the open unit disk and

$$F'(z) = -\sum_{n=1}^{\infty}(-1)^n z^{n-1} = \frac{1}{1+z}.$$

Set $G(z) = (1+z)\exp(-F(z))$. The function G is analytic in the open unit disk and we have

$$G'(z) = -(1+z)F'(z)\exp(-F(z)) + \exp(-F(z)) = 0, \quad |z| < 1.$$

Thus $G(z) = G(0) = 1$ and $\exp(F(z)) = 1 + z$ in the open unit disk.

(2) Since $F'(z) = \frac{1}{1+z}$ and $F(0) = 0$, we have

$$\ln(1+z) = \int_{[0,z]} \frac{ds}{1+s} = \int_0^1 \frac{z}{1+tz}\,dt$$

and

$$\ln(1+z) - z = \int_0^1 \left(\frac{z}{1+tz} - z\right)dt = -\int_0^1 \frac{tz^2}{1+tz}\,dt.$$

Therefore, we can write for $|z| \le 1/2$:

$$\big|\ln(1+z) - z\big| \le \left|\int_0^1 \frac{tz^2}{1+tz}\,dt\right|$$

$$\le \left|\int_0^1 \frac{tz^2}{1-\frac{1}{2}}\,dt\right|$$

$$= |z|^2 \int_0^1 2t = |z|^2.$$

To prove (4.4.18) we use (1.1.35) and get

$$\big||\ln(1+z)| - |z|\big| \le |z|^2,$$

and hence

$$|z| - |z|^2 \le |\ln(1+z)| \le |z| + |z|^2.$$

The result follows since $|z| \le \frac{1}{2}$ and so

$$|z| + |z|^2 \le |z| + \frac{|z|}{2} = \frac{3|z|}{2} \quad \text{and} \quad |z| - |z|^2 \ge |z| - \frac{|z|}{2} = \frac{|z|}{2}. \qquad \square$$

Solution of Exercise 4.4.16. For every $z \in \mathbb{D}$ we have

$$(1-z)\left(\sum_{n=1}^{\infty}\left(\sum_{\ell=1}^{n}\frac{1}{\ell}\right)z^n\right) = \sum_{n=1}^{\infty}\left(\sum_{\ell=1}^{n}\frac{1}{\ell}\right)z^n - \sum_{n=1}^{\infty}\left(\sum_{\ell=1}^{n}\frac{1}{\ell}\right)z^{n+1}$$

$$= z + \sum_{n=2}^{\infty}\left(\sum_{\ell=1}^{n}\frac{1}{\ell}\right)z^n - \sum_{n=1}^{\infty}\left(\sum_{\ell=1}^{n}\frac{1}{\ell}\right)z^{n+1}$$

$$= z + \sum_{n=1}^{\infty}\left(\sum_{\ell=1}^{n+1}\frac{1}{\ell}\right)z^{n+1} - \sum_{n=1}^{\infty}\left(\sum_{\ell=1}^{n}\frac{1}{\ell}\right)z^{n+1}$$

$$= \sum_{n=1}^{\infty}\frac{z^n}{n} = -\ln(1-z).$$

Assume now that a function ψ exists with the required property. Then, for $|z| < 1$, applying Weierstrass' theorem on interchanging the order of summation and integration (Theorem 14.4.1) we have

$$\int_0^{2\pi}\frac{\psi(e^{it})dt}{e^{it}-z} = \sum_{n=0}^{\infty}z^n c_n,$$

where

$$c_n = \frac{1}{2\pi i}\int_0^{2\pi}\psi(e^{it})e^{-i(n+1)t}dt.$$

Thus, by uniqueness of the coefficients of the power series expansion at the origin, we would have

$$\frac{1}{2\pi i}\int_0^{2\pi}\psi(e^{it})e^{-i(n+1)t}dt = -\sum_{\ell=1}^{n}\frac{1}{\ell}.$$

But the sequence on the left of this equality is bounded in modulus:

$$|c_n| \leq \max_{t\in[0,2\pi]}|\psi(e^{it})|,$$

while the sequence on the right is unbounded in modulus:

$$\lim_{n\to\infty}\sum_{\ell=1}^{n}\frac{1}{\ell} = \infty,$$

and this concludes the proof. □

Solution of Exercise 4.4.17. For $|z| < 1 \leq R$ and $t \leq 1$ we have $|zt| < 1 \leq R$ and so by the theorem on differentiability of a power series we have

$$F'(tz) = \sum_{n=1}^{\infty}na_n t^{n-1}z^{n-1}.$$

for $|z| < 1$. Set $f_n(t) = na_{n-1}t^{n-1}z^{n-1}$ and $M_n = n|a_{n-1}||z|^{n-1}$. Since the radius of convergence of the series of the derivative is also R we have

$$\sum_{n=1}^{\infty} M_n < \infty \quad \text{for} \quad |z| < 1 \leq R.$$

Using Weierstrass' theorem we then have

$$\int_0^1 F'(tz)dt = \int_0^1 \left(\sum_{n=1}^{\infty} na_n t^{n-1} z^{n-1} \right) dt$$

$$= \sum_{n=1}^{\infty} na_n z^{n-1} \int_0^1 t^{n-1} dt$$

$$= \sum_{n=1}^{\infty} na_n z^{n-1} \frac{1}{n}$$

$$= \sum_{n=1}^{\infty} a_n z^{n-1}$$

$$= \begin{cases} \dfrac{F(z) - F(0)}{z}, & \text{if } z \neq 0, \\ a_1 = F'(0), & \text{for } z = 0. \end{cases} \qquad \square$$

Solution of Exercise 4.4.18. The idea is the same as in Exercise 4.4.17. We just give the outline of the proof. For a fixed z we set $f_n(t) = a_n z^n t^n f(t)$. Then with $M = \max_{t \in [0,a]} |f(t)|$ we have

$$|f_n(t)| \leq M|a_n||az|^n = M_n.$$

By definition of the radius of convergence, $\sum_{n=0}^{\infty} M_n < \infty$ for $|z| < R/a$. For such z one can apply Weierstrass' theorem and write

$$G(z) = \int_0^a \left(\sum_{n=0}^{\infty} a_n z^n t^n \right) f(t)dt = \sum_{n=0}^{\infty} z^n A_n,$$

where

$$A_n = a_n \int_0^a t^n f(t)dt.$$

Thus G has a development in power series in $|z| < R/a$ and is analytic there. Its derivative is (recall the theorem on differentiability of power series)

$$G'(z) = \sum_{n=1}^{\infty} nA_n z^{n-1}.$$

Another application of Weierstrass' theorem leads to

$$G'(z) = \int_0^a tF'(zt)f(t)dt.$$

The arguments for the function H are similar. Set now $f_n(z) = a_n \dfrac{z^n}{t^n} f(t)$. With $M = \max_{t \in [1,a]} |f(t)|$ and since $1/a \le 1/t \le 1$ for $1 \le t \le a$ we have, for $z \in \mathbb{C}$,

$$|f_n(t)| \le |a_n||z|^n M = M_n,$$

and $\sum_{n=0}^\infty M_n < \infty$ for $|z| < R$. For such z one can apply Weierstrass' theorem and write

$$H(z) = \int_1^a \left(\sum_{n=0}^\infty a_n \frac{z^n}{t^n} \right) f(t)dt = \sum_{n=0}^\infty z^n A_n,$$

where

$$A_n = a_n \int_1^a \frac{f(t)}{t^n} dt.$$

Thus H has a development in power series in $|z| < R$ and is analytic there. Its derivative is (recall the theorem on differentiability of power series)

$$H'(z) = \sum_{n=1}^\infty n A_n z^{n-1}.$$

Another application of Weierstrass' theorem leads to

$$H'(z) = -\int_0^a F'(z/t) \frac{f(t)}{t^2} dt.$$

If $R = \infty$ the above arguments show that G and H are entire functions. □

Solution of Exercise 4.4.19. It follows from (3.4.18) in Exercise 3.4.13 that f is equal to a power series centered at the origin, and convergent for every complex number z. Thus f is analytic in $\mathbb{C}$, that is, is an entire function. Its derivative is equal to

$$f'(z) = \int_0^1 te^{tz} m(t)dt$$

$$= \sum_{n=1}^\infty nz^{n-1} \frac{\int_0^1 t^n m(t)dt}{n!}.$$

The first formula is done by computing the limit

$$\lim_{h \to 0} \frac{f(z+h) - f(z)}{h} = \int_0^1 \lim_{h \to 0} \frac{e^{t(z+h)} - e^{tz}}{h} m(t)dt = \int_0^1 te^{tz} m(t)dt.$$

We justify the interchange of limit and integral in the above chain of equalities by using the dominated convergence theorem. Indeed, it is enough to consider a sequence $(h_n)_{n \in \mathbb{N}}$ with limit 0 to compute the limit. Let

$$f_n(t) = \frac{e^{t(z+h_n)} - e^{tz}}{h_n} m(t), \quad n \in \mathbb{N}.$$

For $|h_n| \leq 1$ we have

$$|f_n(t)| = |e^{tz}| \cdot \left| \frac{e^{th_n} - 1}{h_n} \right| \cdot |m(t)|$$

$$= |e^{tz}| \cdot \left| \sum_{p=1}^{\infty} \frac{t^p h_n^{p-1}}{p!} \right| \cdot |m(t)|$$

$$\leq t|e^{tz}| \cdot |m(t)| \cdot \sum_{p=0}^{\infty} \frac{|t|^p}{(p+1)!},$$

and the function

$$t \mapsto t|e^{tz}| \cdot |m(t)| \cdot \sum_{p=0}^{\infty} \frac{|t|^p}{(p+1)!}$$

is continuous on $[0, 1]$ and in particular absolutely summable on $[0, 1]$. The dominated convergence theorem is therefore applicable, and we obtain a formula for $f'(z)$ in term of an integral. The expression of $f'(z)$ as a power series is a direct consequence of Theorem 4.4.1.   □

We remark that the above argument, slightly adapted, also works if m is only assumed to be in $\mathbf{L}_1(0, 1)$, and in particular in $\mathbf{L}_2(0, 1)$ since by the Cauchy–Schwarz inequality

$$\mathbf{L}_2(0, 1) \subset \mathbf{L}_1(0, 1).$$

The function

$$t \mapsto t|e^{tz}| \cdot |m(t)| \cdot \sum_{p=0}^{\infty} \frac{|t|^p}{(p+1)!}$$

is not anymore continuous, but it is in $\mathbf{L}_1(0, 1)$, and the dominated convergence theorem can still be applied.

Solution of Exercise 4.4.20. We first consider z in the open unit disk. The idea is to show that φ is equal to a power series centered at the origin and with radius of convergence at least 1 (in fact, for m not identically equal to 0, the radius of convergence is exactly 1, but we will not prove it here). We then conclude with the theorem on the analyticity of power series in their disk of convergence. The key is to use Weierstrass' theorem to interchange integration and infinite sum at an appropriate place of the argument. For $|z| < 1$ we can write

$$\frac{e^{it} + z}{e^{it} - z} = \frac{e^{it} - z + 2z}{e^{it} - z}$$

$$= 1 + \frac{2z}{e^{it} - z}$$

$$= 1 + 2ze^{-it}\frac{1}{1 - e^{-it}z}$$

$$= 1 + 2ze^{-it}\left(\sum_{n=0}^{\infty}(e^{-it}z)^n\right)$$

$$= 1 + 2\sum_{n=0}^{\infty} z^{n+1}e^{-i(n+1)t},$$

and so

$$\frac{e^{it} + z}{e^{it} - z}m(t) = m(t) + \sum_{n=0}^{\infty} f_n(t),$$

where

$$f_n(t) = 2z^{n+1}e^{-i(n+1)t}m(t), \quad n \in \mathbb{N}_0.$$

Then, f_n is a continuous function of t in the interval $[0, 2\pi]$ and

$$|f_n(t)| \leq M|z|^{n+1},$$

where $M = 2\sup_{t\in[0,2\pi]} |m(t)|$. Set $M_n = M|z|^{n+1}$. We have

$$\sum_{n=0}^{\infty} M_n = M\sum_{n=0}^{\infty} |z|^{n+1} < +\infty,$$

and so,

$$\varphi(z) = \int_0^{2\pi} \frac{e^{it} + z}{e^{it} - z} m(t)dt$$

$$= \int_0^{2\pi} \left(m(t) + \sum_{n=0}^{\infty} f_n(t)\right) dt$$

$$= \int_0^{2\pi} m(t)dt + \int_0^{2\pi} \left(\sum_{n=0}^{\infty} f_n(t)\right) dt$$

$$= \int_0^{2\pi} m(t)dt + \sum_{n=0}^{\infty} \int_0^{2\pi} f_n(t)dt,$$

where we have used Weierstrass' theorem to go from the penultimate line to the last line.

By definition of f_n we have

$$\int_0^{2\pi} f_{n-1}(t)dt = 2z^n a_n, \quad n = 1, 2, \dots,$$

where, for $n = 0, 1, \dots$,

$$a_n = \int_0^{2\pi} e^{-int} m(t)dt$$

is called the nth trigonometric moment. Then,

$$\varphi(z) = a_0 + 2\sum_{n=1}^{\infty} a_n z^n. \tag{4.5.8}$$

The argument above was made for any $|z| < 1$. Thus (4.5.8) converges for all $|z| < 1$ and by the theorem on the analyticity of convergent power series in their disk of convergence, φ is analytic in the open unit disk.

The case of $|z| > 1$ is done by the change of variable $z \mapsto 1/z$. Assume first m real-valued. The function

$$\varphi(1/z) = -\int_0^{2\pi} \frac{1 + ze^{it}}{1 - ze^{it}} m(t)dt = -\overline{\varphi(\overline{z})} \tag{4.5.9}$$

is analytic in the open unit disk. Therefore φ is analytic in $|z| > 1$. The case of complex-valued m is easily adapted. $\qquad\qquad\square$

Solution of Exercise 4.4.21. For z_0 and z in $\mathbb{C} \setminus [0, 1]$, and since

$$(t - z_0)^2 - (t - z)^2 = (2t - z - z_0)(z - z_0),$$

we have

$$\frac{F(z) - F(z_0)}{z - z_0} - \int_0^1 \frac{2m(t)dt}{(t - z_0)^3} = \int_0^1 \left\{ \frac{\dfrac{1}{(t-z)^2} - \dfrac{1}{(t-z_0)^2}}{z - z_0} - \frac{2}{(t-z_0)^3} \right\} m(t)dt$$

$$= \int_0^1 \left\{ \frac{(2t - z - z_0)}{(t-z)^2 (t-z_0)^2} - \frac{2}{(t-z_0)^3} \right\} m(t)dt$$

$$= \int_0^1 \frac{(2t - z - z_0)(t - z_0) - 2(t - z)^2}{(t-z)^2 (t-z_0)^3} m(t)dt.$$

For $z \in \mathbb{C}$, let $d(z)$ be the distance from z to $[0, 1]$:

$$d(z) = \min_{t \in [0,1]} |t - z|.$$

We have that $d(z_0) > 0$. Moreover, for $|z - z_0| < d(z_0)/2$ we have

$$|t - z| \geq |t - z_0| - |z - z_0| \geq d(z_0) - d(z_0)/2 = d(z_0)/2,$$

and so $d(z) > d(z_0)/2$. Therefore,

$$\left| \frac{F(z) - F(z_0)}{z - z_0} - \int_0^1 \frac{2m(t)dt}{(t - z_0)^3} \right| = \left| \int_0^1 \frac{(2t - z - z_0)(t - z_0) - 2(t - z)^2}{(t - z)^2(t - z_0)^3} m(t)dt \right|$$

$$\leq \frac{4}{d(z_0)^5} \int_0^1 |(2t - z - z_0)(t - z_0) - 2(t - z)^2| \cdot |m(t)|dt.$$

To end the proof we show that

$$\lim_{z \to z_0} \int_0^1 |(2t - z - z_0)(t - z_0) - 2(t - z)^2| \cdot |m(t)|dt = 0. \qquad (4.5.10)$$

But

$$(2t - z - z_0)(t - z_0) - 2(t - z)^2 = (z - z_0)(3t - z_0 - 2z),$$

and so (say, for $|z| \leq 2|z_0|$),

$$\lim_{z \to z_0} \int_0^1 |(2t - z - z_0)(t - z_0) - 2(t - z)^2| \cdot |m(t)|dt$$

$$\leq \lim_{z \to z_0} |z - z_0| \cdot (3 + |z_0| + 2|z|) \int_0^1 |m(t)|dt$$

$$\leq \lim_{z \to z_0} |z - z_0| \cdot (3 + 5|z_0|) \int_0^1 |m(t)|dt$$

$$= 0. \qquad \qquad \square$$

We note that, pointwise,

$$\lim_{z \to z_0} (2t - z - z_0)(t - z_0) - 2(t - z)^2 = 0,$$

and one could also justify interchanging the integral and the limit in (4.5.10) using the dominated convergence theorem (see Theorem 17.5.2) since (say, for $|z| \leq 2|z_0|$)

$$|(2t - z - z_0)(t - z_0) - 2(t - z)^2| \cdot |m(t)| \leq (2 + 3|z_0|)(1 + |z_0|) + 2(1 + 2|z_0|)^2 \cdot M,$$

where $M = \max_{t \in [0,1]} |m(t)|$. In the case at hand, this method is an overkill, but there are cases where it will prove useful.

Solution of Exercise 4.4.22. Recall that

$$E_p(z) = (1 - z)e^{\left(z + \frac{z^2}{2} + \cdots + \frac{z^p}{p}\right)}.$$

We note that

$$(1 - E_p(z))' = e^{\left(z + \frac{z^2}{2} + \cdots + \frac{z^p}{p}\right)} - (1 - z)(1 + z + \cdots + z^{p-1})e^{\left(z + \frac{z^2}{2} + \cdots + \frac{z^p}{p}\right)}$$
$$= z^p e^{\left(z + \frac{z^2}{2} + \cdots + \frac{z^p}{p}\right)},$$

and therefore the coefficients in the power series expansion at the origin of the function

$$(1 - E_p(z))'$$

are positive numbers. Thus there exist positive numbers $b_{n,p}, n > p$ such that

$$1 - E_p(z) = \sum_{n=p+1}^{\infty} z^n b_{n,p}, \quad |z| < 1.$$

The radius of convergence of this series is infinity. Setting $z = 1$ we obtain that

$$\sum_{n=p+1}^{\infty} b_{n,p} = 1.$$

Hence, for $|z| \le 1$,

$$|1 - E_p(z)| = \left| \sum_{n=p+1}^{\infty} z^n b_{n,p} \right|$$
$$\le |z|^{p+1} \left\{ \sum_{n=p+1}^{\infty} |z|^{n-p-1} b_{n,p} \right\}$$
$$\le |z|^{p+1} \left\{ \sum_{n=p+1}^{\infty} b_{n,p} \right\}$$
$$= |z|^{p+1}. \qquad\qquad \square$$

Remark 4.5.1. The bound,

$$|1 - E_p(z)| \le e^{|z|^{p+1}} - 1, \quad z \in \mathbb{C}, \tag{4.5.11}$$

can be found in [60, Problem 7, p. 13]. When $|z| \le 1$ it leads to

$$|1 - E_p(z)| \le (e - 1)|z|^{p+1}$$

Solution of Exercise 4.4.23. The infinite product can be rewritten as

$$\prod_{n=1}^{\infty} E_2(z/\sqrt{n}),$$

where E_2 is defined by (3.7.14) with $p = 2$. Let $R > 0$ and consider $|z| \leq R$. Take $n_0(R) \in \mathbb{N}$ such that $R/\sqrt{n_0(R)} < 1$. Then

$$|z/\sqrt{n}| < 1, \quad \forall |z| \leq R \quad \text{and} \quad n \geq n_0(R).$$

For such z and n, the previous exercise gives us

$$|1 - E_2(z\sqrt{n})| \leq \frac{R^3}{n^{3/2}}.$$

Taking into account the bound (3.7.3) we see that the infinite product

$$\prod_{n=n_0(R)}^{\infty} E_2(z/\sqrt{n})$$

(and hence the infinite product (4.4.22)) converges uniformly in $|z| \leq R$. The claims on the analyticity of the infinite product and on its zeros follow.   □

Chapter 5

Cauchy's Theorem

In this chapter we need the simplest version of Cauchy's theorem, and not the homological or homotopic versions. Furthermore, in the computations of Section 5.1, the weaker form of Cauchy's theorem proved using Green's theorem is enough.

We begin with a couple of exercises on the computation of path integrals, and then focus on exercises related to Cauchy's theorem.

5.1 Line integrals

In topology, and in particular in algebraic topology, the emphasis is on *continuous* rather than differentiable functions. A path (we will also say arc, or curve) will be any subset of $\mathbb{C}$ homeomorphic to $[0, 1]$, and a simple closed path, or Jordan curve, will be any subset of $\mathbb{C}$ homeomorphic to the unit circle. See for instance [118, p. 19]. In the setting of complex integration, we need to change a bit the above definition of a path, and consider appropriate equivalent classes. Two continuous complex-valued functions defined on compact intervals

$$\gamma_\ell(t) = x_\ell(t) + iy_\ell(t), \quad t \in [a_\ell, b_\ell], \quad \ell = 1, 2,$$

are called equivalent if there is an increasing homeomorphism φ of class C_1 from $[a_1, b_1]$ onto $[a_2, b_2]$ such that

$$\gamma_1(t) = \gamma_2(\varphi(t)), \quad t \in [a_1, b_1]. \tag{5.1.1}$$

An equivalence class of such functions will be called a continuous path (or arc, or curve), and the elements of an equivalence class C are called the parametrizations of the arc; it will be called a smooth path when the components of one (and hence all) of the elements in the equivalence class are continuously differentiable. All the elements in a given equivalence class have the same geometric image in the complex plane, but the path and its image are completely different objects. This is illustrated in the next examples, where the two given functions have for image the unit circle, but are not equivalent.

Exercise 5.1.1. *Prove that the smooth paths defined by the parametrizations*

$$\gamma_1(t) = e^{it}, \quad t \in [0, 2\pi],$$

and

$$\gamma_2(t) = e^{it}, \quad t \in [0, 4\pi]$$

are not equivalent.

Let C be a smooth path. It is called *closed* if for one (and hence for every) parametrization $\gamma(t), t \in [a, b]$ of C, it holds that

$$\gamma(a) = \gamma(b).$$

It is called *simple* if one (and hence every) parametrization γ of C is one-to-one onto its image. It is called *simple and closed* if

$$\gamma(t) = \gamma(s) \iff t, s \in \{a, b\}.$$

Let C_1 and C_2 be two smooth paths, with parametrizations $\gamma_1(t), t \in [a_1, b_1]$ and $\gamma_1(t), t \in [a_2, b_2]$, and assume that

$$\gamma_1(b_1) = \gamma_2(a_2).$$

The *concatenation* of γ_1 and γ_2 is the continuous arc with a parametrization given by

$$\gamma(t) = \begin{cases} \gamma_1(t), & t \in [a_1, b_1], \\ \gamma_2(t + a_2 - b_1), & t \in [b_1, b_1 + b_2 - a_2], \end{cases}$$

and will be denoted by $C = C_1 C_2$.[4] We will usually use the simple representation

$$\gamma(t) = \begin{cases} \gamma_1(t), & t \in [a_1, b_1], \\ \gamma_2(t), & t \in [a_2, b_2]. \end{cases}$$

A piecewise smooth arc C is given by the concatenation $C = C_1 C_2 \cdots C_N$ of a finite number of smooth arcs $C_1, C_2, \ldots, C_N$ such that

$$\gamma_j(b_j) = \gamma_{j+1}(a_{j+1}), \quad j = 1, \ldots, N - 1$$

where $\gamma_j : [a_j, b_j] \mapsto \mathbb{C}$ is a parametrization of C_j. The piecewise smooth arc is called closed if moreover,

$$\gamma_N(b_N) = \gamma_1(a_1).$$

---

[4]Sometimes the notation $C = C_2 C_1$ is used. Here we will stick to the notation $C = C_1 C_2$.

The closed piecewise smooth path will be called a closed contour (or a simple closed Jordan curve; see [53, p. 54]) if the continuous path obtained by successive concatenation of the arcs $C_\ell, \ell = 1, \ldots, N$ is closed and simple.

Given a function f continuous on the image of the path, one defines the line integral

$$\int_C f(z)dz = \int_a^b f(\gamma(t))\gamma'(t)dt, \tag{5.1.2}$$

where γ is a parametrization of the path; the integral does not depend on the choice of the parametrization, as follows from the change of variable theorem. The integral on a piecewise smooth path is defined as the sum of the integrals on the smooth paths which compose it.

Exercise 5.1.2. *Compute* $\int_C (x^2 - iy^2)dz$ *where* C *is the upper semicircle:* $z(t) = \cos t + i \sin t$ *with* $0 \le t \le \pi$ *([156, Exercice 13, p. 175]).*

It is well to recall the formula

$$\left| \int_C f(z)dz \right| \le \max_{t \in [a,b]} |f(\gamma(t))| L(C), \tag{5.1.3}$$

for a piecewise smooth path C of length $L(C)$.

The formula is used in particular in Exercise 5.1.4 below, taken from the paper [99]. There, the result serves as a tool in a proof of a simple version of Riemann's mapping theorem. A follow-up of the exercise is given in Exercise 5.5.2. In formula (5.1.5) in Exercise 5.1.4, the notation $|dz|$ stands for the integral with respect to $|\gamma'(t)|dt$: If C is a smooth curve with parametrization $\gamma(t), t \in [a, b]$ and h is a continuous function on C, we have

$$\int_C h(z)|dz| = \int_a^b h(\gamma(t))|\gamma'(t)|dt. \tag{5.1.4}$$

Definition 5.1.3. (5.1.4) is called the line integral with respect to arc length (see, e.g., [22, Section 10.7]).

Exercise 5.1.4. *Let* f *be continuous in the open set* Ω *and let* C *be a smooth curve in* Ω*, not containing the origin. Show that*

$$\left| \int_C \frac{f(z)}{z}dz \right|^2 \le \left(\max_{z \in C} \frac{1}{|z|^2} \right) \cdot L(C) \cdot \int_C |f(z)|^2 |dz|. \tag{5.1.5}$$

The result presented in the following exercise is also taken from [99, p. 825]; see also [98, pp. 397–398] and [145, p. 528].

Exercise 5.1.5. *Let* C *be a smooth Jordan curve, and let* z_0 *and* z_1 *be two points symmetric with respect to a common given normal of* C*. Let* $\gamma(t), t \in [a, b]$*, be a parametrization of* C*. Show that*

$$\lim_{|z_0 - z_1| \to 0} \left(\max_{t \in [a,b]} \frac{|\gamma(t) - z_0|}{|\gamma(t) - z_1|} \right) = 1. \tag{5.1.6}$$

Hint. Use Exercise 2.2.1 and the family of coaxal circles based on z_0 and z_1, and consider the set of points z such that, for $\lambda \in (0,1)$ given,

$$\lambda < \frac{|z - z_0|}{|z - z_1|} < \frac{1}{\lambda}. \tag{5.1.7}$$

The next exercise is taken from Cartan's book; see [45, Exercice 1, p. 76]. In the statement we use the notation $|z| = 1$ for the path

$$\gamma(t) = e^{2\pi i t}, \quad t \in [0,1].$$

This is of course an abuse of notation, since a path and its image are two different objects.

Exercise 5.1.6. *Let C be a path with parametrization $\gamma(t) = x(t) + iy(t)$ and let C^* be the path with parametrization*

$$\gamma^*(t) \stackrel{\text{def.}}{=} \overline{\gamma(t)} = x(t) - iy(t),$$

where $t \in [a,b]$. Assume that f is defined on the images of both C and C^*. Show that

$$\overline{\left(\int_C f(z)dz \right)} = \int_{C^*} \overline{f(\bar{z})}dz, \tag{5.1.8}$$

and

$$\overline{\left(\int_{|z|=1} f(z)dz \right)} = - \int_{|z|=1} \overline{f(z)} \frac{dz}{z^2}. \tag{5.1.9}$$

Let C be a piecewise smooth closed path, and let $z_0 \notin \operatorname{Ran} C$. The number

$$W(C, z_0) = \frac{1}{2\pi i} \int_C \frac{dz}{z - z_0} \tag{5.1.10}$$

is an integer. The number $W(C, z_0)$ is called the *winding number* of the closed curve around the point z_0 (see for instance [144, p. 134]), or the index of z_0 with respect to the closed curve C. We note that various notation and variations on the above terminology appear in the literature, and sometimes one can find the notation $n(z_0, C)$ (that is, z_0 appears before C). For instance, Cartan in [45, p. 62] speaks of the index of C with respect to the point z_0, and uses the notation $I(C, z_0)$. Ahlfors, [4, p. 114], uses the notation $n(C, z_0)$, and Dieudonné in [63, p. 228] uses both $j(z_0, C)$ and $j(C, z_0)$, and speaks of the index of the curve with respect to the point z_0, or of the point z_0 with respect to the curve. Andersson, see [19, p. 18], speaks of the index of the curve with respect to the point z_0, and uses the notation $\operatorname{Ind}_C(z_0)$

Exercise 5.1.7. *Prove that indeed the winding number is an integer.*

For an extension of the preceding result, see Exercise 5.5.22.

The next result is also important when computing definite integrals using the residue theorem. See [1, Exercise 8, p. 81] for a particular case. In the statement, $|z| = R$ denotes the path

$$\gamma(t) = Re^{2\pi it}, \quad t \in [0, 1].$$

Exercise 5.1.8. *Let $p(z)$ and $q(z)$ be two polynomials and assume that $\deg q \geq \deg p + 2$. Show that*

$$\lim_{R \to \infty} \int_{|z|=R} \frac{p(z)}{q(z)} dz = 0.$$

Jordan's lemma plays an important role in the sequel. It reads as follows:

Exercise 5.1.9 (Jordan's lemma). *It holds that*

$$\lim_{R \to \infty} \int_0^{\frac{\pi}{2}} e^{-R \sin t} dt = 0 \quad and \quad \lim_{R \to \infty} \int_0^{\frac{\pi}{2}} e^{-R \cos t} dt = 0. \tag{5.1.11}$$

Integrals in (5.1.11) are integrals on a closed interval, and as such, are particular instances of line integrals. The following exercise requires a different more geometric interpretation.

Exercise 5.1.10. *Write the integrals in (5.1.11) as line integrals on some arc of a circle.*

Exercise 5.1.11. *Let Ω be an open connected subset of $\mathbb{R}^2$, and let u admit continuous first-order partial derivatives in Ω. Show that, for every closed piecewise smooth path C in Ω,*

$$\int_C \frac{\partial u}{\partial x} dx + \frac{\partial u}{\partial y} dy = 0. \tag{5.1.12}$$

Remark 5.1.12. When u admits continuous second-order derivatives and when moreover C is simple, (5.1.12) is a direct consequence of Green's theorem. Compare with (9.2.2).

5.2 The fundamental theorem of calculus for holomorphic functions

Assume now that the function f is $\mathbb{C}$-differentiable at the point $\gamma(t)$. Then, the complex-valued function of a real variable $f(\gamma(t))$ is differentiable, and its derivative is given by the formula

$$f(\gamma(t))' = \gamma'(t)f'(\gamma(t)). \tag{5.2.1}$$

This important and non trivial formula calls for some comments. The function on the left side of (5.2.1) is the derivative of the complex-valued function of a real variable. Similarly, the term $\gamma'(t)$ is the derivative at the point t of the complex-valued function of a real variable $\gamma(t)$. On the other hand, $f'(\gamma(t))$ is the complex derivative of the function f at the point $\gamma(t)$.

Formula (5.2.1) leads to the *fundamental theorem of calculus for holomorphic functions*. In the statement, the derivative $f'(z)$ is assumed continuous. This hypothesis is in fact superfluous, since a holomorphic function has derivatives of all orders. But this fact is proved at a later stage, following the Cauchy–Goursat theorem.

Theorem 5.2.1. *Let Ω be an open connected set, and let C be a smooth path with parametrization $\gamma(t), t \in [a, b]$. Let f be holomorphic in Ω, and assume that f' is continuous in Ω. Then,*

$$\int_C f'(z)dz = f(\gamma(b)) - f(\gamma(a)). \tag{5.2.2}$$

Formula (5.2.2) is also called the Newton–Leibniz formula. See [74, p. 37]. At this stage it is well to recall the following definition: A function f is a primitive of a function g in an open set Ω if we have

$$f'(z) = g(z), \quad \forall z \in \Omega.$$

Note that g is necessarily continuous since differentiability at a point implies continuity at that point. In fact, g is necessarily holomorphic, since f is holomorphic and since a holomorphic function has derivatives of all orders. Every power series has a primitive in its domain of convergence. This is a local result. One important difference with the real case is the following: A given function g may lack a primitive in a set Ω but admits one in some open subset of Ω. The geometry of the set plays an important role; see Exercise 5.2.3 for a first illustration of this fact. Exercises on existence of primitives will be given in Section 5.7. Here, as a corollary of Theorem 5.2.1 we get the proof of the direct statement in the following theorem:

Theorem 5.2.2. *Let Ω be an open connected subset of $\mathbb{C}$ and let g be continuous in Ω. A necessary and sufficient condition for g to have a primitive in Ω is that*

$$\int_C g(z)dz = 0 \tag{5.2.3}$$

holds for every closed path C in Ω.

Exercise 5.2.3. *The function $f(z) = 1/z$ has no primitive in $\mathbb{C} \setminus \{0\}$.*

We recall that the function $1/z$ does have a primitive in the set $\mathbb{C} \setminus (-\infty, 0]$. This last set is star-shaped. In fact it follows from the Cauchy–Goursat theorem that any function holomorphic in a star-shaped domain has a primitive there. This leads us to the following definition of a simply-connected set:

Definition 5.2.4. An open connected set $\Omega \subset \mathbb{C}$ is *simply-connected* if every function holomorphic in Ω has a primitive there, or, equivalently, if (5.2.3) holds for every closed path C in Ω.

Let us go back now to Theorem 5.2.1. Integration by parts in the present setting reads as follows:

Exercise 5.2.5 (see [219, p. 58]). *In the notation of Theorem 5.2.1, let f and g be holomorphic in Ω, and such that f' and g' are continuous on C. Let $\gamma : [a,b] \mapsto \Omega$ be a parametrization of C. Then,*

$$\int_C f(z)g'(z)dz = (fg)(\gamma(b)) - (fg)(\gamma(a)) - \int_C f'(z)g(z)dz. \tag{5.2.4}$$

Theorem 5.2.1 allows us to get a simpler solution to Exercise 3.4.15.

Exercise 5.2.6. *Give a proof of* (3.4.20):

$$|e^{z_1} - e^{z_2}| \le |z_1 - z_2|,$$

where z_1 and z_2 are in the left closed half-plane, using Theorem 5.2.1.

A non-trivial application of Theorem 5.2.1 allows us to compute the Fresnel integrals

$$\int_{\mathbb{R}} \cos(t^2)dt \quad \text{and} \quad \int_{\mathbb{R}} \sin(t^2)dt. \tag{5.2.5}$$

These last integrals play an important role in optics. They can be computed by direct methods, without using complex analysis; see for instance [83], [147]. Theorem 5.2.1 allows us to compute them directly. The existence of the Fresnel integrals follows from the proof itself, but it is also a simple, but instructive, calculus exercise to check directly that they converge. The idea is as follows: It is enough to check that the limit

$$\lim_{R \to \infty} \int_1^R \cos(t^2)dt$$

exists since the integral $\int_0^1 \cos(t^2)dt$ exists. But

$$\int_1^R \cos(t^2)dt = \int_1^R \frac{1}{2t}(2t\cos(t^2))dt$$

$$= \left(\frac{\sin(t^2)}{2t}\right)_1^R + \int_1^R \frac{\sin(t^2)}{2t^2}dt.$$

The first term tends to $-\dfrac{\sin 1}{2}$ as $R \to \infty$ and the integral $\int_1^R \dfrac{\sin(t^2)}{2t^2}dt$ is absolutely convergent since

$$\left|\frac{\sin(t^2)}{2t^2}\right| \le \frac{1}{2t^2}.$$

Thus $\int_0^\infty \cos(t^2)\,dt$ converges. The proof of the convergence of the other Fresnel integral is of course similar. For another proof of convergence, see [61, Solution to 1.5.21, p. 209].

Usually, these integrals are computed in books using the weak version of Cauchy's theorem (see the discussion before Exercise 5.3.2 for the latter). Saks and Zygmund in [192, p. 103] compute the Fresnel integral in an even easier way: They remark that e^{-z^2} has a primitive in $\mathbb{C}$ (since it is given by a power series with infinite radius of convergence), and resort directly to Theorem 5.2.1 to show that the integral in the closed contour in Exercise 5.2.7 below is equal to 0.

To compute the Fresnel integrals we need the Gaussian integral

$$\int_{\mathbb{R}} e^{-t^2}\,dt = \sqrt{\pi}. \tag{5.2.6}$$

More generally, the moments

$$\int_{\mathbb{R}} e^{-t^2} t^{2u}\,dt, \quad u = 1, 2, \ldots \tag{5.2.7}$$

could be computed using integration by part starting from (5.2.6). In Exercise 13.5.1, and also using (5.2.6), these moments are computed using the Fourier transform.

Recall that the integral (5.2.6) may be computed in the following way. Set $K = \int_{\mathbb{R}} e^{-t^2}\,dt$. Then,

$$K^2 = \left(\int_{\mathbb{R}} e^{-t^2}\,dt \right) \left(\int_{\mathbb{R}} e^{-s^2}\,ds \right)$$

$$= \iint_{\mathbb{R}^2} e^{-(t^2+s^2)}\,dt\,ds$$

$$= \iint_{[0,\infty)\times[0,2\pi]} e^{-r^2} r\,dr\,d\theta = \pi,$$

where one has made the change of variables $t = r\cos\theta$ and $s = r\sin\theta$. As recalled in [27], one can also compute this integral using the residue theorem. See Exercise 8.5.1.

Exercise 5.2.7.

(a) *Show that the function e^{-z^2} has a primitive in $\mathbb{C}$.*

(b) *For $R > 0$ consider the closed contour $\Gamma_R = \gamma_{1,R} + \gamma_{2,R} + \gamma_{3,R}$ where:*

    (i) *$\gamma_{1,R}$ is the interval $[0, R]$.*

    (ii) *$\gamma_{2,R}$ is the arc of the circle of radius R and centered at the origin, with angle varying between 0 and $\pi/4$.*

(iii) $\gamma_{3,R}$ *is the interval linking the origin to the point* $Re^{\frac{i\pi}{4}}$.

Compute the Fresnel integrals by computing the integral of the function e^{-z^2} *over the contour* Γ_R *and letting* $R \to \infty$.

Similar arguments allow us to compute the integrals in the next exercise; see [42, p. 386] (see also [83, 147] and Remark 5.9.1 after the solution of the exercise).

Exercise 5.2.8. *Show that*

$$\int_0^\infty e^{-t^2} \cos t^2 dt = \frac{\sqrt{\pi}\sqrt{\sqrt{2}+1}}{4},$$

$$\int_0^\infty e^{-t^2} \sin t^2 dt = \frac{\sqrt{\pi}\sqrt{\sqrt{2}-1}}{4}.$$

We now give an application of the fundamental theorem of calculus for line integrals to prove an injectivity result.

Exercise 5.2.9. *Assume that the complex numbers* $a_2, a_3, \ldots$ *are such that*

$$\sum_{n=2}^\infty n|a_n| < 1. \tag{5.2.8}$$

Show that

$$f(z) = z + \sum_{n=2}^\infty a_n z^n$$

defines a function f holomorphic in the open unit disk $\mathbb{D}$, *and that f is one-to-one in* $\mathbb{D}$.

Another proof of the above result is the topic of Exercise 10.2.3. The result itself can be found in [191], where the following question is also added:

Let $c > 1$. *There exists a function such that*

$$\sum_{n=2}^\infty n|a_n| \le c,$$

and which is not one-to-one in the open unit disk.

5.3 Computations of integrals

Let f be continuous in a convex open set Ω, and holomorphic in $\Omega \setminus \{z_0\}$, for some point $z_0 \in \Omega$. The Cauchy–Goursat theorem for triangles states that, for every triangle in Ω, with boundary $\partial\Delta$, it holds that

$$\int_{\partial\Delta} f(z)dz = 0.$$

It follows that f has a primitive in Ω and therefore (5.2.3)

$$\int_C f(z)dz = 0$$

for every closed path C inside Ω. The same conclusion holds when Ω is star-shaped.

It is also useful to consider open sets which are not simply-connected, or curves which are not connected. For the present setting we will only need the following result:

Theorem 5.3.1. *Let Ω be a convex set and let $C, C_1, \ldots, C_N$ be simple closed non-intersecting curves inside Ω. Assume that the interiors $R_1, \ldots, R_N$ of the curves $C_1, \ldots, C_N$ are inside the interior R of C. Let f be analytic in a neighbourhood of $R - \bigcup_{n=1}^{N} C_j$. Then,*

$$\int_C f(z)dz = \sum_{n=1}^{N} \int_{C_n} f(z)dz, \tag{5.3.1}$$

where all the curves have the positive orientation.

The proof of (5.2.3) is very involved when no continuity hypothesis is made on the derivative of f in Ω. On the other hand, when one assumes f' continuous in Ω, and when γ is a closed simple path whose interior is contained in Ω, (5.2.3) is a simple consequence of Green's theorem; see [53, p. 60]. This is the version of Cauchy's theorem which we need in this section to compute certain definite integrals. If you want to avoid Green's theorem and use Cauchy's theorem, one has, for instance, to choose $\Omega = \mathbb{C} - i(-\infty, 0]$ in the following exercises. It is star-shaped with respect to any point on $i(0, \infty)$, and therefore the Cauchy–Goursat theorem insures that the integral on a closed curve of any function holomorphic in Ω vanishes.

Exercise 5.3.2. *Compute $\int_0^\infty \frac{\sin t}{t} dt$ using Cauchy's theorem as follows: Integrate the function $\frac{e^{iz}}{z}$ on the closed contour defined below and let $R \to \infty$ and $\epsilon \to 0$. The contour is built from four parts (both ϵ and R are strictly positive numbers):*

(i) *$\gamma_{1,R,\epsilon}$ is the real interval $[\epsilon, R]$.*

(ii) *$\gamma_{2,R}$ is the half-circle of radius R, centered at the origin, which lies in the upper half-plane, and positively oriented.*

(iii) *$\gamma_{3,R,\epsilon}$ is the real interval $[-R, -\epsilon]$.*

(iv) *$\gamma_{4,\epsilon}$ is the half-circle of radius ϵ, centered at the origin, and which lies in the upper half-plane, and with negative orientation.*

The above integral is called the *Dirichlet integral*. In the following exercise the formula

$$\lim_{\epsilon \to 0} \int_{\gamma_{4,\epsilon}} \frac{g(z)}{z} dz = -i\pi g(0), \tag{5.3.2}$$

which can also be written, if you already know the notions of residue and of simple pole, as

$$\lim_{\epsilon \to 0} \int_{\gamma_{4,\epsilon}} h(z)dz = -i\pi \operatorname{Res}(h,0),$$ (5.3.3)

when h has a simple pole at $z = 0$, will prove useful.

The above formula still holds when 0 is a pole of h of order possibly bigger than one, when in the Laurent expansion of h at the origin only odd powers occur. See [45, Lemma 4, p. 105] (for a simple pole) and Exercise 7.3.2.

Exercise 5.3.3. *Using the function*

$$f(z) = \frac{1 - e^{2iz}}{z^2}$$

and the same path as in the previous exercise, compute

$$\int_{\mathbb{R}} \left(\frac{\sin x}{x}\right)^2 dx.$$

Exercise 5.3.4. *Show that*

$$\int_{\mathbb{R}} \left(\frac{\sin x}{x}\right)^3 dx = \frac{3\pi}{4}.$$

We want to compute the integrals

$$\int_0^\infty \frac{1 - \cos x}{x^2} x^{1-2H} dx,$$

where $H \in (0,1)$. Such integrals appear in computations related to the fractional Brownian motion.

Exercise 5.3.5. *Show that, for $H \in (0,1)$ different from $1/2$,*

$$\int_0^\infty \frac{1 - \cos x}{x^2} x^{1-2H} dx = \frac{\cos(\pi H)\Gamma(1 - 2H)}{2H} = \frac{\cos(\pi H)\Gamma(2 - 2H)}{(1 - 2H)2H},$$ (5.3.4)

where Γ denotes the Gamma function (3.1.11).

Hint. Distinguish the cases $H \in (0, 1/2)$ and $H \in (1/2, 1)$, and in each case integrate an appropriate function along the contour $\gamma_{\epsilon,R}$ constructed as follows: $\gamma_{\epsilon,R}$ consists of four components:

(1) The interval $[\epsilon, R]$.
(2) The quarter of circle C_R of radius R linking the points R and iR.
(3) The interval $[iR, i\epsilon]$.
(4) The quarter of circle c_ϵ of radius ϵ and linking the points ϵ and $i\epsilon$.

The Gamma function is studied in further detail in Exercise 13.4.2. It is the Mellin transform (see (13.4.1)) of the function e^{-t}, $t > 0$.

5.4 Riemann's removable singularities theorem (Hebbarkeitssatz)

We note that an important consequence of the Cauchy–Goursat theorem is Riemann removable singularity theorem (Riemann Hebbarkeitsseitz):

Theorem 5.4.1. *Let f be bounded in $B(z_0, r)$ and holomorphic in $B(z_0, r) \setminus \{z_0\}$. Then*

$$\lim_{z \to z_0} f(z) \stackrel{\text{def.}}{=} \ell$$

exists, and the function h defined by

$$h(z) = \begin{cases} f(z), & z \neq z_0, \\ \ell, & z = z_0, \end{cases}$$

is holomorphic in $B(z_0, r)$.

See for instance [81, Satz 4.3, p. 79]. We briefly recall its proof here. One first assumes f continuous rather than only bounded. From the Cauchy–Goursat theorem we have that $\int_{\partial \Delta} f(z) dz = 0$ for every triangle $\Delta \subset B(z_0, r)$. Morera's theorem implies then that f has a primitive in $B(z_0, r)$ and hence is analytic in $B(z_0, r)$. To deal with the bounded case, it suffices to consider the function

$$g(z) = (z - z_0) f(z).$$

It is continuous at the point z_0 since f is bounded. By the preceding argument it is analytic in $B(z_0, r)$, and it is a power series expansion there:

$$(z - z_0) f(z) = a_1(z - z_0) + a_2(z - z_0)^2 + \cdots .$$

It follows that f has also a power expansion in $B(z_0, r)$, and therefore is analytic there.

The function $f(x) = |x|$ shows, if need be, that an analog of Theorem 5.4.1 does not hold in the real case. The function $|x|$ is of class C^∞ in $\mathbb{R} \setminus \{0\}$, and is continuous at the origin. It is not differentiable at the origin.

Exercise 5.4.2. *Let $z_0 \in \mathbb{C}$ and let φ be analytic and with positive real part in $\Omega = \{z \,;\, 0 < |z - z_0| < 1\}$. Show that z_0 is a removable singularity.*

For the following two exercises we recall that the zeroes of an analytic function cannot accumulate at a point of the domain of analyticity of the given function. This is a consequence of the existence of a power series expansion at every point of the domain of analyticity.

Exercise 5.4.3. *Let f be a function continuous in an open set $\Omega \subset \mathbb{C}$ and assume that f^2 is holomorphic in Ω. Show that f is holomorphic in Ω.*

The example $f(x) = |x|$ shows that the claim in the previous exercise does not have a direct counterpart in the real case. Furthermore, the example

$$f(z) = \begin{cases} 1, & \text{if } z \neq 0, \\ -1, & \text{if } z = 0, \end{cases}$$

shows that one cannot assume f to be only bounded in Ω. In a similar vein, we have:

Exercise 5.4.4. *Let f be defined in the open set Ω and assume that both f^2 and f^3 are analytic there. Show that f is analytic in Ω. More generally, let $n_1, n_2 \in \mathbb{N}$ be relatively prime, and assume that f^{n_1} and f^{n_2} are analytic in Ω. Show that f is analytic in Ω.*

Exercise 5.4.5. *Let f be analytic in Ω. For $a \in \Omega$ define*

$$R_a f(z) = \begin{cases} \frac{f(z)-f(a)}{z-a}, & z \neq a, \\ f'(a), & z = a. \end{cases} \tag{5.4.1}$$

Show that $R_a f$ is still analytic in Ω and that the resolvent identity

$$R_a f - R_b f = (a-b) R_a R_b f, \quad \forall a, b \in \Omega. \tag{5.4.2}$$

holds.

We note that the resolvent identity (5.4.2) also appears in algebra and in analysis: Let M be an $n \times n$ matrix with complex entries, let I_n denote the $n \times n$ identity matrix, and let

$$R(a) = (M - aI_n)^{-1}$$

for a in the resolvent set of M. Then we have

$$R(a) - R(b) = (a-b)R(a)R(b).$$

Indeed,

$$\begin{aligned} R(a) - R(b) &= (M - aI_n)^{-1} - (M - bI_n)^{-1} \\ &= (M - aI_n)^{-1}\left(M - bI_n - (M - aI_n)\right)(M - bI_n)^{-1} \\ &= (a-b)(M - aI_n)^{-1}(M - bI_n)^{-1} \\ &= (a-b)R(a)R(b). \end{aligned}$$

The resolvent identity, and in particular the backward-shift operator

$$R_0 f(z) = \frac{f(z) - f(0)}{z},$$

play a key role in operator theory. See Section 16.1. An important problem is the study of all closed R_0 invariant subspaces of the Hardy space $\mathbf{H}_2(\mathbb{D})$ (for the definition of this space, see Definition 5.6.11). Another example of operators satisfying (5.4.2) is presented in Exercise 12.3.3. To discuss these various points would lead us too far astray.

5.5   Cauchy's formula and applications

We will begin with a result which is used in the proof of Malgrange's theorem on the existence of a fundamental solution of a partial differential equation. See [29, Lemma 7.3, p. 205]. We send the reader to [29, (7.14) p. 214] to see how the lemma is used. We will not recall the statement of the theorem here. Recall that a monic polynomial is a polynomial whose highest power has coefficient 1:

$$p(z) = z^n + a_{n-1}z^{n-1} + \cdots + a_0. \tag{5.5.1}$$

Exercise 5.5.1. *Let f be a function analytic in $|z| < 1 + \epsilon$ for some $\epsilon > 0$. Then for every monic polynomial p it holds that*

$$|f(0)| \le \frac{1}{2\pi} \int_0^{2\pi} |f(e^{it})p(e^{it})| dt. \tag{5.5.2}$$

The following result, a bit in the spirit of (5.5.2), is a follow-up of Exercise 5.1.4. The result in fact holds for any simply connected set containing the origin, and C any simple Jordan curve with interior containing the origin. See [99].

Exercise 5.5.2. *In the notation of Exercise 5.1.4, take $\Omega = \mathbb{D}$ and C to be a circle of radius $r < 1$. Show that the minimum of the expression*

$$\int_C |f(z)|^2 |dz|$$

over all functions analytic in Ω and such that $f(0) = 1$ is strictly positive.

Consider the function $f(t) = e^{-it}$ on the interval $[0, 2\pi]$. By Weierstrass' approximation theorem, there exists a sequence of polynomials $p_n(t)$ *in the variable t, such that*

$$\lim_{n \to \infty} \max_{t \in [0,2\pi]} |e^{-it} - p_n(t)| = 0.$$

The result is not true if one takes polynomials in e^{it}, as is required to be shown in the following exercise, taken from [184, p. 127].

Exercise 5.5.3. *Show that there is a constant M such that, for every polynomial $p(z)$,*

$$\max_{z \in \mathbb{T}} |z^{-1} - p(z)| \ge M. \tag{5.5.3}$$

Exercise 5.5.4. *Let Ω be a star-shaped open set and let C be a closed simple smooth curve in Ω. Let z_0 be a point not on the image of C, and let f be analytic in Ω. Prove that*

$$\int_C \frac{f'(z)}{z - z_0} dz = \int_C \frac{f(z)}{(z - z_0)^2} dz. \tag{5.5.4}$$

As usual, in the following exercise, $|z| = r$ is an abuse of notation for the closed curve with parametrization

$$\gamma(t) = re^{it}, \quad t \in [0, 2\pi].$$

Exercise 5.5.5. *Compute the integral*

$$\int_{|z|=r} \frac{e^{\sin z^2} dz}{(z^2 + 1)(z - 2i)^3}$$

for strictly positive r different from 1 and 2.

For the next two exercises, it is well to use the formula

$$\int_0^{2\pi} f(e^{it}) dt = \int_{|z|=1} f(z) \frac{dz}{iz}. \tag{5.5.5}$$

Exercise 5.5.6. *Compute*

$$\int_0^{2\pi} e^{e^{2it} - 3it} dt.$$

Exercise 5.5.7. *Solve Exercise* 3.1.8 *using Cauchy's formula.*

Exercise 5.5.8. *Let f be analytic in $|z| < 1 + \epsilon$ for some $\epsilon > 0$. Show that*

$$f(z) = i \operatorname{Im} f(0) + \frac{1}{2\pi} \int_0^{2\pi} \frac{e^{it} + z}{e^{it} - z} \left(\operatorname{Re} f(e^{it}) \right) dt, \quad z \in \mathbb{D}. \tag{5.5.6}$$

Prove that for z in the open unit disk,

$$f(z) - f(0) = \frac{1}{2\pi} \int_0^{2\pi} \frac{2z}{e^{it} - z} \operatorname{Re} f(e^{it}) dt,$$

$$\frac{f^{(n)}(z)}{n!} = \frac{1}{2\pi} \int_0^{2\pi} \frac{2e^{it}}{(e^{it} - z)^{n+1}} \left(\operatorname{Re} f(e^{it}) \right) dt, \quad n = 1, 2, \ldots \tag{5.5.7}$$

$$\operatorname{Re} f(z) = \frac{1 - |z|^2}{2\pi} \int_0^{2\pi} \frac{\operatorname{Re} f(e^{it})}{|e^{it} - z|^2} dt.$$

The proof of Exercise 5.5.8 presented in this section is a direct application of Cauchy's formula. Another proof for polynomials is asked for in Exercise 7.3.11.

It is interesting to remark that the right side of (5.5.6) is analytic in the non-connected set $\mathbb{C} \setminus \mathbb{T}$, while f is analytic in a neighborhood of the closed unit disk. The formula (5.5.6) *does not* coincide with f outside the open unit disk.

We note that formulas (5.5.6) and (5.5.7) express the analytic function as an integral of its real part. Analogous formulas exist for functions analytic in $|z| < R + \epsilon$. This is the topic of the next exercise.

Exercise 5.5.9 (see [139, Hilfssatz 2, z. 242]). *Let f be analytic in $|z| < R + \epsilon$, with $R > 1$. Show that for $|z| < R$ we have*

$$f(z) = i \operatorname{Im} f(0) + \frac{1}{2\pi} \int_0^{2\pi} \frac{2Rz}{Re^{it} - z} \left(\operatorname{Re} f(Re^{it}) \right) dt,$$

$$\frac{f^{(n)}(z)}{n!} = \frac{R^n}{2\pi} \int_0^{2\pi} \frac{2 \left(\operatorname{Re} f(Re^{it}) \right) e^{it} dt}{(Re^{it} - z)^{n+1}}.$$

(5.5.8)

We remark that formula (5.5.8) for $n = 1$ allows us to show directly that an entire function with a bounded real part is a constant.

Exercise 5.5.10. *Assume in Exercise 5.5.8 that $\operatorname{Re} f(e^{it}) \geq 0$ Then, f has a positive real part in the open unit disk.*

The general Herglotz representation formula states that a function analytic and with a positive real part in the open unit disk can be written as

$$f(z) = ia + \int_0^{2\pi} \frac{e^{it} + z}{e^{it} - z} d\mu(t),$$

(5.5.9)

where $d\mu$ is a positive and finite measure on $[0, 2\pi)$. To prove this formula, one first notes that the function $f(rz)$ with $r < 1$ is analytic in $|z| < 1/r$, with $1/r > 1$, and so one can apply to it formula (5.5.6):

$$f(rz) = i \operatorname{Im} f(0) + \frac{1}{2\pi} \int_0^{2\pi} \frac{e^{it} + z}{e^{it} - z} \left(\operatorname{Re} f(re^{it}) \right) dt.$$

The positive measures

$$d\mu_r(t) = \operatorname{Re} f(re^{it}) dt$$

are such that

$$\int_0^{2\pi} d\mu_r(t) = \operatorname{Re} f(0).$$

We pick up a sequence of numbers $(r_n)_{n \in \mathbb{N}}$ in $(0, 1)$ such that

$$\lim_{n \to \infty} r_n = 1.$$

At this stage, one has to resort to a deep result of functional analysis to ensure that the family $(d\mu_{r_n})$ has a convergent subsequence which tends to a positive and finite measure $d\mu$ in the following sense: For every continuous complex-valued function g defined on $[0, 2\pi]$ it holds that

$$\lim_{n \to \infty} \int_0^{2\pi} g(t) d\mu_{r_n}(t) = \int_0^{2\pi} g(t) d\mu(t).$$

This is Helly's theorem. For discussions, see [59, p. 158], [85, Proposition 7.19, p. 223] and [165, p. 220].

The definition of a non-negative (or positive) matrix, used in the following question, is recalled in Section 16.3. See Definition 16.3.1 there.

Exercise 5.5.11. *Let f be analytic in the open unit disk, and with a positive real part there. Prove that, for every $N \in \mathbb{N}$ and every choice of (not necessarily distinct) points in $\mathbb{D}$, the $N \times N$ matrix with (ℓ, j) entry equal to*

$$\frac{f(w_\ell) + \overline{f(w_j)}}{1 - w_\ell \overline{w_j}} \tag{5.5.10}$$

is non-negative.

In fact, the claim in the preceding exercise can be made much stronger: If f is a function *defined* in the open unit disk and such that all $N \times N$ matrices with (ℓ, j) entry (5.5.10) are non-negative, then φ is analytic in the open unit disk. One can also replace the open unit disk by a uniqueness set inside $\mathbb{D}$; see [6, Theorem 2.6.5, p. 39] for the case of contractive functions. The present case is deduced using the Cayley transform. In other words *positivity* implies analyticity. For more information we send the interested student to [67].

The previous discussion and formula (5.5.9) hint at deep connections between the theory of functions of a complex variable and functional analysis. A fascinating fact is that functions for which the condition in Exercise 5.5.11 holds, play an important role in electrical engineering. These connections go beyond the one variable case.

Exercise 5.5.12. *Assume in Exercise 5.5.8 that $\operatorname{Re} f(e^{it}) \geq 0$ for $t \in [0, 2\pi]$. Let*

$$f(z) = f_0 + 2 \sum_{\ell=1}^{\infty} f_\ell z^\ell$$

be the power expansion of f centered at the origin. Show that

$$|f_\ell| \leq \operatorname{Re} f_0, \quad \ell = 1, 2, \ldots. \tag{5.5.11}$$

As a consequence of the previous exercise we have the following result, which is still true when the function f in the statement is not assumed analytic across the unit circle. The proof requires then the general Herglotz representation formula (5.5.9). The result was proved by Harald Bohr in 1914, and is called Bohr's inequality.

Exercise 5.5.13. *Let f be analytic in $|z| < 1 + \epsilon$ for some $\epsilon > 0$ and assume that $|f(z)| \leq 1$ for $|z| < 1$. Let*

$$f(z) = \sum_{\ell=0}^{\infty} f_\ell z^\ell$$

be the power expansion of s centered at the origin. Show that

$$\sum_{\ell=0}^{\infty} |f_\ell z^\ell| \leq 1 \tag{5.5.12}$$

for $|z| \leq 1/3$.

Hint. Consider the function $1 - f$ and apply to it (5.5.11).

Remark 5.5.14. By considering functions of the form $f(z) = c\frac{z-z_0}{1-z\overline{z_0}}$, where $|z_0| < 1$ and $|c| = 1$, one shows that 1 is optimal in (5.5.12).

Also a consequence of Exercise 5.5.8, we have a special case of Harnack's inequalities (see [42, p. 143]). The result itself is valid without the hypothesis that f is analytic in $|z| < 1 + \epsilon$, but merely in $\mathbb{D}$.

Exercise 5.5.15. Let f be analytic in $|z| < 1 + \epsilon$ for some $\epsilon > 0$ and assume that $\operatorname{Re} f(e^{it}) \geq 0$ for $t \in [0, 2\pi]$. Show that

$$\frac{1-|z|}{1+|z|} \operatorname{Re} f(0) \leq \operatorname{Re} f(z) \leq \frac{1+|z|}{1-|z|} \operatorname{Re} f(0), \quad \forall z \in \mathbb{D}. \tag{5.5.13}$$

We now turn to an exercise which has a long history and can be found in numerous places (see for instance [207, p. 10]).

Exercise 5.5.16. Let $f = u + iv$ be analytic in $|z| < 1 + \epsilon$ with $\epsilon > 0$, and assume that $f(0) = 0$. Show that

$$\int_0^{2\pi} u(\cos t, \sin t)^4 dt \leq 36 \int_0^{2\pi} v(\cos t, \sin t)^4 dt \tag{5.5.14}$$

and

$$\int_0^{2\pi} v(\cos t, \sin t)^4 dt \leq 36 \int_0^{2\pi} u(\cos t, \sin t)^4 dt.$$

Hint. Apply Cauchy's formula to f^4 and $z_0 = 0$.

The next exercise is [122, Lemma 2.6.9, p. 61].

Exercise 5.5.17. Let f be an analytic function in $|z| < R$, with power series

$$f(z) = \sum_{k=0}^{\infty} a_k z^k, \quad |z| < R,$$

and let $r < R$. Let

$$M = \max_{|z| \leq r} |f(z)|.$$

Show that

$$|a_k z^k| \leq M, \quad k = 0, 1, 2, \ldots \quad \text{and} \quad |z| \leq r.$$

In a similar vein one has the next exercise, which can be found in [75, Exercise 10.37, p. 120].

Exercise 5.5.18. Let f be analytic in the open disk $|z| < R$ and assume that $|f'(z)| \leq M < \infty$ for $|z| < R$. Show that in the expansion $f(z) = \sum_{n=0}^{\infty} f_n z^n$ one has

$$|f_n| \leq \frac{M}{nR^{n-1}} \quad n = 1, 2, \ldots. \tag{5.5.15}$$

The next exercise can be found in [75, Exercise 10.38, p. 120].

Exercise 5.5.19. *Let f be analytic in the open disk $|z| < R$ and assume that $|f(z)| \leq Me^{|z|}$. Show that in the expansion $f(z) = \sum_{n=0}^{\infty} f_n z^n$ one has*

$$|f_n| \leq M \left(\frac{n}{e}\right)^{-n}. \tag{5.5.16}$$

The next exercise is taken from [62]. It is quite difficult if given without hints towards the solution.

Exercise 5.5.20. *Let f be an analytic function in the open unit disk and assume that*

$$\sup_{r \in [0,1)} \int_0^{2\pi} |f'(re^{it})| dt < M \quad \text{for some} \quad M > 0. \tag{5.5.17}$$

Show that $\int_0^1 |f(x)| dx < \infty$.

Hints. Write $f(z) = \sum_{n=0}^{\infty} a_n z^n$ and give an upper bound on $|a_n|$ using (5.5.17) and then give an upper bound to $\int_0^1 |f(x)| dx$ using

$$|f(x)| \leq \sum_{n=0}^{\infty} |a_n| x^n, \quad x > 0.$$

Exercise 5.5.21. *Show that, for $|z| < 1$,*

$$\frac{1}{2\pi i} \int_{\mathbb{T}} \frac{\zeta + z}{\zeta - z} \zeta^{n-1} d\zeta = \begin{cases} 2z^n, & n \geq 1, \\ 1, & n = 0, \\ 0, & n < 0. \end{cases} \tag{5.5.18}$$

Exercise 5.5.22. *Let f be analytic and not vanishing in $r_0 < |z - z_0| < r_1$. Show that, for $r_0 < r < r_1$,*

$$\frac{1}{2\pi i} \int_{|z-z_0|=r} \frac{f'(z)}{f(z)} dz \in \mathbb{Z}.$$

Exercise 5.5.23. *Show that there is no function f analytic in $\Omega = \mathbb{C} \setminus [-1, 1]$ such that $f(z)^2 = z$ there.*

Remark 5.5.24. The arguments of Exercise 6.3.6 cannot be applied in the exercise below, since $z = 0$ is not assumed to be an isolated singularity. The exercise could be seen as a consequence of the stronger statement in Exercise 4.1.10, but a reasoning using analyticity is asked for here.

There is an analog of formula (5.5.6) for functions analytic in the open upper half-plane. Its proof uses the residue theorem, and therefore the statement is postponed to Exercise 8.5.3. More generally, formula (5.5.9) has a counterpart when the open unit disk is replaced by the open upper half-plane $\mathbb{C}_+$: A function

f is holomorphic in $\mathbb{C}_+$ and has a real positive part there if and only if it can be written as

$$f(z) = a - ibz - i \int_{\mathbb{R}} \left\{ \frac{1}{t-z} - \frac{t}{t^2+1} \right\} d\mu(t), \qquad (5.5.19)$$

where $a \in \mathbb{R}$, $b \geq 0$ and $d\mu$ is a positive Borel measure on the real line subject to

$$\int_{\mathbb{R}} \frac{d\mu(t)}{t^2+1} < \infty. \qquad (5.5.20)$$

It is convenient to rewrite (5.5.19) as

$$f(z) = a - ibz - i \int_{\mathbb{R}} \frac{tz+1}{t-z} \cdot \frac{d\mu(t)}{t^2+1}. \qquad (5.5.21)$$

When a real imaginary part rather than a real positive part is considered (i.e., when one replaces f by if), one obtains the Pick class (see [67, Chapter II]). Pick functions are also called Nevanlinna functions, although this may create confusion with another, and completely different, class of analytic functions.

Exercise 5.5.25. *Let m be a continuous positive function on the real line, subject to (3.1.8)*

$$\int_{\mathbb{R}} \frac{m(t)dt}{t^2+1} < \infty.$$

Show that the function (3.1.9)

$$f_m(z) = -i \int_{\mathbb{R}} \left\{ \frac{1}{t-z} - \frac{t}{t^2+1} \right\} m(t)dt,$$

is holomorphic in $\mathbb{C}_+$ and has a positive real part there. Show that the real part is strictly positive in $\mathbb{C}_+$ unless $m(t) \equiv 0$.

It is of interest to compute f_m for various choices of m. See Exercise 8.3.8 for instance. For more information on functions of the form f_m (and, more generally, of the form (5.5.21)), we refer to [67].

Among other exercises involving Cauchy's formula appearing in the present book, we mention in particular Exercise 7.3.12, where the sum

$$\sum_{n=0}^{\infty} \frac{\binom{2n}{n}}{7^n} = \sqrt{\frac{7}{3}}$$

is computed. There the residue theorem is invoked, but Cauchy's formula could have been used just as well.

We do not present a solution of the following question. You can assume that Ω is the open unit disk if you forgot what a simply-connected domain is. Formula (5.5.22) is called the *inhomogeneous Cauchy formula*.

Question 5.5.26. *Let Ω be a simply-connected domain, which is bounded, and with smooth boundary Γ. Let $f(x,y)$ be smooth in an open neighborhood of Ω. Show that, for every $z \in \Omega$,*

$$f(z) = \frac{1}{2\pi i} \int_\Gamma \frac{f(\zeta)}{\zeta - z} d\zeta + \frac{1}{2\pi i} \iint_\Omega \frac{\frac{\partial f}{\partial \overline{\zeta}}}{\zeta - z} d\zeta \wedge d\overline{\zeta}. \tag{5.5.22}$$

5.6 Power series expansions of analytic functions

We begin with an exercise taken from [208, p. 100]. There, the variable is taken to be real, and the proof is based on the remark that $\cos x \cosh x = \operatorname{Re} \cos(1 + i)x$, and allows also to compute directly the primitive of $\cos x \cosh x$.

Exercise 5.6.1. *Find the power series expansion at the origin of the function*

$$f(z) = \cos z \cosh z.$$

We also note that the function $f(z)$ is a solution of the differential equation

$$f^{(4)} + 4f = 0,$$

as is easily verified, for instance using formula (4.2.2) for the Nth derivative of a product.

The next exercise is taken from [35, p. 118].

Exercise 5.6.2. *Let f be analytic in a neighborhood of the point z_0, and assume that $f'(z_0) \neq 0$. Show that, for $\epsilon \in \mathbb{C}$ where the expression are defined,*

$$\frac{f(z_0 + i\epsilon) - f(z_0 + \epsilon)}{f(z_0 + \epsilon + i\epsilon) - f(z_0)} = i + O(\epsilon^2) \tag{5.6.1}$$

and

$$\frac{f(z_0 + \epsilon) - f(z)}{f(z_0 + i\epsilon) - f(z_0)} \cdot \frac{f(z_0 + i\epsilon + \epsilon) - f(z_0 + i\epsilon)}{f(z_0 + \epsilon + i\epsilon) - f(z_0 + \epsilon)} = -1 + O(\epsilon^2). \tag{5.6.2}$$

It is well to recall that the radius of convergence of the power series centered at $z = z_0$ of a function f is the radius of the largest open disk centered at $z = z_0$ and in which f is analytic.

Exercise 5.6.3. *Let*

$$g(z) = \frac{1}{(z^2 + 1)(z - (1 + i))}.$$

What is the radius of convergence of the Taylor series of g centered at $z_0 = 1/2$?

The following exercise is taken from [81, Exercise 2, p. 88]. We thank Yarden Sharabi for pointing out a mistake in the solution in the first edition of the book.

Exercise 5.6.4. *Let $z_1, \ldots, z_m$ be complex numbers different from 0. Find the radius of convergence of the Taylor expansion at the origin of the function*

$$f(z) = \sum_{\ell=1}^{m} \frac{1}{1 - z_\ell z},$$

and prove (3.2.2)

$$\limsup_{n \to \infty} |\sum_{\ell=1}^{m} z_\ell^n|^{1/n} = \max_{\ell=1,\ldots,m} |z_\ell|.$$

Related to the following exercise, see also Exercise 6.3.10.

Exercise 5.6.5. *Let $N \in \mathbb{N}$ and let $\epsilon = e^{\frac{2\pi i}{N}}$. Let f be analytic in the open unit disk $\mathbb{D}$ and such that*

$$f(z) = f(\epsilon z), \quad \forall z \in \mathbb{D}.$$

(a) *Show that there is a function g analytic in $\mathbb{D}$ such that $f(z) = g(z^N)$.*
(b) *Let $k \in \{1, \ldots, N-1\}$. Is there a function g analytic in $\mathbb{D}$ such that*

$$z^k = g(z^N), \quad z \in \mathbb{D}? \tag{5.6.3}$$

The following exercise is taken from [62]:

Exercise 5.6.6. *Let F be analytic in $|z| < R$, and assume that $F(z)$ is real for $z = \rho$ and $z = \rho \exp(i\pi\sqrt{2})$ when ρ varies in $(0, R)$. Show that F is a constant.*

Exercise 5.6.7. *Show that there are polynomials $h_0(t), h_1(t), \ldots$ such that*

$$e^{tz - z^2/2} = \sum_{n=0}^{\infty} h_n(t) z^n. \tag{5.6.4}$$

The polynomials h_n in the preceding exercise are called the *Hermite polynomials*.

Related to the following question, see also Exercise 7.2.20.

Exercise 5.6.8. *Let f be analytic in $|z| < 1 + \epsilon$ for some $\epsilon > 0$, with power series expansion $f(z) = \sum_{n=0}^{\infty} f_n z^n$. Define, for $|z| < 1$,*

$$A(z) = \frac{\int_{[0,z]} f(s) ds}{1 - z}.$$

Show that the function $A(z)$ is analytic in the open unit disk, and show that its power series expansion at the origin is equal to

$$A(z) = \sum_{n=0}^{\infty} \left(\sum_{j=0}^{n} \frac{f_j}{j+1} \right) z^{n+1}. \tag{5.6.5}$$

In the following two exercises, one has to compute integrals of the form

$$\int_0^{2\pi} |g(re^{it})|^2 dt,$$

where $r > 0$ and where g is analytic in $|z| < R$ with $R > r$. One can use Parseval's identity from the theory of Fourier series, or proceed directly as is done in the proofs presented here.

Exercise 5.6.9. *Let*

$$\mathcal{A} = \{(u(x,y), v(x,y)) ; (x,y) \in \mathbb{D}\}$$

and assume that $f(z) = u(x,y) + iv(x,y)$ is analytic and one-to-one in $\mathbb{D}$. Assume for simplicity that f is moreover analytic in a neighborhood of the closed unit disk. Compute the area of $\mathcal{A}$.

Exercise 5.6.10.

(1) *Let f be analytic in $B(0, R)$ and set*

$$M_2(f, r) = \frac{1}{2\pi} \int_0^{2\pi} |f(re^{it})|^2 dt, \quad r \in (0, R).$$

Show that M_2 is strictly increasing, unless f is a constant.

(2) *Find all entire functions such that*

$$\iint_{\mathbb{R}^2} |f(z)|^2 dx dy < \infty. \tag{5.6.6}$$

Definition 5.6.11. The set of functions analytic in $\mathbb{D}$ and such that

$$\sup_{r \in (0,1)} M_2(f, r) < \infty \tag{5.6.7}$$

is a Hilbert space, called the *Hardy space* $\mathbf{H}_2(\mathbb{D})$ (of order 2, of the disk).

Exercise 5.6.12. *Let f be analytic in the open unit disk, with power series expansion $f(z) = \sum_{n=0}^{\infty} f_n z^n$. Then, f is in the Hardy space $\mathbf{H}_2(\mathbb{D})$ if and only if*

$$\sum_{n=0}^{\infty} |f_n|^2 < \infty. \tag{5.6.8}$$

This last expression is then equal to (5.6.7).

The Hardy space $\mathbf{H}_2(\mathbb{D})$ is the reproducing kernel Hilbert space with reproducing kernel $\frac{1}{1-z\overline{w}}$, meaning that, for every $w \in \mathbb{D}$ the function

$$k_w : z \mapsto \frac{1}{1 - z\overline{w}}$$

belongs to $\mathbf{H}_2(\mathbb{D})$, and that moreover, for every $f \in \mathbf{H}_2(\mathbb{D})$,

$$\langle f, k_w \rangle_{\mathbf{H}_2(\mathbb{D})} = f(w),$$

where we have denoted by $\langle \cdot, \cdot \rangle_{\mathbf{H}_2(\mathbb{D})}$ the inner product in $\mathbf{H}_2(\mathbb{D})$ defined from the norm (5.6.7).

More generally, for $p \in [1, \infty)$, the function

$$M_p(f, r) = \left(\frac{1}{2\pi} \int_0^{2\pi} |f(re^{it})|^p dt \right)^{1/p} \tag{5.6.9}$$

is strictly increasing (unless f is a constant). To prove this fact requires results from the theory of subharmonic functions. See for instance [189, Chapitre 17]. The space of functions analytic in the open unit disk and such that $\sup_{r \in (0,1)} M_p(f, r) < \infty$ is a Banach space (called the Hardy space H_p), when endowed with the norm

$$\|f\|_{H_p} = \sup_{r \in (0,1)} M_p(f, r).$$

When a weight is allowed in (5.6.6) one obtains very interesting spaces.

Definition 5.6.13. The *Fock space* $\mathcal{F}$ is the space of entire functions such that

$$\frac{1}{\pi} \iint_{\mathbb{R}^2} e^{-|z|^2} |f(z)|^2 dxdy < \infty \tag{5.6.10}$$

Exercise 5.6.14.

(a) *Show that for every $n \in \mathbb{N}_0$ the function z^n belongs to the Fock space, and compute the inner products*

$$\langle z^n, z^m \rangle_{\mathcal{F}}$$

for the inner product associated to the norm (5.6.10).

(b) *Show that an entire function $f(z) = \sum_{n=0}^{\infty} f_n z^n$ belongs to the Fock space if and only if*

$$\sum_{n=0}^{\infty} n! |f_n|^2 < \infty. \tag{5.6.11}$$

The Fock space is the reproducing kernel Hilbert space with reproducing kernel $e^{z\overline{w}}$, meaning that:

(a) For every complex number w, the function

$$k_w(z) = e^{z\overline{w}}$$

belongs to $\mathcal{F}$, and

(b) for every $f \in \mathcal{F}$,

$$\langle f, k_w \rangle_{\mathcal{F}} = f(w).$$

Exercise 5.6.15. *Prove claims* (a) *and* (b) *above.*

For the following result, see for instance [203, Theorem 5.16, p. 53].

Exercise 5.6.16. *Let*

$$f(z) = \sum_{n=0}^{\infty} a_n z^n$$

be a convergent power series, with radius of convergence $R > 0$. *Let* $z_0 \in B(0, R)$
(that is, z_0 *arbitrary when* $R = +\infty$). *Show that*

$$f(z) = \sum_{n=0}^{\infty} a_n(z_0)(z - z_0)^n,$$

for $|z - z_0| < R - |z_0|$ *(resp. for any* z *when* $R = +\infty$), *with (see* [203, (5.18),
p. 53]):

$$a_n(z_0) = \sum_{m=0}^{\infty} a_{n+m} \binom{n+m}{n} z_0^m.$$

See Exercise 7.2.20 for another exercise of interest related to power series.

Let us conclude this section with a remark. A function analytic in a neigh-borhood of the origin has a power series expansion

$$f(z) = \sum_{n=0}^{\infty} a_n z^n$$

convergent in some open disk $B(0, r)$, with $r > 0$. An *inverse* problem would be as follows: Given a series $(a_n)_{n \in \mathbb{N}_0}$ of complex numbers, does it correspond to a function analytic f in a neighborhood of the origin with

$$a_n = \frac{f^{(n)}(0)}{n!}, \quad n \in \mathbb{N}_0. \tag{5.6.12}$$

The answer in general is of course *no*, as the example $a_n = n!$ shows. On the other hand, and this is a result of Borel, discussed for instance, in the form of an exercise, in [88, pp. 263–267], for *any* sequence $(a_n)_{n \in \mathbb{N}_0}$ there exists a function in $C^\infty(\mathbb{R})$ such that (5.6.12) holds. The result is also true for C^∞ functions of N real variables. See [63, p. 195].

5.7 Primitives and logarithm

Recall the discussion at the end of Section 5.2. A function analytic in an open connected set Ω has a primitive if and only if

$$\int_C f(z)dz = 0 \tag{5.7.1}$$

for every closed contour C in Ω. Recall also that a function analytic and non-vanishing in an open connected set Ω has an analytic logarithm (that is, there exists a function g analytic in Ω and such that

$$f(z) = \exp g(z), \quad z \in \Omega,$$

if and only if

$$\int_C \frac{f'(z)}{f(z)} dz = 0 \tag{5.7.2}$$

for every closed contour in Ω.

Equation (5.7.2) has a nice interpretation in terms of zeroes and poles of the function f when C is a smooth Jordan curve. See Remark 7.3.6 after Exercise 7.3.5.

Remark 5.7.1. When one works in a convex set, or, more generally, in a star-shaped set, the situation is easier: A function analytic in a star-shaped domain has always a primitive, and a non-vanishing function analytic in a star-shaped domain has always a logarithm.

The case of general connected open sets is much more involved. To disprove the existence of a primitive or of an analytic logarithm, it is enough to find one closed contour in Ω for which (5.7.1) or (5.7.2) fail. On the other hand to prove the existence of an analytic logarithm, one has *a priori* a very difficult task, that is, to check (5.7.2) (and, similarly, to check (5.7.1) to prove the existence of a primitive). It is of interest to find a minimum set of closed contours on which to check (5.7.2) or (5.7.1). For instance, if Ω is an open convex connected set from which a finite number of compact sets have been removed (for instance, the open unit disk from which are removed a finite number of points), it is enough to check conditions such as (5.7.2) on non-overlapping contours around these holes; see for instance [95, p. 126]. Furthermore, the fact that conditions (5.7.1) or (5.7.2) do not hold for some closed contour C is not the end of the story, but rather the beginning of a fascinating other story, related to the homology group of Ω. For the punctured disk mentioned above, its homology group will be generated by non-overlapping circles around these points.

When the function f has an analytic logarithm g in the open set Ω it obviously has analytic roots of any order: For every $N \in \mathbb{Z} \setminus \{0\}$ there exists a function h analytic in Ω such that

$$f(z) = (h(z))^N, \quad \forall z \in \Omega.$$

It suffices to take $h(z) = \exp \frac{g(z)}{N}$. On the other hand, a function may have an analytic square root and no analytic logarithm: For instance, the function $f(z) = z^2$ has an analytic square root in $\mathbb{C}$ (and all the more in $\mathbb{C} \setminus \{0\}$), but no analytic logarithm in $\mathbb{C} \setminus \{0\}$). See for instance [42] for more information on the differences between square roots (or, more generally Nth roots) and logarithms. Analytic square roots are the topic of Section 5.8.

In the first two exercises the sets under consideration are star-shaped. A very special case of a result of Montel is presented in the first exercise (see [42, Theorem 12.20, p. 433], and the bibliographic note page 458 of that same book). The relation with logarithms is not that clear on a first reading!

Exercise 5.7.2. *Let f and g be entire functions. Show that*

$$f(z)^2 + g(z)^2 = 1, \quad \forall z \in \mathbb{C} \tag{5.7.3}$$

if and only if there is an entire function $E(z)$ such that

$$f(z) = \cos E(z) \quad \text{and} \quad g(z) = \sin E(z).$$

Exercise 5.7.3. *Does the function $f(z) = |z|$ have a primitive in $\mathbb{C}$. Is there a function g of class C_1 in $\mathbb{R}^2 \setminus \{(0,0)\}$ such that*

$$\partial_z f = |z|.$$

Exercise 5.7.4. *Does the function*

$$f(z) = \frac{1}{z^2 + 1}$$

have a primitive in Ω where:

(i) $\Omega = \mathbb{C} \setminus \{-i, i\}$.

(ii) $\Omega = \mathbb{C} \setminus [-i, i]$, *where $[-i, i]$ denotes the closed interval $[-i, i]$, that is:*

$$[-i, i] = \{-i + 2ti, \, t \in [0, 1]\}.$$

(iii) $\Omega = \mathbb{C} \setminus \{z = iy, \, y \in \mathbb{R} \quad \text{and} \quad |y| \geq 1\}$. *In this last case, show that on the real line the primitive is $F(x) = \arctan(x)$ when we fix $F(0) = 0$, and find the power expansion of F at the origin.*

(iv) *Let L be the strip defined by (1.2.21). Show that*

$$F(\tan z) = z, \quad z \in L. \tag{5.7.4}$$

Exercise 5.7.5. *Same questions as in Exercise 5.7.4 for the function $\dfrac{z}{z^2 + 1}$. In case (iii), show that on the real line $F(x) = \dfrac{1}{2} \ln(x^2 + 1)$.*

We now give an exercise which plays a role in the proof (due to L. Fejér and F. Riesz) of Riemann's mapping theorem presented in [45]; see [45, (3.2), p. 190]. See [42, p. 239] for more historical background on, and for a different proof (originating with the works of Koebe and Carathéodory) of Riemann's mapping theorem.

Exercise 5.7.6. *Let f be a function analytic in the open unit disk $\mathbb{D}$, and assume that its range is strictly included in $\mathbb{D}$. Let $a \in \mathbb{D} \setminus f(\mathbb{D})$. Show that there exists a function F analytic in $\mathbb{D}$ and such that*

$$e^{F(z)} = \frac{f(z) - a}{1 - \overline{a} f(z)}, \quad z \in \mathbb{D}.$$

Compute $F'(z)$.
  Show that

$$\operatorname{Re} F(z) < 0, \quad z \in \mathbb{D}.$$

Exercise 5.7.7. *Compute in closed form the sum of the series* (3.4.7)

$$\sum_{n=0}^{\infty} n^{(-1)^n} z^n.$$

Exercise 5.7.8. *Let $a \in (0,1)$ and let*

$$\Omega = \mathbb{C} \setminus \{[-1, -a] \cup [a, 1]\}.$$

Show that the function $f(z) = (z^2 - 1)(z^2 - a^2)$ has no analytic logarithm in Ω but show that it has an analytic square root.

Exercise 5.7.9. *The function $\dfrac{\sin z}{z^2}$ has no primitive in $\mathbb{C} \setminus \{0\}$.*

  For related questions, see also Exercise 7.1.15.

Exercise 5.7.10. *Let $a, b \in \mathbb{C}$ and $p, q \in \mathbb{N}_0$. Find a necessary and sufficient condition for the function*

$$f(z) = a\frac{\sin z}{z^{p+1}} - b\frac{\cos z}{z^{q+1}}$$

to have a primitive in $\mathbb{C} \setminus \{0\}$.

Exercise 5.7.11. *Let $a, b \in \mathbb{C}$ and $q \in \mathbb{N}$, and let $Q(z)$ be a polynomial of degree less than or equal to q. Find a necessary and sufficient condition for the function*

$$f(z) = a\frac{\sin z}{z^6} - b\frac{\exp z - Q(z)}{z^{q+2}}$$

to have a primitive in $\mathbb{C} \setminus \{0\}$.

  The following is taken from [5, Exercise 6, p. 108]

Exercise 5.7.12. *Let Ω be an open connected open set and let f be analytic in Ω and such that*

$$|1 - f(z)| < 1 \quad \forall z \in \Omega.$$

(a) *Show that*

$$\int_\gamma \frac{f'(z)}{f(z)} dz = 0$$

for all closed contours γ in Ω.

(b) *Show that f has an analytic logarithm in Ω.*

Exercise 5.7.13. *Let Ω be the complex plane from which are removed the half-lines $x = k$, $y \geq 0$ for $k = 0, 1, 2, 3, 4, \ldots, 2016$. Show that there exists a function analytic in Ω such that*

$$f(z)^{2016} = z(z-1)(z-2)(z-3)(z-4)\cdots(z-2016).$$

Exercise 5.7.14. *Let f be analytic and not vanishing in $r_0 < |z - z_0| < r_1$. Show that, for $r_0 < r < r_1$,*

$$\frac{1}{2\pi i} \int_{|z-z_0|=r} \frac{f'(z)}{f(z)} \in \mathbb{Z}.$$

The next exercise is taken from [62].

Exercise 5.7.15. *Let f be analytic in the annulus $1 < |z| < 2$ and not vanishing there. Show that there exist an integer $n \in \mathbb{Z}$ and a function g analytic in $1 < |z| < 2$ such that*

$$f(z) = z^n e^{g(z)}.$$

Similarly:

Exercise 5.7.16. *Let f be analytic in the domain Ω which consists of the plane, from which are removed the closed unit disk and the closed disk of center 5 and radius 1, and not vanishing there. Show that there exist numbers n_1 and n_2 in $\mathbb{Z}$ and a function g analytic in Ω such that*

$$f(z) = z^{n_1}(z-5)^{n_2} e^{g(z)}, \quad z \in \Omega.$$

Exercise 5.7.17. *Let m be a strictly positive and continuous function on $[0, 1]$. Do the functions*

$$F_n(z) = \int_0^1 \frac{m(t)dt}{(t-z)^n}, \quad n = 1, 2, \ldots$$

have primitives in $\mathbb{C} \setminus [0, 1]$.

Exercise 5.7.18. *Let $a > 0$. Is there a function analytic in $\mathbb{C} \setminus [e^{-a}, e^a]$ and which coincides with the function $\ln(x^2 - 2x \cosh a + 1)$ on $\mathbb{R} \setminus [e^{-a}, e^a]$?*

5.8   Analytic square roots

As illustrated by Exercise 5.7.8, there are subtle differences between existence
of analytic square roots and analytic logarithms; see [42, Exercise 4.60, p. 111,
Exercise 10.5, p. 346]. For instance, there is no analytic logarithm to z^2 in $\mathbb{C}\backslash\{0\}$.
But it has an analytic square root, namely $f(z) = z$ in $\mathbb{C}$. Similarly there is no
analytic logarithm to the function $f(z) = 1 - z^2$ in $\mathbb{C}\backslash[-1,1]$ since

$$\int_{|z|=2} \frac{f'(z)}{f(z)}\,dz = \int_{|z|=2} \frac{2z}{z^2 - 1}\,dz$$

$$= \int_{|z|=2} \left(\frac{1}{z-1} + \frac{1}{z+1} \right) dz = 4\pi i \neq 0.$$

Still, we have:

Exercise 5.8.1. *Show that there is a function analytic in $\mathbb{C}\backslash[-1,1]$ such that*

$$f(z)^2 = 1 - z^2.$$

Exercise 5.8.2.

(a) *Show that the function $\frac{1}{1-z^2}$ has an analytic square root in*

$$\Omega = \mathbb{C} - \{(-\infty, -1] \cup [1, \infty)\},$$

   which takes the value 1 for $z = 0$. We denote by $\frac{1}{\sqrt{1-z^2}}$ this square root.

(b) *Define*

$$\arcsin z = \int_{C_z} \frac{d\zeta}{\sqrt{1 - \zeta^2}},$$

   *where C_z is any smooth path joining the origin to z. Show that $\arcsin z$ is well
   defined, and is the analytic extension to Ω of the function $\arcsin x$ defined on
   the interval $[-1, 1]$.*

(c) *Compute the power expansion of $\arcsin z$ at the origin. What is its radius of
   convergence?*

(d) *Using analytic continuation, compute $\sin(\arcsin z)$ for $z \in \Omega$.*

Exercise 5.8.3. *Let $[\alpha_\ell, \beta_\ell]$, $\ell = 1, \ldots, N$, be N non-intersecting closed intervals.
Show that the function*

$$f(z) = \frac{\prod_{\ell=1}^{N}(z - \alpha_\ell)}{\prod_{\ell=1}^{N}(z - \beta_\ell)} \tag{5.8.1}$$

has an analytic square root in $\Omega = \mathbb{C}\backslash\bigcup_{\ell=1}^{N}[\alpha_\ell, \beta_\ell]$.

Remark 5.8.4. Let $\alpha \neq \beta \in \mathbb{C}$, and $\Omega = \mathbb{C}\backslash[\alpha, \beta]$. The function $f(z) = (z-\alpha)(z-\beta)$ has no analytic logarithm in Ω. It follows from the previous exercise that f has an analytic square root, as is seen by writing

$$(z-\alpha)(z-\beta) = (z-\beta)^2 \frac{z-\alpha}{z-\beta}.$$

Exercise 5.8.5. *There is no analytic square root of z in the annulus $1 < |z| < 2$.*

5.9 Solutions

Solution of Exercise 5.1.1. Assume by contradiction that the functions are equivalent, and let $\varphi : [0, 2\pi] \longrightarrow [0, 4\pi]$ be such that (5.1.1) holds. Then,

$$e^{it} = e^{i\varphi(t)}, \quad t \in [0, 2\pi],$$

and taking the derivative with respect to t we get (see Exercise 3.1.4 if need be)

$$ie^{it} = i\varphi'(t)e^{i\varphi(t)}, \quad t \in [0, 2\pi].$$

Thus

$$\varphi'(t) = 1, \quad t \in [0, 2\pi],$$

and

$$4\pi = \varphi(2\pi) - \varphi(0) = \int_0^{2\pi} \varphi'(t)dt = \int_0^{2\pi} 1dt = 2\pi,$$

which is impossible. $\qquad \square$

Solution of Exercise 5.1.2. By definition of the path integral,

$$\int_C (x^2 - iy^2)dz = \int_0^\pi (\cos^2 t - i\sin^2 t)(-\sin t + i\cos t)dt$$

$$= \int_0^\pi (-\cos^2 t \sin t + \sin^2 t \cos t)dt + i\int_0^\pi (\sin^3 t + \cos^3 t)dt$$

$$= \frac{\cos^3 t + \sin^3 t}{3}\Big|_{t=0}^{t=\pi} + i\left(-\cos t + \frac{\cos^3 t}{3}\right)\Big|_{t=0}^{t=\pi}$$

$$+ i\left(\sin t - \frac{\sin^3 t}{3}\right)\Big|_{t=0}^{t=\pi}$$

$$= \frac{-2 + 4i}{3},$$

where we have used that the primitives of $\sin^3 t$ and $\cos^3 t$ are

$$-\cos t + \frac{\cos^3 t}{3} \quad \text{and} \quad \sin t - \frac{\sin^3 t}{3}$$

respectively. $\qquad \square$

Solution of Exercise 5.1.4. Let $\gamma(t), t \in [a, b]$ be a parametrization of C. By definition of the line integral we have

$$\left| \int_C \frac{f(z)}{z} dz \right|^2 = \left| \int_a^b \frac{f(\gamma(t))}{\gamma(t)} \gamma'(t) dt \right|^2$$

and, using (3.1.4),

$$\leq \left| \int_a^b \frac{|f(\gamma(t))|}{|\gamma(t)|} |\gamma'(t)| dt \right|^2$$

$$\leq \left(\max_{t \in [a,b]} \frac{1}{|\gamma(t)|^2} \right) \cdot \left| \int_a^b |f(\gamma(t))| \sqrt{|\gamma'(t)|} \sqrt{|\gamma'(t)|} dt \right|^2$$

and, using the Cauchy-Schwarz inequality (see (3.1.5)),

$$\leq \left(\max_{t \in [a,b]} \frac{1}{|\gamma(t)|^2} \right) \cdot \left(\int_a^b |f(\gamma(t))|^2 |\gamma'(t)| dt \right) \left(\int_a^b |\gamma'(t)| dt \right),$$

which is the required result.   □

Solution of Exercise 5.1.5. Let $t_0 \in [a, b]$ and let M_0 be the point on the curve defined by $z = \gamma(t)$. Let z_0 and z_1 be on the normal line of the curve at the point M, and symmetric with respect to M_0. We want to show that

$$\lim_{|z_0 - z_1| \to 0} \frac{|\gamma(t) - z_0|}{|\gamma(t) - z_1|} = 1$$

uniformly in $t \in [a, b]$. To that purpose we will show the following: For every $\lambda \in (0, 1)$ there exists $\eta > 0$ such that

$$|z_0 - z_1| < \eta \implies \lambda < \frac{|\gamma(t) - z_0|}{|\gamma(t) - z_1|} < \frac{1}{\lambda}. \tag{5.9.1}$$

At this stage we recall that, for $u \in (0, 1)$ the set of points z such that

$$\frac{|z - z_0|}{|z - z_1|} = u \quad \text{or} \quad \frac{|z - z_0|}{|z - z_1|} = \frac{1}{u} \tag{5.9.2}$$

form two circles, symmetric with respect to M, and with same radius

$$R = \frac{u}{1 - u^2} |z_0 - z_1|, \tag{5.9.3}$$

and centers Ω and Ψ given respectively by

$$\Omega = \frac{z_0 - u^2 z_1}{1 - u^2} = z_0 + \frac{u^2}{1 - u^2}(z_0 - z_1) \quad \text{and} \quad \Psi = \frac{z_1 - u^2 z_0}{1 - u^2} = z_1 + \frac{u^2}{1 - u^2}(z_1 - z_0).$$
$$\tag{5.9.4}$$

These circles, say C_u and $C_{\frac{1}{u}}$, form a coaxal family, and it is useful to remark that when $u_1 < u_2$ the circle C_{u_1} lies in the interior of the disk defined by the circle C_{u_2}, and $C_0 = \{z_0\}$.

Using formulas (5.9.3)–(5.9.4) we see that for any $\lambda \in (0,1)$, there exists η such that, for $|z_0 - z_1| < \eta$, the circle C_λ lies inside the curve and $C_{\frac{1}{\lambda}}$ lies outside the curve. Thus the curve lies in the set (5.1.7), and the claim follows. $\qquad\square$

Solution of Exercise 5.1.6. We only prove (5.1.9):

$$\overline{\left(\int_{|z|=1} f(z)dz\right)} = \overline{\left(\int_0^{2\pi} f(e^{it})e^{it}idt\right)}$$

$$= -\int_0^{2\pi} \overline{f(e^{it})}e^{-it}idt$$

$$= -\int_0^{2\pi} \frac{\overline{f(e^{it})}}{e^{2it}}e^{it}idt$$

$$= -\int_{|z|=1} \frac{\overline{f(z)}}{z^2}dz. \qquad\square$$

Solution of Exercise 5.1.7. Let $\gamma(t)$, $t \in [a,b]$ be a parametrization of the closed and piecewise smooth path C. Set

$$g(s) = (\gamma(s) - z_0)\exp-\left\{\int_a^s \frac{\gamma'(t)}{\gamma(t) - z_0}dt\right\}, \quad s \in [a,b].$$

We have

$$g'(s) = \gamma'(s)\exp-\left\{\int_a^s \frac{\gamma'(t)}{\gamma(t) - z_0}dt\right\}$$

$$- (\gamma(s) - z_0)\frac{\gamma'(s)}{\gamma(s) - z_0}\exp-\left\{\int_a^s \frac{\gamma'(t)}{\gamma(t) - z_0}dt\right\}$$

$$\equiv 0,$$

and so $g(a) = g(b)$. Thus

$$\gamma(a) - z_0 = (\gamma(b) - z_0)\exp-\left\{\int_a^b \frac{\gamma'(t)}{\gamma(t) - z_0}dt\right\}.$$

We have $\gamma(a) = \gamma(b) \neq z_0$. Thus,

$$\int_a^b \frac{\gamma'(t)}{\gamma(t) - z_0}dt \in 2\pi i\mathbb{Z}. \qquad\square$$

Solution of Exercise 5.1.8. We write

$$p(z) = p_n z^n + p_{n-1} z^{n-1} + \cdots + p_0 \quad \text{and} \quad q(z) = q_m z^m + q_{m-1} z^{m-1} + \cdots + q_0,$$

with p_n and q_m different from 0 and $m \geq n+2$. For $z \neq 0$ we can write

$$q(z) = q_m z^m (1 + r(z))$$

where

$$r(z) = \frac{q_{m-1}}{q_m} \frac{1}{z} + \cdots + \frac{q_0}{q_m} \frac{1}{z^m}.$$

Since $\lim_{z \to \infty} r(z) = 0$, there is $R_0 > 0$, which we will assume greater than 1, such that

$$|z| > R_0 \implies |r(z)| < \frac{1}{2},$$

and hence, still for $|z| > R_0$,

$$\frac{1}{|q(z)|} \leq \frac{1}{|q_m z^m|(1 - |r(z)|)} < \frac{1}{|q_m z^m|(1 - 1/2)} = \frac{2}{|q_m z^m|}.$$

Furthermore, for $|z| > 1$,

$$|p(z)| \leq K|z|^n, \quad \text{with} \quad K = \sum_{\ell=0}^{n} |p_\ell|.$$

Thus, for $|z| = R > R_0$,

$$\left| \frac{p(z)}{q(z)} \right| \leq \frac{2KR^n}{|q_m|R^m} = \frac{K_1}{R^{m-n}}, \quad \text{with} \quad K_1 = \frac{2K}{|q_m|}.$$

Using formula (5.1.3) we thus have

$$\left| \int_{|z|=R} \frac{p(z)}{q(z)} dz \right| \leq 2\pi R \frac{K_1}{R^{m-n}} = \frac{2\pi K_1}{R^{m-n-1}},$$

which goes to 0 as $R \to \infty$ since $m - n - 1 > 0$. □

Solution of Exercise 5.1.9. The change of variable $t \mapsto \frac{\pi}{2} - t$ shows that both integrals coincide. The claim (5.1.11) is a direct consequence of Jordan's inequality (see for instance [184, §19.5, p. 224], [175, p. 114]):

$$\frac{2}{\pi} \leq \frac{\sin t}{t} \leq 1 \quad \text{where} \quad 0 < t \leq \frac{\pi}{2},$$

which leads to

$$\int_0^{\frac{\pi}{2}} e^{-R \sin t} dt < \frac{\pi}{R}. \tag{5.9.5}$$

See [49, p. 187]. □

We note that a shorter, but not elementary proof, consists in invoking the dominated convergence theorem for sequences of positive numbers $(R_n)_{n\in\mathbb{R}}$ which tend to ∞.

Solution of Exercise 5.1.10. Let γ_{++} denote the first quarter of the unit circle, with the positive orientation. Then

$$\int_0^{\frac{\pi}{2}} e^{-R\sin t}dt = \int_{\gamma_{++}} e^{-R\frac{z-z^{-1}}{2i}}\frac{dz}{iz}.$$

The second integral is treated in the same way. $\square$

Solution of Exercise 5.1.11. Let

$$\gamma(t) = (x(t), y(t)), \ t \in [a, b],$$

be a parametrization of C. We have

$$\frac{du(\gamma(t))}{dt} = \frac{\partial u}{\partial x}(\gamma(t))x'(t) + \frac{\partial u}{\partial y}(\gamma(t))y'(t)$$

and so

$$\int_C \frac{\partial u}{\partial x}dx + \frac{\partial u}{\partial y}dy = \int_a^b \frac{du(\gamma(t))}{dt}dt = u(\gamma(b)) - u(\gamma(a)) = 0,$$

since C is closed. $\square$

Solution of Exercise 5.2.3. It suffices to take as path γ the closed unit circle:

$$\gamma(t) = e^{it}, \quad t \in [0, 2\pi].$$

Then $\gamma'(t) = i\gamma(t)$ and we have

$$\int_\gamma \frac{dz}{z} = \int_0^{2\pi} \frac{\gamma'(t)}{\gamma(t)}dt = \int_0^{2\pi} i dt = 2\pi i \neq 0. \qquad \square$$

Solution of Exercise 5.2.5. The rule of differentiation for a product holds for $\mathbb{C}$-differentiable functions, and thus

$$(fg)'(z) = f'(z)g(z) + f(z)g'(z).$$

Taking into account this formula and applying the Newton–Leibniz formula (5.2.2) to fg we obtain

$$(fg)(\gamma(b)) - (fg)(\gamma(a)) = \int_C (fg)'(z)dz$$
$$= \int_C f'(z)g(z)dz + \int_C f(z)g'(z)dz,$$

and hence the result. $\square$

Solution of Exercise 5.2.6. Take z_1 and z_2 to be in the left closed half-plane, and assume $z_1 \neq z_2$ (if $z_1 = z_2$ the result is trivial). The interval $[z_1, z_2]$ is also included in the left closed half-plane. Consider the parametrization

$$\gamma(t) = z_1 + t(z_2 - z_1), \quad t \in [0,1]$$

of the interval. The function e^z is its own derivative, and therefore

$$e^{z_2} - e^{z_1} = \int_{[z_1,z_2]} e^z dz.$$

For every $t \in [0,1]$, we have

$$|e^{\gamma(t)}| \leq 1$$

since $\gamma(t)$ belongs to the left closed half-plane. Using (5.1.3) we have:

$$|e^{z_2} - e^{z_1}| = |\int_{[z_1,z_2]} e^z dz| \leq \max_{t \in [0,1]} |e^{\gamma(t)}| \cdot |z_2 - z_1| \leq |z_2 - z_1|. \qquad \square$$

Solution of Exercise 5.2.7. We give to Γ_R the positive orientation. We then have the following parametrizations for the components of Γ_R:

$$\gamma_{1,R}(t) = t, \qquad t \in [0, R],$$
$$\gamma_{2,R}(\theta) = Re^{i\theta}, \qquad \theta \in [0, \pi/4],$$
$$\gamma_{3,R}(t) = (R - t)e^{\frac{i\pi}{4}}, \quad t \in [0, R].$$

Since e^{-z^2} is defined by a power series centered at the origin, and converging in all of $\mathbb{C}$, it has a primitive in $\mathbb{C}$ and we can write

$$\int_{\Gamma_R} e^{-z^2} dz = 0, \quad \forall R > 0,$$

that is,

$$\int_{\gamma_{1,R}} e^{-z^2} dz + \int_{\gamma_{2,R}} e^{-z^2} dz + \int_{\gamma_{3,R}} e^{-z^2} dz = 0, \quad \forall R > 0. \tag{5.9.6}$$

We now show that $\lim_{R\to\infty} \int_{\gamma_{2,R}} e^{-z^2} dz = 0$. Indeed, for $\theta \in \left[0, \dfrac{\pi}{4}\right]$ we have

$$\cos(2\theta) \geq 1 - \frac{4}{\pi}\theta. \tag{5.9.7}$$

Thus

$$\left| \int_{\gamma_{2,R}} e^{-z^2} dz \right| = \left| \int_0^{\frac{\pi}{4}} e^{-\{R^2(\cos(2\theta)+i\sin(2\theta))\}} iRe^{i\theta} d\theta \right|$$

$$\leq R \int_0^{\frac{\pi}{4}} e^{-\{R^2\cos(2\theta)\}} d\theta$$

$$\leq R \int_0^{\frac{\pi}{4}} e^{\{-R^2(1-\frac{4\theta}{\pi})\}} \quad \text{(where we use (5.9.7))}$$

$$= Re^{-R^2} \int_0^{\frac{\pi}{4}} e^{\left(R^2 \frac{4\theta}{\pi}\right)} d\theta$$

$$= Re^{-R^2} \frac{\pi}{4R^2} \left(e^{\frac{4R^2\theta}{\pi}} \right)_{\theta=0}^{\theta=\frac{\pi}{4}}$$

$$= Re^{-R^2} \frac{\pi}{4R^2} \left(e^{R^2} - 1 \right)$$

$$= \frac{\pi}{4R} \left(1 - e^{-R^2} \right)$$

$$\to 0 \quad \text{as} \quad R \to \infty.$$

Since $\lim_{R\to\infty} \int_{\gamma_{1,R}} e^{-z^2} dz = \int_0^\infty e^{-t^2} dt < \infty$, the limit

$$\lim_{R\to\infty} \int_{\gamma_{3,R}} e^{-R^2} dz$$

also exists and we have

$$\lim_{R\to\infty} \int_{\gamma_{1,R}} e^{-z^2} dz + \lim_{R\to\infty} \int_{\gamma_{3,R}} e^{-z^2} dz = 0,$$

i.e.,

$$\int_{\gamma_{3,R}} e^{-z^2} dz = -\frac{\sqrt{\pi}}{2}.$$

But

$$\int_{\gamma_{3,R}} e^{-z^2} dz = \int_0^R e^{\left\{-e^{\frac{i\pi}{2}}(R-t)^2\right\}} (-1) e^{\frac{i\pi}{4}} dt$$

$$= \left(\frac{1+i}{\sqrt{2}} \right) \int_0^R e^{-i(R-t)^2} (-1) dt$$

$$\to - \left(\frac{1+i}{\sqrt{2}} \right) \int_0^\infty (\cos(t^2) - i\sin(t^2)) dt \quad \text{as} \quad R \to \infty,$$

where to go from the penultimate line to the last line we made the change of variable $t \mapsto R - t$. Hence

$$- \left(\frac{1+i}{\sqrt{2}} \right) \left(\int_0^\infty \cos(t^2) dt - i \int_0^\infty \sin(t^2) dt \right) = -\frac{\sqrt{\pi}}{2}.$$

Thus

$$\int_0^\infty \cos(t^2)dt = \int_0^\infty \sin(t^2)dt = \sqrt{\frac{\pi}{8}},$$

and the Fresnel integrals are equal to twice this number, i.e., $\sqrt{\dfrac{\pi}{2}}$.   □

For a proof which uses complex analysis but not the Jordan lemma see the paper of C. Olds [172]. This paper, as well as the papers of Flanders and Leonard quoted in the introduction of Section 5.3, can be obtained from the site:

$$\mathtt{http://www.jstor.org}.$$

Solution of Exercise 5.2.8. Both integrals are absolutely convergent since

$$\left| e^{-t^2}\cos t^2 \right| \le e^{-t^2} \quad \text{and} \quad \left| e^{-t^2}\sin t^2 \right| \le e^{-t^2}.$$

We note that

$$\int_0^\infty e^{-t^2}\cos t^2 dt + i \int_0^\infty e^{-t^2}\sin t^2 dt = \int_0^\infty e^{-t^2(1-i)}dt.$$

This suggests taking the following closed contour $\Gamma_R = \gamma_{1,R} + \gamma_{2,R} + \gamma_{3,R}$ where:

(i) $\gamma_{1,R}$ is the interval $[0,R]$.

(ii) $\gamma_{2,R}$ is the arc of the circle of radius R and centered at the origin, with angle varying between 0 and $\pi/8$.

(iii) $\gamma_{3,R}$ is the interval linking the point $R\exp\frac{i\pi}{8}$ to the origin.

We now remark that the functions

$$f(z) = \cos z^2 e^{-z^2} \quad \text{and} \quad \sin z^2 e^{-z^2}$$

are equal to power series centered at the origin and with radius of convergence ∞. Therefore they admit primitives in $\mathbb{C}$. By Theorem 5.2.1, we have

$$\int_{\Gamma_R} e^{-z^2(1-i)}dz = 0 \quad \text{for all} \quad R > 0,$$

and in a way similar to the computations of the Fresnel integrals

$$\lim_{R\to\infty} \int_{\gamma_{2,R}} e^{-z^2(1-i)}dz = 0.$$

Thus

$$\lim_{R\to\infty} \int_{\gamma_{1,R}} e^{-z^2(1-i)}dz = - \lim_{R\to\infty} \int_{\gamma_{3,R}} e^{-z^2(1-i)}dz. \tag{5.9.8}$$

The limit on the left is

$$\int_0^\infty e^{-t^2(1-i)}\,dt = \int_0^\infty e^{-t^2}\cos t^2\,dt + i\int_0^\infty e^{-t^2}\sin t^2\,dt.$$

A parametrization of $\gamma_{3,R}$ is given by

$$\gamma(u) = e^{i\frac{\pi}{8}}(R-u), \quad u \in [0,R],$$

and so the limit on the right side of (5.9.8) is equal to

$$-\lim_{R\to\infty}\int_0^R e^{-\left\{(R-u)^2 e^{i\frac{\pi}{4}}(1-i)\right\}}e^{i\frac{\pi}{8}}(-1)\,du = \lim_{R\to\infty}\int_0^R e^{-\left\{u^2\frac{(1+i)}{\sqrt{2}}(1-i)\right\}}e^{i\frac{\pi}{8}}\,du$$

$$= \int_0^\infty e^{-\sqrt{2}u^2}e^{i\frac{\pi}{8}}\,du$$

$$= \int_0^\infty e^{-v^2}2^{1/4}e^{i\frac{\pi}{8}}\,dv$$

$$= \frac{\sqrt{\pi}}{2}2^{-1/4}\left(\frac{\sqrt{\sqrt{2}+2}}{2} + i\frac{\sqrt{2-\sqrt{2}}}{2}\right)$$

$$= \frac{\sqrt{\pi}}{4}\left(\sqrt{\sqrt{2}+1} + i\sqrt{\sqrt{2}-1}\right),$$

and hence the result.

In the chain of equalities we have used that

$$\cos\frac{\pi}{8} = \sqrt{\frac{\cos\frac{\pi}{4}+1}{2}} = \frac{\sqrt{2+\sqrt{2}}}{2} \quad\text{and}\quad \sin\frac{\pi}{8} = \sqrt{\frac{1-\cos\frac{\pi}{4}}{2}} = \frac{\sqrt{2-\sqrt{2}}}{2},$$

and

$$2^{-\frac{1}{4}}\frac{\sqrt{\sqrt{2}+2}}{2} = 2^{-\frac{1}{4}}\frac{\sqrt{\sqrt{2}(\sqrt{2}+1)}}{2} = \frac{\sqrt{\sqrt{2}+1}}{2},$$

$$2^{-\frac{1}{4}}\frac{\sqrt{2-\sqrt{2}}}{2} = 2^{-\frac{1}{4}}\frac{\sqrt{\sqrt{2}(\sqrt{2}-1)}}{2} = \frac{\sqrt{\sqrt{2}-1}}{2}. \qquad \square$$

Remark 5.9.1. More generally, for $\mathrm{Re}\,z > 0$, it holds that

$$\int_0^\infty e^{-zt^2}\,dt = \frac{1}{2}\sqrt{\frac{\pi}{z}}, \tag{5.9.9}$$

where $\sqrt{z}$ denotes the analytic square root of z in the open right half-plane, which coincides with $\sqrt{x}$ on $(0,\infty)$; see [53, Example 1, p. 113]. Equality (5.9.9) is clear for

$z = x > 0$. This is just a change of variable in the Gaussian integral (5.2.6). On the other hand both sides of (5.9.9) are analytic in $\operatorname{Re} z > 0$, and the equality follows by analytic continuation. We also note that the function $\dfrac{1}{2}\sqrt{\dfrac{\pi}{z}}$ is an analytic extension of the function $\int_0^\infty e^{-zt^2}\,dt$ to $\mathbb{C} \setminus (-\infty, 0]$. See Section 6.3. Finally, setting $z = x - i$ with $x \geq 0$ leads to the formula (see [83], [147]):

$$\int_0^\infty e^{-xt^2}\cos t^2 dt = \sqrt{\frac{\pi}{8}}\sqrt{\frac{\sqrt{x^2+1}+x}{x^2+1}},$$

$$(5.9.10)$$

$$\int_0^\infty e^{-xt^2}\sin t^2 dt = \sqrt{\frac{\pi}{8}}\sqrt{\frac{\sqrt{x^2+1}-x}{x^2+1}},$$

where $x \geq 0$. Indeed, write $\sqrt{\frac{1}{x-i}} = a(x) + ib(x)$. Then we get the system

$$a(x)^2 - b(x)^2 = \frac{x}{x^2+1} \quad \text{and} \quad a(x)b(x) = \frac{1}{x^2+1},$$

which has a unique solution such that $a(0) = b(0) > 0$.

We note also the two formulas

$$\int_0^\infty \sin x^n dx = \frac{1}{n}\Gamma(1/n)\sin(\frac{\pi}{2n}),$$

$$(5.9.11)$$

$$\int_0^\infty \cos x^n dx = \frac{1}{n}\Gamma(1/n)\cos(\frac{\pi}{2n}),$$

where Γ denotes Euler's Gamma function. See [204, 18.54 and 18.55].

Solution of Exercise 5.2.9. The function f is analytic in the open unit disk since it is the sum of a convergent power series there. Let now z_1 and z_2 be in $\mathbb{D}$, and let $[z_1, z_2]$ be the interval linking z_1 and z_2. We have $[z_1, z_2] \subset \mathbb{D}$, and a parametrization of $[z_1, z_2]$ is given by

$$\gamma(t) = z_1 + t(z_2 - z_1), \quad t \in [0, 1].$$

By the fundamental theorem of calculus for analytic functions,

$$f(z_2) - f(z_1) = \int_{[z_1, z_2]} f'(z)dz$$

$$= \int_0^1 f'(\gamma(t))\gamma'(t)dt$$

$$= (z_2 - z_1)\int_0^1 f'(\gamma(t))dt$$

$$= (z_2 - z_1)\left\{1 + \int_0^1 \left(\sum_{n=2}^\infty na_n(\gamma(t))^{n-1}\right)dt\right\},$$

since

$$f'(z) = 1 + \sum_{n=2}^{\infty} a_n z^{n-1}.$$

Thus

$$|f(z_2) - f(z_1)| = |z_2 - z_1| \cdot \left| \left\{ 1 + \int_0^1 \left(\sum_{n=2}^{\infty} na_n(\gamma(t))^{n-1}\right) dt \right\} \right|$$

$$\geq |z_2 - z_1| \cdot \left| 1 - \left| \int_0^1 \left(\sum_{n=2}^{\infty} na_n(\gamma(t))^{n-1} \right) dt \right| \right|. \tag{5.9.12}$$

But we have

$$\left| \int_0^1 \left(\sum_{n=2}^{\infty} na_n(\gamma(t))^{n-1} \right) dt \right| \leq \int_0^1 \left(\sum_{n=2}^{\infty} n|a_n| \cdot |(\gamma(t))^{n-1}| \right) dt$$

$$\leq \int_0^1 \left(\sum_{n=2}^{\infty} n|a_n| \right) dt$$

$$= \sum_{n=2}^{\infty} n|a_n| < 1.$$

Hence:

$$\left| 1 - \left| \int_0^1 \left(\sum_{n=2}^{\infty} na_n(\gamma(t))^{n-1} \right) dt \right| \right| = 1 - \left| \int_0^1 \left(\sum_{n=2}^{\infty} na_n(\gamma(t))^{n-1} \right) dt \right|$$

$$\geq 1 - \sum_{n=2}^{\infty} n|a_n| > 0,$$

and (5.9.12) leads to

$$|f(z_2) - f(z_1)| \geq |z_2 - z_1| \cdot \left| 1 - \left| \int_0^1 \left(\sum_{n=2}^{\infty} na_n(\gamma(t))^{n-1} \right) dt \right| \right|$$

$$\geq |z_2 - z_1| \left(1 - \sum_{n=2}^{\infty} n|a_n| \right).$$

Thus, in view of (5.2.8),

$$|f(z_2) - f(z_1)| = 0 \quad \Longrightarrow \quad |z_2 - z_1| = 0,$$

that is,

$$f(z_2) = f(z_1) \quad \Longrightarrow \quad z_2 = z_1. \qquad \square$$

Solution of Exercise 5.3.2. An abridged solution is as follows: First consider $\int_{\gamma_{2,R}} f(z)dz$. We have

$$\left| \int_{\gamma_{2,R}} f(z)dz \right| = \left| \int_0^\pi \left(e^{iR\cos t - R\sin t} \right) \frac{iRe^{it}}{Re^{it}} dt \right|$$

$$\leq \int_0^\pi e^{-R\sin t} dt$$

$$= 2 \int_0^{\frac{\pi}{2}} e^{-R\sin t} dt$$

$$\longrightarrow 0$$

by Jordan's lemma. On the other hand,

$$\lim_{\epsilon \to 0} \int_{\gamma_{4,\epsilon}} f(z)dz = -\lim_{\epsilon \to 0} \int_0^\pi \left(e^{i\epsilon\cos t - \epsilon\sin t} \right) \frac{i\epsilon e^{it}}{\epsilon e^{it}} dt$$

$$= -i\pi.$$

Thus

$$\lim_{\substack{\epsilon \to 0 \\ R \to \infty}} \int_{-\epsilon}^{-R} \frac{e^{it}}{t} dt + \int_{\epsilon}^R \frac{e^{it}}{t} dt = -i\pi. \tag{5.9.13}$$

But

$$\int_{-R}^{-\epsilon} \frac{e^{it}}{t} dt + \int_{\epsilon}^R \frac{e^{it}}{t} dt = 2i \int_{\epsilon}^R \frac{\sin t}{t} dt,$$

and hence the result is

$$\int_0^\infty \frac{\sin t}{t} dt = \frac{\pi}{2}. \tag{5.9.14}$$

Solution of Exercise 5.3.3. With the notation of the previous exercise,

$$\lim_{R \to \infty} \int_{\gamma_{2,R}} f(z)dz = 0,$$

and, using (5.3.2) with $g(z) = \frac{1 - e^{2iz}}{z}$ we have:

$$\lim_{\epsilon \to 0} \int_{\gamma_{4,\epsilon}} f(z)dz = -\pi \frac{4}{2} = -2\pi,$$

and hence

$$\int_{-\infty}^\infty \frac{1 - e^{2ix}}{x^2} dx = 2\pi.$$

Taking into account that $1 - \cos(2x) = 2\sin^2 x$ we have

$$\int_{\mathbb{R}} \frac{\sin^2 x}{x^2} dx = \pi. \tag{5.9.15}$$

For a more detailed proof we suggest [80, pp. 159–161]. The previous result can be checked in [204, p. 107]. The definite integrals on pages 107–109 of Schaum's *Mathematical handbook of formulas and tables* form a nice source of integrals which you can try to compute using Cauchy's theorem (or, for some of them, the residue theorem). For instance, you might want to check that

$$\int_{\mathbb{R}} \left(\frac{\sin x}{x} \right)^4 dx = \frac{2\pi}{3},$$

see [204, 18.59, p. 108]. To evaluate the integral first note that

$$\sin^4 x = \frac{3 - 4\cos(2x) + \cos(4x)}{8}.$$

One calculates on the same contour as in the previous two exercises the integral of the function

$$f(z) = \frac{3 - 4e^{2iz} + e^{4iz}}{8z^4}.$$

We now check that

$$\lim_{\epsilon \to 0} \int_{\gamma_{4,\epsilon}} f(z)dz = -\frac{2\pi}{3}.$$

We have

$$f(z) = \frac{3 - 4(1 + 2iz - \frac{4z^2}{2} - \frac{8iz^3}{3!} + \cdots) + 1 + 4iz - \frac{16z^2}{2} - \frac{64iz^3}{3!} + \cdots}{8z^4}$$

$$= -\frac{i}{2z^3} - \frac{2i}{3z} + g(z),$$

where g is analytic in a neighborhood of the origin. The important point is that there is no term in $1/z^2$. We have

$$\int_{\gamma_{4,\epsilon}} f(z)dz = \int_{\pi}^{0} f(\epsilon e^{it})\epsilon i e^{it} dt$$

$$= -\frac{i}{2} \int_{\pi}^{0} \frac{\epsilon i e^{it}}{\epsilon^3 e^{3it}} dt - \frac{2i}{3} \int_{\pi}^{0} \frac{\epsilon i e^{it}}{\epsilon e^{it}} dt + \int_{\pi}^{0} g(\epsilon e^{it}) i \epsilon e^{it} dt$$

$$= T_1(\epsilon) + T_2(\epsilon) + T_3(\epsilon)$$

where

$$T_1(\epsilon) = -\frac{i}{2} \int_{\pi}^{0} \frac{\epsilon i e^{it}}{\epsilon^3 e^{3it}} dt = \frac{1}{2\epsilon^2} \int_{\pi}^{0} e^{-2it} dt \equiv 0,$$

$$T_2(\epsilon) = -\frac{2i}{3} \int_{\pi}^{0} \frac{\epsilon i e^{it}}{\epsilon e^{it}} dt \equiv -\frac{2\pi}{3},$$

$$T_3(\epsilon) = \int_{\pi}^{0} g(\epsilon e^{it}) i \epsilon e^{it} dt \to 0 \quad \text{as} \quad \epsilon \to 0.$$

Thus, as $\epsilon \to 0$ and $R \to \infty$ we have

$$\int_{-R}^{-\epsilon} f(x)dx + \int_{\epsilon}^{R} f(x)dx - \frac{2\pi}{3} \to 0.$$

Thus

$$\int_{\epsilon}^{R} (f(x) + f(-x))dx - \frac{2\pi}{3} \to 0.$$

Since $f(-x) = \overline{f(x)}$, we have

$$\int_{0}^{\infty} 2\operatorname{Re} f(x)dx = \frac{2\pi}{3}.$$

Since $\operatorname{Re} f(x) = \frac{\sin^4 x}{x^4}$ the result follows.   □

Understanding this method leads without too much difficulty to the computation of the integrals

$$\int_{\mathbb{R}} \left(\frac{\sin x}{x}\right)^{2p} dx, \quad p \in \mathbb{N}, \tag{5.9.16}$$

as we now explain. Newton's binomial formula (1.3.6) applied to

$$(e^{ix} - e^{-ix})^{2p} = e^{2ixp}(1 - e^{-2ix})^{2p}$$

leads to a sum which contains only even powers of $e^{\pm ix}$, and which moreover is an even function of x. Thus, we can write

$$\left(\frac{e^{ix} - e^{-ix}}{2i}\right)^{2p} = \frac{\sum_{k=-p}^{p} c_{p,k} e^{2ikx}}{2},$$

where the numbers $c_{p,k}$, $k = -p, \ldots, p$ are (real) rational numbers such that

$$c_{p,-k} = c_{p,k}, \quad k = 0, \ldots, p.$$

Thus,

$$\sin^{2p} x = \left(\frac{e^{ix} - e^{-ix}}{2i}\right)^{2p}$$

$$= \frac{\sum_{k=-p}^{p} c_{p,k} e^{2ikx}}{2}$$

$$= \frac{c_{p,0}}{2} + \operatorname{Re} \sum_{k=1}^{p} c_{p,k} e^{2ikx} \tag{5.9.17}$$

$$= \frac{c_{p,0}}{2} + \sum_{k=1}^{p} c_{p,k} \cos(2kx). \tag{5.9.18}$$

In particular,

$$\frac{c_{p,0}}{2} + \sum_{k=1}^{p} c_{p,k} = 0. \tag{5.9.19}$$

We note that

$$c_{p,k} = \frac{\binom{2p}{p-k}(-1)^{k-p}}{2^{2p-1}}, \quad k = 0, \dots, p. \tag{5.9.20}$$

Differentiating (5.9.18) $2p - 2$ times, and setting each time $x = 0$ we obtain that

$$\sum_{k=1}^{p} c_{p,k} k^{2t} = 0, \quad t = 1, \dots, p-1. \tag{5.9.21}$$

To compute (5.9.16) we integrate, along the same contour as above, the function

$$g(z) = \frac{\frac{c_{p,0}}{2} + \sum_{k=1}^{p} c_{p,k} e^{2ikz}}{z^{2p}}. \tag{5.9.22}$$

Let $z = \epsilon e^{i\theta}$. To compute the line integral $\int_{\gamma_{4,\epsilon}} f(z) dz$ we first write

$$
\begin{aligned}
g(\epsilon e^{i\theta}) i\epsilon e^{i\theta} &= \frac{c_{p,0} i\epsilon e^{i\theta}}{2\epsilon^{2p} e^{2pi\theta}} + \sum_{k=1}^{p} c_{p,k} \frac{\sum_{\ell=0}^{\infty} \frac{(2i\epsilon)^\ell}{\ell!} k^\ell e^{i\ell\theta}}{\epsilon^{2p} e^{2pi\theta}} i\epsilon e^{i\theta} \\
&= \frac{c_{p,0} i\epsilon e^{i\theta}}{2\epsilon^{2p} e^{2pi\theta}} + \sum_{\ell=0}^{\infty} \left(\sum_{k=1}^{p} c_{p,k} k^\ell \right) \frac{2^\ell i^{\ell+1} \epsilon^{(\ell-2p+1)}}{\ell!} e^{i(\ell-2p+1)\theta}.
\end{aligned} \tag{5.9.23}
$$

In the computation of $\int_{\gamma_{4,\epsilon}} f(z) dz$ the terms with a strictly positive power of ϵ in (5.9.23) do not play a role because their sum goes to 0 as $\epsilon \to 0$. We thus focus on

$$\sum_{\ell=0}^{2p-1} \left(\sum_{k=1}^{p} c_{p,k} k^\ell \right) \frac{2^\ell i^{\ell+1} \epsilon^{(\ell-2p+1)}}{\ell!} e^{i(\ell-2p+1)\theta}.$$

The terms corresponding to even values of ℓ vanish in view of (5.9.19) (for $\ell = 0$) and (5.9.21) (for $\ell \neq 0$). On the other hand, the integral of $e^{i(\ell-2p+1)\theta}$ on $\gamma_{4,\epsilon}$ is equal to 0 when ℓ is odd and different from $2p-1$. Hence the only contribution from the sum (5.9.23) to $\lim_{\epsilon \to 0} \int_{\gamma_{4,\epsilon}} g(z) dz$ is the term corresponding to $\ell = 2p - 1$, and we conclude that

$$\int_{\mathbb{R}} \left(\frac{\sin x}{x} \right)^{2p} dx = \pi \frac{(-1)^p 2^{2p-1} \left(\sum_{k=1}^{p} c_{p,k} k^{2p-1} \right)}{(2p-1)!}. \tag{5.9.24}$$

Taking into account (5.9.20), we have

$$\int_{\mathbb{R}} \left(\frac{\sin x}{x} \right)^{2p} dx = \pi \frac{\sum_{k=1}^{p} \binom{2p}{p-k} (-1)^k k^{2p-1}}{(2p-1)!}. \tag{5.9.25}$$

In the following two examples we use the first formula. When $p = 2$ we have

$$c_{2,0} = \frac{3}{8}, \quad c_{2,1} = -\frac{4}{8}, \quad c_{2,2} = \frac{1}{8},$$

and

$$\int_{\mathbb{R}} \left(\frac{\sin x}{x}\right)^4 dx = \pi \frac{(-1)^2 2^3 \left(\sum_{k=1}^{2} c_{2,k} k^3\right)}{3!} = \pi \frac{8\left(\frac{-4+2^3}{8}\right)}{6} = \frac{2\pi}{3}.$$

When $p = 3$, we have

$$c_{3,0} = \frac{10}{2^5}, \quad c_{3,1} = -\frac{15}{2^5}, \quad c_{3,2} = \frac{6}{2^5}, \quad \text{and} \quad c_{3,3} = -\frac{1}{2^5},$$

and

$$\int_{\mathbb{R}} \left(\frac{\sin x}{x}\right)^6 dx = \pi \frac{(-1)^3 2^5 (-3^5 + 6 \cdot 2^5 - 15)}{2^5 \cdot 5!} = \frac{11\pi}{20}.$$

We now turn to the computation of the integrals

$$\int_{\mathbb{R}} \left(\frac{\sin x}{x}\right)^{2p+1} dx, \quad p \in \mathbb{N}.$$

Exercise 5.3.4 considers the case $p = 1$. The general case is considered after the solution.

Solution of Exercise 5.3.4. We have

$$\sin^3 x = \left(\frac{e^{ix} - e^{-ix}}{2i}\right)^3 = \frac{3e^{ix} - 3e^{-ix} + e^{3ix} - e^{-3ix}}{8i} = \frac{3\sin x - \sin 3x}{4}.$$

These equalities suggest integrating the function

$$f(z) = \frac{3e^{iz} - e^{3iz}}{4z^3}$$

along the above contour, and letting $\epsilon \to 0$ and $R \to \infty$. Cauchy's theorem gives

$$\int_{[-R,-\epsilon]\cup[\epsilon,R]} \frac{3e^{ix} - e^{3ix}}{4x^3} dx + \int_{\gamma_{2,R}} f(z)dz + \int_{\gamma_{4,\epsilon}} f(z)dz = 0. \qquad (5.9.26)$$

It is easy to check that

$$\lim_{R \to \infty} \int_{\gamma_{2,R}} f(z)dz = 0.$$

We now prove that

$$\lim_{\epsilon \to 0} \int_{\gamma_{4,\epsilon}} f(z)dz = i\frac{3\pi}{4}.$$

Indeed, we have

$$f(z) = \frac{3 + 3iz - \frac{3}{2}z^2 + \cdots - 1 - 3iz + \frac{9}{2}z^2 + \cdots}{4z^3}$$

$$= \frac{2 + 3z^2 + z^3 O(z)}{4z^3}$$

$$= \frac{2}{4z^3} + \frac{3}{4z} + O(z),$$

where $O(z)$ denotes a quantity which stays bounded as z goes to 0. But (with $\gamma_{4,\epsilon}$ defined as in Exercise 5.3.2)

$$\int_{\gamma_{4,\epsilon}} \frac{dz}{z^3} = \int_\pi^0 \frac{\epsilon i e^{it}}{\epsilon^3 e^{3it}} dt = -\frac{i}{\epsilon^2} \int_0^\pi e^{-2it} dt = 0,$$

$$\int_{\gamma_{4,\epsilon}} \frac{dz}{z} \equiv -i\pi$$

and

$$\lim_{\epsilon \to 0} \int_{\gamma_{4,\epsilon}} O(z) dz = 0.$$

Hence letting $\epsilon \longrightarrow 0$ and $R \longrightarrow \infty$ in (5.9.26) we have that

$$\int_{\mathbb{R}} \frac{3e^{ix} - e^{3ix}}{4x^3} dx = \frac{3\pi i}{4}.$$

Taking imaginary parts on both sides leads to the result.   □

A different and shorter proof, based on formula (5.3.3), appears in [207, p. 132].

We now present the analog of formula (5.9.24). We only outline the arguments. The first step consists in finding the rational (real) numbers such that

$$\left(\frac{e^{ix} - e^{-ix}}{2i} \right)^{2p+1} = \frac{\sum_{k=0}^p c_{p,k} (e^{i(2k+1)x} - e^{-i(2k+1)x})}{2i}.$$

We have therefore

$$\sin^{2p+1}(x) = \sum_{k=0}^p c_{p,k} \sin((2k+1)x).$$

As above we note that

$$c_{p,k} = \frac{\binom{2p+1}{p-k} (-1)^{k-p}}{2^{2p}}, \quad k = 0, \ldots, p. \qquad (5.9.27)$$

Differentiating this equation $2p + 1$ times and setting $x = 0$ we obtain the counterpart of (5.9.21):

$$\sum_{k=0}^{p} c_{p,k}(2k+1)^{2t+1} = 0, \quad t = 0, \ldots, p. \tag{5.9.28}$$

We now integrate the function

$$f(z) = \frac{\sum_{k=0}^{p} c_{p,k} z^{2k+1}}{z^{2p+1}}$$

along the same contour as above. Since

$$f(\epsilon e^{i\theta})i\epsilon e^{i\theta} = \sum_{\ell=0}^{\infty}\left(\sum_{k=0}^{p} c_{p,k}(2k+1)^{\ell}\right)\frac{i^{\ell+1}\epsilon^{\ell-2p}}{\ell!}e^{i(\ell-2p)\theta}, \tag{5.9.29}$$

the terms corresponding to odd values of ℓ vanish in view of (5.9.28). On the other hand, the integral of $e^{i(\ell-2p)\theta}$ on $\gamma_{4,\epsilon}$ is equal to 0 when ℓ is even and different from $2p$. Hence the only contribution from the sum (5.9.29) to $\lim_{\epsilon\to 0}\int_{\gamma_{4,\epsilon}} f(z)dz$ is the term corresponding to $\ell = 2p$, and we conclude that

$$\int_{\mathbb{R}}\left(\frac{\sin x}{x}\right)^{2p+1} dx = \pi\frac{(-1)^p\left(\sum_{k=0}^{p} c_{p,k}(2k+1)^{2p}\right)}{(2p)!}. \tag{5.9.30}$$

Taking into account (5.9.27) this formula can be rewritten as

$$\int_{\mathbb{R}}\left(\frac{\sin x}{x}\right)^{2p+1} dx = \pi\frac{\sum_{k=0}^{p}(-1)^k\binom{2p+1}{2k}(2k+1)^{2p}}{2^{2p}(2p)!}$$
$$= \pi\frac{\sum_{k=0}^{p}(-1)^k\binom{2p+1}{2k}(k+\frac{1}{2})^{2p}}{(2p)!}. \tag{5.9.31}$$

As a check, let us use formula (5.9.30) and consider the cases $p = 0$, $p = 1$ and $p = 2$. For $p = 0$ we trivially have $c_{0,1} = 1$ and so

$$\int_{\mathbb{R}}\frac{\sin x}{x}dx = \pi,$$

and we get back the value of the Dirichlet integral (5.9.13). When $p = 1$ we have

$$c_{1,0} = \frac{3}{4}, \quad c_{1,1} = -\frac{1}{4},$$

and formula (5.9.30) gives

$$\int_{\mathbb{R}}\frac{\sin^3 x}{x^3}dx = \frac{3\pi}{4},$$

and we get back the integral computed in Exercise 5.3.4. For $p = 2$, we have

$$c_{2,0} = \frac{10}{16}, \quad c_{2,1} = -\frac{5}{16}, \quad c_{2,2} = \frac{1}{16},$$

and we get the integral

$$\int_{\mathbb{R}} \left(\frac{\sin x}{x}\right)^5 dx = \frac{115\pi}{192}.$$

Solution of Exercise 5.3.5.
Case 1: $H \in (0, 1/2)$. We take the function

$$f(z) = \frac{1 - e^{iz}}{z^2} \exp\{(1 - 2H) \ln z\},$$

where $\ln z$ is the logarithm function defined on $\mathbb{C} \setminus \mathbb{R}_-$, and equal to $\ln x$ on the positive line. By Cauchy's theorem the integral of f along $\gamma_{\epsilon,R}$ is equal to 0. We now show that

$$\lim_{\epsilon \to 0} \int_{C_\epsilon} f(z) dz = \lim_{R \to \infty} \int_{C_R} f(z) dz = 0. \tag{5.9.32}$$

Indeed, $\ln(\epsilon e^{it}) = \ln \epsilon + it$ for $t \in [0, \pi/2]$, and we have

$$\int_{C_\epsilon} f(z) dz = \int_{\pi/2}^0 \frac{1 - e^{i\epsilon e^{it}}}{\epsilon^2 e^{2it}} \exp\{(1 - 2H)(\ln \epsilon + it)\} i\epsilon e^{it} dt$$

$$= -i \int_0^{\pi/2} \frac{1 - e^{i\epsilon e^{it}}}{\epsilon e^{it}} \epsilon^{1-2H} e^{(1-2H)it} dt.$$

For $|z| \le 1$ (and in particular when $z = i\epsilon e^{it}$ with $\epsilon < 1$), we have that

$$\left|\frac{1 - e^z}{z}\right| = \left|\sum_{n=1}^{\infty} \frac{z^{n-1}}{n!}\right| \le \sum_{n=1}^{\infty} \frac{1}{n!} = e - 1.$$

Moreover, for $H \in (0, 1/2)$ we have $1 - 2H > 0$ and so $\lim_{\epsilon \to 0} \epsilon^{1-2H} = 0$. Hence, for $\epsilon < 1$,

$$\left|\int_{C_\epsilon} f(z) dz\right| \le (e - 1)\frac{\pi}{2} \epsilon^{1-2H} \to 0,$$

as $\epsilon \to 0$. Similarly,

$$\int_{C_R} f(z) dz = i \int_0^{\pi/2} \frac{1 - e^{iRe^{it}}}{Re^{it}} R^{1-2H} e^{(1-2H)it} dt$$

$$= i \int_0^{\pi/2} \frac{1 - e^{iRe^{it}}}{e^{it}} R^{-2H} e^{(1-2H)it} dt.$$

Since

$$|e^{iRe^{it}}| = e^{-R\sin t} \leq 1, \quad t \in [0, \pi/2],$$

we have $|1 - e^{iRe^{it}}| \leq 2$ for $t \in [0, \pi/2]$, and

$$\left| \int_{C_R} f(z)dz \right| \leq \pi R^{-2H} \to 0,$$

as $R \to \infty$. Thus we have

$$\int_0^\infty f(x)dx = -\lim_{\substack{\epsilon \to 0, \\ R \to \infty}} \int_R^\epsilon i \frac{1 - e^{-y}}{-y^2} y^{1-2H} e^{i(1-2H)\pi/2} dy$$

$$= (ie^{i(1-2H)\pi/2}) \int_0^\infty \frac{1 - e^{-y}}{-y^2} y^{1-2H} dy$$

$$= e^{-i\pi H} \int_0^\infty \frac{1 - e^{-y}}{y^2} y^{1-2H} dy,$$

so that, equating real parts of both sides, we have

$$\int_0^\infty \frac{1 - \cos x}{x^2} x^{1-2H} dx = \cos(\pi H) \int_0^\infty \frac{1 - e^{-y}}{y^2} y^{1-2H} dy.$$

Finally, integration by parts gives

$$\int_0^\infty (1 - e^{-y}) y^{-1-2H} dy = \frac{(1 - e^{-y})y^{-2H}}{-2H} \Big|_0^\infty - \int_0^\infty \frac{e^{-y}y^{-2H}}{-2H} dy$$

$$= -\int_0^\infty \frac{e^{-y}y^{-2H}}{-2H} dy$$

$$= \frac{\Gamma(1 - 2H)}{2H}.$$

To go from the first line to the second, note that $\frac{(1-e^{-y})y^{-2H}}{-2H} \Big|_0^\infty = 0$ since $H \in (0, 1/2)$. So

$$\int_0^\infty \frac{1 - \cos x}{x^2} x^{1-2H} dx = \cos(\pi H) \frac{\Gamma(1 - 2H)}{2H},$$

which can also be rewritten as

$$\int_0^\infty \frac{1 - \cos x}{x^2} x^{1-2H} dx = \cos(\pi H) \frac{\Gamma(2 - 2H)}{2H(1 - 2H)},$$

since $\Gamma(1 + z) = z\Gamma(z)$.

Case 2: $H \in (1/2, 1)$. We first integrate twice by parts to obtain

$$
\int_0^\infty (1 - \cos x) x^{-1-2H} \, dx = (1 - \cos x) \frac{x^{-2H}}{-2H} \Big|_0^\infty + \frac{1}{2H} \int_0^\infty \frac{\sin x}{x^{2H}} \, dx
$$

$$
= \frac{1}{2H} \int_0^\infty \frac{\sin x}{x^{2H}} \, dx
$$

$$
= \sin x \frac{x^{1-2H}}{2H(1 - 2H)} \Big|_0^\infty - \frac{1}{2H(1 - 2H)} \int_0^\infty x^{1-2H} \cos x \, dx
$$

$$
= -\frac{1}{2H(1 - 2H)} \int_0^\infty x^{1-2H} \cos x \, dx.
$$

(5.9.33)

Note that

$$
(1 - \cos x) \frac{x^{-2H}}{-2H} \Big|_0^\infty = \sin x \frac{x^{1-2H}}{2H(1 - 2H)} \Big|_0^\infty = 0,
$$

since $H \in (1/2, 1)$.

To compute the last integral we use the same contour as in the first case, but with the function $f(z) = e^{iz} e^{(1-2H) \ln z}$. We first show that (5.9.32) still holds here with the present choice of f. We have

$$
\int_{C_R} f(z) \, dz = \int_0^{\pi/2} e^{iRe^{it}} R^{1-2H} e^{i(1-2H)t} i Re^{it} \, dt,
$$

and so

$$
\left| \int_{C_R} f(z) \, dz \right| \leq \int_0^{\pi/2} e^{-R \sin t} R^{2-2H} \, dt.
$$

From the proof of Jordan's lemma, or, equivalently, checking that on $[0, \pi/2]$, it holds that

$$
\sin t \geq \frac{2t}{\pi},
$$

we obtain

$$
\int_0^{\pi/2} e^{-R \sin t} \, dt \leq \int_0^{\pi/2} e^{-2tR/\pi} \, dt = \frac{\pi}{2R} (1 - e^{-R}) \leq \frac{\pi}{R}.
$$

Hence

$$
\left| \int_{C_R} f(z) \, dz \right| \leq \int_0^{\pi/2} e^{-R \sin t} R^{2-2H} \, dt \leq \pi R^{1-2H} \to 0,
$$

as $R \to \infty$ since $1 - 2H < 0$. In a similar way,

$$
\int_{C_\epsilon} f(z) \, dz = \int_{\pi/2}^0 e^{i\epsilon e^{it}} \epsilon^{1-2H} e^{i(1-2H)t} i\epsilon e^{it} \, dt.
$$

We now use that

$$
|e^{i\epsilon e^{it}}| = e^{-\epsilon \sin t} \leq 1, \quad t \in \left[0, \frac{\pi}{2}\right],
$$

and so

$$\left| \int_{C_\epsilon} f(z)dz \right| \leq \frac{\pi}{2} \epsilon^{2-2H} \to 0,$$

as $\epsilon \to 0$ since $2 - 2H > 0$. Hence,

$$\int_0^\infty f(x)dx = - \lim_{\substack{\epsilon \to 0, \\ R \to \infty}} \int_R^\epsilon e^{-y} e^{(1-2H)(i\pi/2 + \ln y)} idy$$

$$= (ie^{(1-2H)i\pi/2}) \int_0^\infty e^{-y} y^{1-2H} dy$$

$$= -e^{-i\pi H} \Gamma(2 - 2H).$$

Taking real parts we obtain

$$\int_0^\infty x^{1-2H} \cos x dx = -\cos(\pi H)\Gamma(2 - 2H).$$

Comparing with (5.9.33), we obtain:

$$\int_0^\infty \frac{1 - \cos x}{x^2} x^{1-2H} dx = \frac{\cos(\pi H)\Gamma(2 - 2H)}{2H(1 - 2H)}. \qquad \square$$

Solution of Exercise 5.4.2. Let $z \in \Omega$. Then

$$|1 + \varphi(z)| \geq |1 + \operatorname{Re} \varphi(z)| \geq 1$$

since $\operatorname{Re} \varphi(z) \geq 0$. Thus $(1 + \varphi(z)) \neq 0$ and the function

$$s(z) = \frac{1 - \varphi(z)}{1 + \varphi(z)}$$

is well defined in Ω. Furthermore,

$$1 - |s(z)|^2 = \frac{2 \operatorname{Re} \varphi(z)}{|1 + \varphi(z)|^2} \geq 0.$$

Thus $s(z)$ is bounded in a punctured neighborhood of z_0, and hence, by Riemann's removable singularity theorem (see Theorem 5.4.1), z_0 is a removable singularity.
$\qquad \square$

Solution of Exercise 5.4.3. From the arguments of Exercise 4.2.18 the function f has a derivative at every point where it does not vanish, and so f is analytic in $\Omega \setminus \{w \in \Omega \; ; \; f(w) = 0\}$. The points $\{w \in \Omega \; ; \; f(w) = 0\}$ are *a priori* isolated singularities of f. Since f is continuous at these points, an application of Riemann's removable singularity Theorem (see Theorem 5.4.1) shows that f is analytic in all of Ω.
$\qquad \square$

Solution of Exercise 5.4.4. Let $\mathcal{Z}(f)$ denote the set of zeroes of f in Ω. We have $\mathcal{Z}(f) = \mathcal{Z}(f^2)$. Hence, $\mathcal{Z}(f)$ has only isolated points in Ω since f^2 is analytic. The function

$$f = \frac{f^3}{f^2} \tag{5.9.34}$$

is a quotient of two functions analytic in $\Omega \setminus \mathcal{Z}(f)$, and hence is analytic there. From expression (5.9.34) we see that the points in $\mathcal{Z}(f)$ are isolated singularities of f. We will now show that they are removable singularities. Let $z_0 \in \mathcal{Z}(f)$. We write

$$f^2(z) = (z - z_0)^N g(z),$$
$$f^3(z) = (z - z_0)^M h(z),$$

where $N, M \in \mathbb{N}$ and where g and h are analytic in $B(z_0, r)$ for some $r > 0$, and do not vanish in $B(z_0, r)$. Therefore, for $z \in B(z_0, r) \setminus \{z_0\}$ we have

$$f(z) = (z - z_0)^{M-N} \frac{h(z)}{g(z)}. \tag{5.9.35}$$

The function f^2 is analytic in Ω and so it is bounded in modulus in a neighborhood of z_0, and so f is also bounded in modulus in a neighborhood of z_0. This forces $M \geq N$, and (5.9.35) expresses that z_0 is a removable singularity of f. Since $f(z_0) = 0$ we have in fact $M > N$.

To prove the last claim, let p_1 and p_2 in $\mathbb{Z}$ be such that

$$p_1 n_1 + p_2 n_2 = 1.$$

One, and only one, of the numbers p_1 or p_2 is negative. Without loss of generality we assume $p_1 > 0$. It suffices to redo the preceding analysis with $f^{p_1 n_1}$ and $f^{-p_2 n_2}$ instead of f^3 and f^2. $\square$

Solution of Exercise 5.4.5. The first claim is a direct application of Riemann's removable singularity theorem (see Theorem 5.4.1 for the latter). To prove the second claim, one first takes $a \neq b$, and z different from a and b. One then has:

$$\frac{(R_a f)(z) - (R_b f)(z)}{a - b} = \frac{\dfrac{f(z) - f(a)}{z - a} - \dfrac{f(z) - f(b)}{z - b}}{a - b}$$
$$= \frac{(f(z) - f(a))(z - b) - (f(z) - f(b))(z - a)}{(z - a)(z - b)(a - b)},$$

and

$$R_a R_b f(z) = \frac{\dfrac{f(z) - f(b)}{z - b} - \dfrac{f(a) - f(b)}{a - b}}{z - a}$$
$$= \frac{(f(z) - f(b))(a - b) - (f(a) - f(b))(z - b)}{(z - a)(z - b)(a - b)}.$$

The equality follows (still for z different from a and b, and $a \neq b$) since

$$(f(z) - f(a))(z - b) - (f(z) - f(b))(z - a)$$
$$= (f(z) - f(b))(a - b) - (f(a) - f(b))(z - b)$$
$$= (a - b)f(z) - (z - b)f(a) + (z - a)f(b).$$

Therefore, for $a \neq b$, and using (5.4.1), we see that the function

$$\frac{R_a f(z) - R_b f(z)}{a - b}$$

can be extended analytically to all of Ω. Hence $R_a R_b f$ can also be extended analytically to all of Ω, and we obtain the resolvent equation. Finally, in the case $a = b$, equation (5.4.2) is trivial.   □

Solution of Exercise 5.5.1. We follow [29, p. 206]. Let p be as in (5.5.1). The function

$$p^\sharp(z) = 1 + \overline{a_{n-1}}z + \cdots + \overline{a_0}z^n \qquad\qquad (5.9.36)$$

is still a polynomial, and Cauchy's formula applied to $p^\sharp(z)f(z)$ leads to

$$p^\sharp(0)f(0) = \frac{1}{2\pi} \int_0^{2\pi} p^\sharp(e^{it})f(e^{it})\,dt.$$

Since $p^\sharp(0) = 1$, using (3.1.4) we have

$$|f(0)| \leq \frac{1}{2\pi} \int_0^{2\pi} |p^\sharp(e^{it})f(e^{it})|\,dt.$$

To conclude we note that, for $z \neq 0$,

$$p^\sharp(z) = z^n \overline{p(1/\overline{z})}.$$

In particular, for $z = e^{it}$ (with $t \in \mathbb{R}$) we have $1/\overline{z} = z$ and so

$$|p^\sharp(e^{it})| = |p(e^{it})|,$$

and hence we get the required inequality.   □

We note that the operation $p \mapsto p^\sharp$ appears also in the solution of Exercise 6.4.3.

Solution of Exercise 5.5.2. In view of Cauchy's formula, (5.1.5) can be rewritten as

$$4\pi^2 |f(0)|^2 \leq \left(\max_{z \in C} \frac{1}{|z|^2} \right) \cdot L(C) \cdot \int_C |f(z)|^2 |dz|,$$

and so the infimum is strictly positive when we fix $f(0) = 1$.   □

Solution of Exercise 5.5.3. Let p be a polynomial. Then, by Cauchy's formula,

$$\frac{1}{2\pi i}\int_{\mathbb{T}}\frac{1-zp(z)}{z}dz=1. \qquad (5.9.37)$$

Therefore, using (5.1.3),

$$1=\left|\frac{1}{2\pi i}\int_{\mathbb{T}}\frac{1-zp(z)}{z}dz\right|$$

$$\leq \max_{z\in\mathbb{T}}|1-zp(z)|$$

$$=\max_{z\in\mathbb{T}}|z^{-1}-p(z)|,$$

where we have used formula (5.1.3) to go from the first to the second line. This proves the claim with $M=1$. $\qquad\square$

Solution of Exercise 5.5.4. If z_0 is in the exterior of C, both functions

$$\frac{f'(z)}{z-z_0}\quad\text{and}\quad\frac{f(z)}{(z-z_0)^2}$$

are analytic in a neighborhood of the interior of C, and Cauchy's theorem insures that both sides of (5.5.4) vanish.

If z_0 is in the interior of C, Cauchy's formula applied to f' shows that

$$\int_C\frac{f'(z)}{z-z_0}dz=2\pi i f'(z_0),$$

while Cauchy's formula applied to f shows that

$$\int_C\frac{f(z)}{(z-z_0)^2}dz=2\pi i f'(z_0),$$

and hence the result. $\qquad\square$

Another proof, using the notion of removable singularity, is presented in Exercise 7.2.3.

Solution of Exercise 5.5.5. The function

$$\frac{e^{\sin z^2}}{(z^2+1)(z-2i)^3}$$

is analytic in $|z|<1$, and therefore for every $r<1$, Cauchy's theorem implies that

$$\int_{|z|=r}\frac{e^{\sin z^2}\,dz}{(z^2+1)(z-2i)^3}=0.$$

Let now $1 < r < 2$. Let C_i and C_{-i} be two circles around the points i and $-i$, with (say common) radius ρ such that $B(i, \rho)$ and $B(-i, \rho)$ are both inside $B(0, r)$. By Theorem 5.3.1, we have

$$\int_{|z|=r} \frac{e^{\sin z^2} dz}{(z^2+1)(z-2i)^3} = \int_{C_i} \frac{e^{\sin z^2} dz}{(z^2+1)(z-2i)^3} + \int_{C_{-i}} \frac{e^{\sin z^2} dz}{(z^2+1)(z-2i)^3},$$

where all the curves are taken with the positive orientation. The function

$$\frac{e^{\sin z^2}}{(z+i)(z-2i)^3}$$

is analytic in an open neighborhood of $|z - i| \le \rho$. By Cauchy's formula,

$$\int_{C_i} \frac{e^{\sin z^2} dz}{(z^2+1)(z-2i)^3} = \int_{C_i} \frac{\frac{e^{\sin z^2}}{(z+i)(z-2i)^3}}{z-i} dz$$

$$= 2\pi i \frac{e^{\sin z^2}}{(z+i)(z-2i)^3}\Big|_{z=i}$$

$$= -i\pi e^{-\sin 1}.$$

The integral

$$\int_{C_{-i}} \frac{e^{\sin z^2} dz}{(z^2+1)(z-2i)^3} = \int_{C_{-i}} \frac{\frac{e^{\sin z^2}}{(z-i)(z-2i)^3}}{z+i} dz$$

is computed in a similar way.

For the case $r > 2$ one takes ρ such that $B(2i, \rho)$ is inside $B(0, r)$. Theorem 5.3.1 now gives

$$\int_{|z|=r} \frac{e^{\sin z^2} dz}{(z^2+1)(z-2i)^3} = \int_{C_i} \frac{e^{\sin z^2} dz}{(z^2+1)(z-2i)^3} + \int_{C_{-i}} \frac{e^{\sin z^2} dz}{(z^2+1)(z-2i)^3}$$

$$+ \int_{C_{2i}} \frac{e^{\sin z^2} dz}{(z^2+1)(z-2i)^3},$$

where C_{2i} denotes the positively oriented circle C_{2i} around $z = 2i$ with radius ρ. The first two integrals on the right side of the above equality have already been computed using Cauchy's formula. The third one is computed in the same way, namely

$$\int_{C_{2i}} \frac{e^{\sin z^2} dz}{(z^2+1)(z-2i)^3} = \int_{C_{2i}} \frac{\frac{e^{\sin z^2}}{z^2+1}}{(z-2i)^3} dz = 2\pi i \cdot 2! \cdot \left(\frac{e^{\sin z^2}}{z^2+1}\right)^{(2)}\Big|_{z=2i}. \qquad \square$$

Solution of Exercise 5.5.6. Using formula (5.5.5), we have

$$\int_0^{2\pi} e^{e^{2it}-3it}\,dt = \int_{|z|=1} \frac{e^{z^2}}{iz^4}\,dz$$

$$= 2\pi\frac{1}{2\pi i}\int_{|z|=1} \frac{e^{z^2}}{z^4}\,dz$$

$$= 2\pi\,\frac{(e^{z^2})^{(3)}}{3!}\bigg|_{z=0}$$

$$= 0,$$

where we have used Cauchy's formula. $\qquad\square$

Solution of Exercise 5.5.7. We begin as for Exercise 3.1.8 and write $\cos t = \dfrac{e^{it}+e^{-it}}{2}$ and

$$\int_0^{2\pi} \cos^{2p} t\,dt = \int_0^{2\pi}\left(\frac{e^{it}+e^{-it}}{2}\right)^{2p}\,dt = \int_0^{2\pi} \frac{(e^{2it}+1)^{2p}}{2^{2p}e^{2pit}}\,dt.$$

Using formula (5.5.5) we see that this equation in turn is equal to

$$\frac{1}{4^p i}\int_{|z|=1} \frac{(z^2+1)^{2p}}{z^{2p+1}}\,dz.$$

Let $f(z) = (z^2+1)^{2p}$. Cauchy's formula implies that

$$\frac{1}{2\pi i}\int_{|z|=1} \frac{(z^2+1)^{2p}}{z^{2p+1}}\,dz = \frac{f^{(2p)}(0)}{(2p)!}.$$

But

$$f^{(2p)}(0) = (2p)!\binom{2p}{p},$$

since

$$f(z) = \sum_{\ell=0}^{2p} z^{2\ell}\binom{2p}{\ell},$$

and so

$$\int_0^{2\pi} \cos^{2p} t\,dt = \frac{2\pi i}{4^p i}\frac{1}{2\pi i}\int_{|z|=1} \frac{(z^2+1)^{2p}}{z^{2p+1}}\,dz$$

$$= \frac{2\pi i}{4^p i}\frac{f^{(2p)}(0)}{(2p)!}$$

$$= \frac{2\pi}{4^p}\binom{2p}{p}. \qquad\square$$

The dominated convergence theorem leads to

$$\lim_{p\to\infty} \int_0^{2\pi} \cos^{2p} t\, dt = 0.$$

This can also be checked directly from the formula for $\int_0^{2\pi} \cos^{2p} t\, dt$ using Stirling's formula.

We also note that $\int_0^{2\pi} \cos^{2p+1} t\, dt = 0$. Indeed, the change of variable $t \mapsto u = t + \pi$ leads to

$$\int_0^{2\pi} \cos^{2p+1} t\, dt = \int_\pi^{3\pi} (-1)^{2p+1} \cos^{2p+1} u\, du$$

$$= -\int_\pi^{3\pi} \cos^{2p+1} u\, du$$

$$= -\int_0^{2\pi} \cos^{2p+1} u\, du,$$

where the last equality holds since the function $\cos^{2p+1} u$ has period 2π, and so its integral is the same on any interval of length 2π.

Finally we remark that one can apply the residue theorem to compute

$$\int_{|z|=1} \frac{(z^2+1)^{2p}}{z^{2p+1}}\, dz.$$

Solution of Exercise 5.5.8. We have

$$\frac{1}{4\pi}\int_0^{2\pi} \frac{e^{it}+z}{e^{it}-z} f(e^{it})\, dt = \frac{1}{4\pi}\int_0^{2\pi} \frac{2e^{it}-e^{it}+z}{e^{it}-z} f(e^{it})\, dt$$

$$= \frac{1}{2\pi}\int_0^{2\pi} \frac{e^{it}}{e^{it}-z} f(e^{it})\, dt - \frac{1}{4\pi}\int_0^{2\pi} f(e^{it})\, dt$$

$$= f(z) - \frac{f(0)}{2},$$

since, by Cauchy's formula,

$$f(z) = \frac{1}{2\pi i}\int_{|\zeta|=1} \frac{f(\zeta)}{\zeta - z}\, d\zeta = \frac{1}{2\pi}\int_0^{2\pi} \frac{e^{it}}{e^{it}-z} f(e^{it})\, dt$$

and in particular

$$f(0) = \frac{1}{2\pi}\int_0^{2\pi} \frac{e^{it}}{e^{it}} f(e^{it})\, dt = \frac{1}{2\pi}\int_0^{2\pi} f(e^{it})\, dt.$$

On the other hand, for a given $z \in \mathbb{D}$, Cauchy's formula applied to the function

$$g(\zeta) = f(\zeta)\frac{1+\zeta\bar{z}}{1-\zeta\bar{z}}$$

leads to

$$f(0) = g(0) = \frac{1}{2\pi}\int_0^{2\pi} g(e^{it})dt = \frac{1}{2\pi}\int_0^{2\pi} f(e^{it})\frac{1+e^{it}\overline{z}}{1-e^{it}\overline{z}}dt,$$

and so

$$\frac{1}{2\pi}\int_0^{2\pi} f(e^{it})\frac{1+e^{it}\overline{z}}{1-e^{it}\overline{z}}dt = f(0).$$

Therefore,

$$\frac{1}{4\pi}\int_0^{2\pi} \overline{f(e^{it})}\frac{e^{it}+z}{e^{it}-z}dt = \overline{\left(\frac{1}{4\pi}\int_0^{2\pi} f(e^{it})\frac{e^{-it}+\overline{z}}{e^{-it}-\overline{z}}dt\right)}$$

$$= \overline{\left(\frac{1}{4\pi}\int_0^{2\pi} f(e^{it})\frac{1+e^{it}\overline{z}}{1-e^{it}\overline{z}}dt\right)}$$

$$= \frac{\overline{f(0)}}{2},$$

and hence the result since

$$\frac{1}{2\pi}\int_0^{2\pi} \frac{e^{it}+z}{e^{it}-z}\operatorname{Re} f(e^{it})dt$$

$$= \frac{1}{2}\left(\frac{1}{2\pi}\int_0^{2\pi} \frac{e^{it}+z}{e^{it}-z}f(e^{it})dt + \frac{1}{2\pi}\int_0^{2\pi} \frac{e^{it}+z}{e^{it}-z}\overline{f(e^{it})}dt\right).$$

To prove the first equality in (5.5.7), it suffices to subtract $f(0)$ from each side of (5.5.6). One obtains

$$f(z) - f(0) = i\operatorname{Im} f(0) - f(0) + \frac{1}{2\pi}\int_0^{2\pi} \frac{e^{it}+z}{e^{it}-z}\operatorname{Re} f(e^{it})dt$$

$$= \frac{1}{2\pi}\int_0^{2\pi} \frac{e^{it}+z}{e^{it}-z}\operatorname{Re} f(e^{it})dt - \operatorname{Re} f(0)$$

$$= \frac{1}{2\pi}\int_0^{2\pi} \frac{e^{it}+z}{e^{it}-z}\operatorname{Re} f(e^{it})dt - \frac{1}{2\pi}\int_0^{2\pi} \operatorname{Re} f(e^{it})dt$$

$$= \frac{1}{2\pi}\int_0^{2\pi} \frac{2z}{e^{it}-z}\operatorname{Re} f(e^{it})dt.$$

Differentiating n times with respect to z both sides of the first equality in (5.5.7) we obtain the second claim. Interchanging integration and differentiation is legitimate thanks to Theorem 14.6.1. The third claim is a direct computation, done as follows:

$$\operatorname{Re} f(z) = \operatorname{Re}\left(i \operatorname{Im} f(0) + \frac{1}{2\pi} \int_0^{2\pi} \frac{e^{it} + z}{e^{it} - z} \operatorname{Re} f(e^{it}) dt \right)$$

$$= \frac{1}{2\pi} \int_0^{2\pi} \left(\operatorname{Re} \frac{e^{it} + z}{e^{it} - z} \right) \operatorname{Re} f(e^{it}) dt$$

$$= \frac{1}{2\pi} \int_0^{2\pi} \frac{1 - |z|^2}{|e^{it} - z|^2} \operatorname{Re} f(e^{it}) dt. \qquad \square$$

Solution of Exercise 5.5.9. It suffices to define the function $g(z) = f(Rz)$ for $z \in \mathbb{D}$ and apply Exercise 5.5.8 to g. $\qquad \square$

Solution of Exercise 5.5.10. This is a direct consequence of the third equality in (5.5.7). $\qquad \square$

Solution of Exercise 5.5.11. We use the representation (5.5.9) for f:

$$f(z) = ia + \int_0^{2\pi} \frac{e^{it} + z}{e^{it} - z} d\mu(t),$$

and write for $z, w \in \mathbb{D}$,

$$\frac{f(z) + \overline{f(w)}}{1 - z\overline{w}} = \frac{\int_0^{2\pi} \left\{ \frac{e^{it} + z}{e^{it} - z} + \frac{e^{-it} + \overline{w}}{e^{-it} - \overline{w}} \right\} d\mu(t)}{1 - z\overline{w}}$$

$$= \frac{\int_0^{2\pi} \left\{ \frac{2 - 2z\overline{w}}{(e^{it} - z)(e^{-it} - \overline{w})} \right\} d\mu(t)}{1 - z\overline{w}}$$

$$= 2 \int_0^{2\pi} \frac{d\mu(t)}{(e^{it} - z)(e^{-it} - \overline{w})}.$$

Therefore for $N \in \mathbb{N}$, $w_1, \ldots, w_N \in \mathbb{D}$ and $c_1, \ldots, c_N \in \mathbb{C}$ we have

$$\sum_{\ell,j=1}^N \overline{c_\ell} c_j \frac{f(w_\ell) + \overline{f(w_j)}}{1 - w_\ell \overline{w_j}} = 2 \int_0^{2\pi} \sum_{\ell,j=1}^N \left(\frac{\overline{c_\ell} c_j}{(e^{it} - w_\ell)(e^{-it} - \overline{w_j})} \right) d\mu(t)$$

$$= 2 \int_0^{2\pi} \left| \sum_0^N \frac{\overline{c_\ell}}{e^{it} - w_\ell} \right|^2 d\mu(t)$$

$$\geq 0. \qquad \square$$

Solution of Exercise 5.5.12. Using formula (5.5.6) and the equality

$$\frac{e^{it} + z}{e^{it} - z} = 1 + 2 \sum_{n=0}^\infty z^{n+1} e^{-i(n+1)t}, \quad |z| < 1,$$

we have for $|z| < 1$,

$$f(z) = i \operatorname{Im} f(0) + \frac{1}{2\pi} \int_0^{2\pi} \operatorname{Re} f(e^{it})dt + 2 \sum_{\ell=1}^{\infty} z^{\ell} \frac{1}{2\pi} \int_0^{2\pi} e^{-i\ell t} \operatorname{Re} f(e^{it})dt.$$

Thus

$$f_0 = \operatorname{Re} f(0) = \frac{1}{2\pi} \int_0^{2\pi} \operatorname{Re} f(e^{it})dt,$$

and

$$f_\ell = \frac{1}{2\pi} \int_0^{2\pi} e^{-i\ell t} \operatorname{Re} f(e^{it})dt, \quad \ell = 1, 2, \ldots.$$

Since the real part of f is positive on $\mathbb{T}$, it follows from these expressions that

$$|f_\ell| \leq \frac{1}{2\pi} \int_0^{2\pi} \operatorname{Re} f(e^{it})dt = 2 \operatorname{Re} f(0), \quad \ell = 1, 2, \ldots. \qquad \square$$

Solution of Exercise 5.5.13. We will assume $f_0 \in \mathbb{R}$ and positive. This can always be achieved by multiplying f by a constant of modulus 1, and this operation does not affect (5.5.12). The function $g = 1 - f$ is analytic in $|z| < 1 + \epsilon$ and has a real positive part in the open unit disk. By (5.5.11)

$$|f_\ell| \leq 2 \operatorname{Re}(1 - f_0) = 2(1 - f_0), \quad \ell = 1, 2, \ldots,$$

and thus, for $|z| \leq 1/3$ we have:

$$\sum_{\ell=0}^{\infty} |f_\ell z^{\ell}| \leq f_0 + \sum_{\ell=1}^{\infty} \frac{|f_\ell|}{3^{\ell}}$$

$$\leq f_0 + \sum_{\ell=1}^{\infty} \frac{2(1 - f_0)}{3^{\ell}}$$

$$= f_0 + 2(1 - f_0) \sum_{\ell=1}^{\infty} \frac{1}{3^{\ell}}$$

$$= f_0 + (1 - f_0) = 1. \qquad \square$$

Solution of Exercise 5.5.15. From (5.5.6) we have:

$$\operatorname{Re} f(z) = \frac{1}{2\pi} \int_0^{2\pi} \operatorname{Re} \left(\frac{e^{it} + z}{e^{it} - z} \right) \operatorname{Re} f(e^{it})dt = \frac{1}{2\pi} \int_0^{2\pi} \frac{1 - |z|^2}{|e^{it} - z|^2} \operatorname{Re} f(e^{it})dt.$$

But, for $|z| < 1$ we have

$$\frac{1 - |z|^2}{(1 + |z|)^2} \leq \frac{1 - |z|^2}{|e^{it} - z|^2} \leq \frac{1 - |z|^2}{(1 - |z|)^2}$$

and so (since $1 - |z|^2 = (1 - |z|)(1 + |z|)$)

$$\frac{1 - |z|}{1 + |z|} \leq \frac{1 - |z|^2}{|e^{it} - z|^2} \leq \frac{1 + |z|}{1 - |z|}.$$

Therefore

$$\frac{1 - |z|}{1 + |z|} \operatorname{Re} f(e^{it}) \leq \frac{1 - |z|^2}{|e^{it} - z|^2} \operatorname{Re} f(e^{it}) \leq \frac{1 + |z|}{1 - |z|} \operatorname{Re} f(e^{it}).$$

Integrating these inequalities we obtain (5.5.13) since

$$\operatorname{Re} f(0) = \frac{1}{2\pi} \int_0^{2\pi} \operatorname{Re} f(e^{it}) dt. \qquad \square$$

Solution of Exercise 5.5.16. By Cauchy's formula applied to f^4 and $z_0 = 0$ we obtain:

$$0 = f(0)^4 = \frac{1}{2\pi i} \int_{|z|=1} \frac{f^4(z)}{z} dz = \frac{1}{2\pi} \int_0^{2\pi} f^4(e^{it}) dt.$$

In particular,

$$\operatorname{Re} \int_0^{2\pi} f^4(e^{it}) dt = 0.$$

Since

$$\operatorname{Re} f^4 = \operatorname{Re}(u + iv)^4 = u^4 + v^4 - 6u^2 v^2,$$

we obtain

$$\int_0^{2\pi} (u(\cos t, \sin t)^4 + v(\cos t, \sin t)^4) dt = 6 \int_0^{2\pi} u(\cos t, \sin t)^2 v(\cos t, \sin t)^2 dt,$$

and in particular

$$\int_0^{2\pi} u(\cos t, \sin t)^4 dt \leq 6 \int_0^{2\pi} u(\cos t, \sin t)^2 v(\cos t, \sin t)^2 dt.$$

Taking squares of both sides and applying the Cauchy–Schwarz inequality to the expression on the right, we obtain

$$\left(\int_0^{2\pi} u(\cos t, \sin t)^4 dt \right)^2 \leq 36 \left(\int_0^{2\pi} u(\cos t, \sin t)^2 v(\cos t, \sin t)^2 dt \right)^2$$

$$\leq 36 \left(\int_0^{2\pi} u(\cos t, \sin t)^4 dt \right) \left(\int_0^{2\pi} v(\cos t, \sin t)^4 dt \right).$$

If $\int_0^{2\pi} u(\cos t, \sin t)^4 dt > 0$ we divide both sides of the last inequality by this expression and obtain (5.5.14). If $\int_0^{2\pi} u(\cos t, \sin t)^4 dt = 0$, then (5.5.14) is trivially satisfied. The second inequality follows since the function $if = -v + iu$ is still analytic in $|z| < 1 + \epsilon$ and vanishes at the origin. $\qquad \square$

Solution of Exercise 5.5.17. By Cauchy's formula we have

$$a_k = \frac{f^{(k)}(0)}{k!} = \frac{1}{2\pi i} \int_{|z|=r} \frac{f(z)dz}{z^{k+1}},$$

and so

$$|a_k| = \left| \int_0^{2\pi} \frac{f(re^{it})ire^{it}dt}{r^{k+1}e^{i(k+1)t}} \right| \leq \frac{M}{r^k}.$$

Thus, for $|z| \leq r$,

$$|a_k z^k| \leq r^k \frac{M}{r^k} = M. \qquad \square$$

Solution of Exercise 5.5.18. Take $R_1 < R$. The coefficient of the power z^{n-1} in the power expansion of f' is nf_n. By Cauchy's formula applied to f',

$$\begin{aligned}
|nf_n| &= \left| \frac{1}{2\pi i} \int_{|z|=R_1} \frac{f'(z)dz}{z^n} \right| \\
&= \left| \frac{1}{2\pi i} \int_0^{2\pi} \frac{f'(R_1 e^{it})R_1 i e^{it}}{R_1^n e^{int}} dt \right| \\
&\leq \frac{1}{2\pi} \int_0^{2\pi} \left| \frac{f'(R_1 e^{it})R_1 i e^{it}}{R_1^n e^{int}} \right| dt \\
&\leq \frac{1}{2\pi} \int_0^{2\pi} \left| \frac{MR_1 i e^{it}}{R_1^n e^{int}} \right| dt \\
&= \frac{1}{2\pi} \int_0^{2\pi} \left| \frac{MR_1}{R_1^n} \right| dt \\
&= \frac{M}{R_1^{n-1}}.
\end{aligned}$$

Since this estimate holds for all $R_1 \leq R$, we obtain (5.5.15). $\qquad \square$

Solution of Exercise 5.5.19. Take $r > 0$. Cauchy's formula gives us

$$\begin{aligned}
|f_n| &= \left| \frac{1}{2\pi i} \int_{|z|=r} \frac{f(z)dz}{z^{n+1}} \right| \\
&= \left| \frac{1}{2\pi i} \int_0^{2\pi} \frac{f(re^{it})rie^{it}}{r^{n+1}e^{i(n+1)t}} dt \right| \\
&\leq \frac{1}{2\pi} \int_0^{2\pi} \left| \frac{f(re^{it})rie^{it}}{r^{n+1}e^{i(n+1)t}} \right| dt \\
&\leq \frac{1}{2\pi} \int_0^{2\pi} \left| \frac{Mre^r}{r^{n+1}} \right| dt \\
&= M \frac{e^r}{r^n}.
\end{aligned}$$

Thus
$$|f_n| \leq M \inf_{r>0} \frac{e^r}{r^n}.$$

The minimum of the function $r \mapsto \frac{e^r}{r^n}$ is for $r = n$. Plugging $r = n$ in the formula one obtains the required estimate.   □

Solution of Exercise 5.5.20. From
$$f'(z) = \sum_{n=1}^{\infty} n a_n z^{n-1},$$

we have
$$n a_n = \frac{1}{2\pi i} \int_{|z|=r} \frac{f'(z)}{z^n} dz, \quad n = 1, 2, \ldots$$

for any $0 < r < 1$. Thus,
$$n a_n = \frac{1}{2\pi} \int_0^{2\pi} \frac{f'(re^{it})}{r^{n-1} e^{i(n-1)t}} dt.$$

In view of (5.5.17) we have
$$n |a_n| \leq M.$$

Now we can write
$$\int_0^1 |f(x)| dx \leq \int_0^1 \sum_{n=0}^{\infty} |a_n| x^n dx$$
$$\leq |a_0| + \int_0^1 \sum_{n=1}^{\infty} \frac{M}{n} x^n dx$$
$$\leq |a_0| + M \int_0^1 (-\ln(1-x)) dx < \infty.$$ □

Solution of Exercise 5.5.21. For $n \geq 1$, this is just Cauchy's formula applied to the function
$$g(\zeta) = (\zeta + z)\zeta^{n-1}.$$

For $n \leq 0$, and using (5.1.9) if need be, we have that
$$\frac{1}{2\pi i} \int_{\mathbb{T}} \frac{\zeta + z}{\zeta - z} \zeta^{n-1} d\zeta = \frac{1}{2\pi i} \int_{\mathbb{T}} \frac{1 + \zeta \bar{z}}{1 - \zeta \bar{z}} \zeta^{-1-n} d\zeta,$$

which is equal to -1 for $n = 0$ thanks to Cauchy's formula, and 0 for $n < 0$ thanks to Cauchy's theorem since $|z| < 1$.   □

Solution of Exercise 5.5.22. Without loss of generality and to keep the notation simple we set $z_0 = 0$ and $r = 1$. The integral to compute is

$$\frac{1}{2\pi} \int_0^{2\pi} \frac{f'(e^{it})}{f(e^{it})} e^{it} dt.$$

Recall that (see (5.2.1))

$$f(e^{is})' = ie^{is} f'(e^{is}). \tag{5.9.38}$$

Set, for $0 \le s \le 2\pi$,

$$g(s) = f(e^{is}) \exp -i\left\{ \frac{1}{2\pi} \int_0^s \frac{f'(e^{it})}{f(e^{it})} e^{it} dt \right\}.$$

Then, in view of (5.9.38),

$$g'(s) \equiv 0$$

and so $g(s) = g(0) = g(2\pi)$. Hence

$$1 = \exp -i\left\{ \frac{1}{2\pi} \int_0^{2\pi} \frac{f'(e^{it})}{f(e^{it})} e^{it} dt \right\},$$

and hence the result. □

Solution of Exercise 5.5.23. Assume that such a function exists. Then for $z \in \Omega$,

$$2f(z)f'(z) = 1.$$

Dividing this expression by $f(z)^2 = z$ we obtain

$$\frac{f'(z)}{f(z)} = \frac{1}{2z}.$$

Integrating both sides of this equality along the circle of radius 2 and using the preceding exercise, we obtain on the left side a multiple of 2π, while on the right side we get π, and hence a contradiction. □

Solution of Exercise 5.5.25. We saw in Exercise 3.1.12 that the integral (3.1.9) is well defined for z off the real line and that

$$\operatorname{Re} f_m(z) = -i(z - \bar{z}) \int_{\mathbb{R}} \frac{m(t)dt}{|t - z|^2}.$$

Furthermore, the integral will vanish if and only if

$$\frac{m(t)}{|t - z|^2} \equiv 0,$$

that is, if and only if $m(t) \equiv 0$. We now turn to the proof that f_m is holomorphic in $\mathbb{C} \setminus \mathbb{R}$. We focus on the case where z is in the open upper half-plane $\mathbb{C}_+$, and will show that

$$f'_m(z) = -i \int_{\mathbb{R}} \frac{m(t)dt}{(t-z)^2}. \tag{5.9.39}$$

The case of the open lower half-plane is treated in the same way. Let z and w in $\mathbb{C}_+$. We have

$$\frac{f_m(z) - f_m(w)}{z - w} = -i \int_{\mathbb{R}} \frac{m(t)dt}{(t-z)(t-w)},$$

and therefore,

$$\frac{f_m(z) - f_m(w)}{z - w} + i \int_{\mathbb{R}} \frac{m(t)dt}{(t-z)^2} = -i \int_{\mathbb{R}} \left\{ \frac{1}{(t-z)(t-w)} - \frac{1}{(t-z)^2} \right\} m(t)dt$$

$$= -i(w - z) \int_{\mathbb{R}} \frac{m(t)dt}{(t-z)^2(t-w)}.$$

Let

$$d = \operatorname{dist}(z, \mathbb{R}) = \min_{t \in \mathbb{R}} |t - z| = \operatorname{Im} z.$$

We chose w such that $|z - w| < \frac{d}{2}$. Then, for $t \in \mathbb{R}$ we have

$$\left| \frac{w - z}{t - z} \right| \le \frac{|w - z|}{d} < \frac{1}{2}. \tag{5.9.40}$$

In view of (5.9.40) we have, for w as above,

$$\left| \frac{1}{t - w} \right| = \left| \frac{1}{t - z + z - w} \right|$$

$$= \left| \frac{1}{(t-z)(1 - \frac{w-z}{t-z})} \right|$$

$$\le \frac{1}{|t-z|(1 - |\frac{w-z}{t-z}|)}$$

$$\le \frac{1}{|t-z|(1 - 1/2)} = \frac{2}{|t-z|}.$$

Therefore, for w such that $|z - w| < \frac{d}{2}$ we have

$$\left| \frac{f_m(z) - f_m(w)}{z - w} + i \int_{\mathbb{R}} \frac{m(t)dt}{(t-z)^2} \right| \le |w - z| \cdot 2 \int_{\mathbb{R}} \frac{m(t)dt}{|t-z|^3},$$

from which we obtain that f_m is holomorphic in the open upper half-plane, and that its derivative is given by (5.9.39). □

Solution of Exercise 5.6.1. We have

$$\cos z \cosh z = \frac{e^{iz} + e^{-iz}}{2} \frac{e^z + e^{-z}}{2}$$

$$= \frac{e^{(1+i)z} + e^{(i-1)z} + e^{(-i+1)z} + e^{-(1+i)z}}{4}.$$

Thus

$$4 \cos z \cosh z = \sum_{n=0}^{\infty} \frac{z^n}{n!} \left\{ (i+1)^n + (i-1)^n + (-i+1)^n + (-i-1)^n \right\}.$$

But

$$c_n \overset{\text{def.}}{=} (i+1)^n + (i-1)^n + (-i+1)^n + (-i-1)^n$$
$$= (i+1)^n (1 + (-1)^n) + (i-1)^n (1 + (-1)^n).$$

Thus $c_n = 0$ when n is odd, as it should be since the function $\cos z \cosh z$ is even. For $n = 2p$ we have

$$c_{2p} = 2((i+1)^{2p} + (i-1)^{2p})$$
$$= 2((2i)^p + (-2i)^p)$$
$$= \begin{cases} 0, & \text{if } p \text{ is odd,} \\ 4 \cdot 2^{2\ell}(-1)^\ell, & \text{if } p = 2\ell \text{ is even.} \end{cases}$$

Hence

$$\cos z \cosh z = \sum_{\ell=0}^{\infty} \frac{(-1)^\ell 2^{2\ell}}{(4\ell)!} z^{4\ell}. \qquad \square$$

Solution of Exercise 5.6.2. We prove only (5.6.1). The proof of (5.6.2) is similar. For small enough ϵ we can write:

$$\frac{f(z_0 + i\epsilon) - f(z_0 + \epsilon)}{f(z_0 + \epsilon + i\epsilon) - f(z_0)} = \frac{A}{B}$$

where

$$A = f(z_0) + i\epsilon f'(z_0) - \frac{\epsilon^2}{2} f''(z_0) + \epsilon^2 O(\epsilon) - f(z_0) - \epsilon f'(z_0) - \frac{\epsilon^2}{2} f''(z_0) + \epsilon^2 O(\epsilon)$$
$$= \epsilon \left((i-1)f'(z_0) - \epsilon f''(z_0) + \epsilon O(\epsilon) \right)$$

and (since $(1+i)^2 = 2i$)

$$B = f(z_0) + \epsilon(1+i)f'(z_0) + i\epsilon^2 f''(z_0) + \epsilon^2 O(\epsilon) - f(z_0)$$
$$= \epsilon \left((1+i)f'(z_0) + i\epsilon f''(z_0) + \epsilon O(\epsilon) \right).$$

Hence,

$$\frac{f(z_0 + i\epsilon) - f(z_0 + \epsilon)}{f(z_0 + \epsilon + i\epsilon) - f(z_0)} - i$$

$$= \frac{(i-1)f'(z_0) - \epsilon f''(z_0) + \epsilon O(\epsilon) - i\left((1+i)f'(z_0) + i\epsilon f''(z_0) + \epsilon O(\epsilon)\right)}{(1+i)f'(z_0) + i\epsilon f''(z_0) + \epsilon O(\epsilon)}$$

$$= \frac{\epsilon O(\epsilon)}{(1+i)f'(z_0) + i\epsilon f''(z_0) + \epsilon O(\epsilon)}$$

from which the result follows since $f'(z_0) \neq 0$. $\quad\square$

Solution of Exercise 5.6.3. By definition of the radius of convergence,

$$R = \min(|1/2 - \pm i|, |1/2 - (1+i)|) = \frac{\sqrt{5}}{2}. \quad\square$$

Solution of Exercise 5.6.4. The radius of convergence of the power series expansion at the origin is

$$R = \min_{\ell=1,2,\ldots,m} 1/|z_\ell| = \frac{1}{\max_{\ell=1,2,\ldots,m} |z_\ell|},$$

since this is the radius of the largest open disk centered at the origin where the power expansion exists. Moreover, for $z < 1/\max_{\ell=1,2,\ldots,m} |z_\ell|$, we have

$$f(z) = \sum_{\ell=1}^{m} \sum_{n=0}^{\infty} z^n z_\ell^n = \sum_{n=0}^{\infty} z^n \left(\sum_{\ell=1}^{m} z_\ell^n\right). \tag{5.9.41}$$

The interchange of summations makes sense since one of the sums is finite. The radius of convergence of (5.9.41) is

$$\frac{1}{\limsup_{n\to\infty} |\sum_{\ell=1}^{m} z_\ell^n|^{1/n}}.$$

Thus

$$\frac{1}{\limsup_{n\to\infty} |\sum_{\ell=1}^{m} z_\ell^n|^{1/n}} = \frac{1}{\max_{\ell=1,2,\ldots,m} |z_\ell|},$$

and hence the result. $\quad\square$

Solution of Exercise 5.6.5. (a) Let $f(z) = \sum_{n=0}^{\infty} a_n z^n$ be the power series expansion of f in $\mathbb{D}$ centered at the origin. We have

$$\sum_{n=0}^{\infty} a_n z^n = \sum_{n=0}^{\infty} a_n \epsilon^N z^n, \quad \forall z \in \mathbb{D}.$$

By uniqueness of the coefficients in the power series expansion, we have

$$a_n = a_n \epsilon^N, \quad n \in \mathbb{N}_0.$$

It follows that $a_n = 0$ when n is not a multiple of N, that is

$$f(z) = \sum_{k=0}^{\infty} a_{kN} z^{kN}, \quad z \in \mathbb{D}.$$

Set $g(z) = \sum_{k=0}^{\infty} a_{kN} z^k$, and let R denote the radius of convergence of g. If $R = \infty$, we have $f(z) = g(z^N)$ in particular in $\mathbb{D}$. Assume now $R < \infty$. Then

$$R^N = \frac{1}{\limsup_{k\to\infty} |a_{kN}|^{1/kN}} \geq \frac{1}{\limsup_{n\to\infty} |a_n|^{1/n}} \geq 1,$$

since f is analytic in the open unit disk. Thus $R \geq 1$ and g is analytic in the open unit disk, and satisfies $f(z) = g(z^N)$ there.

(b) Assume that (5.6.3) holds in the open unit disk. Then replacing z by ϵz we have

$$z^k = g(z^N) = g((\epsilon z)^N) = (\epsilon z)^k,$$

and hence $\epsilon^k = 1$. But this is impossible for $k \in \{1, \ldots, N-1\}$. $\qquad \square$

Solution of Exercise 5.6.6. Let $F(z) = \sum_{\ell=0}^{\infty} a_\ell z^\ell$ be the power series expansion of F centered at the origin. By hypothesis,

$$\sum_{\ell=0}^{\infty} a_\ell \rho^\ell \in \mathbb{R} \quad \forall \rho \in (0, R).$$

We prove by induction that $a_n \in \mathbb{R}$ for all n. For $n = 0$ the claim is true, as is seen by letting $\rho \to 0$ (recall that the power series is continuous in $|z| < R$ and in particular at the origin). Assume now that $a_0, \ldots, a_n$ are real. Then the expression

$$\sum_{\ell=n+1}^{\infty} a_\ell \rho^\ell = \sum_{\ell=0}^{\infty} a_\ell \rho^\ell - \sum_{\ell=0}^{n} a_\ell \rho^\ell$$

is real for all $\rho \in (0, R)$. Dividing by ρ^{n+1} we obtain that

$$\sum_{\ell=n+1}^{\infty} a_\ell \rho^{\ell-n-1} \in \mathbb{R}, \quad \forall \rho \in (0, R).$$

Letting $\rho \to 0$ we get that $a_{n+1} \in \mathbb{R}$. We now use the second hypothesis

$$\sum_{n=0}^{\infty} a_n \rho^n \exp(in\pi\sqrt{2}) \in \mathbb{R}, \quad \forall \rho \in (0, R).$$

In particular the imaginary part of the above expression is 0 for all $\rho \in (0, R)$, that is

$$\sum_{n=1}^{\infty} a_n \rho^n \sin(n\pi\sqrt{2}) = 0, \quad \forall \rho \in (0, R).$$

By the same argument as above,

$$a_n \sin(n\pi\sqrt{2}) = 0, \quad n = 1, 2, \dots.$$

But $\sqrt{2}$ is irrational and so $n\pi\sqrt{2}$ is never a multiple of π and so $\sin(n\pi\sqrt{2}) \neq 0$ for $n = 1, 2, \dots$. It follows that $a_n = 0$ for $n = 1, 2, \dots$ and so $F(z) \equiv a_0$. $\square$

Solution of Exercise 5.6.7. For a given $t \in \mathbb{C}$ the function $z \mapsto e^{tz - z^2/2}$ is entire and therefore admits a power series expansion of the form (5.6.4). We now show that the h_n are polynomial functions of the variable t. By Cauchy's formula

$$h_n(t) = \frac{1}{2\pi i} \int_{|z|=1} \frac{e^{tz - z^2/2}}{z^{n+1}} dz$$

$$= \sum_{\ell=0}^{\infty} t^\ell \frac{1}{2\pi i} \int_{|z|=1} \frac{z^\ell}{\ell!} \frac{e^{-z^2/2}}{z^{n+1}} dz.$$

By Cauchy's theorem, the integral $\int_{|z|=1} z^{(\ell-n-1)} e^{-z^2/2} dz = 0$ for $\ell > n$. By Cauchy's formula, it is equal to $2\pi i$ for $\ell = n$. Thus h_n is a polynomial of degree n. $\square$

Solution of Exercise 5.6.8. The formula for $A(z)$ involves an integral along the interval $[0, z]$. Since the function f is analytic in the open unit disk, and since the open unit disk is convex (and in particular star-shaped) the integral does not depend in fact on the path within $\mathbb{D}$ linking 0 to z. We have

$$\int_{[0,z]} f(s) ds = \sum_{n=0}^{\infty} \frac{f_n}{n+1} z^{n+1}.$$

Thus, for $|z| < 1$,

$$A(z) = \left(\sum_{n=0}^{\infty} \frac{f_n}{n+1} z^{n+1} \right) \left(\sum_{n=0}^{\infty} z^n \right)$$

$$= z \left(\sum_{n=0}^{\infty} \frac{f_n}{n+1} z^n \right) \left(\sum_{n=0}^{\infty} z^n \right),$$

and (5.6.5) is a direct consequence of the formula (4.4.14) for the product of power series. $\square$

Solution of Exercise 5.6.9. By definition, the area $S(\mathcal{A})$ is equal to

$$S(\mathcal{A}) = \iint_{\mathcal{A}} dx\, dy.$$

We make the change of variable

$$\varphi(x, y) = (u(x, y), v(x, y)).$$

Since f is analytic in $\mathbb{D}$, the Jacobian matrix is equal to (4.2.8), and its determinant is equal to $|f'(z)|^2$. By the theorem of change of variables for double integrals we have

$$S(A) = \iint_{\mathbb{D}} |f'(z)|^2 \, dxdy. \tag{5.9.42}$$

Let $f(z) = \sum_{n=0}^{\infty} a_n z^n$ be the power series expansion centered at the origin of f in the open unit disk. With the change of variable

$$x = r\cos t \quad y = r\sin t,$$

this integral becomes

$$S(A) = \int_0^{2\pi} \int_0^1 \sum_{n,m=1}^{\infty} nmr^{n+m-2} e^{i(n-m)t} \, drdt. \tag{5.9.43}$$

The family of functions

$$f_{n,m}(r,t) = nmr^{n+m-2} e^{i(n-m)t}, \quad n, m = 1, \ldots$$

converges uniformly and absolutely in $[0,1] \times [0,2\pi]$. This allows us to exchange the integral and the double sum in (5.9.43). Indeed, let $\epsilon > 0$. There exists $N \in \mathbb{N}$ such that

$$\sum_{n,m=N+1}^{\infty} |f_{n,m}(r,t)| < \epsilon,$$

and thus

$$\int_0^{2\pi} \int_0^1 \sum_{n,m=1}^{\infty} nmr^{n+m-2} e^{i(n-m)t} r drdt$$

$$= \int_0^{2\pi} \int_0^1 \left(\sum_{n,m=1}^{N} nmr^{n+m-2} e^{i(n-m)t} \right) r drdt$$

$$+ \int_0^{2\pi} \int_0^1 \left(\sum_{n,m=N+1}^{\infty} nmr^{n+m-2} e^{i(n-m)t} \right) r drdt.$$

The second integral is bounded by $2\pi\epsilon$, and the first integral is equal to

$$\pi \sum_{n=1}^{N} n|a_n|^2.$$

The formula

$$S(A) = \pi \sum_{n=1}^{N} n|a_n|^2$$

for the area follows. □

We note the following: Starting from (5.9.42), we can use the Parseval equality, and obtain

$$\int_0^{2\pi} \left| \sum_{n=1}^{\infty} n a_n r^{n-1} e^{i(n-1)t} \right|^2 dt = 2\pi \sum_{n=1}^{\infty} n^2 |a_n|^2 r^{2n-2}$$

and proceed as follows (recall that we assume f analytic in a neighborhood of the closed unit disk):

$$S(\mathcal{A}) = \iint_{\mathbb{D}} |f'(z)|^2 dx dy$$

$$= \int_0^1 \left(\int_0^{2\pi} |f'(re^{it})|^2 dt \right) r dr$$

$$= 2\pi \int_0^1 \sum_{n=1}^{\infty} n^2 |a_n|^2 r^{2n-1} dr$$

$$= \pi \sum_{n=1}^{\infty} n |a_n|^2,$$

where the interchange of integral and summation to get to the last line is done using for instance the monotone convergence theorem.

Solution of Exercise 5.6.10. (1) Let $f(z) = \sum_{n=0}^{\infty} a_n z^n$ be the Maclaurin expansion of f in $B(0, R)$. As in the solution of the preceding exercise we have

$$\frac{1}{2\pi} \int_0^{2\pi} |f(re^{it})|^2 dt = \sum_{n=0}^{\infty} r^{2n} |a_n|^2, \quad r \in [0, R).$$

Therefore $M_2(f, r)$ is increasing, and is strictly increasing unless

$$a_1 = a_2 = \cdots = 0.$$

(2) Let f be entire and let $R > 0$. We have

$$\iint_{|z| \le R} |f(z)|^2 dx dy = \int_0^R \int_0^{2\pi} |f(re^{it})|^2 r dr dt = 2\pi \int_0^R r M_2(f, r) dr. \quad (5.9.44)$$

Since $M_2(f, r)$ is increasing, this last integral cannot converge as $R \to \infty$, unless $M_2(f, r) \equiv 0$, that is, unless $f(z) \equiv 0$. Another proof using subharmonic functions goes as follows: The function $|f|^2$ is subharmonic (see (9.3.3)), and so (9.3.6) implies that

$$|f(0)|^2 \le M_2(f, r).$$

If $f(0) \ne 0$, we see from (5.9.44) that

$$\iint_{|z| \le R} |f(z)|^2 dx dy \ge \pi R^2 |f(0)|^2 \longrightarrow +\infty$$

as $R \longrightarrow +\infty$. If $f(0) = 0$, we have $f(z) = z^n g(z)$ where $n \in \mathbb{N}$ and where g is an entire function not vanishing at the origin. We have

$$\iint_{|z|>1} |g(z)|^2 \, dx dy \le \iint_{|z|>1} |z^{2n}| \cdot |g(z)|^2 \, dx dy \le \iint_{\mathbb{C}} |f(z)|^2 \, dx dy < \infty,$$

and therefore

$$\iint_{\mathbb{C}} |g(z)|^2 \, dx dy < \infty,$$

and we are back to the preceding case.   □

Here also we could use Parseval's identity to compute

$$\int_0^{2\pi} |f(re^{it})|^2 \, dt = 2\pi \left(\sum_{n=0}^{\infty} |a_n|^2 r^{2n} \right), \tag{5.9.45}$$

and proceed.

Solution of Exercise 5.6.12. This is a direct consequence of (5.9.45).   □

Solution of Exercise 5.6.14. (a) Let $n, m \in \mathbb{N}_0$. Going to polar coordinates we have:

$$\frac{1}{\pi} \iint_{\mathbb{R}^2} e^{-|z|^2} z^n (\bar{z})^m \, dx dy = \frac{1}{\pi} \int_0^{2\pi} \int_0^{\infty} e^{-r^2} r^{n+m+1} e^{i(n-m)t} \, dt dr$$

$$= \begin{cases} 2 \int_0^{\infty} e^{-r^2} r^{2n+1} \, dr, & \text{if } n = m, \\ 0, & \text{if } n \ne m. \end{cases}$$

But

$$2 \int_0^{\infty} e^{-r^2} r^{2n+1} \, dr = \int_0^{\infty} e^{-r} r^n \, dr = n!. \tag{5.9.46}$$

(b) Using polar coordinates and using (5.9.45) we have:

$$\frac{1}{\pi} \int_{\mathbb{R}^2} e^{-|z|^2} |f(z)|^2 \, dx dy = \int_0^{\infty} e^{-r^2} \left(\int_0^{2\pi} |f(re^{it})|^2 \, dt \right) r dr$$

$$= 2 \int_0^{\infty} e^{-r^2} \left(\sum_{n=0}^{\infty} r^{2n} |f_n|^{2n} \right) r dr$$

$$= 2 \sum_{n=0}^{\infty} |f_n|^2 \int_0^{\infty} e^{-r^2} r^{2n+1} \, dr$$

$$= \sum_{n=0}^{\infty} n! |f_n|^2$$

in view of (5.9.46). The interchange of sum and integral above can be done for instance using the monotone convergence theorem.   □

Solution of Exercise 5.6.15. (a) We have

$$e^{z\overline{w}} = \sum_{n=0}^{\infty} c_n z^n \quad \text{with} \quad c_n = \frac{\overline{w}^n}{n!}, \; n = 0, 1, \dots.$$

Since

$$\sum_{n=0}^{\infty} n! |c_n|^2 = \sum_{n=0}^{\infty} n! \frac{|w|^{2n}}{(n!)^2} = \sum_{n=0}^{\infty} \frac{|w|^{2n}}{n!} = e^{|w|^2} < \infty,$$

the function

$$k_w : z \mapsto e^{z\overline{w}} \in \mathcal{F}.$$

Furthermore, for $f \in \mathcal{F}$ with power series expansion $f(z) = \sum_{n=0}^{\infty} f_n z^n$, we have

$$\langle f, k_w \rangle_{\mathcal{F}} = \sum_{n=0}^{\infty} n! f_n \frac{\overline{w}^n}{n!} = \sum_{n=0}^{\infty} f_n w^n = f(w). \qquad \square$$

Solution of Exercise 5.6.16. We have

$$a_n(z_0) = \frac{f^{(n)}(z_0)}{n!}.$$

Hence

$$
\begin{aligned}
a_n(z_0) &= \sum_{v=n}^{\infty} \frac{v(v-1)\cdots(v-n+1)}{nF!} a_v z_0^{v-n} \\
&= \sum_{m=0}^{\infty} \frac{(n+m)(n+m-1)\cdots(m+1)}{n!} a_{n+m} z_0^m \\
&= \sum_{m=0}^{\infty} a_{n+m} z_0^m \binom{n+m}{n}. \qquad \square
\end{aligned}
$$

Solution of Exercise 5.7.2. Assume that a solution exists. Then,

$$(f(z) - ig(z))(f(z) + ig(z)) = 1.$$

In particular, the function $(f(z) + ig(z))$ is entire and does not vanish in $\mathbb{C}$. It has therefore a logarithm, which is analytic in $\mathbb{C}$ (that is, which is an entire function): We denote this logarithm by $iE(z)$.

$$f(z) + ig(z) = e^{iE(z)}.$$

Then,

$$f(z) - ig(z) = (f(z) + ig(z))^{-1} = e^{-iE(z)},$$

and we get

$$f(z) = \cos E(z) \quad \text{and} \quad g(z) = \sin E(z). \qquad \square$$

As a consequence of the previous exercise, we check that (5.7.3) has no non-constant polynomial solutions. Indeed, if there are polynomials $p(z)$ and $q(z)$ such that (5.7.3) holds, then

$$p(z) + iq(z) = \cos H(z) + i \sin H(z) = e^{iH(z)}$$

for some entire function $H(z)$. The right side of this equation does not vanish in $\mathbb{C}$, while the left side has roots if $p(z) + iq(z)$ is not equal identically to a constant. But $p(z) + iq(z)$ constant implies that $p(z) - iq(z)$ is also a constant since

$$(p(z) + iq(z))(p(z) - iq(z)) = p(z)^2 + q(z)^2 = 1,$$

and hence $p(z)$ and $q(z)$ are constant polynomials.

Finally try the following question: Find all entire solutions to the equation

$$f(z)^2 + g(z)^2 = h(z)^2,$$

where $h(z) \not\equiv 0$.

Hint. Assume first that f, g and h have no common zero. Then the functions f/h and g/h have entire extensions.

Solution of Exercise 5.7.3. If there was a function g such that $g'(z) = f(z)$, then f would be analytic and with identically vanishing imaginary part. By the Cauchy–Riemann equations, this forces f to be a constant, and thus leads to a contradiction.

Using (4.5.4) and Exercise 4.2.20 we now show that

$$\partial_z \left(\frac{2z|z|}{3} \right) = |z|, \quad z \neq 0.$$

Indeed, in a way similar to Exercise 4.2.22,

$$\partial_z |z| = \frac{1}{2} \left(\frac{2x}{2|z|} - i \frac{2y}{2|z|} \right) = \frac{\bar{z}}{2|z|}$$

and so

$$\partial_z \left(\frac{2z|z|}{3} \right) = \frac{2|z|}{3} + \frac{2z}{3} \cdot \frac{\bar{z}}{2|z|} = |z|, \quad \text{for} \quad z \neq 0. \qquad \square$$

Solution of Exercise 5.7.4. In case (i), we compute (using Cauchy's formula)

$$\int_{|z-i|=1/2} \frac{dz}{z^2 + 1} = \int_{|z-i|=1/2} \frac{dz}{(z+i)(z-i)} = 2\pi i \times \frac{1}{2i} = \pi \neq 0,$$

and hence the function has no primitive in $\mathbb{C} \setminus \{-i, i\}$.

In case (ii), we compute

$$\int_{|z|=2} \frac{dz}{z^2+1} = \int_{|z-i|=1/2} \frac{dz}{(z+i)(z-i)} + \int_{|z+i|=1/2} \frac{dz}{(z+i)(z-i)} = 0,$$

and in this case there is a primitive in the given open (non-simply-connected) set.

In case (iii), the set

$$\Omega = \mathbb{C} \setminus \{z = iy, \text{ with } y \in \mathbb{R} \quad \text{and} \quad |y| \geq 1\}$$

is star-shaped, and thus a primitive exists. The primitive F which takes value $F(0) = 0$ is given by the formula

$$F(z) = \int_{\gamma_z} \frac{ds}{s^2+1},$$

where γ_z is any path linking 0 and z in Ω. For $z = x$ on the real line, one can takes γ_x to be

$$\gamma_x(t) = t, \quad t \in [0, x]$$

and we get $F(x) = \arctan(x)$. We have in particular $\tan(F(z)) = z$ on the real line, and so on the whole of Ω by analytic continuation. The Taylor expansion

$$\frac{1}{1+z^2} = \sum_{n=0}^{\infty} (-1)^n z^{2n}$$

has radius of convergence $R = 1$, and so

$$F(z) = \sum_{n=0}^{\infty} \frac{(-1)^n}{2n+1} z^{2n+1}, \quad |z| < 1.$$

(iv) The image of L under tan is the set Ω (see Exercise 1.2.8), and therefore $F(\tan z)$ is well defined for $z \in L$. We have

$$(F(\tan z))' = (1 + \tan^2 z) \frac{1}{1 + \tan^2 z} = 1,$$

and thus

$$F(\tan z) = z, \quad z \in L. \tag{5.9.47}$$

$\square$

Solution of Exercise 5.7.5. In case (i) there is no primitive since, using Cauchy's formula for the function $z/(z+i)$ and the point i we have

$$\int_{|z-i|=1/2} \frac{zdz}{z^2+1} = \int_{|z-i|=1/2} \frac{zdz}{(z+i)(z-i)} = 2\pi i i/2ii \neq 0.$$

In case (ii), we have

$$\int_{|z|=2} \frac{z dz}{z^2+1} = \int_{|z-i|=1/2} \frac{z dz}{(z+i)(z-i)} + \int_{|z+i|=1/2} \frac{z dz}{(z-i)(z+i)}$$

$$= 2\pi i \left\{ \frac{i}{2i} + \frac{i}{-2i} \right\} = 0.$$

Hence a primitive exists in the asserted set.

In case (iii) a primitive exists since the set is star-shaped. The argument to show that the primitive which vanishes at the origin is the analytic extension of the function $\frac{1}{2} \ln(x^2+1)$ to $\mathbb{C} \setminus \{z = iy,\, y \in \mathbb{R} \text{ and } |y| \geq 1\}$ is done as in the previous exercise. $\qquad\square$

The function F in (iii) in the preceding exercise is the analytic extension of $\arctan x$ to $\Omega = \mathbb{C} \setminus \{z = iy,\, y \in \mathbb{R} \text{ and } |y| \geq 1\}$. Note that the left side of (5.9.47) is periodic, with period π, while the right side is not periodic.

Solution of Exercise 5.7.6. The function

$$\frac{f(z) - a}{1 - \overline{a} f(z)}$$

has no zeros in the open unit disk since $a \in \mathbb{D} \setminus f(\mathbb{D})$. Therefore there exists an analytic logarithm $F(z)$.

We have

$$F'(z) = \frac{\left(\frac{f(z)-a}{1-\overline{a}f(z)} \right)'}{\frac{f(z)-a}{1-\overline{a}f(z)}} = \frac{(1 - |a|^2) f'(z)}{(1 - \overline{a}f(z))(f(z) - a)}.$$

Finally, in view of (1.1.41), we have that

$$\frac{f(z) - a}{1 - \overline{a} f(z)} \in \mathbb{D}, \quad \forall z \in \mathbb{D}.$$

It follows that

$$e^{\operatorname{Re} F(z)} < 1, \quad \forall z \in \mathbb{D},$$

and so $\operatorname{Re} F(z) < 0$. $\qquad\square$

Solution of Exercise 5.7.7. We go back to the solution of Exercise 3.4.4. The sum (3.9.14) creates no special problem, and is not related to the topic of the present section. Using formula (3.4.12) we obtain

$$\sum_{p=0}^{\infty} (2p)(-1)^{2p} z^{2p} = \sum_{p=0}^{\infty} 2p z^{2p} = \frac{2z^2}{(1-z^2)^2}.$$

The second sum,

$$\sum_{p=0}^{\infty}(2p+1)(-1)^{2p+1}z^{2p+1} = \sum_{p=0}^{\infty}\frac{z^{2p+1}}{2p+1},$$

is the unique primitive of the function

$$\frac{1}{1-z^2} = 1 + z^2 + z^4 + \cdots$$

$$= \frac{1}{2}\left(\frac{1}{1-z} + \frac{1}{1+z}\right)$$

in the open unit disk, with value 0 at the origin. This primitive is the analytic extension of the function $\ln\sqrt{\frac{1+x}{1-x}}$ to the open unit disk, and will still be denoted by $\ln\sqrt{\frac{1+z}{1-z}}$.   □

Solution of Exercise 5.7.8. Using the logarithmic derivative (see formula (4.2.3)), or by direct computation, we have

$$\frac{f'(z)}{f(z)} = \frac{2z}{z^2-1} + \frac{2z}{z^2-a^2} = \left\{\frac{1}{z-1} + \frac{1}{z-a}\right\} + \left\{\frac{1}{z+1} + \frac{1}{z+a}\right\}.$$

Let now γ be a simple closed contour such that $[a,1]$ lies in the interior of γ and $[-1,-a]$ lies in the exterior of γ. We have

$$\frac{1}{2\pi i}\int_\gamma \frac{f'(z)}{f(z)}dz = \frac{1}{2\pi i}\int_\gamma\left\{\frac{1}{z-1} + \frac{1}{z-a}\right\}dz + \frac{1}{2\pi i}\int_\gamma\left\{\frac{1}{z+1} + \frac{1}{z+a}\right\}dz.$$

The first integral is equal to 2 while the second integral is equal to 0. Therefore, $\int_\gamma \frac{f'(z)}{f(z)}dz \neq 0$ and f has no analytic logarithm. On the other hand we can write

$$f(z) = (z^2-a^2)^2\frac{z^2-1}{z^2-a^2}.$$

Let $q(z) = \frac{z^2-1}{z^2-a^2}$. We have now

$$\frac{q'(z)}{q(z)} = \frac{2z}{z^2-1} - \frac{2z}{z^2-a^2} = \left\{\frac{1}{z-1} - \frac{1}{z-a}\right\} + \left\{\frac{1}{z+1} - \frac{1}{z+a}\right\},$$

and so q has an analytic logarithm, and so an analytic square root, in the asserted set. Hence f has an analytic square root in Ω.   □

Solution of Exercise 5.7.9. Indeed, by Cauchy's formula,

$$\int_{|z|=1}\frac{\sin z}{z^2}dz = 2\pi i(\cos 0) \neq 0.$$ □

We note that, for $p \in \mathbb{N}$ the function $\dfrac{\sin z}{z^p}$ will have a primitive in $\mathbb{C} \setminus \{0\}$ if and only if p is odd.

Solution of Exercise 5.7.10. A necessary and sufficient condition is that

$$\int_{|z|=1} f(z)dz = 0.$$

Using Cauchy's formula, we obtain

$$2\pi i a p! (\sin z)^{(p)}\big|_{z=0} = 2\pi i b q! (\cos z)^{(q)}\big|_{z=0}. \tag{5.9.48}$$

Depending on p and q one can get specific conditions. For instance, if p is even and q is odd, condition (5.9.48) reduces to $0 = 0$, and thus is met for all $a, b \in \mathbb{C}$. □

Solution of Exercise 5.7.11. Here too, as in the preceding exercise, a necessary and sufficient condition for the existence of a primitive is that the integral of the function along the unit circle vanishes. We have

$$\int_{|z|=1} \frac{\sin z}{z^6} = -\frac{2\pi i}{5!},$$

and, since $\deg Q \leq q$,

$$\int_{|z|=1} \frac{e^z - Q(z)}{z^{q+2}} = 2\pi i \frac{1}{(q+1)!}.$$

Hence the condition for the existence of a primitive in $\mathbb{C} \setminus \{0\}$ is

$$\frac{a}{5!} + \frac{b}{(q+1)!} = 0. \qquad\qquad □$$

Solution of Exercise 5.7.12. For $z \in \Omega$ we have

$$\frac{f'(z)}{f(z)} = \frac{f'(z)}{1 - (1 - f(z))}$$

$$= \sum_{n=0}^{\infty} f'(z)(1 - f(z))^n$$

$$= -\sum_{n=0}^{\infty} (1 - f)'(z)(1 - f(z))^n.$$

Let now γ be a closed contour in Ω. Using Weierstrass' theorem we can write

$$\int_{\gamma} \frac{f'(z)}{f(z)}dz = -\sum_{n=0}^{\infty} \int_{\gamma} (1 - f)'(z)(1 - f(z))^n dz.$$

Each of the integrals

$$\int_\gamma (1 - f)'(z)(1 - f(z))^n dz = 0,$$

since the function $(1 - f)'(z)(1 - f(z))^n$ is the derivative of the function

$$(1 - f(z))^{n+1}/(n + 1).$$

Therefore, f has a logarithm in Ω. Fix $z_0 \in \Omega$. The logarithm function g which vanishes at $z = z_0$ is given by

$$g(z) = \int_{z_0}^z \frac{f'(s)}{f(s)} ds$$

$$= -\sum_{n=0}^\infty \int_{z_0}^z (1 - f'(s))(1 - f(s))^n ds$$

$$= -\sum_{n=0}^\infty \frac{(1 - f(z))^{n+1}}{n+1} + \sum_{n=0}^\infty \frac{(1 - f(z_0))^{n+1}}{n+1}. \qquad \square$$

Solution of Exercise 5.7.13. The set Ω is simply-connected (consider the complement of its image on the Riemann sphere). The function $\prod_{\ell=0}^{2016}(z - \ell)$ does not vanish there. Thus, it has an analytic logarithm and analytic roots of any order. $\qquad \square$

Solution of Exercise 5.7.14. Without loss of generality, and to keep the notation simple, we set $z_0 = 0$ and $r = 1$. The integral to compute is

$$\frac{1}{2\pi} \int_0^{2\pi} \frac{f'(e^{it})}{f(e^{it})} e^{it} dt.$$

Recall that

$$f(e^{is})' = ie^{is} f'(e^{is}). \qquad (5.9.49)$$

Set, for $0 \le s \le 2\pi$,

$$g(s) = f(e^{is}) \exp -i \left\{ \int_0^s \frac{f'(e^{it})}{f(e^{it})} e^{it} dt \right\}.$$

Then, in view of (5.9.49),

$$g'(s) \equiv 0$$

and so $g(s) = g(0) = g(2\pi)$. Hence

$$1 = \exp -i \left\{ \int_0^{2\pi} \frac{f'(e^{it})}{f(e^{it})} e^{it} dt \right\},$$

and hence the result. $\qquad \square$

Solution of Exercise 5.7.15. By the preceding exercise,

$$n = \frac{1}{2\pi i} \int_{|z|=3/2} \frac{f'(z)}{f(z)} dz \in \mathbb{Z}.$$

Set $h(z) = z^{-n} f(z)$. Then

$$\frac{h'(z)}{h(z)} = -\frac{n}{z} + \frac{f'(z)}{f(z)}.$$

In particular,

$$\frac{1}{2\pi i} \int_{|z|=3/2} \frac{h'(z)}{h(z)} dz = -n + n = 0,$$

and the function h has an analytic logarithm in the given annulus: $h(z) = \exp g(z)$, where g is analytic in $1 < |z| < 2$. It follows that $f(z) = z^n \exp g(z)$. □

Solution of Exercise 5.7.16. Let $\gamma_1(t) = 2e^{it}$, and let $\gamma_2(t) = 5 + 2e^{it}$, where in both cases $t \in [0, 2\pi]$. The numbers

$$n_\ell = \frac{1}{2\pi i} \int_{\gamma_\ell} \frac{f'(z)}{f(z)} dz, \quad \ell = 1, 2,$$

belong to $\mathbb{Z}$. Let

$$h(z) = \frac{f(z)}{z^{n_1}(z-5)^{n_2}}.$$

Then, for any closed curve γ in Ω,

$$\int_\gamma \frac{h'(z)}{h(z)} dz = 0,$$

and so there exists a function g analytic in Ω such that $h(z) = e^{g(z)}$ there. This ends the proof. □

See [42, p. 97] for related results.

Solution of Exercise 5.7.17. We see that

$$F_n'(z) = -nF_{n+1}(z),$$

and therefore the function F_n has a primitive for $n \geq 2$. We now show that F_1 has no primitive in $\mathbb{C} \setminus [0,1]$. Let C denote the circle of center 0 and radius 4, positively oriented. For $t \in [0,1]$ we have

$$\int_C \frac{dz}{t-z} = 2\pi i$$

and so

$$\int_C F(z)dz = \int_0^1 \left(\int_C \frac{dz}{t-z}\right) m(t)dt$$

$$= 2\pi i \int_0^1 m(t)dt \neq 0 \quad \text{since} \quad m(t) > 0, \ t \in [0,1],$$

where we have used Theorem 14.6.1 to interchange integration and derivation. It follows that F_1 has no primitive in the asserted domain.   □

Solution of Exercise 5.7.18. We first remark that the function $\ln(x^2 - 2x \cosh a + 1)$ is defined in $\mathbb{R} \setminus [e^{-a}, e^a]$. Assume that a function f exists. Then, on $\mathbb{R} \setminus [e^{-a}, e^a]$, its complex derivative is given by

$$f'(x) = \frac{2(x - \cosh a)}{x^2 - 2x \cosh a + 1}$$

as is seen from formula (4.2.6). By analytic continuation, we have that

$$f'(z) = \frac{2(z - \cosh a)}{z^2 - 2z \cosh a + 1}, \quad \forall z \in \mathbb{C} \setminus [e^{-a}, e^a].$$

But the function

$$\frac{2(z - \cosh a)}{z^2 - 2z \cosh a + 1} = \frac{1}{z - e^a} + \frac{1}{z - e^{-a}}$$

has no primitive in $\mathbb{R} \setminus [e^{-a}, e^a]$, as is seen by computing its integral on a circle of center 0 and radius $R > e^a$.   □

Solution of Exercise 5.8.1. The function

$$h(z) = \frac{1-z}{1+z}$$

has an analytic logarithm in the asserted domain. Indeed, we have

$$\frac{h'(z)}{h(z)} = \frac{1}{z-1} - \frac{1}{z+1}$$

and the integral along a closed simple curve whose interior contains the interval $[-1, 1]$ is therefore equal to 0. We conclude by noting that

$$f(z) = 1 - z^2 = \frac{1-z}{1+z}(z+1)^2.$$ □

Solution of Exercise 5.8.2. (a) We first note that Ω is star-shaped with respect to the origin. Therefore the function $\frac{1}{1-z^2}$ has an analytic logarithm in Ω, and hence an analytic square root there. By possibly multiplying this square root by a constant of modulus 1, one can always suppose that it takes the value 1 at the origin. For $|z| < 1$ we then have (with $f_{-1/2}$ defined as in Exercise 4.4.4)

$$\frac{1}{\sqrt{1-z^2}} = f_{-1/2}(z^2) = 1 + \frac{1}{2}z^2 + \frac{1\cdot3}{2\cdot4}z^4 + \frac{1\cdot3\cdot5}{2\cdot4\cdot6}z^6 + \cdots .$$

See, e.g., [204, p. 135] for the latter.

(b) The function $\arcsin z$ is well defined since Ω is star-shaped. For $z = x \in (-1,1)$ we take as C_z the interval on the real line defined by 0 and x. We have

$$\arcsin x = \int_0^x \frac{ds}{\sqrt{1-s^2}},$$

which indeed coincides with the classical function $\arcsin$ of calculus.

(c) In view of the power series expansion in (a) we have

$$\arcsin z = z + \frac{1}{2\cdot3}z^3 + \frac{1\cdot3}{2\cdot4\cdot5}z^5 + \frac{1\cdot3\cdot5}{2\cdot4\cdot6\cdot7}z^7 + \cdots , \quad z \in \mathbb{D}.$$

(d) We have $\sin(\arcsin x) = x$ for $x \in (-1,1)$. By analytic continuation, this identity holds in Ω. $\qquad\square$

Solution of Exercise 5.8.3. Let γ_ℓ be for $\ell = 1,\ldots,N$ a simple closed path in Ω, whose interior contains only the interval $[\alpha_\ell, \beta_\ell]$. To check the existence of a square root, it is enough to check that

$$\int_{\gamma_\ell} \frac{f'(z)}{f(z)} dz = 0, \quad \ell = 1,\ldots,N.$$

Using the formula (4.2.3) for the logarithmic derivative we have:

$$\frac{f'(z)}{f(z)} = \sum_{\ell=1}^N \left(\frac{1}{z-\alpha_\ell} - \frac{1}{z-\beta_\ell} \right),$$

and thus for $\ell_0 \in \{1,\ldots,N\}$ we have

$$\int_{\gamma_{\ell_0}} \frac{f'(z)}{f(z)} = \int_{\gamma_{\ell_0}} \left(\frac{1}{z-\alpha_{\ell_0}} - \frac{1}{z-\beta_{\ell_0}} \right) dz + \sum_{\substack{\ell=1 \\ \ell\neq\ell_0}}^N \int_{\gamma_{\ell_0}} \left(\frac{1}{z-\alpha_j} - \frac{1}{z-\beta_j} \right) dz.$$

The first integral on the right side is equal to

$$\int_{\gamma_{\ell_0}} \left(\frac{1}{z-\alpha_{\ell_0}} - \frac{1}{z-\beta_{\ell_0}} \right) dz = \int_{\gamma_{\ell_0}} \frac{1}{z-\alpha_{\ell_0}} dz - \int_{\gamma_{\ell_0}} \frac{1}{z-\beta_{\ell_0}} dz = 2\pi i - 2\pi i = 0.$$

The second integral vanishes thanks to Cauchy' theorem. Thus the function f has an analytic square root in the asserted domain. $\qquad\square$

Solution of Exercise 5.8.5. Assume by contradiction that such a function exists. Then,

$$2f(z)f'(z) = 1$$

in $1 < |z| < 2$, and dividing by $2\pi i f(z)^2$ on both sides we have

$$2\frac{1}{2\pi i}\frac{f'(z)}{f(z)} = \frac{1}{2\pi i z}. \qquad (5.9.50)$$

Let

$$\gamma : \gamma(t) = 1.5e^{it}, \quad t \in [0, 2\pi].$$

Integrating both sides of (5.9.50) on γ, and taking into account that

$$\frac{1}{2\pi i}\int_\gamma \frac{f'(z)}{f(z)}\,dz \in \mathbb{Z},$$

we obtain that there is an integer N such that

$$2N = 1,$$

which is a contradiction. $\qquad\qquad\qquad\qquad\qquad\qquad\qquad\qquad\qquad\square$

Chapter 6

Morera, Liouville, Schwarz, et les autres: First Applications

Cauchy's formula is the key to most, if not all, important results in complex variables, and in particular to the existence of a power series expansion around each point of analyticity, the maximum modulus principle and the fact that the zeros of a non-identically vanishing analytic function are isolated. In this chapter we present exercises on these topics.

6.1 Zeroes of analytic functions

The zeros of a non-identically vanishing analytic function cannot accumulate at an inner point, but they can of course accumulate at a boundary point of the domain of analyticity. This is the essence of item (a) in the next exercise. Item (c) illustrates the fact that one should always be careful with conditions on functions given in terms of integrals. In relation with this exercise, see also Exercise 6.8.10.

Exercise 6.1.1.

(a) *Let f be a function analytic in $|z| < 1$ and such that*

$$\int_{|z|=1} \frac{f(z)}{(n+1)z - 1} dz = 0, \quad \forall n \in \mathbb{N}. \tag{6.1.1}$$

Show that $f(z) \equiv 0$ in $|z| < 2$.

(b) *What can be said if (6.1.1) is replaced by*

$$\int_{|z|=1} \frac{f(z)}{((n+1)z - 1)^2} dz = 0, \quad \forall n \in \mathbb{N}? \tag{6.1.2}$$

(c) *Assume that f is analytic in $0 < |z| < 2$ and satisfies (6.1.1). Is f necessarily identically equal to 0?*

Hint. Look at $f(z) = \sin \frac{\pi}{z}$.

The following exercise appears in [190].

Exercise 6.1.2. *Let f be an entire function and assume that for every Maclaurin expansion*

$$f(z) = \sum_{n=0}^{\infty} c_n(z_0)(z - z_0)^n,$$

there is an n such that $c_n(z_0) = 0$. Show that f is a polynomial.

Exercise 6.1.3. *Let $a_0, a_1, \ldots$ be a sequence of complex numbers such that*

$$\sum_{n=0}^{\infty} |a_n| < \infty. \tag{6.1.3}$$

Assume that for all integers $k \geq 2$,

$$\sum_{n=0}^{\infty} \frac{a_n}{k^n} = 0. \tag{6.1.4}$$

Show that all the a_n are equal to 0.

Exercise 6.1.4. *Find all functions f analytic in the open unit disk and such that $f(\frac{1}{n}) = 0$ for $n = 2, 3, \ldots$.*

Exercise 6.1.5 (see, e.g., [75, Exercice 13.03, p. 141]). *Same question as in the previous exercise for each of the following conditions:*

$$f\left(\frac{1}{n}\right) = e^{-n}, \tag{6.1.5}$$

$$\left| f\left(\frac{1}{n}\right) \right| \leq e^{-n}. \tag{6.1.6}$$

Is there a function analytic in the punctured disk such that (6.1.5) hold? Explain the difference with the first question.

Exercise 6.1.6. *Is there a function analytic in the open unit disk and such that*

$$f(1/(2n)) = f(1/(2n-1)) = \frac{1}{n}, \quad n = 2, 3, \ldots?$$

Exercise 6.1.7 (see [45, Exercise 11, p. 78]). *Let f and g be analytic and not vanishing in the open unit disk $\mathbb{D}$. Assume that for all $n \geq 2$,*

$$\frac{f'(\frac{1}{n})}{f(\frac{1}{n})} = \frac{g'(\frac{1}{n})}{g(\frac{1}{n})}. \tag{6.1.7}$$

Show that there is a constant c such that $f(z) = cg(z)$ for all $z \in \mathbb{D}$.

Exercise 6.1.8.

(a) *Find all functions analytic in a neighborhood of the origin, and for which there exists $n_0 \in \mathbb{N}$ such that*

$$f(1/n) + f''(1/n) = 0$$

for $n \geq n_0$.

(b) *Same question for*

$$f(1/n) = f''(1/n).$$

The following exercise considers Exercise 4.1.13 when a and b are analytic in the given set Ω.

Exercise 6.1.9. *In Exercise 4.1.13 we moreover assume that a and b are analytic in Ω. Prove the following claims:*

(a) *There are at most countably many points $z \in \Omega$ for which*

$$a(z) = b(z) = 0, \tag{6.1.8}$$

and the set of such points has no accumulation points in Ω.

(b) *There is a point $z \in \Omega$ such that*

$$|a(z)| = |b(z)| \neq 0. \tag{6.1.9}$$

(c) *The set $\Omega \setminus \Omega_0$ is not connected.*

Hint. Use Theorem 15.6.2.

In Exercise 10.2.7 we show that in fact there are uncountably many points for which (6.1.9) holds.

The properties of the zeroes of analytic functions lead to the following fascinating questions: Given a connected set Ω, a set $\{z_1, z_2, \ldots\}$ with an accumulation point in Ω, and numbers $w_1, w_2, \ldots$, can we reconstruct a function analytic in Ω and such that

$$f(z_j) = w_j, \ldots \quad j = 1, 2, \ldots?$$

More generally, can we reconstruct, if possible in a unique way, an analytic function from its values at given points? This type of question is called an interpolation problem. When the points do not accumulate in Ω, one can always find such an f if no metric structure is imposed on the function. See for instance [189, Chapitre 15]. In engineering and signal theory, interpolation problems are often set under additional constraints on the function (for instance, such as being analytic and bounded by 1 in modulus in the open unit disk). Such questions will be considered in two special instances in Chapter 11. We will meet simple instances of the Nevanlinna–Pick and Carathéodory–Fejér interpolation problems in Problems 6.5.9 and 6.5.10. These problems will also be discussed in Section 11.5, and the sampling theorem in Section 11.2.

6.2   Morera's theorem

The following theorem, called Morera's theorem, is used in particular to prove that certain functions defined by integrals are analytic; see Theorem 6.2.3. In the statement, Δ denotes a triangle defined by points in Ω (and, in particular, not on the boundary of Ω) and $\partial\Delta$ denotes the contour defined by the boundary of the triangle.

Theorem 6.2.1 (Morera's theorem). *Let Ω be an open convex subset of $\mathbb{C}$, and let f be continuous in Ω. Assume that*

$$\int_{\partial\Delta} f(z)dz = 0, \tag{6.2.1}$$

for every triangle in Ω. Then, f is analytic in Ω.

A first, and key, application of this result is:

Theorem 6.2.2. *Let $(f_n)_{n\in\mathbb{N}}$ be a sequence of functions analytic in the open subset Ω of the complex plane, and converging uniformly on compact subsets of Ω to f. Then, f is analytic in Ω.*

The proof goes as follows. The function f is continuous since the limit is uniform on compact sets. Take an open convex subset U of Ω. For any triangle in U, the uniform convergence implies that

$$\lim_{n\to\infty} \int_{\partial\Delta} f_n(z)dz = \int_{\partial\Delta} f(z)dz.$$

By Cauchy's theorem, $\int_{\partial\Delta} f_n(z)dz = 0$ for every n, and we obtain that (6.2.1) holds, and so f is analytic in U. This ends the proof since Ω can be covered by open convex sets (for instance, open balls).

Among the other applications of Morera's theorem we mention the Schwarz reflection principle (see Exercise 6.3.9 in the next section). Another important application is the following result, which gives a useful criterium to prove analyticity for functions defined by integrals.

Theorem 6.2.3. *Let Ω be an open connected set, and let $F(z,s)$ be a function continuous for $(z,s) \in \Omega \times [c,d]$ such that for every $s \in [c,d]$ the function $z \mapsto F(z,s)$ is analytic in Ω. Then the function*

$$G(z) = \int_c^d F(z,s)ds \tag{6.2.2}$$

is analytic in Ω.

Proof. Since the property is local, we can assume that Ω is convex. The result is then a consequence of Morera's theorem and of Theorem 14.5.2. Indeed for every $s \in [c, d]$ and every triangle $\Delta \subset \Omega$, Cauchy's theorem gives

$$\int_{\partial\Delta} F(z, s)dz = 0,$$

where we have denoted by $\partial\Delta$ the boundary of the triangle, and hence

$$\int_c^d \left(\int_{\partial\Delta} F(z, s)dz \right) ds = 0.$$

By (14.5.1) we have

$$\int_{\partial\Delta} \left(\int_c^d F(z, s)ds \right) dz = 0.$$

Morera's theorem implies that the function $z \mapsto \int_c^d F(z, s)ds$ is analytic in Ω. $\qquad\square$

This last theorem does not give the power series expansion of the function (6.2.2). For the latter, one needs more (Weierstrass' theorem, for instance).

The following result is used in particular in Dixon's proof of the global Cauchy's theorem; see [64] and [144, p. 147]. There a closed chain homologous to 0 is considered. We here consider only the case of a closed contour. The statement of the question speaks of star-shaped domain, but the true setting is that of simply-connected domains.

Question 6.2.4. *Let Ω be an open star-shaped domain, and let Γ be a closed smooth curve inside Ω. Let f be analytic in Ω. Show that the function*

$$F(z) = \int_\Gamma g(s, z)ds \tag{6.2.3}$$

where

$$g(s, z) = \begin{cases} \dfrac{f(s) - f(z)}{s - z}, & z, s \in \Omega, \ z \neq s, \\ f'(s), & z, s \in \Omega, \ z = s, \end{cases} \tag{6.2.4}$$

is analytic in Ω.

6.3   Analytic continuation

This notion has already been used earlier in the book. See for instance item (d) of Exercise 5.8.2, and the discussion following (5.9.9).

For the first exercise, see also Exercise 4.4.13.

Exercise 6.3.1. *Define* $F(z) = -\sum_{n=1}^{\infty}(-1)^n \dfrac{z^n}{n}$ *(we denote the function $F(z) = $* $\ln(1+z)$*).*

(a) *Show that F is analytic in the open unit disk and that*

$$\exp F(z) = 1 + z, \quad for \quad |z| < 1.$$

(b) *Does F have an analytic extension across a sub-arc of the unit circle?*

In a similar way to Exercise 6.3.1, and in view of Exercise 10.3.9, we mention:

Exercise 6.3.2. *For $x \in (-1,1)$ and $\alpha \in \mathbb{R}$ define*[5]

$$(1+x)^\alpha = \exp(\alpha \ln(1+x)). \qquad\qquad (6.3.1)$$

(1) *Show that the function f_α defined in (4.4.9) is the analytic extension of* (6.3.1) *to $\mathbb{D}$ (resp. to $\mathbb{C}$) when $\alpha \in \mathbb{R} \setminus \mathbb{N}_0$ (resp. $\alpha \in \mathbb{N}_0$).*
(2) *Let $\alpha \in (-1,0)$. Show that (with the notation (4.4.8))*

$$(-1)^n \binom{\alpha}{n} > 0, \quad n = 0,1,\ldots$$

and

$$\sum_{n=0}^{\infty}(-1)^n \binom{\alpha}{n} \; diverges.$$

(3) *Let $\alpha \in (0,1)$. Show that*

$$(-1)^{n-1}\binom{\alpha}{n} > 0, \quad n = 0,1,\ldots$$

and

$$\sum_{n=0}^{\infty}(-1)^{n-1}\frac{\binom{\alpha}{n}}{n+1} < \infty.$$

In the following exercise, the natural boundary means that the function cannot be continued analytically across the unit circle. We refer to [195, Chapter VI] for a survey on natural boundaries.

Exercise 6.3.3.

(a) *Show that the unit circle is the natural boundary of the series*

$$\sum_{n=0}^{\infty} z^{n!} \quad and \quad \sum_{n=0}^{\infty} z^{2^n}.$$

---

[5] One could take $\alpha \in \mathbb{C}$; see Remark 6.9.3.

(b) *Does the power series*

$$\sum_{n=1}^{\infty} \frac{z^n}{n^2}$$

have an analytic extension across some sub-arc of the unit circle?

We now want to find all rational functions which send the unit circle onto itself. Because of *a priori* possible poles or zeros on the unit circle, we set the question as follows (see also [42, Example 11.14, p. 367] for instance):

Exercise 6.3.4. *Find all rational functions which map an open arc of the unit circle into the unit circle.*

For the next exercise, see [53, p. 109] and see also Exercise 8.4.1.

Exercise 6.3.5.

(a) *Prove that for $z \in \mathbb{C} \setminus \{(-\infty - 1] \cup [1, \infty)\}$, there exists an analytic square root to the function $1 - z^2$ such that $f(0) = 1$.*

(b) *With f as in (a), show that*

$$\int_0^{2\pi} \frac{dt}{1 + z \sin t} = \frac{2\pi}{f(z)} \tag{6.3.2}$$

for all $z \in \mathbb{C} \setminus \{(-\infty - 1] \cup [1, \infty)\}$.

We already saw (see Exercise 4.1.10) that there does not exist a continuous square root to the function $f(z) = z$ in the plane. Another proof of a weaker statement (one requires analyticity rather than continuity) is presented now (see also Exercise 5.5.23):

Exercise 6.3.6. *Let $r > 0$. Show that there is no analytic square root to the function z in $0 < |z| < r$.*

Let $\Omega = \mathbb{C} \setminus (-\infty, 0]$. In view of the following exercise, we recall that the function defined by

$$\sqrt{z} = \sqrt{\rho} e^{i \frac{\theta}{2}}, \tag{6.3.3}$$

where $z = \rho e^{i\theta}$ and $\theta \in (-\pi, \pi)$ is an analytic square root of z in Ω. We also remark that $1 - z \in \mathbb{C} \setminus (-\infty, 0]$ if and only if $z \in \mathbb{C} \setminus [1, \infty)$.

Exercise 6.3.7. *Let $U = \mathbb{C} \setminus [1, \infty)$. The function $\sqrt{1 - z}$ obtained by composition of the function $1 - z$ with the square root function (6.3.3) is an analytic extension in U of the function $f_{1/2}$ defined by (4.4.9) with $\alpha = 1/2$.*

As we have already discussed, Riemann's removable singularity theorem states that if f is analytic in $B(z_0, r) \setminus \{z_0\}$ and continuous in $B(z_0, r)$, then it is analytic in $B(z_0, r)$. The following exercise considers what happens when a point is replaced by a closed interval:

Exercise 6.3.8. *Let Ω be an open and convex subset of $\mathbb{C}$ and let I be a closed interval in Ω. Let f be continuous in Ω and analytic in $\Omega \setminus I$. Show that f is analytic in Ω.*

Hint. Use Morera's theorem (see Theorem 6.2.1).

The next related result is called Schwarz' reflection principle, and also follows from Morera's theorem. It has numerous applications. An important one is the derivation of the Schwarz–Christoffel formula which gives the conformal mapping from the open upper half-plane onto the interior of a non-intersecting (but not necessarily convex) polygon.

Exercise 6.3.9. *Let $x_0 \in \mathbb{R}$ and f be holomorphic in the open half-disk*

$$B_+(x_0, r) = \{z \in \mathbb{C};\ \mathrm{Im}\, z > 0 \quad and \quad |z - x_0| < r\}.$$

Assume that f is continuous up to the real line and that it takes real values on the real line. Show that the function

$$h(z) = \begin{cases} f(z), & \text{if } z \in B_+(z_0, r) \cup (x_0 - r, x_0 + r), \\ \overline{f(\bar{z})}, & \text{if } \bar{z} \in B_+(z_0, r) \cup (x_0 - r, x_0 + r) \end{cases}$$

is holomorphic in $B(x_0, r)$.

Other related exercises, whose proofs require the residue theorem, are Exercise 8.2.4 and 8.5.2. We conclude this section with an exercise which has its importance in multiscale signal processing. Recall that $B(0, r) = \{z \in \mathbb{C};\ |z| < r\}$. See also Exercise 3.4.11 for some related formulas.

Exercise 6.3.10. *Let f be analytic in the open unit disk $\mathbb{D}$, and let $n \in \mathbb{N}$. The purpose of this exercise is to show that the formula*

$$g_n(z) = \frac{1}{n} \sum_{\substack{w \in \mathbb{D} \\ w^n = z}} f(w) \tag{6.3.4}$$

defines a function analytic in $\mathbb{D}$, and expresses its power series expansion at the origin in terms of the power series expansion of f at the origin. The map $f \mapsto g_n$ is called the decimation, or downsampling, operator.

(a) *Let q be the analytic square root of order n in $\mathbb{D} \setminus (-1, 0]$ and which coincides with $x^{1/n}$ on $(0, 1)$. Express g_n in terms of q for $z \in \mathbb{D} \setminus (-1, 0]$.*

(b) *Using the power expansion of f at the origin, show that g_n has an analytic extension to the open unit disk.*

6.4 The maximum modulus principle

The first example provides an application of complex variables to a geometrical question; see [211, p. 401].

Exercise 6.4.1. *Given n points $P_1, \ldots, P_n$ in the plane, show that the product of their distance to a point in the plane has no local extremum, with the exception of the points P_j.*

If p is a polynomial of degree n, the function $z^n p(1/z)$ is also a polynomial (possibly of lesser degree). This is the idea behind the next two exercises.

Exercise 6.4.2 (see [75]). *Let $p(z) = z^n + a_{n-1}z^{n-1} + \cdots + a_0$. Prove that either $p(z) \equiv z^n$ or that there exists a point ζ on the unit circle such that $|p(\zeta)| > 1$.*

Exercise 6.4.3. *Let p be a polynomial of degree n and let, for $r > 0$,*

$$M(r, p) = \sup_{|z|=r} |p(z)|.$$

Show that the function $r \mapsto M(r, p)$ is increasing and that the function $r \mapsto \dfrac{M(r, p)}{r^n}$ is decreasing.

Exercise 6.4.4. *Is there a function analytic in the open unit disk and such that $|f(z)| = e^{|z|}$ there?*

Using Exercise 5.5.15 one can prove the following surprising result; it appears as an exercise (without any hint) in [121, Exercise 4, p. 40]. The proof is given in [42, Exercise 5.36, p. 146]. We give the exercise and the solution with the hypothesis that f is analytic in a neighborhood of the closed unit disk. This can be dispensed with.

Exercise 6.4.5. *Let h be analytic in $|z| < 1 + \epsilon$ for some $\epsilon > 0$ and assume that h does not vanish in $|z| < 1 + \epsilon$ and is bounded by 1 in modulus in $\mathbb{D}$. Show that*

$$\sup_{|z|\leq 1/5} |h(z)|^2 \leq \inf_{|z|\leq 1/7} |h(z)|.$$

Hint. Since h does not vanish in $|z| < 1 + \epsilon$, there is a function f analytic there and such that $h = e^{-f}$. Since $|h(z)| \leq 1$ for z in the open unit disk, we have that $\operatorname{Re} f(z) \geq 0$ there. Then apply (5.5.13) to f for appropriate values of R.

6.5 Schwarz' lemma

Schwarz' lemma states that if a function f is analytic and contractive in the open unit disk, and if moreover $f(0) = 0$, then

$$|f(z)| \leq |z|, \quad \forall z \in \mathbb{D}. \tag{6.5.1}$$

Moreover, equality holds if and only if $f(z) = cz$ for some $c \in \mathbb{T}$. This lemma is of utmost importance. It is the key to the Schur algorithm (see [6] for a survey, and Section 11.5 below), which itself is related to fast algorithms in signal processing. See [131].

It follows from (6.5.1) that

$$|f'(0)| \leq 1. \tag{6.5.2}$$

Usually, it appears in textbooks after the notion of singularities, since $z = 0$ is a removable singularity of the function $f(z)/z$. A shorter way to prove Schwarz' lemma is to use Riemann's Hebbarkeitssatz (see Theorem 5.4.1).

We send to the paper [173] for a survey and history of the Schwarz lemma. As explained in that paper, considering a function analytic in $B(0, R_1)$ and bounded in modulus by R_2 there, we obtain the bound

$$|f(z)| \leq \frac{R_2}{R_1}|z|, \quad |z| \leq R_1, \tag{6.5.3}$$

from which one get Liouville's theorem; see Section 6.8 for the latter.

The next exercise is [45, Exercice 1, p. 171].

Exercise 6.5.1. *Let f be a function analytic in the open unit disk, bounded by 1 in modulus there and vanishing at the origin. Show that the series*

$$\sum_{n=0}^{\infty} f(z^n)$$

converges uniformly in the closed disks $|z| \leq r < 1$.

The following exercise gives an extension of (6.5.2), called Pick's inequality, to the whole open unit disk. Recall that the functions b_a were defined in (1.1.44):

$$b_a(z) = \frac{z - a}{1 - z\overline{a}}, \quad a \in \mathbb{D}.$$

Exercise 6.5.2. *Let f be analytic in the open unit disk and bounded by 1 in modulus. Then,*

$$|f'(z)| \leq \frac{1 - |f(z)|^2}{1 - |z|^2}, \quad \forall z \in \mathbb{D}. \tag{6.5.4}$$

Show that equality holds if and only if f is of the form

$$f(z) = c\frac{z - a}{1 - z\overline{a}} = cb_a(z),$$

where $c \in \mathbb{T}$ and $a \in \mathbb{D}$.

We note that, more generally, for any choice of different complex numbers z_1 and z_2 in $\mathbb{D}$, the matrix

$$P = \begin{pmatrix} \frac{1-|f(z_1)|^2}{1-|z_1|^2} & \frac{f(z_1)-f(z_2)}{z_1-z_2} \\ \frac{\overline{f(z_1)-f(z_2)}}{\overline{z_1-z_2}} & \frac{1-|f(z_2)|^2}{1-|z_2|^2} \end{pmatrix} \tag{6.5.5}$$

is non-negative. Pick's inequality is a limiting case of this inequality, by considering the determinant of P, and letting z_1 tend to z_2. The methods to prove that P is non-negative are far beyond the scope of this book. See [9] for instance for a survey and references.

Another application of Schwarz' lemma is given in [45, Exercice 10, p. 112] and [42, Exercise 6.5, p. 192]: If f is analytic from the open unit disk into itself and has two fixed points, then $f(z) \equiv z$. The result is called Cartan's theorem. As a simpler case, solve the following:

Exercise 6.5.3. *If f sends $\mathbb{D}$ into itself and if*

$$f(0) = 0 \quad and \quad f\left(\frac{1}{2}\right) = \frac{1}{2},$$

then $f(z) \equiv z$.

Exercise 6.5.4. *Prove the general case of the preceding question: If f is analytic from the open unit disk into itself and has two fixed points, then $f(z) \equiv z$.*

Exercise 6.5.5. *Let f be a function analytic from the open unit disk $\mathbb{D}$ into itself. Assume that for some point $z_0 \in \mathbb{D}$ it holds that*

$$f(z_0) = z_0 \quad and \quad f'(z_0) = 1.$$

Find f.

The case of the boundary is more involved. The first positive result seems to be due to Burns and Krantz, see [43], but is beyond the scope of this book.

Exercise 6.5.6. *Show by a counterexample that the preceding result does not remain valid when z_0 is on the unit circle. To avoid any problem with boundary values, assume that f is analytic in a neighborhood of the closed unit disk; you can look for a Moebius map.*

The following exercise originates with Schur's celebrated paper [194]. We denote by $\mathcal{S}$ the family of functions analytic in the open unit disk and with values in the closed unit disk. These functions bear various names, and are in particular called Schur functions. They are the transfer functions of discrete time-invariant dissipative systems. See Section 11.4 for more information. Blaschke products (see (3.7.17)) are special cases of Schur functions.

Exercise 6.5.7.

(a) *Let $f \in S$. Assume that $|f(0)| < 1$. Show that the function*

$$
f_1(z) = \begin{cases} \dfrac{f(z) - f(0)}{z(1 - \overline{f(0)}f(z))}, & z \in \mathbb{D} \setminus \{0\}, \\[3mm] \dfrac{f'(0)}{1 - |f(0)|^2}, & z = 0, \end{cases}
$$

is analytic and contractive in the open unit disk.

(b) *Let $\rho_0 \in \mathbb{D}$. Show that the formula*

$$
f(z) = \frac{\rho_0 + zg(z)}{1 + z\overline{\rho_0}g(z)} \tag{6.5.6}
$$

describes all functions in S such that $f(0) = \rho_0$ when g runs through S.

(c) *Find $f'(0)$ in terms of $g(0)$ and ρ_0 in (6.5.6). Conclusion?*

In his 1917 paper [194] Schur associates to $f \in S$ a series of functions $f_1, f_2, \ldots$ of S via the recursion

$$
f_0(z) = f(z),
$$

$$
f_{n+1}(z) = \begin{cases} \dfrac{f_n(z) - f_n(0)}{z(1 - \overline{f_n(0)}f_n(z))}, & z \in \mathbb{D} \setminus \{0\}, \\[3mm] f'_n(0), & z = 0, \end{cases} \qquad n = 0, 1, \ldots \tag{6.5.7}
$$

called the Schur algorithm. The recursion stops at rank n if $|f_n(0)| = 1$. This algorithm, originally developed to solve classical interpolation problems such a the trigonometric moment problem, has been shown to have numerous applications in signal processing. In Exercise 11.5.3 in Section 11.5 we will see that the Schur algorithm ends after a finite number of iterations if and only if f is a finite Blaschke product.

Remark 6.5.8. It is interesting to note that G. Hamel, at the same time but independently of Schur, developed a very closely related algorithm (in the words of Szegö in his review for Zentralblatt MATH, *dasselbe Hauptresultat*). See [113, 128].

The following two exercises are solved using Exercise 6.5.7, and are (very) special instances of the Carathéodory–Fejér and Nevanlinna–Pick interpolation problems. See Section 11.5 for more on these interpolation problems.

Exercise 6.5.9. *Given a and b in $\mathbb{C}$, find a necessary and sufficient condition for $f \in S$ to exist such that*

$$
f(0) = 0 \quad and \quad f'(0) = b, \tag{6.5.8}
$$

and describe the set of all solutions.

Exercise 6.5.10. *Given two pairs of numbers* (z_1, w_1) *and* (z_2, w_2) *in* $\mathbb{D}^2$, *find a necessary and sufficient condition for a function* $f \in \mathcal{S}$ *to exist such that*

$$f(z_1) = w_1 \quad and \quad f(z_2) = w_2, \tag{6.5.9}$$

and describe the set of all solutions.

Exercise 6.5.11. *Let* f *be analytic from* $\mathbb{D}$ *into* $\mathbb{D}$ *and assume that* $f(a) = f(b) = 0$ *for two different numbers* a *and* b *in* $\mathbb{D}$. *Show that*

$$|f(z)| \le \left| \frac{z-a}{1-z\bar{a}} \right| \cdot \left| \frac{z-b}{1-z\bar{b}} \right|.$$

As a variation on this theme, find a bound on f when f (which is analytic and map $\mathbb{D}$ into itself) vanishes at the point a and at its first N_a derivatives at a and at the point b and at its first N_b derivatives at b.

Exercise 6.5.12. *Consider the function* (4.4.20) *and assume that* $m(t) \ge 0$ *and that* $\int_0^{2\pi} m(t)dt = 1$. *Show that*

$$\left| \frac{\varphi(z) - 1}{\varphi(z) + 1} \right| \le |z|, \quad \forall z \in \mathbb{D}.$$

Exercise 6.5.13.

(a) *Let* z_0 *be in the open upper half-plane* $\mathbb{C}_+$. *Show that the map* $\mathscr{B}_{z_0}$ *defined by* (1.1.46),

$$\mathscr{B}_{z_0}(z) = \frac{z - z_0}{z - \bar{z}_0},$$

is one-to-one from the closed upper half-plane onto the closed unit disk. It is one-to-one onto from the real line onto the unit circle and from $\mathbb{C}_+$ *onto the open unit disk* $\mathbb{D}$.

Let f *be a function analytic in* $\mathbb{C}_+$ *and bounded by* 1 *in modulus there.*

(b) *Assume that* $f(z_0) = 0$. *Show that*

$$|f(z)| \le \left| \frac{z - z_0}{z - \bar{z}_0} \right|, \quad \forall z \in \mathbb{C}_+.$$

(c) *Assume that* $f(z_0) = f'(z_0) = 0$. *Show that*

$$|f(z)| \le \left| \frac{z - z_0}{z - \bar{z}_0} \right|^2, \quad \forall z \in \mathbb{C}_+.$$

The book of Burckel [42] contains a whole chapter on applications of Schwarz' lemma (pp. 191–217).

6.6   Series of analytic functions

Series and infinite products of functions have appeared already in the book, for instance as power series, or as series of complex numbers depending on a parameter. In the present section we focus on examples of series of analytic functions which converge uniformly on compact subsets of their common domain of analyticity. The limit is then an analytic function. See Theorem 6.2.2.

Remark 6.6.1. Uniform convergence on compact subsets defines a topology, which is moreover metrizable. These aspects are reviewed in Section 10.1.

Exercise 6.6.2. *Show that*

$$|\sin z| \le \sinh |z|, \quad z \in \mathbb{C}.$$

Show that, for $x \in [0,1]$,

$$\sinh x \le x(\cosh 1).$$

Show that the series

$$\sum_{n=0}^{\infty} \sin(z^n)$$

converges absolutely in the open unit disk, uniformly on compact sets.

Exercise 6.6.3. *Show that the series*

$$\sum_{n=1}^{\infty} \frac{z(z+1)\cdots(z+n-1)}{n^n}$$

defines an entire function.

See [23, Exercice 1.6, p. 278] for the following problem.

Exercise 6.6.4. *Show that the function f defined in Exercise 3.7.8 is entire, and that*

$$f(z) = 1 + \sum_{n=1}^{\infty} \frac{q^{\frac{n(n+1)}{2}}}{(1-q)\cdots(1-q^n)}(-1)^n z^n.$$

Exercise 6.6.5. *Show that the domain of convergence of the sum*

$$\sum_{n=0}^{\infty} e^{-n^2 z} \tag{6.6.1}$$

is the open right half-plane, and that the convergence is uniform in every set of the form $\operatorname{Re} z \ge \epsilon > 0$.

Another example of series is given in Exercise 7.2.22, where, using Liouville's theorem, one is asked to prove that

$$\frac{1}{z} + \sum_{n=1}^{\infty} \left(\frac{1}{z-n} + \frac{1}{n} \right) = \pi \cot(\pi z).$$

6.7 Analytic functions as infinite products

We have already met in Exercise 3.7.16 the representation of $\sin z$ as an infinite product. Note that item (a) of Exercise 6.7.2 is Exercise 3.7.12.

Exercise 6.7.1. *Show that the Blaschke products defined in Exercise 3.7.12 are analytic in the open unit disk.*

Exercise 6.7.2.

(a) *Show that, for* $|z| < 1$,
$$|(1-z)e^z - 1| \le |z|^2.$$

(b) *Given a sequence of non-zero complex numbers* z_n, $n = 1, 2, \ldots$ *such that*
$$\sum_{n=1}^{\infty} \frac{1}{|z_n|^2} < \infty.$$

Show that the function
$$\prod_{n=1}^{\infty} \left(1 - \frac{z}{z_n}\right) e^{z/z_n}$$

is entire and vanishes exactly at the points z_n.

Exercise 6.7.3. *Consider the infinite product representation (3.7.25) of* $\sin z$:
$$\sin z = z \prod_{k=1}^{\infty} \left(1 - \frac{z^2}{k^2 \pi^2}\right), \quad z \in \mathbb{C}.$$

Comparing with the power series expansion of $\sin \pi z$, *prove (1.3.14), that is:*
$$\sum_{k=1}^{\infty} \frac{1}{k^2} = \frac{\pi^2}{6}.$$

(see [28, p. 233]). *Similarly, using the infinite product representation of* $\cos z$ (see (3.7.26)), *prove that*
$$\sum_{k=0}^{\infty} \frac{1}{(2k+1)^2} = \frac{\pi^2}{8}. \tag{6.7.1}$$

Remark 6.7.4. Knowing the value of (1.3.14) and of (6.7.1) is equivalent. Another proof of (6.7.1) is given after Theorem 11.2.1, and gives a proof of (1.3.14). In [56] another elementary proof of (6.7.1) can be found, leading to yet another proof of (1.3.14).

Remark 6.7.5. We take this opportunity to mention an open problem, consisting in getting any information, let alone compute explicitly, the Catalan constant

$$\sum_{k=0}^{\infty} \frac{(-1)^k}{(2k+1)^2},$$

that is, of the value at $z = 2$ of Dirichlet beta function $\beta(z) = \sum_{k=0}^{\infty} \frac{(-1)^k}{(2k+1)^z}$.

6.8   Liouville's theorem and the fundamental theorem of algebra

Liouville's theorem states that a bounded entire function is constant. We note that it was first proved by Cauchy; see [42, p. 82]. As an application of Liouville's theorem we will prove in Exercise 16.1.11 an important result from functional analysis. The fundamental theorem of algebra has numerous proofs. One very short proof, due to C. Fefferman, was outlined in Question 1.5.1. One topological proof is presented in Section 15.6. Another proof, usually given in complex analysis classes, uses Liouville's theorem. We begin this section by recalling this theorem and this latter proof.

Theorem 6.8.1 (The fundamental theorem of algebra). *Every non-constant polynomial has at least one complex root.*

Proof. We consider a non-constant polynomial (which we assume monic[6] without loss of generality) $p(z) = z^n + a_{n-1}z^{n-1} + \cdots + a_0$, and assume by contradiction that it does not vanish in $\mathbb{C}$. Then the function $1/p(z)$ is entire and non-constant. We will show that it is bounded in modulus, which will lead to a contradiction, thanks to Liouville's theorem. For $z \neq 0$ we can write

$$p(z) = z^n(1 + g(z)),$$

where

$$g(z) = \frac{a_{n-1}}{z} + \cdots + \frac{a_0}{z^n}$$

is such that $\lim_{z\to\infty} g(z) = 0$. Take $R > 0$ such that

$$|z| > R \Longrightarrow |g(z)| < \frac{1}{2}.$$

Then, using (1.1.38) with $z_1 = 1$ and $z_2 = g(z)$, we have for $|z| > R$,

$$\left|\frac{1}{p(z)}\right| \leq \frac{1}{1 - |g(z)|} \leq 2.$$

---

[6] that is, the coefficient of the highest power is equal to 1.

On the other hand the function $\left|\frac{1}{p(z)}\right|$ is bounded in $|z| \leq R$ since it is continuous there and since $|z| \leq R$ is compact. It follows that the non-constant entire function $1/p(z)$ is bounded in modulus. This cannot be, in view of Liouville's theorem. □

It follows from the previous theorem that a polynomial of degree $n > 0$ has n roots, counting multiplicities. Let $n \geq 2$. The n roots of unity of order n, $z_k = \exp\left(\frac{2\pi i k}{n}\right)$, $k = 0, \ldots, n-1$ have total sum equal to 0, as follows for instance from (1.1.54) with $z = \exp\left(\frac{2\pi i}{n}\right)$. See also Exercise 1.2.3. But (1.1.54) will not be useful to solve the following exercise:

Exercise 6.8.2. *Let $n \geq 3$ and $a, b \in \mathbb{C}$. Show that the sum of the roots of the polynomial equation*

$$z^n + az + b = 0$$

is equal to 0.

Remark 6.8.3. Let $p(z) = a_n z^n + a_{n-1} z^{n-1} + \cdots + a_0$ be a polynomial of degree n, with (possibly repeated) roots $z_1, \ldots, z_n$. We have

$$p(z) = a_n(z - z_1)(z - z_2) \cdots (z - z_n). \tag{6.8.1}$$

Comparing the coefficient of z^{n-1} we get

$$z_1 + \cdots + z_n = -\frac{a_{n-1}}{a_n}. \tag{6.8.2}$$

For instance, the sum of the roots of the equation

$$z^N - 3z^{N-1} + a_{N-2}z^{N-2} + \cdots + a_0 = 0$$

is equal to 3, independently of $N \geq 3$ and of the values of $a_0, \ldots, a_{N-2}$.

More generally, the numbers $(-1)^k \frac{a_{n-k}}{a_n}$ $(k = 0, \ldots, n-1)$ express the symmetric functions of the roots:

$$(-1)^k \frac{a_{n-k}}{a_n} = \sum_{1 \leq i_1 < i_2 < \cdots < i_k \leq n} z_{i_1} \cdots z_{i_k}.$$

For instance, for $k = 2$ one has:

$$z_1 z_2 + \cdots + z_1 z_N + z_2 z_3 + \cdots + z_2 z_N + \cdots + z_{N-1} z_N = \frac{a_{N-2}}{a_N}.$$

Exercise 6.8.4. *Find all entire functions f such that*

(a) $|f(z)| \leq M(1 + \sqrt{|z - i|})$,

(b) $|f'(z)| \leq M(1 + \sqrt{|z|})$,

(c) $|f(z)| \leq M(1 + |z - i|)$,

(d) $|f'(z)| \leq M(1 + |z|)$,

for some strictly positive number M.

Exercise 6.8.5. *Show that the range of a non-constant entire function is dense in $\mathbb{C}$.*

When in the above exercise, f is not a polynomial, then the function $f(1/z)$ has an essential singularity at the origin, and a much more precise statement can be given using Picard's theorem: $f(\mathbb{C})$ is equal to $\mathbb{C}$ or to $\mathbb{C} \setminus \{w\}$ for some $w \in \mathbb{C}$. The example $f(z) = e^z$ illustrates the fact that $f(\mathbb{C})$ can be strictly included in $\mathbb{C}$.

Exercise 6.8.6. *Show that an entire function f such that $\operatorname{Re} f(z) \leq M$ for some $M \in \mathbb{R}$ is constant.*

The following exercise can be found for instance in [150, Exercise 11, p. 123].

Exercise 6.8.7. *Find all entire functions f such that*

$$|f(z)| = 1 \quad for \quad |z| = 1. \tag{6.8.3}$$

The result in the next exercise is called Liouville's first theorem (when a general lattice is considered).

Exercise 6.8.8. *Show that there are no entire non-constant functions f such that*

$$f(z+1) = f(z), \tag{6.8.4}$$
$$f(z+i) = f(z), \quad \forall z \in \mathbb{C}. \tag{6.8.5}$$

Remark. There are biperiodic non-constant *meromorphic* functions. These are the *elliptic functions*. See Exercise 13.1.1.

Exercise 6.8.9. *Is there a polynomial P of degree N such that*

$$\int_{|z|=2} \frac{P(z)}{(n+1)z - 1} dz = 0, \quad n = 0, 1, \ldots, N.$$

Same question for P such that

$$\int_{|z|=1} \frac{P(z)}{(2z-1)^{n+1}} dz = 0, \quad n = 0, \ldots, N.$$

Exercise 6.8.10. *Find all polynomials of degree N such that*

$$\int_{|z|=r} \frac{P(z)}{(n+1)z - 1} dz = 0, \quad n = 0, 1, \ldots, N, \tag{6.8.6}$$

for $r = 3/4$ and $r = 1/(N+2)$ respectively.

Related to these problems see also Exercise 6.1.1.

Exercise 6.8.11. *Find all polynomials p and q such that*

$$p(z) \cos^2 z + q(z) \sin^2 z = 1.$$

Exercise 6.8.12 (see [62, Problem 5.6.7, p. 69]). *Is there a function f analytic in* $\mathbb{C} \setminus \{0\}$ *such that*

$$|f(z)| \geq \frac{1}{\sqrt{|z|}}, \quad \forall z \in \mathbb{C} \setminus \{0\}. \tag{6.8.7}$$

For other exercises using, or related to, Liouville's theorem, see for instance Exercises 6.8.5 and 6.8.7. Finally we note that there exist non-constant entire functions H which go to 0 along every fixed direction:

$$\lim_{r \to \infty} H(re^{i\theta}) = 0$$

for every $\theta \in [0, 2\pi)$. One such function is given by $R_0 f$ (recall that R_0 has been defined in (5.4.1), with

$$f(z) = \int_0^\infty \frac{e^{tz}}{t^t} \, dt.$$

See [28, §12.2, pp. 152–155]. Another, more involved example, is given by the function

$$g(z)e^{-g(z)},$$

where $g(z)$ is the analytic extension to the complex plane of the function

$$\frac{1}{2\pi i} \int_{\partial U} \frac{e^{e^u} \, du}{u - z},$$

where U is the half-strip

$$U = \{z \in \mathbb{C}; \ \mathrm{Re}\, z > 0 \quad \text{and} \quad |\mathrm{Im}\, z| < \pi\}.$$

See [63, Exercice 5, p. 248], [75, Exercice 10.54, p. 124], [189, Exercice 11, p. 314].

The solution of the following exercise using Rouché's theorem was shown to us by Dennis Gulko. See Exercise 7.4.7. What is asked in Exercise 6.8.13 is an elementary solution without this theorem, but using the fact that the polynomial has four roots.

Exercise 6.8.13. *How many solutions has the equation*

$$z^4 + 3z^2 + z + 1$$

in the closed upper half unit disk?

We conclude with a unicity question, first considered in [3]. See [181] for possible extensions to the case of rational functions.

Question 6.8.14. *Let P and Q be two non constant polynomials, with same sets of preimages of 0 and 1, that is*

$$P^{-1}\{0\} = Q^{-1}\{0\} \quad \text{and} \quad P^{-1}\{1\} = Q^{-1}\{1\}.$$

Show that $P = Q$.

6.9   Solutions

Solution of Exercise 6.1.1. (a) By Cauchy's formula, condition (1.6.17) can be rewritten as

$$f\left(\frac{1}{n+1}\right) = 0, \quad \forall n \in \mathbb{N}.$$

It follows that $f \equiv 0$ since the points $\frac{1}{n+1}$ accumulate at the origin.

(b) Cauchy's formula now leads to

$$f'\left(\frac{1}{n+1}\right) = 0, \quad \forall n \in \mathbb{N}.$$

It follows that $f' \equiv 0$ and f is constant.

(c) When we assume that f is analytic only in $0 < |z| < 2$, Cauchy's formula cannot be used, and the preceding argument fails, as is illustrated by the function $f(z) = \sin\frac{\pi}{z}$. For every $a \in \mathbb{D}$ it holds that

$$\int_{|z|=1} \frac{\sin\frac{\pi}{z}}{z-a} dz = 0. \tag{6.9.1}$$

To show that (6.9.1) holds, we first rewrite the integral as a line integral

$$\int_{|z|=1} \frac{f(z)}{z-a} dz = i \int_0^{2\pi} \frac{\sin(\pi e^{-it})}{1-e^{-it}a} dt.$$

To prove (6.9.1), we write

$$\frac{\sin(\pi e^{-it})}{1-e^{-it}a} = \sum_{n=0}^{\infty} a^n e^{-int} \sin(\pi e^{-it}),$$

and, using Weierstrass' theorem, we see that

$$\int_0^{2\pi} \frac{\sin(\pi e^{-it})}{1-e^{-it}} dt = \sum_{n=0}^{\infty} a^n \int_0^{2\pi} e^{-int} \sin(\pi e^{-it}) dt.$$

Another application of Weierstrass' theorem leads to

$$\int_0^{2\pi} e^{-int} \sin(\pi e^{-it}) dt = 0, \quad n = 0, 1, 2, \ldots. \qquad \square$$

Remark 6.9.1. A much quicker proof, but which is beyond the scope of this book, goes as follows: The function $\dfrac{\sin(\pi e^{-it})}{1-e^{-it}a}$ is the boundary value of the function

$$\frac{\sin(\frac{\pi}{z})}{1-z^{-1}a},$$

which belongs to the Hardy space $\overline{\mathbf{H}_2(\mathbb{D})}$ (see [190] for the definition) and so the integral is equal to 0.

Remark 6.9.2. The function $\dfrac{\sin(\pi e^{-it})}{1 - e^{-it}a}$ belongs to the Wiener algebra $\mathcal{W}_-$. See for instance [104] for the definition.

Solution of Exercise 6.1.2. First a remark: If there is an integer n such that $c_n(z_0) \equiv 0$, then f is a polynomial since

$$c_n(z_0) = \frac{f^{(n)}(z_0)}{n!}.$$

Assume now by contradiction that f is not a polynomial. Then, for every n the coefficient $c_n(z_0)$ does not vanish identically. Since $z \mapsto c_n(z) = f^{(n)}(z)/n!$ is an entire function, its set of zeroes is at most countable (but of course can be empty or finite). Thus the set

$$A = \{z \in \mathbb{C}, \text{there exists } n \in \mathbb{N} \text{ such that } c_n(z) = 0\}$$

is at most countable. On the other hand, this set is equal to $\mathbb{C}$ by hypothesis, which leads to a contradiction. □

The preceding result is still true for C^∞ functions of a real variable. See [217, pp. 278–279] for two proofs, one involving Baire's theorem. As mentioned in Burckel's book [42, p. 166], the result (proved using Baire's theorem) originates with [55], and can be found in the book [66].

Solution of Exercise 6.1.3. Let z be such that $|z| < 1$. We have $|a_n z^n| \le |a_n|$ and so, in view of (6.1.3), the power series

$$F(z) = \sum_{n=0}^{\infty} a_n z^n$$

converges absolutely for all z in the open unit disk, and defines an analytic function there. Condition (6.1.4) reads

$$F(1/k) = 0, \quad k = 2, 3, \ldots.$$

Thus the zeroes of F have a limit point in the open unit disk, and so $F(z) \equiv 0$, from which we obtain that all the coefficients

$$a_n = \frac{F^{(n)}(0)}{n!}, \quad n \in \mathbb{N}_0,$$

vanish. □

Solution of Exercise 6.1.4. Since the zeros accumulate at the point $z = 0$, the only function is $f(z) \equiv 0$. □

Solution of Exercise 6.1.5. We consider the first case and show by contradiction that no such function exists. If it exists, we note that it is not identically vanishing, since the values at $1/n$ are different from 0. From (6.1.5) we obtain $f(0) = \lim_{n\to\infty} f(1/n) = 0$. Thus there is an integer $p > 0$ and a function g analytic in the open unit disk such that $g(0) \neq 0$ and

$$f(z) = z^p g(z). \tag{6.9.2}$$

Thus

$$f(1/n) = n^{-p} g(1/n), \quad n = 2, 3, \ldots.$$

Since $f(1/n) = e^{-n}$ we obtain

$$g(1/n) = n^p e^{-n}, \quad n = 2, 3, \ldots,$$

and thus

$$g(0) = \lim_{n\to\infty} n^p e^{-n} = 0.$$

But $g(0) \neq 0$, and thus we obtain a contradiction and there is no function with the required property.

We now turn to the second case. The function $f(z) \equiv 0$ answers the question. We show that it is the only solution. The proof is quite similar as in the first case. Assume that a solution which is not identically vanishing exists. Condition (6.1.6) implies that

$$|f(0)| = \lim_{n\to\infty} |f(1/n)| = 0,$$

so that $f(0) = 0$ and we have the representation (6.9.2). Thus,

$$|f(1/n)| = \frac{|g(1/n)|}{n^p} \leq e^{-n}$$

and so

$$|g(1/n)| \leq n^p e^{-n},$$

once more leading to $g(0) = 0$, and hence to a contradiction.

In the last case we cannot assume that f is continuous at the origin. The function $f(z) = e^{-1/z}$ meets conditions (6.1.5). □

Solution of Exercise 6.1.6. Assume such a function exists. The condition

$$f(1/(2n)) = \frac{1}{n}, \quad n = 2, 3, \ldots$$

forces

$$f(z) = 2z,$$

since both functions coincide at the points $1/2, 1/3, \ldots.$. Similarly, the second condition forces

$$f(z) = \frac{2z}{z+1},$$

since both functions coincide at the points $1/2, 1/3, \ldots.$. Hence we obtain a contradiction, and no such function exists. $\qquad\square$

Solution of Exercise 6.1.7. The function $h(z) = \frac{f(z)}{g(z)}$ is analytic in the open unit disk (since g does not vanish there). The derivative of h is

$$h'(z) = \frac{f'(z)g(z) - f(z)g'(z)}{g^2(z)}$$

$$= h(z)\left(\frac{f'(z)}{f(z)} - \frac{g'(z)}{g(z)}\right).$$

Since f does not vanish in $\mathbb{D}$, (6.1.7) expresses that $h'(\frac{1}{n}) = 0$ for $n \geq 2$. Since the sequence $\frac{1}{n}$ converges to 0 which is an interior point of $\mathbb{D}$, it follows that $h'(z)$ is identically equal to 0 and hence h is a constant. $\qquad\square$

The previous exercise is [45, Exercice, 11 p. 78].

Solution of Exercise 6.1.8. (a) The analytic function $f + f''$ vanishes at the points $1/n$ for $n \geq n_0$ and so its zeros accumulate at $z = 0$. It follows that

$$f(z) + f''(z) = 0$$

in a neighborhood of the origin. Thus

$$f(z) = a\cos z + b\sin z$$

for some complex numbers a and b (use analytic continuation from a real neighborhood of the origin to a complex neighborhood of the origin). The formula for f shows moreover that it extends to an entire function.

In case (b),

$$f(z) = a\cosh z + b\sinh z. \qquad\square$$

Solution of Exercise 6.1.9. We denote by $\mathscr{Z}(a)$ and $\mathscr{Z}(b)$ the sets of zeros of a and b, and by $\Omega_{00} = \mathscr{Z}(a) \cap \mathscr{Z}(b)$ the set of points in Ω for which (6.1.8) holds. We first remark the following: The sets of zeros of a and b are at most countable, and without accumulation points in Ω. Otherwise, one of the functions a and b would be identically vanishing in Ω, and one of the sets Ω_+ and Ω_- would be empty. We now prove the claims of the exercise.

(a) By the above, $\Omega_{00} = \mathscr{Z}(a) \cap \mathscr{Z}(b)$ is at most countable, and has no accumulation points in Ω.

(b) By Theorem 15.6.2, the set $\Omega \setminus \Omega_{00}$ is connected. It is open since Ω_{00} is closed, and hence $\Omega \setminus \Omega_{00}$ is path-connected. See Lemma 15.4.7. Let $z_+ \in \Omega_+$ and $z_- \in \Omega_-$, and let C be a path linking z_+ to z_-, with parametrization $\gamma(t)$, $t \in [a, b]$. The function

$$t \mapsto l(t) = \frac{|b(\gamma(t))|}{|a(\gamma(t))|}$$

is continuous and

$$l(a) < 1 \quad \text{and} \quad l(b) > 1.$$

By continuity, there exists a point $c \in (a, b)$ such that $l(c) = 1$. This point c belongs to $\Omega_0 \setminus \Omega_{00}$.

(c) The set Ω_0 is closed, and so $\Omega \setminus \Omega_0$ is open. Assume that $\Omega \setminus \Omega_0$ is connected. Then, it is path-connected, see Lemma 15.4.7, and any pair of points $z_\pm$ in $\Omega_\pm$ is connected by a path inside $\Omega \setminus \Omega_0$. But, as above, this implies the existence of a μ where $|a(\mu)| = |b(\mu)|$. This point μ is thus in Ω_0, and this contradicts the fact that the path lies inside $\Omega \setminus \Omega_0$. □

Solution of Exercise 6.3.1. (a) The function F is analytic in the open unit disk (this was shown in the proof of Exercise 4.4.13 using a theorem on power series). The functions $\exp F(z)$ and $1+z$ are both analytic in the open unit disk. In calculus classes one proves (basically, the same proof as the one in Exercise 4.4.13) that

$$\exp F(x) = 1 + x, \quad x \in (-1, 1).$$

By analytic continuation this identity extends to the open unit disk.

(b) Let $\ln(z)$ be the usual analytic extension of $\ln(x)$ to $\mathbb{C} \setminus (-\infty, 0)$:

$$\ln(z) = \ln \rho + i\theta, \quad z = re^{i\theta}, \quad \theta \in (-\pi, \pi).$$

Then, for $z \in \mathbb{C} \setminus (-\infty, 0)$ we have

$$e^{\ln(z)} = z.$$

Thus, for $z \in \mathbb{C} \setminus (-\infty, -1)$, we have

$$e^{\ln(1+z)} = 1 + z.$$

It follows that $\ln(1+z)$ is the analytic extension of F to $\mathbb{C} \setminus (-\infty, 0)$. For (b), use item (b) of the previous exercise. □

Solution of Exercise 6.3.2. (1) The power series f_α restricted to $(-1, 1)$ and the function (6.3.1) have the same value at the origin and solve the same differential equation, namely

$$y'(x) = \frac{\alpha}{1+x} y(x),$$

Both functions are thus equal in $(-1, 1)$.

(2) The first claim follows from the definition (4.4.8). Furthermore, from the previous item we have for $x \in (-1, 1)$,

$$e^{\alpha \ln(1-x)} = \sum_{n=0}^{\infty} (-1)^n \binom{\alpha}{n} x^n.$$

See (4.4.8) for the definition of $\binom{\alpha}{n}$. The result follows by letting $x \uparrow 1$.

(3) The proof is similar to that of (2), and will be omitted.   □

Remark 6.9.3. The previous exercise also holds for complex $\alpha = a + bi$ provided we defined for real $x \in (-1, 1)$

$$(1 + x)^{\alpha} = \exp(a \ln(1 + x))(\cos(b \ln(1 + x)) + i \sin(b \ln(1 + x))).$$

Solution of Exercise 6.3.3. For the first function in (a), consider a rational number p/q. For $n \geq q$,

$$\frac{p}{q} n! \in 2\mathbb{N}$$

and hence

$$(re^{2\pi ip/q})^{n!} = r^{n!} e^{2\pi i \frac{pn!}{q}} = r^{n!}.$$

It follows that

$$\lim_{r \uparrow 1} \sum_{n=q}^{\infty} r^{n!} e^{2\pi i \frac{pn!}{q}} = \sum_{n=q}^{\infty} \lim_{r \uparrow 1} r^{n!} e^{2\pi i \frac{pn!}{q}}$$

$$= \sum_{n=q}^{\infty} \lim_{r \uparrow 1} r^{n!} = +\infty, \tag{6.9.3}$$

where the interchange of limit and summation is done using the monotone convergence theorem if you know measure theory, and by direct estimates, omitted here, if you want an elementary proof. Since the numbers $e^{2\pi ip/q}$ are dense on the unit circle, the function f cannot be continued in any open sub-arc of the unit circle.

For the second function, we first remark that f satisfies the functional equation

$$f(z) = -z + f(z^2).$$

Therefore

$$\lim_{\rho \uparrow 1} f(\rho)$$

does not exist. Since

$$f(-\rho) = \rho + f(\rho^2),$$

it follows that

$$\lim_{\rho \uparrow 1} f(-\rho)$$

does not exist either. Since

$$f(i\rho) = -i\rho + f(-\rho^2),$$

it then follows that

$$\lim_{\rho \uparrow 1} f(i\rho)$$

does not exist. The same argument will show that the limit

$$\lim_{\rho \uparrow 1} f(\rho e^{i\theta})$$

does not exist at the points $\theta = 2\pi \frac{k}{2^n}$, with $n = 0, 1, \ldots$ and $k = 0, 1, \ldots, 2^n - 1$. $\square$

Solution of Exercise 6.3.4. We assume that f is rational and that $|f(z)| = 1$ when $z \in I$, where I is some open arc of a circle. Thus,

$$f(z) = \frac{1}{\left(f\left(\frac{1}{\bar{z}}\right)\right)} = 1 \quad \text{when} \quad z \in I. \tag{6.9.4}$$

By analytic continuation, this equation holds for all complex numbers where both $f(z)$ and $f(1/\bar{z})$ are defined. From (6.9.4) we note that f cannot have a zero or a pole on the unit circle. Moreover, if w is a zero (resp. a pole) of f different from 0, then $1/\bar{w}$ is a pole (resp. a zero) of f. Let $w_1, \ldots, w_N$ be the zeros of f different from 0. The function

$$\frac{f(z)}{\prod_{n=1}^{N} b_{w_n}(z)}$$

has at most a zero or a pole at the origin. Therefore there exists $M \in \mathbb{Z}$ such that

$$\frac{f(z)}{z^M \prod_{n=1}^{N} b_{w_n}(z)}$$

is a constant. This constant is unitary since the function takes unitary values on I (and in fact is defined on all of $\mathbb{T}$ and takes unitary values there). $\square$

Solution of Exercise 6.3.5. We note the following. The left side of (6.3.2) defines a function analytic in the open unit disk. Indeed, using Weierstrass' theorem, we have that for $|z| < 1$,

$$\int_0^{2\pi} \frac{dt}{1 + z \sin t} = \sum_{n=0}^{\infty} z^n (-1)^n \int_0^{2\pi} \sin^n t \, dt.$$

On the other hand, for $z = a \in (-1, 1)$, the right-hand side is equal to

$$\frac{2\pi}{\sqrt{1 - a^2}}.$$

See Exercise 8.4.1 below. Thus both sides of (6.3.2) coincide on the interval $(-1, 1)$, and by analytic continuation they coincide in the open unit disk. Thus, the right side is an analytic continuation of the left side. $\square$

Other examples of analytic extensions appear in Exercises 5.7.4 and 5.7.10.

Solution of Exercise 6.3.6. Assume by contradiction that such a function f exists. The function f is in particular bounded in a neighborhood of the origin, and thus, by Riemann's Hebbarkeitssatz (see Theorem 5.4.1), the function F defined by

$$F(z) = \begin{cases} f(z), & \text{if } 0 < |z| < r \\ 0, & \text{if } z = 0 \end{cases}$$

is analytic in $B(0,r)$. The point $z = 0$ is a zero of F, say of order $N \geq 1$. Thus we have

$$z^{2N} G(z)^2 = z, \quad z \in B(0,r),$$

where G is analytic in $B(0,r)$ and such that $G(0) \neq 0$. We thus get

$$z^{2N-1} G(z)^2 = 1, \quad z \in B(0,r).$$

Setting $z = 0$ in the above equation we obtain a contradiction. □

Solution of Exercise 6.3.7. It suffices to remark that the two functions coincide on the real interval $(-1,1)$ with the function $\sqrt{1-x}$. □

Solution of Exercise 6.3.8. Let Δ be a triangle inside Ω, with boundary $\partial\Delta$. We take Δ to be closed (that is, the interior of the triangle and its boundary). To check that $\int_{\partial\Delta} f(z)dz = 0$. Four cases have to be distinguished. More precisely we need to consider:

(a) $I \cap \partial\Delta \neq \emptyset$ and $I \cap \overset{\circ}{\Delta} = \emptyset$,

   which itself is divided into:

   (a1) $I \cap \partial\Delta$ consists of one point.

   (a2) $I \cap \partial\Delta$ consists of an interval.

(b) $I \cap \partial\Delta \neq \emptyset$ and $I \cap \overset{\circ}{\Delta} \neq \emptyset$.

(c) $I \cap \partial\Delta = \emptyset$ and $I \cap \overset{\circ}{\Delta} = \emptyset$.

(d) $I \cap \partial\Delta = \emptyset$ and $I \cap \overset{\circ}{\Delta} \neq \emptyset$.

In the last case, I is in the outside of Δ, and Morera's theorem insures that $\int_{\partial\Delta} f(z)dz = 0$. In the other cases, the continuity of f insures that we still have $\int_{\partial\Delta} f(z)dz = 0$. In case (a1), the continuity of f insures that the integral of f on one of the segments of $\partial\Delta$ does not change if one removes one point from that segment. In cases (b), (c) and (d), we triangularize Δ in such a way that $I \cap \Delta$ is a side of part of the triangles which make the triangularization, and then use the continuity of f to approximate the integral of f on this segment. □

Solution of Exercise 6.3.9. We remark first that the function h is continuous in $B(x_0, r)$. Let $f(z) = u(x,y) + iv(x,y)$. We note that

$$\overline{f(\bar{z})} = u(x,-y) - iv(x,-y),$$

and it is clear that the functions

$$U(x, y) = u(x, -y) \quad \text{and} \quad V(x, y) = -v(x, -y)$$

satisfy the Cauchy–Riemann equations when u and v do. See Exercise 4.2.6 if need be. Therefore, $\overline{f(\bar{z})}$ is analytic in the image of $B_+(z_0, r)$ under conjugation. To prove that the function h is holomorphic in $B(x_0, r)$, we use Morera's theorem. Consider a triangle Δ in $B(x_0, r)$. We want to show that

$$\int_{\partial \Delta} h(z)dz = 0. \tag{6.9.5}$$

Various cases occur:

(a) The triangle does not intersect the real interval $(x_0 - r, x_0 + r)$. Then, the integral is zero because h is analytic in $B(z_0, r) \setminus (x_0 - r, x_0 + r)$.

(b) The triangle intersects $(x_0 - r, x_0 + r)$ at one point only.

(c) The triangle intersects $(x_0 - r, x_0 + r)$ at two points.

(d) One of the edges is $(x_0 - r, x_0 + r)$.

We discuss the case (d). Two nodes, say A and B, of the triangle are on the real axis. Without loss of generality we may assume that the remaining node, say C, is in the open upper half-plane, that is in $B_+(x_0, r)$. Let $\epsilon > 0$. The intersection of the line $-i\epsilon + x$, $x \in \mathbb{R}$ with the triangle consists of two points. We denote by $A(\epsilon)$ the one on the interval $[A, C]$ and by $B(\epsilon)$ the one on the interval $[B, C]$. The integral of h on the triangle defined by $A(\epsilon), B(\epsilon)$ and C is equal to 0 since h is analytic in the open convex set $B_+(x_0, r)$. Furthermore, in view of the continuity at the points of $(x_0 - r, x_0 + r)$, the integral of f on the quadrilateral defined by $A, B, A(\epsilon)$ and $B(\epsilon)$ goes to 0 as ϵ goes to 0. This shows that (6.9.5) holds. Cases (b) and (c) are treated in a similar way. By Morera's theorem, h is analytic in $B(z_0, r)$. □

Solution of Exercise 6.3.10. (a) For $z = \rho e^{i\theta} \in \mathbb{D} \setminus (-1, 0]$ we have $q(z) = \rho^{1/n} e^{\frac{i\theta}{n}}$. Let $\epsilon = e^{\frac{2\pi i}{n}}$. We have $w^n = z$ if and only if

$$w = w_j = \epsilon^j q(z), \quad j = 0, \ldots, n - 1.$$

The formula

$$g_n(z) = \frac{1}{n} \sum_{j=0}^{n-1} f(\epsilon^j q(z))$$

expresses that g_n is analytic in $\mathbb{D} \setminus (-1, 0]$.

(b) Let

$$f(z) = \sum_{m=0}^{\infty} a_m z^m$$

be the Taylor expansion of f in $\mathbb{D}$ centered at the origin. We have

$$g_n(z) = \frac{1}{n} \sum_{j=0}^{n-1} \sum_{m=0}^{\infty} a_m \rho^{m/n} e^{\frac{im\theta}{n}}$$

$$= \frac{1}{n} \sum_{m=0}^{\infty} a_m \rho^{m/n} e^{\frac{im\theta}{n}} \left(\sum_{j=0}^{n-1} \epsilon^{jm} \right).$$

By formula (1.1.17) we have

$$\sum_{j=0}^{n-1} \epsilon^{jm} = \begin{cases} n, & \text{if } m = kn, \ k \in \mathbb{N}_0, \\ 0, & \text{otherwise.} \end{cases}$$

It follows that

$$g_n(z) = \sum_{k=0}^{\infty} a_{kn} z^k.$$

Hence g_n has an analytic continuation to $\mathbb{D}$. $\qquad\square$

Solution of Exercise 6.4.1. We translate the problem in the complex plane. Let $z_1, \ldots, z_n$ be the complex numbers corresponding to $P_1, \ldots, P_n$. Let $P = (x, y)$ be in the plane, and let $z = x + iy$. We have

$$\prod_{\ell=1}^{n} \|PP_\ell\| = \left| \prod_{\ell=1}^{n} (z - z_\ell) \right|.$$

The function $\prod_{\ell=1}^{n}(z - z_\ell)$ is a polynomial, and in particular is an entire function. By the maximum modulus principle, its modulus has no local extrema, besides at the points z_ℓ. $\qquad\square$

Solution of Exercise 6.4.2. Assume that

$$|p(e^{i\theta})| \leq 1 \quad \forall \theta \in [0, 2\pi],$$

and consider

$$q(z) = z^n p(1/z) = 1 + a_{n-1}z + \cdots + a_0 z^n.$$

The function q is also a polynomial and it holds that

$$\max_{|z| \leq 1} |q(z)| = \max_{|z|=1} |q(z)| = \max_{\theta \in [0, 2\pi]} |e^{in\theta} p(e^{-i\theta})| \leq 1,$$

where we used the maximum modulus principle and the hypothesis that $|p(e^{i\theta})| \leq 1$ for all $\theta \in [0, 2\pi]$. But $q(0) = 1$ and so $q(z) \equiv 1$ by the maximum modulus principle. This forces $a_{n-1} = \cdots = a_0 = 0$ and so $p(z) = z^n$. The converse is trivial.   □

Solution of Exercise 6.4.3. Let $r_1 < r_2$. We have

$$\{z \; ; \; |z| \leq r_1\} \subset \{z \; ; \; |z| \leq r_2\},$$

and so $\sup_{|z| \leq r_1} |f(z)| \leq \sup_{|z| \leq r_2} |f(z)|$. By the maximum modulus principle,

$$M(r, p) \overset{\text{def.}}{=} \sup_{|z|=r} |p(z)| = \sup_{|z| \leq r} |p(z)|,$$

and hence $M(r, p)$ is increasing.

Let $p(z) = a_n z^n + a_{n-1} z^{n-1} + \cdots + a_0$. To prove the second claim, we consider the function $p^\sharp(z)$ given by (5.9.36):

$$p^\sharp(z) = \begin{cases} z^n p\left(\dfrac{1}{z}\right), & z \neq 0, \\ a_n, & z = 0. \end{cases}$$

We have:

$$M(r, p^\sharp) = \sup_{|z|=r} |p^\sharp(z)| = \sup_{|z|=r} r^n \left| p\left(\frac{1}{z}\right) \right| = r^n M(\frac{1}{r}, p), \qquad (6.9.6)$$

and so

$$M\left(\frac{1}{r}, p^\sharp\right) = \frac{M(r, p)}{r^n}. \qquad (6.9.7)$$

The function $p^\sharp(z)$ is a polynomial and so the function $z \mapsto M(r, p^\sharp)$ is increasing and so the map $z \mapsto M(\frac{1}{r}, p^\sharp)$ is decreasing. This concludes the proof, thanks to (6.9.7).   □

Solution of Exercise 6.4.4. Assume by contradiction that such a function f exists. It does not vanish and $|f^{-1}(z)| = e^{-|z|} \leq 1$ for $z \in \mathbb{D}$. But $|f(0)| = e^0 = 1$. The maximum modulus principle implies that f is a constant. But no constant function can satisfy $|f(z)| = e^{|z|}$ for all $|z| < 1$.   □

Solution of Exercise 6.4.5. Since h does not vanish in $|z| < 1+\epsilon$, there is a function f analytic there and such that $h = e^{-f}$; in particular

$$|h(z)| = e^{-\operatorname{Re} f(z)}.$$

Since $|h(z)| \leq 1$ for z in the open unit disk we have that $\operatorname{Re} f(z) \geq 0$ there. Since f is analytic in $|z| < 1+\epsilon$ we can use Harnack's inequalities (as proved in the present set of exercises; recall that (5.5.13) is a special case of Harnack's inequalities). Fix $R < 1$ and $\zeta \in [0, 2\pi]$. We have then

$$\frac{1 - |R|}{1 + |R|} \operatorname{Re} f(0) \leq \operatorname{Re} f(Re^{i\zeta}) \leq \frac{1 + |R|}{1 - |R|} \operatorname{Re} f(0), \quad \forall \zeta \in [0, 2\pi].$$

The cases $R = 1/7$ and $R = 1/5$ lead to

$$\frac{3}{4} \operatorname{Re} f(0) \leq \operatorname{Re} f\left(\frac{e^{i\zeta}}{7}\right) \leq \frac{4}{3} \operatorname{Re} f(0), \quad \forall \zeta \in [0, 2\pi],$$

and

$$\frac{2}{3} \operatorname{Re} f(0) \leq \operatorname{Re} f\left(\frac{e^{i\zeta}}{5}\right) \leq \frac{3}{2} \operatorname{Re} f(0), \quad \forall \zeta \in [0, 2\pi].$$

In particular,

$$\operatorname{Re} f\left(\frac{e^{i\zeta}}{7}\right) \leq \frac{4}{3} \operatorname{Re} f(0) \leq 2 \operatorname{Re} f\left(\frac{e^{i\zeta}}{5}\right), \quad \forall \zeta \in [0, 2\pi].$$

Thus

$$\sup_{|z|=1/5} |h(z)|^2 = \inf_{|z|=1/5} e^{-2 \operatorname{Re} f(z)}$$

$$\leq e^{-\frac{4}{3} \operatorname{Re} f(0)}$$

$$\leq \inf_{|z|=1/7} e^{-\operatorname{Re} f(z)}$$

$$= \inf_{|z|=1/7} |h(z)|,$$

and thus, using the maximum modulus principle for h in $|z| \leq 1/5$ and $1/h$ in $|z| \leq 1/7$ we obtain the result. $\qquad \square$

Solution of Exercise 6.5.1. By Schwarz' lemma, we have $|f(z)| \leq |z|$ in the open unit disk $\mathbb{D}$. If z is in $\mathbb{D}$ so is z^n for any positive integer n, and so $|f(z^n)| \leq |z^n|$. Thus for $|z| \leq r$ we have

$$\left| \sum_{n=0}^{\infty} f(z^n) \right| \leq \sum_{n=0}^{\infty} |f(z^n)|$$

$$\leq \sum_{n=0}^{\infty} |z^n|$$

$$\leq \sum_{n=0}^{\infty} r^n$$

$$= \frac{1}{1 - r}.$$

Hence, the series $\sum_{n=0}^{\infty} f(z^n)$ converges absolutely and uniformly in $|z| \leq r$ for $r < 1$. $\qquad \square$

We recall that, for $a \in \mathbb{D}$, the map

$$b_a(z) = \frac{z - a}{1 - z\overline{a}}$$

denotes the elementary Blaschke factor (1.1.44), and that b_a maps conformally $\mathbb{D}$ onto itself, with inverse b_{-a}.

Solution of Exercise 6.5.2. If f is a unitary constant, (6.5.4) is trivial. If f is not a unitary constant, the maximum modulus principle implies that $|f(z)| < 1$ for all $z \in \mathbb{D}$. Let $z_0 \in \mathbb{D}$. The idea is to reduce the situation to the case where $z = 0$ using appropriate maps of the form b_a. The function

$$g(z) = \frac{f(b_{-z_0}(z)) - f(z_0)}{1 - f(b_{-z_0}(z))\overline{f(z_0)}}$$

is analytic and contractive in $\mathbb{D}$, and vanishes at the origin. Schwarz' lemma implies that

$$|g'(0)| \leq 1.$$

An easy computation leads to

$$g'(z) = \frac{1 - |z_0|^2}{(1 + z\overline{z_0})^2} f'(b_{-z_0}(z)) \frac{1 - |f(z_0)|^2}{(1 - f(b_{-z_0}(z))\overline{f(z_0)})^2}.$$

Thus,

$$g'(0) = \frac{1 - |z_0|^2}{1 - |f(z_0)|^2} f'(z_0),$$

and the result follows. The claim on the case of equality follows from the fact that $g'(0)$ has modulus 1 if and only if $g(z) = cz$ for some $c \in \mathbb{T}$. $\square$

Solution of Exercise 6.5.3. By Schwarz' lemma we can write $f(z) = zg(z)$ where g is analytic and bounded by 1 in modulus in $\mathbb{D}$. The condition $f(1/2) = 1/2$ leads to $\frac{g(1/2)}{2} = \frac{1}{2}$, so that $g(1/2) = 1$, and the maximum modulus principle leads to $g(z) \equiv 1$. $\square$

Solution of Exercise 6.5.4. Let z_1 and z_2 in $\mathbb{D}$ such that $f(z_\ell) = z_\ell$ for $\ell = 1, 2$. The function

$$g = b_{z_1} \circ f \circ b_{-z_1},$$

sends the open unit disk into itself and is such that $g(0) = 0$. By Schwarz' lemma, $g(z) = z\sigma(z)$, where the function σ is analytic and bounded by 1 in modulus in the open unit disk.

The second condition $f(z_2) = z_2$ can be rewritten as

$$f \circ b_{-z_1} \circ b_{z_1}(z_2) = z_2,$$

and so

$$b_{z_1} \circ f \circ b_{-z_1} \circ b_{z_1}(z_2) = b_{z_1}(z_2).$$

Thus with $w = b_{z_1}(z_2)$, we have $g(w) = w$, i.e., $\sigma(w) = 1$. By the maximum modulus principle, $\sigma(z) \equiv 1$ and so $g(z) = z$, meaning that $f(z) \equiv z$. □

Solution of Exercise 6.5.5. The function $F = b_{z_0} \circ f \circ b_{-z_0}$ is analytic from $\mathbb{D}$ into itself and

$$F(0) = b_{z_0}(f(b_{-z_0}(0)) = b_{z_0}(f(z_0)) = b_{z_0}(z_0) = 0.$$

By Schwarz' lemma,

$$F(z) = zG(z),$$

where G is analytic and contractive in $\mathbb{D}$. Hence

$$(f \circ b_{-z_0})(z) = (b_{-z_0})(zG(z))$$
$$= \frac{zG(z) + z_0}{1 + zG(z)\overline{z_0}}.$$

But

$$(f \circ b_{-z_0})'(z) = \frac{1 - |z_0|^2}{(1 + z\overline{z_0})^2} f'(b_{-z_0}(z)),$$

and

$$\left(\frac{zG(z) + z_0}{1 + zG(z)\overline{z_0}} \right)' = \frac{1 - |z_0|^2}{(1 + zG(z)\overline{z_0})^2}(zG'(z) + G(z)).$$

Hence,

$$\frac{1 - |z_0|^2}{(1 + z\overline{z_0})^2} f'(b_{-z_0}(z)) = \frac{1 - |z_0|^2}{(1 + zG(z)\overline{z_0})^2}(zG'(z) + G(z)).$$

Setting $z = 0$ and using the fact that $f'(z_0) = 1$ we obtain

$$G(z_0) = 1.$$

By the maximum principle, $G(z) \equiv 1$ and so $f(z) = z$. □

Solution of Exercise 6.5.6. It suffices to take

$$f(z) = \frac{1 + z}{3 - z}.$$

We have $f(1) = f'(1) = 1$. □

Solution of Exercise 6.5.7. (a) We assume $|f(0)| < 1$, and therefore f is not a constant of modulus 1 (it can be a constant of modulus strictly less than 1), and therefore $|f(z)| < 1$ for all $z \in \mathbb{D}$. Therefore (see (1.1.41) if needed with $f(z)$ instead of z and $f(0)$ instead of w),

$$h(z) = \frac{f(z) - f(0)}{1 - \overline{f(0)}f(z)} \in \mathbb{D}, \quad \forall z \in \mathbb{D}.$$

Since $h(0) = 0$, it follows from Schwarz' lemma that the function

$$f_1(z) = \begin{cases} \dfrac{h(z)}{z}, & z \in \backslash \{0\}, \\ h'(0), & z = 0 \end{cases} \tag{6.9.8}$$

is in $\mathcal{S}$.

(b) Given $g \in \mathcal{S}$ we have that $|zg(z)| < 1$ for all $z \in \mathbb{D}$. Using (1.1.41) with $zg(z)$ instead of z and ρ_0 instead of w, we see the function (6.5.6)

$$f(z) = \frac{\rho_0 + zg(z)}{1 + z\overline{\rho_0}g(z)}$$

is in $\mathcal{S}$. It clearly satisfies $f(0) = \rho_0$. Conversely, if f is a solution, the function f_1 defined by (6.9.8) with $f(0) = \rho_0$ belongs to $\mathcal{S}$. Solving f in terms of f_1 we obtain (6.5.6) (with f_1 instead of g). So f is indeed of the form (6.5.6).

(c) Recall that the derivative of the function

$$z \mapsto \frac{z + \rho_0}{z\overline{\rho_0} + 1}$$

is given by

$$z \mapsto \frac{1 - |\rho_0|^2}{(z\overline{\rho_0} + 1)^2}.$$

Furthermore

$$(zg(z))' = zg'(z) + g(z).$$

Hence, the usual rule of differentiation for the composition of two functions leads to

$$f'(z) = (zg'(z) + g(z)) \cdot \frac{1 - |\rho_0|^2}{(zg(z)\overline{\rho_0} + 1)^2}.$$

Thus,

$$f'(0) = g(0)(1 - |\rho_0|^2).$$

It follows that to find $f \in \mathcal{S}$ with preassigned values of $f(0)$ and $f'(0)$, it suffices to first solve the problem of finding all $f \in \mathcal{S}$ such that $f(0)$ is given. If $|f(0)| > 1$ there are no solutions. If $|f(0)| = 1$, then $f(z) \equiv f(0)$, and necessarily $f'(0) = 0$. Assume now that $|f(0)| < 1$. Then finding all f (if any) such that $f'(0)$ is given amounts to find all $g \in \mathcal{S}$ such that

$$g(0) = \frac{f'(0)}{1 - |\rho_0|^2}.$$

This problem has no solution if $\rho_1 \stackrel{\text{def.}}{=} \frac{f'(0)}{1 - |\rho_0|^2}$ has modulus greater than 1; it has a unique solution if $|\rho_1| = 1$, and it has an infinity of solutions, described in a way similar to (6.5.6) if $\rho_1 \in \mathbb{D}$. $\square$

Solution of Exercise 6.5.9. By Schwarz' lemma, we have $f(0) = 0$ if and only if

$$f(z) = zg(z)$$

for some $g \in \mathcal{S}$. We have

$$f'(z) = zg'(z) + g(z),$$

and therefore

$$g(0) = f'(0) = b.$$

If $|b| > 1$ there is no solution to (6.5.8). If $|b| = 1$, the maximum modulus principle implies that g is unique and equal to $g(z) \equiv b$. Thus the interpolation problem (6.5.8) has a unique solution, namely $f(z) = zb$. If $|b| < 1$, it follows from (6.5.6) that g is of the form

$$g(z) = \frac{b + zh(z)}{1 + z\bar{b}h(z)},$$

where h varies in $\mathcal{S}$. The solutions of the interpolation problem (6.5.8) are therefore exactly the functions of the form

$$f(z) = z\frac{b + zh(z)}{1 + z\bar{b}h(z)},$$

where h varies in $\mathcal{S}$. □

Solution of Exercise 6.5.10. If $|w_1| = 1$, the only function in $\mathcal{S}$ which satisfies $f(z_1) = w_1$ is the constant function $f(z) \equiv w_1$. Thus, if $w_2 \neq w_1$, the interpolation problem at hand has no solution, and has a unique solution $f(z) \equiv w_1$ if $w_1 = w_2$. Assume now $|w_1| < 1$. Then by Exercise 6.5.7, a function $f \in \mathcal{S}$ satisfies $f(z_1) = w_1$ if and only if it is of the form

$$f(z) = \frac{w_1 + \dfrac{z - z_1}{1 - \bar{z}_1 z}g(z)}{1 + \overline{w_1}\dfrac{z - z_1}{1 - \bar{z}_1 z}g(z)}, \qquad g \in \mathcal{S}. \tag{6.9.9}$$

The condition $f(z_2) = w_2$ reads

$$w_2 = \frac{w_1 + \dfrac{z_2 - z_1}{1 - \bar{z}_1 z_2}g(z_2)}{1 + \overline{w_1}\dfrac{z_2 - z_1}{1 - \bar{z}_1 z_2}g(z_2)},$$

that is

$$g(z_2) = \frac{w_2 - w_1}{1 - \overline{w_1}w_2} \cdot \frac{1 - z_1\bar{z}_2}{z_2 - z_1}.$$

As in the previous exercise, three cases are possible depending on the value of

$$\rho \overset{\text{def.}}{=} \frac{w_2 - w_1}{1 - \overline{w_1}w_2} \cdot \frac{1 - z_1\bar{z}_2}{z_2 - z_1}.$$

If $|\rho| > 1$ there is no solution. If $|\rho| = 1$, the unique solution of the interpolation problem is

$$f(z) = \frac{w_1 + \dfrac{z - z_1}{1 - z\overline{z_1}} \dfrac{w_2 - w_1}{1 - \overline{w_1}w_2} \cdot \dfrac{1 - z_1\overline{z_2}}{z_2 - z_1}}{1 + \overline{w_1} \dfrac{z - z_1}{1 - z\overline{z_1}} \dfrac{w_2 - w_1}{1 - \overline{w_1}w_2} \cdot \dfrac{1 - z_1\overline{z_2}}{z_2 - z_1}}.$$

If $|\rho| < 1$, in view of Exercise 6.5.7, the set of all solutions g is of the form

$$g(z) = \frac{\dfrac{w_2 - w_1}{1 - \overline{w_1}w_2} \cdot \dfrac{1 - z_1\overline{z_2}}{z_2 - z_1} + \dfrac{z - z_2}{1 - z\overline{z_2}} h(z)}{1 + \dfrac{\overline{w_2} - \overline{w_1}}{1 - w_1\overline{w_2}} \cdot \dfrac{1 - \overline{z_1}z_2}{\overline{z_2} - \overline{z_1}} \dfrac{z - z_2}{1 - z\overline{z_2}} h(z)}, \qquad h \in \mathcal{S}. \qquad \square$$

To conclude it suffices to plug this expression in (6.9.9).

It can be shown that a necessary and sufficient condition for Problem 6.5.9 to have a solution is that the matrix

$$\begin{pmatrix} \dfrac{1 - |w_1|^2}{1 - |z_1|^2} & \dfrac{1 - w_1\overline{w_2}}{1 - z_1\overline{z_2}} \\ \dfrac{1 - w_2\overline{w_1}}{1 - z_2\overline{z_1}} & \dfrac{1 - |w_2|^2}{1 - |z_2|^2} \end{pmatrix}$$

is non-negative. Recall that a hermitian matrix is called non-negative if all its eigenvalues are greater than or equal to 0. See Definition 16.3.1.

Solution of Exercise 6.5.11. Recall that

$$b_{-a}(z) = \frac{z + a}{1 + z\overline{a}}.$$

The function $F(z) = f(b_{-a}(z))$ is analytic in the open unit disk $\mathbb{D}$, and maps $\mathbb{D}$ into itself. Furthermore

$$F(0) = f(b_{-a}(0)) = f(a) = 0.$$

By Schwarz' lemma,
$$F(z) = zG(z),$$

where G is analytic and contractive in the open unit disk. Replacing in this equation z by $b_a(z)$ we obtain

$$F(b_a(z)) = f(b_{-a}(b_a(z))) = f(z) \quad \text{on the one hand,}$$
$$= b_a(z)G(b_a(z)) \quad \text{on the other hand,}$$

and so
$$f(z) = \frac{z - a}{1 - z\overline{a}} g(z),$$

where the function $g(z) = G(b_a(z))$ is analytic and contractive in $\mathbb{D}$. Furthermore, $g(b) = 0$ since $f(b) = 0$ and $a \neq -b$. We reiterate the same argument on g and obtain

$$g(z) = \frac{z - b}{1 - z\overline{b}} h(z)$$

for some function h analytic and contractive in $\mathbb{D}$. Thus

$$f(z) = \frac{z - a}{1 - z\overline{a}} g(z) = \frac{z - a}{1 - z\overline{a}} \frac{z - b}{1 - z\overline{b}} h(z),$$

and so

$$|f(z)| = \left| \frac{z - a}{1 - z\overline{a}} \right| \left| \frac{z - b}{1 - z\overline{b}} \right| |h(z)| \leq \left| \frac{z - a}{1 - z\overline{a}} \right| \left| \frac{z - b}{1 - z\overline{b}} \right|. \qquad \square$$

Solution of Exercise 6.5.12. We have

$$\operatorname{Re} \varphi(z) = \int_0^{2\pi} \frac{1 - |z|^2}{|e^{it} - z|^2} m(t) dt \geq 0.$$

Thus the function

$$s(z) = \frac{\varphi(z) - 1}{\varphi(z) + 1}$$

is bounded by one in modulus in the open unit disk. The condition

$$\int_0^{2\pi} m(t) dt = 1$$

forces $\varphi(0) = 1$ and so $s(0) = 0$. It suffices then to apply Schwarz' lemma to obtain the result. $\qquad \square$

Solution of Exercise 6.5.13. The first claim follows from the identity

$$1 - \left| \frac{z - z_0}{z - \overline{z_0}} \right|^2 = \frac{2 \operatorname{Re} z(\overline{z_0} - z_0)}{|z - \overline{z_0}|^2} = \frac{4 y y_0}{|z - \overline{z_0}|^2},$$

where $y = \operatorname{Im} z$ and $y_0 = \operatorname{Im} z_0$. See formula (1.1.53) with $w = z_0$ and $v = z$ if need be. The inverse map

$$\mathscr{B}_{z_0}^{-1}(z) = \frac{z_0 - z\overline{z_0}}{1 - z}$$

maps then in a one-to-one way the closed unit disk onto the closed upper half-plane. To prove the second claim, consider the function

$$s(z) = f(\mathscr{B}_{z_0}^{-1}(z)) = f\left(\frac{z_0 - z\overline{z_0}}{1 - z} \right).$$

It is analytic and contractive in the open unit disk. Furthermore, $s(0) = f(z_0) = 0$. Applying Schwarz' lemma to s we get

$$|s(z)| \leq |z|, \quad z \in \mathbb{D}.$$

Replacing in this inequality z by $\mathscr{B}_{z_0}(z)$, with now $z \in \mathbb{C}_+$ we obtain the claim.

To prove the last claim, we see now that $s(0) = s'(0) = 0$. By Schwarz' lemma, the first condition implies that $s(z) = z\sigma(z)$, where σ is analytic and contractive in the open unit disk. Since

$$s'(z) = \sigma(z) + z\sigma'(z),$$

the second condition implies that $\sigma(z) = z\sigma_1(z)$, where σ_1 is analytic and contractive in the open unit disk. Thus $s(z) = z^2\sigma_1(z)$, and in particular,

$$|s(z)| \le |z|^2, \quad z \in \mathbb{D}.$$

Once more replacing in this inequality z by $\mathscr{B}_{z_0}(z)$, with now $z \in \mathbb{C}_+$ we obtain the last claim.   □

Solution of Exercise 6.6.2. From the power series defining $\sin z$ we have

$$|\sin z| = \left| \sum_{n=0}^{\infty} \frac{(-1)^n z^{2n+1}}{(2n+1)!} \right| \le \sum_{n=0}^{\infty} \frac{|z|^{2n+1}}{(2n+1)!} = \sinh|z|.$$

The second claim is an easy calculus exercise. Consider the function

$$f(x) = x\cosh 1 - \sinh x.$$

Then

$$f'(x) = \cosh 1 - \cosh x \ge 0 \quad \text{for} \quad x \in [0,1].$$

Thus f is increasing on $[0,1]$ and $f(x) \ge f(0) = 0$, that is $\sinh x \le x(\cosh 1)$ on $[0,1]$. The last claim follows easily from

$$|\sin z^n| \le |z|^n \cosh 1$$

for $z \in \mathbb{D}$.   □

Solution of Exercise 6.6.3. Let

$$f_n(z) = \frac{z(z+1)\cdots(z+n-1)}{n^n}, \quad n = 1, 2, \ldots.$$

We have

$$f_{n+1}(z) = \frac{z+n}{n+1} \cdot \left(\frac{n}{n+1}\right)^n f_n(z) = \frac{1 + \dfrac{z}{n}}{1 + \dfrac{1}{n}} \cdot \left(1 + \frac{1}{n}\right)^{-n} f_n(z).$$

For $R > 0$ we have

$$\lim_{n \to \infty} \frac{1 + \dfrac{R}{n}}{1 + \dfrac{1}{n}} \cdot \left(1 + \frac{1}{n}\right)^{-n} = e^{-1}.$$

Therefore, for any $R > 0$ there exists $n_0(R) \in \mathbb{N}$ such that:

$$n \geq n_0(R) \implies |f_{n+1}(z)| \leq \frac{2}{e}|f_n(z)|, \quad \forall |z| \leq R. \tag{6.9.10}$$

Since $2/e < 1$, it follows that the series $\sum_{n=1}^{\infty} f_n(z)$ converges absolutely and uniformly on every closed set $|z| \leq R$, and its sum is therefore analytic. □

Solution of Exercise 6.6.4. It suffices to notice that the series $q^\ell z$ converges uniformly on compact subsets of $\mathbb{C}$. We now turn to the formula for the coefficients $(a_n)_{n \in \mathbb{N}}$ of the Taylor series

$$f(z) = 1 + \sum_{n=1}^{\infty} a_n z^n$$

of f centered at the origin. From formula (3.7.12) (see Exercise 3.7.8) we get (with $a_0 = 1$)

$$a_n = -\frac{q^n}{1 - q^n} a_{n-1}, \quad n = 1, 2, \ldots,$$

and hence the result since

$$1 + 2 + \cdots + n = \frac{n(n+1)}{2}. \qquad \square$$

Solution of Exercise 6.6.5. Let $z = x + iy$. Then

$$|e^{-n^2 z}| = e^{-n^2 x} = \left(\frac{1}{e^x}\right)^{n^2},$$

and the series is absolutely convergent, and hence convergent, in the open right half-plane. On the other hand, for $x \leq 0$, the absolute value $|e^{-n^2 z}| = e^{-n^2 x}$ does not go to zero, and hence the sum (6.6.1) cannot be convergent.

For $x \geq \epsilon > 0$ we have
$$|e^{-n^2 z}| \leq e^{-n^2 \epsilon},$$

and hence the sum is absolutely and uniformly convergent there. □

Solution of Exercise 6.7.1. From the estimate (3.9.22) it follows that the sequence of finite products converges uniformly on compact subsets of $\mathbb{D}$. Thus the limit is an analytic function. □

Solution of Exercise 6.7.2. It suffices to use (3.9.19) with $a_n = 1/n$. □

Solution of Exercise 6.7.3. Replacing z by $i\pi z$ in the formula (3.7.23) for $\sinh z$ we obtain

$$\frac{\sin \pi z}{\pi z} = \prod_{k=1}^{\infty}\left(1 - \frac{z^2}{k^2}\right). \tag{6.9.11}$$

The infinite product defines an entire function, and it is therefore legitimate to identity the coefficients of the powers of z on both sides of (6.9.11). The coefficient of z^2 on the left is

$$-\frac{\pi^2}{6}.$$

On the right the coefficient of z^2 is

$$-\sum_{k=1}^{\infty} \frac{1}{k^2},$$

and hence the result. The proof of (6.7.1) is done in the same way, using (3.7.26) to obtain

$$\cos z = 1 - \frac{z^2}{2} + \text{higher-order terms}$$

$$= 1 - 4z^2 \left\{ \sum_{k=0}^{\infty} \frac{1}{(2k+1)^2 \pi^2} \right\} + \text{higher-order terms.} \qquad \square$$

Solution of Exercise 6.8.2. The result is a direct application of formula (6.8.2). $\square$

Solution of Exercise 6.8.4. We consider (a). The other cases are treated similarly. Let

$$f(z) = \sum_{n=0}^{\infty} f_n (z - i)^n$$

be the power expansion of f at the point i. It has infinite radius of convergence, and hence

$$f_n = \frac{1}{2\pi i} \int_{|z-i|=R} \frac{f(z)}{(z-i)^{n+1}} dz = \frac{1}{2\pi} \int_0^{2\pi} \frac{f(i + Re^{it})}{R^n e^{int}} dt$$

for every $R > 0$. Taking into account the bound in (a) we have

$$|f_n| \le \frac{1}{2\pi} \int_0^{2\pi} \frac{|f(i + Re^{it})|}{R^n} dt \le \frac{1}{2\pi} \int_0^{2\pi} \frac{M(1 + \sqrt{R})}{R^n} dt = \frac{M(1 + \sqrt{R})}{R^n}.$$

Letting $R \to \infty$ we get that $f_n = 0$ for $n \ge 1$. Thus, f is a constant, of modulus less than or equal to M. $\square$

Solution of Exercise 6.8.5. If the given function f is a polynomial, this is clear from the fundamental theorem of algebra. The image is in fact all of $\mathbb{C}$ since the equation

$$f(z) = w$$

has at least one solution for all $w \in \mathbb{C}$.

Assume now that f is an arbitrary non-constant entire function, and proceed by contradiction. There is then a number a and a positive number r such that

$$|f(z) - a| > r, \quad \forall z \in \mathbb{C},$$

and in particular

$$\left| \frac{1}{f(z) - a} \right| \leq \frac{1}{r}, \quad \forall z \in \mathbb{C}.$$

Hence the function $1/(f(z) - a)$ is entire and bounded. By Liouville's theorem it is constant. This is a contradiction since f itself is not constant. □

Solution of Exercise 6.8.6. The function $F(z) = e^{f(z) - M}$ is entire. But

$$|F(z)| = e^{\operatorname{Re} f(z) - M} \leq 1.$$

By Liouville's theorem, F is a constant function. Thus

$$F'(z) = f'(z) F(z) \equiv 0,$$

and so $f'(z) \equiv 0$ and f is a constant function. □

Solution of Exercise 6.8.7. We note that functions of the form $f(z) = cz^n$ where c is a complex number of modulus 1 and $n \in \mathbb{N}$ answer the question. We show that these are the only entire functions satisfying (6.8.3). Assume f is an entire function satisfying (6.8.3). We can always write $f(z) = z^n f_0(z)$, where f_0 is entire and does not vanish at the origin. Rewriting (6.8.3) as

$$f_0(z) \overline{f_0(1/\overline{z})} = 1, \quad z \neq 0, \tag{6.9.12}$$

we see that

$$\lim_{z \to \infty} f_0(z) = 1 / \overline{f_0(0)}.$$

By Liouville's theorem, f_0 is a constant, which is unitary thanks to (6.9.12), and this concludes the proof. □

Solution of Exercise 6.8.8. By the periodicity conditions, it is enough to know the values of the function f in the closed square with corners

$$(0, 0), (0, 1), (1, 0), (1, 1).$$

In this square, the function is bounded in modulus (by an elementary property of continuous functions on closed bounded sets). Thus, it is bounded in the complex plane, and hence constant by Liouville's theorem. □

Solution of Exercise 6.8.9. Assume that such a polynomial exists. By Cauchy's formula, the first condition reads

$$P(1) = P(1/2) = P(1/3) = \cdots = P(1/(N+1)) = 0.$$

Thus P would be a polynomial of degree N with $(N+1)$ zeroes; this cannot be, and the only polynomial satisfying the condition is the polynomial $P(z) \equiv 0$.

In the second case, we have that all the derivatives of P up to order N vanish at $z = 1/2$. But, by the Taylor expansion at $z = 1/2$ and since P has degree N,

$$P(z) = \sum_{\ell=0}^{N} \frac{P^{(\ell)}(1/2)}{\ell!}(z - 1/2)^\ell,$$

and so $P \equiv 0$ (and in particular no polynomial of degree N meets the requirement). □

Solution of Exercise 6.8.10. We rewrite the vanishing condition (6.8.6) as

$$\int_{|z|=r} \frac{P(z)}{z - \frac{1}{n+1}} dz = 0, \quad n = 0, 1, \ldots, N.$$

For $r = 3/4$, the points $\frac{1}{2}, \frac{1}{3}, \ldots, \frac{1}{N+1}$ are inside the circle $|z| = r$, and (6.8.6) becomes

$$P(1/2) = P(1/3) = \cdots = P(1/(N+1)) = 0.$$

Since P has degree N, these conditions uniquely determine it, up to a multiplicative constant:

$$P(z) = K \cdot \left(z - \frac{1}{2}\right) \cdots \left(z - \frac{1}{N+1}\right).$$

For $r = 1/(N+2)$, no point of the form $1/(n+1)$ with $n = 0, \ldots, N+1$ lies inside the circle $|z| = r$, and every polynomial (not only of degree N) answers the question. □

Solution of Exercise 6.8.11. Set first $z = 2\pi k$ with $k \in \mathbb{Z}$. You get

$$p(2\pi k) \cos^2 2\pi k = 1.$$

Thus

$$p(2\pi k) = 1 \quad k \in \mathbb{Z},$$

the polynomial $p(z) - 1$ has an infinite number of zeros, and so must be equal to 0. So $p(z) \equiv 1$. To see that $q(z) \equiv 1$, put $z = \frac{\pi}{2} + k\pi$. □

Solution of Exercise 6.8.12. Assume that such a function exists. Then f does not vanish in $\mathbb{C} \setminus \{0\}$ and therefore $1/f$ is analytic there. Moreover, $1/f$ is bounded in a neighborhood of the origin, and hence, by Riemann's removable singularity theorem, is analytic in a neighborhood of the origin, and hence the function

$$
h(z) = \begin{cases} \frac{1}{f(z)}, & z \neq 0, \\ 0, & z = 0 \end{cases}
$$

is entire. It can be written as

$$
h(z) = z^N h_1(z),
$$

where $N \in \mathbb{N}$ and h_1 is an entire function not vanishing at the origin (and hence not vanishing in $\mathbb{C}$). From (6.8.7) we get

$$
\left| \frac{1}{f(z)^2} \right| = |z^{2N} h_1^2(z)| \leq |z|, \quad z \in \mathbb{C},
$$

and the entire function $z^{2N-1} h_1^2(z)$ is bounded and hence is equal to a constant, say K. Thus

$$
z^{2N-1} h_1^2(z) = K.
$$

Setting $z = 0$ we obtain $0 = 1$. Hence, no such function f exists. □

Solution of Exercise 6.8.13. The polynomial

$$
p(z) = z^4 + 3z^2 + z + 1
$$

has real coefficients, and so its roots are either real or appear in conjugate pairs. See Exercise 1.5.5. For $z = x \in (-1, 1)$, we have that $x + 1 > 0$ and hence $x^4 + 3x^2 + x + 1 > 0$. Furthermore, ± 1 are not roots of p and hence no roots are in $[-1, 1]$. Let $z_0, \overline{z_0}, z_1, \overline{z_1}$ be the roots of p, with z_0 and z_1 in the open upper half-plane (and *a priori* possibly equal). We have

$$
p(z) = (z - z_0)(z - \overline{z_0})(z - z_1)(z - \overline{z_1}),
$$

and in particular

$$
z_0 \overline{z_0} z_1 \overline{z_1} = 1,
$$

that is $|z_0 z_1| = 1$. It follows that two cases may occur:

(a) z_0 or z_1 has modulus strictly less than 1. Then p has exactly one solution in the closed upper half unit disk.

(b) All roots are of modulus 1. Let $z_0 = e^{i\theta_0}$ and $z_1 = e^{i\theta_1}$. Then,

$$
p(z) = (z^2 - 2z \cos \theta_0 + 1)(z^2 - 2z \cos \theta_1 + 1).
$$

Comparing the coefficients of z and z^3 we obtain:

$$
\cos \theta_0 + \cos \theta_1 = 0 \quad \text{and} \quad -2(\cos \theta_0 + \cos \theta_1) = 1,
$$

which cannot be. □

Chapter 7

Laurent Expansions, Residues, Singularities and Applications

Laurent expansions deal with functions analytic in an open ring

$$r_0 < |z - z_0| < r_1, \qquad (7.0.1)$$

($r_0 = 0$ and $r_1 = +\infty$ are allowed). The result (see Theorem 7.1.1 below) expresses f as the sum $f = f_+ + f_-$ of a part f_+ analytic in the disk $|z - z_0| < r_1$ and of a part f_- analytic in $r_0 < |z - z_0|$. When $r_1 = +\infty$, the function f_+ is entire, while the function $f_-(1/z)$ is entire when $r_0 = 0$. The case $r_0 = 0$ is of particular importance. The point z_0 is then called an *isolated singularity* of f. At this stage of a course on complex variables, the student already knows that a function analytic in a punctured open neighborhood of a point, say z_0, and continuous (or even, bounded) in that neighborhood is analytic in the whole neighborhood. This is called Riemann's removable singularity theorem (also known by its German name Riemann's Hebbarkeitssatz) and its proof follows from the proof of Cauchy's theorem. The point z_0 is then called a *removable (isolated) singularity*. When the function is analytic in a punctured neighborhood of a point z_0, but not assumed bounded there, two possibilities occur:

(a) We have

$$\lim_{z \to z_0} |f(z)| = +\infty. \qquad (7.0.2)$$

Then, and only then, z_0 is a pole.

(b) The limit (7.0.2) does not exist. Note the following: Since f is assumed unbounded near z_0, $|f(z)|$ will go to infinity via a subsequence, but via other subsequences the values of $f(z)$ will stay bounded, and the limit will not exist, or will be finite.

7.1   Laurent expansions

We recall the theorem on Laurent expansion for functions analytic in a ring. In the statement, we use the abuse of notation $|\zeta - z_0| = r$ to denote the path

$$\gamma(t) = z_0 + re^{it}, \quad t \in [0, 2\pi].$$

Theorem 7.1.1. *Let f be a function analytic in $r_0 < |z - z_0| < r_1$. Then, f can be written as the sum of two functions $f(z) = f_+(z) + f_-(z)$, where f_+ is analytic in $|z - z_0| < r_1$ and f_- is analytic in $r_0 < |z - z_0|$. We have*

$$f_+(z) = \sum_{n=0}^{\infty} a_n (z - z_0)^n,$$

where the coefficients a_n are given by

$$a_n = \frac{1}{2\pi i} \int_{|\zeta - z_0| = r} \frac{f(\zeta)}{(\zeta - z_0)^{n+1}} d\zeta, \quad n = 0, 1, \dots$$

with r any number in $(0, r_1)$, and

$$f_-(z) = \sum_{n=1}^{\infty} \frac{b_n}{(z - z_0)^n},$$

where the coefficients b_n are given by

$$b_n = \frac{1}{2\pi i} \int_{|\zeta - z_0| = r} f(\zeta)(\zeta - z_0)^{n-1} d\zeta, \quad n = 1, 2, \dots \tag{7.1.1}$$

with r any number in (r_0, ∞).

When $r_0 = 0$, one has the following classification of isolated singular points:

(a) If all the $b_n = 0$, the point z_0 is a *removable singularity*.

(b) If for some $N \in \mathbb{N}$, $b_N \neq 0$ and $b_{N+1} = b_{N+2} = \cdots = 0$, the point z_0 is a *pole* of order N.

(c) If $b_n \neq 0$ for an infinite number of indices, z_0 is an *essential singularity*.

Before turning to the exercises it is of interest to hint at a connection with another mathematical topic. When in Theorem 7.1.1 we have $z_0 = 0$ and $1 \in (r_0, r_1)$ the function f is analytic in a neighborhood of the unit circle. For $z = e^{it}$ the Laurent expansion becomes

$$f(e^{it}) = \sum_{n=0}^{\infty} a_n e^{int} + \sum_{n=1}^{\infty} \frac{b_n}{e^{int}},$$

and the reader will recognize a Fourier series. In particular, Parseval's identity leads to

$$\frac{1}{2\pi} \int_0^{2\pi} |f(e^{it})|^2 dt = |a_0|^2 + \sum_{n=1}^{\infty} (|a_n|^2 + |b_n|^2).$$

Exercise 7.1.2. *Find the Laurent expansions for the following functions in the indicated domains:*

(a) $\dfrac{1}{z(z-1)}$, $0 < |z-1| < 1$;

(b) $\dfrac{1}{(z^2+1)^2}$, $0 < |z-i| < 2$;

(c) $\dfrac{1}{z} \sin^2 \dfrac{2}{z}$, $0 < |z|$;

(d) $\dfrac{1-e^{-z}}{z^3}$, $0 < |z|$;

(e) $\dfrac{\sin z}{z-2}$, $|z-2| \neq 0$;

(f) $\dfrac{1}{z^2(z^2+1)}$, $0 < |z| < 1$.

Exercise 7.1.3. *Find the Laurent expansion of the function*

$$\frac{e^z}{z(z^2+1)}$$

in the domain $0 < |z| < 1$.

Exercise 7.1.4. *Represent the function*

$$\frac{z+1}{z-1}$$

(a) *as a Maclaurin series and find its convergence radius;*

(b) *as a Laurent series in the domain $\{z : |z| > 1\}$.*

Exercise 7.1.5. *Represent the function*

$$\frac{1}{z^2(z-1)}$$

in all possible series in powers of z centered at $z = 0$ and in powers of $z - 1$ centered at $z = 1$. Find the domains of convergence of these representations.

Exercise 7.1.6. *Check the formula*

$$\frac{1}{4z - z^2} = -\sum_{n=2}^{\infty} \frac{4^{n-2}}{z^n}, \qquad |z| > 4.$$

Exercise 7.1.7. *Build an analytic square root for the function $1+z^2$ in the domains $\{z : |z| < 1\}$ and $\{z : |z| > 1\}$, denoted in either case by $\sqrt{1+z^2}$, and give the Laurent expansions of*

$$\frac{\sqrt{1+z^2}}{z}$$

in $\{z : 0 < |z| < 1\}$ and $\{z : |z| > 1\}$.

The next exercise is taken from [53, Example 1, p. 77].

Exercise 7.1.8. *Show that*

$$\cosh\left(z + \frac{1}{z}\right) = a_0 + \sum_{n=1}^{\infty} a_n \left(z^n + \frac{1}{z^n}\right), \quad z \neq 0,$$

where

$$a_n = \frac{1}{2\pi} \int_0^{2\pi} \cos(nt)\cosh(2\cos t)dt, \quad n = 0,1,2,\ldots.$$

Exercise 7.1.9. *Prove the following theorem of Weierstrass: If f is analytic in $\mathbb{C} \setminus \{z_1,\ldots,z_N\}$, then there are $N+1$ entire functions $f_0,\ldots,f_N$ such that*

$$f(z) = f_0(z) + f_1(1/(z - z_1)) + \cdots + f_N(1/(z - z_N)). \tag{7.1.2}$$

Remarks 7.1.10. When f is rational the functions $f_0,\ldots,f_N$ are polynomials, and (7.1.2) is the partial fraction expansion of f. For an application of Exercise 7.1.9 to the realization theory of rational functions, see Exercise 12.1.2.

Question 7.1.11. *Let $f(z) = \int_0^1 \frac{e^t}{(t-z)^2} dt$.*

(1) *Let $t_0 \in [0,1]$. Explain why there is only one expansion of f centered at t_0.*
(2) *Let $z_0 \in \mathbb{C} \setminus [0,1]$. How may Laurent expansions centered at z_0 are there? Compute them.*

Exercise 7.1.12. *Prove that*

$$f(z) = \int_0^1 \frac{\cos t}{t - z} dt + \int_2^3 \frac{\sin t}{t - z} dt \tag{7.1.3}$$

admits a Laurent expansion in the ring $1 < |z| < 2$ and compute this expansion.

Exercise 7.1.13. *Let $z_0 \in \mathbb{C}$ and f as in (7.1.3).*

(a) *Compute the Laurent expansion of (7.1.3) in $|z - z_0| > D$ where*

$$D = \max_{t \in [0,1] \cup [2,3]} |t - z_0|.$$

(b) *Give a condition on z_0 for Laurent expansions centered at z_0 (possibly degenerating to a Taylor expansion) to exist. Compute these as well.*

Exercise 7.1.14. *Find the Laurent expansion of the function (4.4.23) and of its second derivative in $|z| > 1$.*

Exercise 7.1.15. *Let $p \in \mathbb{N}$. Show that the function $\dfrac{\sin z}{z^p}$ will have primitives in $\mathbb{C} \setminus \{0\}$ if and only if p is odd. Give the Laurent expansion centered at the origin of any of its primitives.*

Another interesting example of Laurent expansion is presented in Exercise 7.3.13.

7.2 Singularities

We first recall the following characterization of poles and zeros:

Theorem 7.2.1. *Let f be analytic in a punctured neighborhood of the complex number z_0. The following are equivalent:*

1. *z_0 is a pole of order N of f.*
2. *z_0 is a zero of order N of $1/f$.*
3. *The limit*
$$\lim_{z \to z_0} (z - z_0)^N f(z)$$
exists, is finite and different from 0.

Exercise 7.2.2. *Assume that z_0 is a zero (resp. a pole) of order N of the function f, and let $M \in \mathbb{N}$. Show that z_0 is a zero (resp. a pole) of order NM of f^M.*

In the next exercise we go back to Exercise 5.5.4.

Exercise 7.2.3. *Let Ω be a star-shaped open set and let C be a closed simple smooth curve in Ω. Let z_0 not belong to the image of C, and let f be analytic in Ω. Show that $z = z_0$ is a removable singularity of the function*

$$h(z) = \frac{f(z) - (z - z_0)f'(z)}{(z - z_0)^2}$$

and, using this fact, give another solution of (5.5.4).

Exercise 7.2.4. *Show that $z = 0$ is a removable singularity of the function*

$$f(z) = \frac{1}{\tan z} - \frac{1}{\sin z}.$$

Exercise 7.2.5. *Find the poles and zeros of the function*

$$f(z) = \frac{(\cos z - 1)^3 \sin(z^2) \sin(\pi z)}{(e^z - 1)(z^2 + 1)}.$$

Exercise 7.2.6. *What are the singularities of the functions*

$$f(z) = \frac{z(z-1)^2}{\sin^2(\pi z)},$$

$$g(z) = \frac{z^4(z-1)}{\sin^2(\pi z)},$$

$$h(z) = \frac{z^3(z-1)^6}{\sin^5(\pi z)}?$$

Hints. For f, $z = 0$ is a pole of order 1, $z = 1$ is a removable singularity (which is not a zero) and all other integers are poles of order 2.

For g, $z = 0$ is a removable singularity which is a zero of order 2, $z = 1$ is a pole of order 1, and all other integers are poles of order 2.

For h, $z = 0$ is pole of order 2, $z = 1$ is a removable singularity which is a zero of order 1, and all other integers are poles of order 5.

Exercise 7.2.7. *What is the point $z = 0$ for the function*

$$f(z) = \frac{\sin z}{z^4}?$$

Exercise 7.2.8. *Nature of $z = 0$ for the function $f(z) = \frac{\sin^3 z}{z}$.*

Exercise 7.2.9. *Show that the function*

$$\cos e^{(\frac{1}{z^2}+z^2)} + \left(\frac{8\sin(z^2)}{z^{50}}\right)^{100}$$

has an essential singularity at the origin.

Exercise 7.2.10. *Show that $z=0$ is an essential singularity of the function $\cos(e^{1/z})$.*

Exercise 7.2.11. *Let f and g be analytic in a punctured neighborhood of the point z_0 and assume that z_0 is an essential singularity of f and a pole of g. What kind of singularity is it for any of the functions $fg, f/g$ and $f + g$.*

Exercise 7.2.12. *Assume that $z = 0$ is an essential singularity of f. Show that it is also an essential singularity of f^2.*

Exercise 7.2.13. *Assume that $z = a$ is a pole of order N of the function f. Show that it is a pole of order $N + 1$ of the function f'.*

Related to Exercise 7.2.9 we have the following general fact:

Exercise 7.2.14. *Assume that f has a pole at the point z_0. Show that z_0 is an essential singularity of e^f.*

Exercise 7.2.15. *Let $f(z)$ be even and analytic in $\Omega = \mathbb{C}\backslash\{z = m + in\,;\, m, n \in \mathbb{Z}\}$, and assume that*

$$f(z) = f(z + m + ni) \tag{7.2.1}$$

for all $m, n \in \mathbb{Z}$ and all $z \in \Omega$. Assume that the only singular point of f modulo the lattice $\mathbb{Z} + i\mathbb{Z}$ is the origin, and that it is a pole of f of order 2. Show that there exist complex numbers g_0, g_1, g_2 and g_3 such that $g_0 \neq 0$ and

$$(f')^2 = g_3 f^3 + g_2 f^2 + g_1 f + g_0. \tag{7.2.2}$$

Hint. The function f has poles at all the points of the lattice

$$L = \{m + in\,;\, m, n \in \mathbb{Z}\}.$$

For any choice of a_0, a_1, a_2, a_3, the function

$$q(z) = (f')^2 - (a_3 f^3 + a_2 f^2 + a_1 f + a_0)$$

is biperiodic, with periods i and 1. Assume there is a choice of the a_i such that $z = 0$ is a removable singularity of q, and moreover is a zero of q. Then, all the points of the lattice L are zeros of q. The function q is therefore entire and biperiodic, and so is a constant (see Exercise 6.8.8).

Recall that ∞ is an isolated singular point of $f(z)$ if, by definition, 0 is an isolated singular point of $f(1/z)$.

Exercise 7.2.16. *Let f be an entire function and assume that ∞ is a pole of f. Find f.*

Exercise 7.2.17. *Show that the origin is a removable singularity of the function*

$$\frac{z}{e^z - 1}.$$

Show that there exist numbers $B_0, B_1, \ldots$ such that

$$\frac{z}{e^z - 1} = \sum_{n=0}^{\infty} \frac{B_n}{n!} z^n \tag{7.2.3}$$

for $|z| < 2\pi$. Show that $B_0 = 1$ and that the recursion

$$\binom{n+1}{0} B_0 + \binom{n+1}{1} B_1 + \cdots + \binom{n+1}{n} B_n = 0, \quad n \geq 1,$$

holds.
Prove that $B_{2k+1} = 0$ for $k \geq 1$.

The numbers $b_n = (-1)^{(n-1)} B_{2n}$, $n = 1, 2, \ldots$ are called the Bernoulli numbers. As topic of exercises, one can find them in numerous places, for instance in [75, p. 132]. Bernoulli numbers appear in various places, and particular in the expressions for the sums

$$\sum_{n=1}^{\infty} \frac{1}{n^{2p}} = \frac{\pi^{2p} b_p 2^{2p-1}}{(2p)!}. \tag{7.2.4}$$

Following [45, pp. 114–115], a proof of (7.2.4) is outlined in the next question.

Question 7.2.18.

(1) *Let $n \in \mathbb{N}$. Show that*

$$\mathrm{Res}\left(\frac{1}{z^{2n}(e^z - 1)}, 0\right) = \frac{B_{2n}}{(2n)!},$$

where $B_2, B_4, \ldots$ are defined as in (7.2.3), and compute

$$\mathrm{Res}\left(\frac{1}{z^{2n}(e^z - 1)}, 2i\pi p\right), \quad p \in \mathbb{Z} \setminus \{0\}.$$

(2) *Prove (7.2.4) by computing the integral of the function $\frac{1}{z^{2n}(e^z-1)}$ along the square with vertices $\pm(2k+1) \pm (2k+1)\pi i$, and letting $k \to \infty$.*

More generally than (7.2.3) one has:

Exercise 7.2.19. *Show that there are polynomials $B_n(t)$, $n = 0, 1, \ldots$ such that*

$$\frac{z e^{tz}}{e^z - 1} = \sum_{n=0}^{\infty} \frac{B_n(t)}{n!} z^n, \quad |z| < 2\pi. \tag{7.2.5}$$

Prove that, with B_ℓ as in the previous exercise,

$$B_n(t) = \sum_{\ell=0}^{n} B_\ell \binom{n}{\ell} t^{n-\ell}, \tag{7.2.6}$$

$$B_n(t+1) = B_n(t) + nt^{n-1}, \tag{7.2.7}$$

$$B_n'(t) = n B_{n-1}(t). \tag{7.2.8}$$

The functions B_n appear in quadratic approximation (one step Euler–McLaurin formula); for a discussion, see for instance [23, pp. 255–256].

For the formulas in the next two exercises, see for instance [201, p. 153].

Exercise 7.2.20. *Let f be analytic in the open unit disk, with power series $f(z) = \sum_{n=0}^{\infty} a_n z^n$. Show that the function*

$$C(z) = \begin{cases} \frac{1}{z} \int_{[0,z]} \frac{f(s)}{1-s} ds, & z \neq 0, \\ a_0, & z = 0, \end{cases}$$

is analytic in the open unit disk, and show that its power series expansion at the origin is equal to

$$C(z) = \sum_{n=0}^{\infty} \left(\frac{\sum_{j=0}^{n} a_j}{n+1} \right) z^n. \tag{7.2.9}$$

The map which to the sequence $(a_n)_{n \in \mathbb{N}_0}$ associates the sequence

$$\left(\frac{\sum_{j=0}^{n} a_j}{n+1} \right)_{n \in \mathbb{N}_0} \tag{7.2.10}$$

is called the Cesàro operator. See Question 16.1.9 for more information on this operator.

We conclude with the following result. We have already proved in fact a stronger result in Exercise 4.1.10. The strategy here is to argue by contradiction and to use the power expansion at the origin of a function f satisfying the claim of the exercise.

Exercise 7.2.21. *Show that there is no function analytic in $\mathbb{C} \setminus \{0\}$ such that $f(z)^2 = z$.*

We conclude this section with the series appearing in Exercise 3.3.2.

Exercise 7.2.22. *Show the formulas*

$$\frac{1}{z} + \sum_{n=1}^{\infty} \left(\frac{1}{z-n} + \frac{1}{n} \right) = \pi \cot(\pi z),$$

and

$$\sum_{n \in \mathbb{Z}} \frac{1}{(z-n)^2} = \frac{\pi^2}{\sin^2(\pi z)}.$$

7.3 Residues and the residue theorem

Suppose that, in Theorem 7.1.1, we have $r_0 = 0$. The coefficient b_1,

$$b_1 = \frac{1}{2\pi i} \int_{|\zeta - z_0| = r} f(\zeta) d\zeta, \tag{7.3.1}$$

is called the residue of f at the point z_0. We set

$$b_1 = \mathrm{Res}(f, z_0).$$

Various formulas are available to compute the residue without computing the Laurent expansion. We mention in particular the following:

Proposition 7.3.1. *Let f and g be analytic in a neighborhood of the point z_0 and assume that z_0 is a simple zero of g. Then,*

$$\operatorname{Res}\left(\frac{f}{g}, z_0\right) = \frac{f(z_0)}{g'(z_0)}. \tag{7.3.2}$$

More generally, assume that z_0 is a zero of order N of g, and write

$$g(z) = (z - z_0)^N c(z), \tag{7.3.3}$$

where c is analytic in a neighborhood of z_0 and does not vanish at z_0. Then formula (7.3.1) for the residue gives (where r is small enough)

$$\begin{aligned}
\operatorname{Res}\left(\frac{f}{g}, z_0\right) &= \frac{1}{2\pi i} \int_{|\zeta - z_0| = r} \frac{\dfrac{f(\zeta)}{c(\zeta)}}{(\zeta - z_0)^N} d\zeta \\
&= \frac{\left(\dfrac{f}{c}\right)^{(N-1)}(z_0)}{(N-1)!} \\
&= \lim_{z \to z_0} \frac{\left(\dfrac{f(z)(z - z_0)^N}{g(z)}\right)^{(N-1)}}{(N-1)!}
\end{aligned} \tag{7.3.4}$$

where we have used Cauchy's formula for the derivative. To express (7.3.4) directly in terms of g we remark that the relation (7.3.3) implies that

$$\frac{g^{(N+n)}(z_0)}{(N+n)!} = \frac{c^{(n)}(z_0)}{n!}, \quad n = 0, 1, \ldots \tag{7.3.5}$$

The case $N = 2$ gives the formula

$$\operatorname{Res}\left(\frac{f}{g}, z_0\right) = -\frac{f'(z_0)\dfrac{g^{(2)}(z_0)}{2} - f(z_0)\dfrac{g^{(3)}(z_0)}{3!}}{\left(\dfrac{g(z_0)}{2!}\right)^2}. \tag{7.3.6}$$

When $c(z) \equiv 1$ we then get from (7.3.4) the formula

$$\operatorname{Res}\left(\frac{f(z)}{(z - z_0)^N}, z_0\right) = \frac{f^{(N-1)}(z_0)}{(N-1)!}. \tag{7.3.7}$$

The following formula is a generalization of (5.3.3), and is used implicitly in the computations of the integrals $\int_{\mathbb{R}} \left(\frac{\sin x}{x}\right)^p dx$ for $p \in \mathbb{N}$; see Exercises 5.3.3, 5.3.4, and the discussions after the proofs of these exercises. In the case of a simple pole it can be found for instance in [45, Lemma 4, p. 105].

Question 7.3.2. *Let* 0 *be a pole of the function* h, *and assume that the principal part in the Laurent expansion of* h *at the origin has only odd powers. Let* c_ϵ *be the half-circle of radius* ϵ, *centered at the origin, and which lies in the upper half-plane, and with negative orientation. Then*

$$\lim_{\epsilon \longrightarrow 0} \int_{c_\epsilon} h(z)dz = -i\pi \operatorname{Res}(h, 0). \tag{7.3.8}$$

Exercise 7.3.3. *Compute the residue at the origin of*

$$\frac{e^z - 1}{\sin^2 z}.$$

Let f be analytic for $|z| > R$ for some $R > 0$. The residue at infinity is defined to be

$$\operatorname{Res}(f, \infty) = -\operatorname{Res}(\frac{1}{z^2} f(1/z), 0).$$

We note that

$$\operatorname{Res}(f, \infty) = -\frac{1}{2\pi i} \lim_{R \to \infty} \int_{|z|=R} f(z)dz. \tag{7.3.9}$$

Indeed, for any r small enough,

$$\operatorname{Res}(f, \infty) = -\operatorname{Res}(1/z^2 f(1/z), 0)$$

$$= -\frac{1}{2\pi} \int_0^{2\pi} f(e^{-it}/r)e^{-it}/rdt$$

$$= -\frac{1}{2\pi} \int_0^{2\pi} f(e^{it}/r)e^{it}/rdt$$

$$= -\lim_{R \to \infty} \frac{1}{2\pi} \int_0^{2\pi} f(Re^{it})Re^{it}dt$$

$$= -\frac{1}{2\pi i} \lim_{R \to \infty} \int_{|z|=R} f(z)dz.$$

Exercise 7.3.4.

(a) *Compute the residues of the function*

$$\frac{nz^{n-1}}{z^n - 1},$$

at its poles, including infinity.

(b) *Prove formula* (3.3.1)

$$\frac{nz^{n-1}}{z^n - 1} = \sum_{\ell=0}^{n-1} \frac{1}{z - z_\ell},$$

where $z_0, \ldots, z_{n-1}$ *are the roots of unity of order* n.

Exercise 7.3.5. *Let Γ be a simple closed contour, and f analytic in and on Γ, with the possible exception of a finite number of poles inside Γ. Show that*

$$\frac{1}{2\pi i}\int_\Gamma \frac{f'(z)}{f(z)}dz = Z - P, \qquad (7.3.10)$$

where Z (resp. P) denotes the number of zeros (resp. poles) of f inside Γ, counting multiplicity.

Hint. Apply the residue theorem to $\frac{f'}{f}$.

Remark 7.3.6. We can now express condition (5.7.2) in terms of zeros and poles when Γ is a smooth Jordan curve: The function f should have the same numbers of poles and zeros (counting multiplicity) inside Γ (of course, these points do not belong to the domain where the logarithm is looked for). For example consider four different complex numbers a, b, c, d, such that

$$[a,c]\cap[b,d] = [a,b]\cap[c,d] = \emptyset,$$

and let

$$F(z) = \frac{(z-a)(z-b)}{(z-c)(z-d)}.$$

Then F has an analytic logarithm in $\mathbb{C}\setminus\{[a,c]\cup[b,d]\}$ but not in $\mathbb{C}\setminus\{[a,b]\cup[c,d]\}$. A simple closed curve around $[a,c]$, and with $[c,d]$ in its exterior (resp. $[b,d]$, and with $[a,b]$ in its exterior) encloses one zero and one pole of F, while a simple closed curve around $[a,b]$ encloses two zeros and no poles of F. The same arguments allow easily to find (non simply connected) domains in which a rational function, that is (with obvious notations) a function of the form

$$\frac{\prod_{i=1}^n (z-a_i)^{n_i}}{\prod_{j=1}^m (z-c_j)^{m_j}}$$

has, or has not, an analytic logarithm.

We now give a small variation on Exercise 7.3.5.

Exercise 7.3.7. *Let Γ be a simple closed contour, and f analytic in and on Γ, with the possible exception of a finite number of poles inside Γ. Let g be analytic in and on Γ. Compute*

$$\frac{1}{2\pi i}\int_\Gamma \frac{g(z)f'(z)}{f(z)}dz. \qquad (7.3.11)$$

As a corollary of Exercise 7.3.5 we have the following important result (the converse statement is the content of Theorem 10.2.2 and of Exercise 10.2.1)

Exercise 7.3.8. *Let f analytic in a neighborhood of the point z_0, and assume that z_0 is a zero of order $M > 1$ of f. Show that there is no neighborhood of z_0 where f is one-to-one.*

For the partial fraction expansion appearing in the following exercise, see the discussion following Exercise 12.1.2.

Exercise 7.3.9.

(a) Let P be a polynomial of degree $n \geq 2$ and let $z_1, \ldots, z_k$ be the distinct roots of P. Let

$$\frac{1}{P(z)} = \frac{A_1}{z - z_1} + \cdots + \frac{A_k}{z - z_k} + \text{terms of the form } \frac{B}{(z - z_j)^{j_\ell}}, \quad (7.3.12)$$

with $j_\ell \geq 2$ being the partial fraction expansion of $1/P$. Show that

$$\sum_{\ell=1}^{k} A_\ell = 0. \quad (7.3.13)$$

(b) *Compute*

$$\int_{|z|=2} \frac{dz}{(z^{1000} + 1)(z - 3)}.$$

Hint for (a). Compute $\int_{|z|=R} \frac{dz}{P(z)}$ and let $R \to \infty$.

More generally we have:

Exercise 7.3.10. *Let f be a rational function. The sum of all the residues, included at infinity, is equal to 0.*

The fact that the sum of all the residues in the previous exercise is equal to 0 (or, as a particular case, (7.3.13)) is called the *exactity relation*; see [91, p. 173].

Exercise 7.3.11. *Compute*

$$\frac{1}{2\pi} \int_0^{2\pi} \frac{e^{it} + z}{e^{it} - z} e^{int} dt, \quad z \in \mathbb{D}, \ n \in \mathbb{Z}, \quad (7.3.14)$$

and deduce a proof of formula (5.5.6) for polynomials.

It seems difficult to compute the sum (7.3.15) below by direct methods, that is, using only real analysis.

Exercise 7.3.12. *Compute*

$$\sum_{n=0}^{\infty} \frac{\binom{2n}{n}}{7^n}. \quad (7.3.15)$$

Exercise 7.3.12 is taken from [176, Example 8.5, p. 195]. In a similar vein, we have (see [28, p. 144], [18, Exercise 8.38, p. 261])

$$\sum_{n=0}^{\infty} \frac{\binom{2n}{n}}{5^n} = \sqrt{5}. \tag{7.3.16}$$

More generally:

Exercise 7.3.13. *Show that the sum*

$$\sum_{n=0}^{\infty} \frac{\binom{2n}{n}}{\zeta^n}$$

is the Laurent expansion of a function analytic in $|\zeta| > 4$ *and compute the sum in closed form.*

The same method allows us to solve the next exercise. That exercise appears in [211, p. 328], and is solved there by a completely different method (using the power series expansion of the real function $x \mapsto \sin(\alpha \arcsin x)$).

Exercise 7.3.14. *Show that*

$$\sum_{n=0}^{\infty} \left(\frac{2}{27}\right)^n \binom{3n}{n} = \frac{\sqrt{3}+1}{2}.$$

We conclude with a question taken from [75, pp. 276–277], and which is conducive to the computation of sums of inverses of trigonometric functions. See Remark 1.3.9.

Question 7.3.15 (see [75, Exercice 30.04, p. 276, Exercice 30.05, p. 277]). *Let* $f(z)$ *be a rational function of* $\cos z$ *and* $\sin z$ *with no poles on the* x *and* y *axis, and going out* 0 *as* $\operatorname{Im} z \to \infty$. *Let* $z_1, \ldots, z_m$ *be the poles of* f *with real part in* $(0, 2\pi)$. *Show that for* $n \in \mathbb{N}$,

$$\sum_{k=0}^{n-1} f\left(\frac{2\pi k}{n}\right) = -\frac{n}{2} \sum_{u=1}^{m} \operatorname{Res}\left(f(z) \cot \frac{nz}{2}, z_u\right). \tag{7.3.17}$$

As an application, compute the sums

$$\sum_{k=1}^{n-1} \frac{1}{\sin^2 \frac{k\pi}{n}} \quad and \quad \sum_{k=0}^{n-1} \frac{1}{1 + \cos^2 \frac{k\pi}{n}}.$$

7.4 Rouché's theorem

Exercise 7.4.1. *Let γ be a simple closed curve. Using Exercise 7.3.5 prove Rouché's theorem: If f and g are analytic in a neighborhood of the interior of γ, and if $|f(z)| > |g(z)|$ on γ, then f and $f + g$ have the same number of zeros inside γ.*

Hint. Compute $\int_\gamma \left(\frac{f'}{f} - \frac{(f+g)'}{f+g} \right) dz$.

For a nice application of Rouché's theorem which is used in the proof of Riemann's mapping theorem, see Exercise 10.2.8. See also [45, Exercice 19, p. 116].

Exercise 7.4.2. *If f is analytic in $|z| \leq 1.2$ and if $|f(z)| < 1$ on $|z| = 1$, the equation $f(z) = z^n$ has exactly n solutions in $|z| < 1$.*

Exercise 7.4.3. *By using Rouché's theorem with $F(z) = z^4$ and $f(z) = z^3 + 1$, show that all the roots of $z^4 + z^3 + 1 = 0$ are of modulus less than $\frac{3}{2}$ (see [84, pp. 302–303]).*

The following exercise is also taken from Flanigan's book [84, p. 303]. It consists in proving, for analytic functions, a very important theorem of topology, *Brouwer's theorem*, which is in fact true for continuous functions.

Exercise 7.4.4. *Let F be analytic in $|z| < 1.2$ and map the closed unit disk into the open unit disk. Show that F has a fixed point, i.e., there is z in the open unit disk such that $F(z) = z$.*

Hint. Apply Rouché's theorem with $f(z) = -2z$ and $g(z) = F(z) + z$.

Exercise 7.4.5. *How many roots has the equation*

$$z^4 - 3z + 1 = 0$$

in the open unit disk?

Exercise 7.4.6. *How many roots has the equation*

$$z^4 + z^3 - 4z + 1 = 0$$

in the ring $1 < |z| < 3$?

Exercise 7.4.7. *Solve Exercise 6.8.13 using Rouché's theorem.*

Exercise 7.4.8. *Prove that for real λ strictly greater than 1, there is a unique solution to the equation*

$$ze^{\lambda-z} = 1$$

in the open unit disk.

We conclude with a result which allows to prove the open mapping theorem (see Theorem 10.2.6 for a statement of the latter).

Exercise 7.4.9. *Let Ω be an open subset of $\mathbb{C}$, and let f be analytic in Ω. Let $z_0 \in \Omega$ be a zero of order N of f.*

(1) *Show that there exists $r_0 > 0$ with the following property: For every $r \in (0, r_0)$ there exists $\epsilon > 0$ such that, for every $a \in B(f(z_0), \epsilon)$, the equation $f(z) = a$ has exactly N solutions.*

(2) *Show that $f(B(z_0, r))$ contains $B(f(z_0), \epsilon)$.*

7.5   Solutions

Solution of Exercise 7.1.2. We consider only (a) and (e).

(a) We have

$$\frac{1}{z(z-1)} = \frac{1}{(1+z-1)(z-1)}$$

$$= \frac{1}{z-1} \cdot \sum_{n=0}^{\infty} (-1)^n (z-1)^n,$$

and hence the result.

(e) We write

$$\frac{\sin z}{z-2} = \frac{\sin(z-2+2)}{z-2}$$

$$= \frac{\sin(z-2)\cos 2 + \cos(z-2)\sin 2}{z-2}$$

$$= \frac{\sum_{p=0}^{\infty} \frac{(-1)^p (z-2)^{2p+1}}{(2p+1)!} \cos 2 + \sum_{p=0}^{\infty} \frac{(-1)^p (z-2)^{2p}}{(2p)!} \sin 2}{z-2}$$

and the rest is smooth sailing.   □

Solution of Exercise 7.1.3. The function $\frac{e^z}{z(z^2+1)}$ is analytic in $\mathbb{C} \setminus \{0, i, -i\}$. Thus the Laurent expansion at the origin converges in $0 < |z| < 1$. The point $z = 0$ is a simple pole, and so there is only one term corresponding to a negative power of z in the Laurent expansion. The terms of the expansion can be computed as follows. For $|z| < 1$ we have

$$\frac{e^z}{1+z^2} = \left(\sum_{n=0}^{\infty} \frac{z^n}{n!} \right) \left(\sum_{p=0}^{\infty} (-1)^p z^{2p} \right)$$

$$= \sum_{j=0}^{\infty} z^j \left(\sum_{\substack{p \in \mathbb{N}_0, \\ 2p \le j}} \frac{(-1)^p}{(j-2p)!} \right),$$

where we have used (4.4.14), and so, for $0 < |z| < 1$,

$$\frac{e^z}{z(1+z^2)} = \sum_{j=0}^{\infty} z^{j-1} \left(\sum_{\substack{p \in \mathbb{N}_0 \\ 2p \leq j}} \frac{(-1)^p}{(j-2p)!} \right). \qquad \square$$

Solution of Exercise 7.1.4. The MacLaurin series has radius of convergence equal to 1 since $z = 1$ is a pole of the function $\frac{z+1}{z-1}$. Moreover,

$$\frac{z+1}{z-1} = 1 - \frac{2}{1-z} = 1 - 2 \sum_{n=0}^{\infty} z^n$$

and hence the result.

In the domain $|z| > 1$ we have

$$\frac{z+1}{z-1} = 1 + \frac{2}{z-1}$$
$$= 1 + \frac{1}{z} \frac{2}{1 - \frac{1}{z}}$$
$$= 1 + \frac{1}{z} \sum_{n=0}^{\infty} \frac{1}{z^n},$$

and hence the result. $\square$

Solution of Exercise 7.1.5. There are four possible cases:

$$|z| < 1, \quad |z| > 1, \quad |z-1| < 1, \quad \text{and} \quad |z-1| > 1.$$

The first and third cases are Taylor expansions. The other two cases are Laurent expansions. We will only compute the Laurent expansion for the last case. We have

$$\frac{1}{z^2(z-1)} = \frac{1}{(z-1+1)^2(z-1)}$$
$$= \frac{1}{(z-1)^3} \frac{1}{\left(1 + \frac{1}{(z-1)}\right)^2}$$
$$= \frac{1}{(z-1)^3} \sum_{n=0}^{\infty} \frac{(-1)^n (n+1)}{(z-1)^{2n}}$$
$$= \sum_{n=0}^{\infty} \frac{(-1)^n (n+1)}{(z-1)^{2n+3}}. \qquad \square$$

Solution of Exercise 7.1.6. It suffices to write

$$\frac{1}{4z - z^2} = -\frac{1}{z^2(1 - \frac{4}{z})} = -\frac{1}{z^2}\left(\sum_{\ell=0}^{\infty}\frac{4^\ell}{z^\ell}\right) = -\sum_{\ell=0}^{\infty}\frac{4^\ell}{z^{\ell+2}}. \qquad \square$$

Solution of Exercise 7.1.7. The function f_α defined by (4.4.9) with $\alpha = 1/2$ is analytic in the open unit disk. In view of (4.4.10) with $\alpha = \beta = 1/2$, it satisfies

$$(f_{1/2}(z))^2 = 1 + z, \quad z \in \mathbb{D},$$

and thus

$$(f_{1/2}(z^2))^2 = 1 + z^2, \quad z \in \mathbb{D}.$$

From the power series expansion of $f_{1/2}$ (see (4.5.7)), the Laurent expansion of the function $(f_{1/2}(z^2))/z$ is equal to

$$\frac{f_{1/2}(z^2)}{z} = \frac{1 + \frac{1}{2}z^2 - \frac{1}{2\cdot4}z^4 + \frac{1\cdot3}{2\cdot4\cdot6}z^6 + \cdots}{z}$$

$$= \frac{1}{z} + \frac{1}{2}z - \frac{1}{2\cdot4}z^3 + \frac{1\cdot3}{2\cdot4\cdot6}z^5 + \cdots, \quad z \in \mathbb{D}\setminus\{0\}.$$

We now turn to the domain $|z| > 1$. Writing

$$1 + z^2 = z^2\left(1 + \frac{1}{z^2}\right),$$

we see that the function $g(z) = zf_{1/2}(1/z^2)$ is analytic in $|z| > 1$ and satisfies $g(z)^2 = 1 + z^2$ there. Using again (4.4.10) we get the Laurent expansion

$$zf_{1/2}(1/z^2) = z + \frac{1}{2z} - \frac{1}{2\cdot4z^3} + \frac{1\cdot3}{2\cdot4\cdot6z^5} + \cdots, \quad |z| > 1. \qquad \square$$

Solution of Exercise 7.1.8. Since the function $f(z) = \cosh\left(z + \frac{1}{z}\right)$ is such that $f(z) = f(1/z)$, the Laurent expansion at the origin is symmetric:

$$a_n = b_{n-1}, \quad n = 1, 2, \ldots$$

and so

$$\cosh\left(z + \frac{1}{z}\right) = a_0 + \sum_{n=1}^{\infty} a_n\left(z^n + \frac{1}{z^n}\right),$$

where

$$a_n = \frac{1}{2\pi i}\int_{|z|=1}\frac{\cosh\left(z + \frac{1}{z}\right)}{z^{n+1}}dz, \quad n = 0, 1, 2, \ldots.$$

But

$$a_n = \frac{1}{2\pi}\int_0^{2\pi}\cosh(e^{it} + e^{-it})e^{-int}dt$$

$$= \frac{1}{2\pi}\int_0^{2\pi}\cosh(2\cos t)\cos(nt)dt$$

since, by the change of variable $t \mapsto t - \pi$,

$$\frac{1}{2\pi} \int_0^{2\pi} \cosh(2 \cos t) \sin(nt) dt = (-1)^n \frac{1}{2\pi} \int_{-\pi}^{\pi} \cosh(2 \cos t) \sin(nt) dt = 0.$$

The last equality follows from the fact that the function $t \mapsto \cosh(2 \cos t) \sin(nt)$ is odd. ☐

It is easy to construct similar exercises for the Laurent expansion of $\exp(z + 1/z)$, or generally, of $g(z + 1/z)$ for appropriate functions g.

Solution of Exercise 7.1.9. Consider the Laurent expansion at one of the singular points, say z_1. There is $R_1 > 0$ such that

$$f(z) = \sum_{\ell=0}^{\infty} a_\ell (z - z_1)^\ell + \sum_{\ell=1}^{\infty} \frac{b_\ell}{(z - z_1)^\ell}, \qquad |z - z_1| < R_1.$$

The series $\sum_{\ell=1}^{\infty} \frac{b_\ell}{(z - z_1)^\ell}$ converges for all $z \neq z_1$. This last fact implies that the function

$$f_1(z) = \sum_{\ell=1}^{\infty} b_\ell z^\ell$$

is entire. The function

$$F_1(z) = f(z) - f_1(1/(z - z_1))$$

has a removable singularity at z_1. If $N = 1$, the function F_1 is entire and the result is proved. Assume that $N > 1$. We reiterate the argument just done for f and z_1 with now F_1 and one of the remaining singularities, say z_2, to obtain an entire function f_2 such that z_2 is a removable singularity of the function

$$F_2(z) = F_1(z) - f_2(1/(z - z_2)) = f(z) - \{f_1(1/(z - z_1)) + f_2(1/(z - z_2))\}.$$

Reiterating this argument a finite number of times, we obtain the result. ☐

For a discussion of the above result, and much more information, see [139, § 7].

Solution of Exercise 7.1.12. The function $\int_0^1 \frac{\cos t}{t - z} dt$ is analytic in $\mathbb{C} \setminus [0, 1]$ while the function $\int_2^3 \frac{\sin t}{t - z} dt$ is analytic in $\mathbb{C} \setminus [2, 3]$. So f admits a Laurent expansion in the asserted ring. Let $1 < |z| < 2$. Then, for every $t \in [0, 1]$ we have that $|t/z| < 1$ while $|z/t| < 1$ for every $t \in [2, 3]$. Thus

$$f(z) = \int_0^1 \frac{\cos t}{t - z} dt + \int_2^3 \frac{\sin t}{t - z} dt$$

$$= -\frac{1}{z} \int_0^1 \frac{\cos t}{1 - t/z} dt + \int_2^3 \frac{\sin t}{t(1 - z/t)} dt$$

$$= -\sum_{n=0}^{\infty} \frac{\int_0^1 t^n \cos t \, dt}{z^{n+1}} + \sum_{n=0}^{\infty} z^n \left(\int_2^3 t^{n-1} \sin t \, dt \right),$$

where we have used Weierstrass' theorem (Theorem 14.4.1) to interchange sums and integrals.   □

Solution of Exercise 7.1.13.

(a) We write

$$\frac{1}{t-z} = \frac{1}{t-z_0-(z-z_0)} = -\frac{1}{(z-z_0)\left(1 - \dfrac{t-z_0}{z-z_0}\right)} = -\sum_{u=0}^{\infty} \frac{(t-z_0)^u}{(z-z_0)^{u+1}}$$

for z such that $|z - z_0| > D$. Plugging this expression in (7.1.3) and using Weierstrass' theorem one gets the required expansion.

(b) Let

$$d_1 = \min_{t\in[0,1]} |z_0 - t| \quad \text{and} \quad D_1 = \max_{t\in[2,3]} |z_0 - t|,$$

and

$$d_2 = \min_{t\in[2,3]} |z_0 - t| \quad \text{and} \quad D_2 = \max_{t\in[0,1]} |z_0 - t|.$$

Note that d_1 or d_2 may be equal to 0 (but not simultaneously). The function $\int_0^1 \frac{\cos t}{t-z}\,dt$ has a Taylor expansion in $B(z_0, d_1)$ when $d_1 > 0$ and a Laurent expansion in $|z - z_0| > D_1$. Similarly, the function $\int_2^3 \frac{\sin t}{t-z}\,dt$ has a Taylor expansion in $B(z_0, d_2)$ when $d_2 > 0$ and a Laurent expansion in $|z - z_0| > D_2$. The number of Laurent expansions centered at z_0 depends on the respective positions of $[d_1, D_1]$ and $[d_2, D_2]$. A number of cases may occur (in the list below we do not mention symmetric cases, where the role of the indices 1 and 2 is interchanged):

(a) $d_1 = 0 < D_1 < d_2 < D_2$. Then, there are Laurent expansions in $D_1 < |z| < d_2$ and $|z| > D_2$.

(b) $d_1 = 0 < d_2 \le D_1 < D_2$. Then, there is a Laurent expansion in $|z| > D_2$.

(c) $d_1 = 0 < d_2 \le D_2 < D_1$. Then, there is a Laurent expansion in $|z| > D_1$.

(d) $0 < d_1 < D_1 < d_2 < D_2$ Then, there is Taylor expansion in $B(0, d_1)$ and Laurent expansions in $D_1 < |z| < d_2$ and $|z| > D_2$.

(e) $0 < d_1 < d_2 \le D_1 < D_2$. Then, there is a Taylor expansion in $B(0, d_1)$ and a Laurent expansion in $|z| > D_2$.

(f) $0 < d_1 < d_2 \le D_2 < D_1$. Then, there is a Taylor expansion in $B(0, d_1)$ and a Laurent expansion in $|z| > D_1$.

We will leave to the reader the computations of these various expansions.   □

Solution of Exercise 7.1.14. For $|z| > 1$ we have that

$$|t/z| < 1 \quad \text{for} \quad t \in [0,1],$$

and therefore,

$$F(z) = \int_0^1 \frac{m(t)dt}{z^2(1 - t/z)^2}$$

$$= \frac{1}{z^2} \int_0^1 m(t) \left(\sum_{\ell=0}^{\infty} (\ell + 1)(t/z)^{\ell} \right) dt$$

$$= \sum_{\ell=0}^{\infty} \frac{F_{\ell}}{z^{\ell+2}},$$

where

$$F_{\ell} = (\ell + 1) \int_0^1 t^{\ell} m(t)dt, \quad \ell = 0, 1, \ldots. \qquad \square$$

Solution of Exercise 7.1.15. The function has a primitive if and only if

$$\int_{\mathbb{T}} \frac{\sin z}{z^p} dz = 0.$$

By Cauchy's formula, we have

$$\int_{\mathbb{T}} \frac{\sin z}{z^p} dz = 2\pi i \frac{\sin^{(p-1)}(z)}{(p-1)!}\Big|_{z=0}.$$

This last number is equal to 0 if and only if $p-1$ is even, that is, if and only if p is odd.

Writing

$$\frac{\sin z}{z^p} = \sum_{\ell=0}^{\infty} \frac{(-1)^{\ell} z^{2\ell+1-p}}{(2\ell+1)!},$$

we see that the primitives of $\frac{\sin z}{z^p}$ in $\mathbb{C} \setminus \{0\}$ can be written as

$$F(z) = K + \sum_{\ell=0}^{\infty} \frac{(-1)^{\ell} z^{2\ell-p+2}}{(2\ell+1)!(2\ell-p+2)}, \quad K \in \mathbb{C}. \qquad \square$$

Solution of Exercise 7.2.2. Assume that z_0 is a zero of order N of f. Then, in a neighborhood V of z_0 we have

$$f(z) = (z - z_0)^N h(z),$$

where h is analytic in V and $h(z_0) \neq 0$. Thus

$$f^M(z) = (z - z_0)^{NM} h^M(z), \quad z \in V. \tag{7.5.1}$$

Since h^M is analytic in V and $h^M(z_0) \neq 0$, equation (7.5.1) expresses exactly that z_0 is a zero of order NM of f^M.

The case of a pole is treated by considering the function $1/f$. $\qquad \square$

Solution of Exercise 7.2.3. We assume that f is not identically equal to 0 in Ω (the case $f(z) \equiv 0$ is trivial). Let $g(z) = f(z) - (z - z_0)f'(z)$. We have $g(z_0) = 0$. Moreover,

$$g'(z) = f'(z) - f'(z) - (z - z_0)f''(z) = -(z - z_0)f''(z).$$

Hence, $g'(z_0) = 0$, and g has a zero of order at least 2 at $z = z_0$. Thus, $h(z) = g(z)/(z - z_0)^2$ has a removable singularity at $z = z_0$, and has an analytic extension to all of Ω. We still call h this extension. By Cauchy's theorem,

$$\int_\gamma h(z)dz = 0$$

for every closed path γ in Ω. When the image of γ does not contain z_0, this last integral can be divided into two integrals to obtain

$$\int_\gamma \frac{f(z)}{(z - z_0)^2}dz - \int_\gamma \frac{(z - z_0)f'(z)}{(z - z_0)^2}dz = 0,$$

which is exactly (5.5.4). □

Solution of Exercise 7.2.4. We have

$$f(z) = \frac{z}{\sin z}\left(\frac{\cos z - 1}{z}\right).$$

The point $z = 0$ is a removable singularity of $z/\sin z$ since

$$\lim_{z \to 0} \frac{\sin z}{z} = 1.$$

(To check this, note that the limit is equal to $\sin'(0)$.) It is also a removable singularity of

$$\frac{\cos z - 1}{z} = \frac{z}{2} - \frac{z^3}{4!} + \cdots, \quad z \neq 0.$$

This last equation also shows that the origin is a first-order zero of $(\cos z - 1)/z$. Thus $z = 0$ is a removable singularity of f, and moreover is a zero of order 1 of this function. □

Solution of Exercise 7.2.5. The various factors composing f vanish at

(i) $z = 2\pi k$, $k \in \mathbb{Z}$ (for $(\cos z - 1)^3$),

(ii) $z = \pm\sqrt{k\pi}$ and $z = \pm i\sqrt{k\pi}$, $k = 0, 1, 2, \ldots$ (for $\sin(z^2)$),

(iii) $z = k$, $k \in \mathbb{Z}$ (for $\sin(\pi z)$),

(iv) $z = 2i\pi k$, $k \in \mathbb{Z}$ (for $(e^z - 1)$),

(v) $z = \pm i$ (for $(z^2 + 1)$).

The only difficulty is really $z = 0$, which appears in (i)–(iv). The origin is a zero of order 2 of the function $\cos z - 1$, a zero of order 2 of the function $\sin z^2$ and a simple zero of the functions $\sin(\pi z)$ and $e^z - 1$. Thus we have

$$(\cos z - 1)^3 = z^6 g_1(z), \quad \sin(z^2) = z^2 g_2(z), \quad \sin(\pi z) = z g_3(z), \quad e^z - 1 = z g_4(z),$$

where the g_j denote functions analytic in a neighborhood of the origin (in fact they are entire functions) not vanishing at the origin. Thus we can write

$$f(z) = \frac{z^6 g_1(z) z^2 g_2(z) z g_3(z)}{z g_4(z)(z^2 + 1)} = z^8 g(z), \tag{7.5.2}$$

where g is analytic in a neighborhood of the origin and not vanishing there. Thus, $z = 0$ is a removable singularity of f, which moreover is a zero of order 8.

For $k = 1, 2, \ldots$, the derivative of $\sin(z^2)$ does not vanish at $z = \pm\sqrt{k\pi}$ or $z = \pm i\sqrt{k\pi}$, and thus the corresponding point is a simple zero of f. For $k = \pm 1, \pm 2, \ldots$, $z = k$ is a simple zero of f and $z = 2i\pi k$ is a simple pole of f. Finally $z = \pm i$ are simple poles. We prove only this last assertion. We have

$$f(z) = \frac{h(z)}{z - i},$$

where

$$h(z) = \frac{(\cos z - 1)^3 \sin(z^2) \sin(\pi z)}{(e^z - 1)(z + i)}$$

is analytic in a neighborhood of i and does not vanish there. So $z = i$ is a simple pole. □

Remark 7.5.1. The function g appearing in (7.5.2) is in fact analytic in $|z| < 1$. Explain why.

Solution of Exercise 7.2.6. We will consider only the function $h(z)$. Its (possibly removable) singularities are at the points where $\sin \pi z$ vanishes, that is for $z \in \mathbb{Z}$. The integers are therefore either poles or removable singularities (these last ones may turn out to be zeroes of h). The numerator in the expression for h,

$$h(z) = \frac{z^3(z - 1)^6}{\sin^5(\pi z)},$$

vanishes at $z = 0$ and $z = 1$. We therefore distinguish three cases:

(a) $z = n$ with $n \notin \{0, 1\}$. Then z is a simple zero of $\sin \pi z$ and therefore (see Exercise 7.2.2), it is a zero of order 5 of $\sin^5(\pi z)$. Since the numerator of h does not vanish at these points, they are poles of order 5 of h.

(b) $z = 0$. We have

$$\lim_{z \to 0} z^2 h(z) = \lim_{z \to 0} \left(\frac{z}{\sin \pi z}\right)^5 = \frac{1}{\pi^5} \neq 0,$$

and thus $z = 0$ is a pole of order 2 of h.

(c) $z = 1$. We have

$$\lim_{z \to 1} \frac{\sin \pi z}{z - 1} = \pi \cos \pi z \big|_{z=1} = -\pi.$$

Therefore,

$$\lim_{z \to 1} \frac{z - 1}{h(z)} = \lim_{z \to 1} \frac{(z-1)^5}{\sin^5(\pi z)} = \frac{1}{(-\pi)^5} \neq 0.$$

Thus $z = 1$ is a pole of order 1 of $1/h$, and a removable singularity, which moreover is a zero of order 1, of h. $\qquad \square$

Solution of Exercise 7.2.7. The function f is analytic in $\mathbb{C} \setminus \{0\}$, and therefore has a Laurent expansion around $z = 0$ in that set. Using the expansion for $\sin z$ we have

$$f(z) = \frac{z - \frac{z^3}{3!} + \frac{z^5}{5!} - \cdots}{z^4}$$

$$= \frac{1}{z^3} - \frac{1}{3!} \cdot \frac{1}{z} + \frac{1}{5!} \cdot z + \cdots . \qquad (7.5.3)$$

Hence, $z = 0$ is a pole of order 3. $\qquad \square$

Note. The function $g(z) = z^3 f(z) = \frac{\sin z}{z}$ has a removable singularity at 0 and $g(0) = 1 \neq 0$.

Solution of Exercise 7.2.8. Since

$$\lim_{z \to 0} \frac{z^2}{f(z)} = \lim_{z \to 0} \left(\frac{z}{\sin z} \right)^3 = 1,$$

the origin is a pole of order 2 of $1/f$ and hence a zero of order 2 of f.

A different and longer proof would go as follows: The function f is analytic in $\mathbb{C} \setminus \{0\}$, and therefore has a Laurent expansion around $z = 0$ in that set. Using the formula

$$\sin^3 z = \frac{3}{4} \sin z - \frac{1}{4} \sin 3z$$

and the power series expansion of $\sin z$ we obtain

$$f(z) = \frac{\frac{3}{4} \left(z - \frac{z^3}{3!} + \frac{z^5}{5!} - \cdots z^4 \right) - \frac{1}{4} \left(3z - \frac{(3z)^3}{3!} + \frac{(3z)^5}{5!} - \cdots \right)}{z}$$

$$= \frac{z^3 + z^5 \frac{1}{4} \frac{1}{5!} (3 - 3^5) + \cdots}{z}$$

$$= z^2 + z^4 \frac{1}{4} \frac{1}{5!} (3 - 3^5) + \cdots . \qquad (7.5.4)$$

Hence $z = 0$ is a removable singularity for f, which has an analytic extension to all of $\mathbb{C}$. By abuse of notation we still denote by f this extension. Furthermore (7.5.4) expresses that $z = 0$ is a zero of order 2. $\qquad \square$

Solution of Exercise 7.2.9. The origin is a pole of the function

$$\left(\frac{8\sin(z^2)}{z^{50}}\right)^{100} = \left(8^{100}\left(\frac{\sin(z^2)}{z^2}\right)^{100}\right)(z^{-48})^{100}$$

since it is a removable singularity of the function $\frac{\sin(z^2)}{z^2}$. It is thus enough to prove that it is an essential singularity of

$$h(z) = \cos e^{\left(\frac{1}{z^2}+z^2\right)}.$$

Let z be such that

$$z^2 + \frac{1}{z^2} = \ln(\pi n), \quad \text{i.e.,} \quad z^4 - z^2(\ln \pi n) + 1 = 0.$$

Thus

$$z^2 = \frac{(\ln \pi n) \pm \sqrt{(\ln \pi n)^2 - 4}}{2}.$$

The choice

$$z^2 = \frac{(\ln \pi n) - \sqrt{(\ln \pi n)^2 - 4}}{2} = \frac{2}{\ln(\pi n) + \sqrt{(\ln(\pi n))^2 - 4}}$$

leads to the sequence

$$z_n = \frac{\sqrt{2}}{\sqrt{\ln(\pi n) + \sqrt{(\ln(\pi n))^2 - 4}}},$$

which goes to 0 as n goes to infinity. Since $h(z_n) = (-1)^n$, the limit does not exist and $z = 0$ is an essential singularity. $\qquad\Box$

Solution of Exercise 7.2.10. We will show that the function has no limit as $z \to 0$. Let

$$z_n = \frac{1}{\ln(n\pi)}, \quad n = 1, 2, \ldots.$$

Then $e^{1/z_n} = n\pi$ and $\sin(e^{1/z_n}) = (-1)^n$. This shows that the limit does not exist, and so $z = 0$ is an essential singularity. $\qquad\Box$

Such examples are classical; see for instance [75, p. 187].

Solution of Exercise 7.2.11. We first note that, in a punctured neighborhood of z_0, we have

$$g(z) = \frac{h(z)}{(z - z_0)^N}, \tag{7.5.5}$$

where $N \in \mathbb{N}$ and h is analytic in a neighborhood of z_0, and such that $h(z_0) \neq 0$. We will now show that, in the three cases, the point z_0 is an essential singularity.

The proof goes by contradiction. Consider first the function fg and assume that z_0 is not an essential singularity of fg. Then, in a neighborhood of z_0, we have, for $z \neq z_0$,

$$f(z)g(z) = H(z)(z - z_0)^M,$$

where $M \in \mathbb{Z}$ and H is analytic in a neighborhood of z_0, and such that $H(z_0) \neq 0$. Taking into account (7.5.5) we have

$$f(z) = \frac{H(z)}{h(z)}(z - z_0)^{M+N},$$

in a punctured neighborhood of z_0. Since H/h is an analytic neighborhood of z_0, the above expression contradicts the fact that z_0 is an essential singularity of f.

The case of f/g is treated in the same way. We now focus on $f + g$. Assuming that z_0 is not an essential singularity of $f + g$ we have, with the same notation as above,

$$f(z) + \frac{h(z)}{(z - z_0)^N} = H(z)(z - z_0)^M$$

for z in a punctured neighborhood of z_0. It follows that

$$f(z) = -\frac{h(z)}{(z - z_0)^N} + H(z)(z - z_0)^M,$$

and z_0 will be, depending on M and N, either a pole or a removable singularity of f, which cannot be by assumption.   □

Solution of Exercise 7.2.12. Assume that 0 is a removable singularity of f^2. Then, $|f|^2$ would be bounded in a punctured neighborhood of 0, and so $|f|$ would be bounded in a punctured neighborhood of 0, and thus 0 would be a removable singularity of f. Assume now that 0 is a pole of f^2. Then $\lim_{z \to 0} |f|^2(z) = +\infty$, and thus also $\lim_{z \to 0} |f|(z) = +\infty$, that is $z = 0$ would be a pole of f.

Another phrasing is as follows: Assume by contradiction that 0 is not an essential singularity of f. Then there exist an integer $n \in \mathbb{Z}$ and a function g analytic in a neighborhood of the origin such that $g(0) \neq 0$ such that

$$f^2(z) = z^n g(z).$$

If $n \geq 0$ it follows that $|f|^2$, and hence $|f|$, is bounded near the origin, and thus $z = 0$ is a removable singularity of f. If $n < 0$, the same argument applies to $z^{-2n} f(z)^2 = z^{-n} g$. Thus $z = 0$ would be a removable singularity of $z^{-n} f$, and thus $z = 0$ would then be a pole of f.   □

Solution of Exercise 7.2.13. By definition of a pole of order N, we can write in a punctured neighborhood of $z = a$,

$$f(z) = \frac{h(z)}{(z - a)^N}, \tag{7.5.6}$$

where h is analytic in a neighborhood of $z = a$, and is such that $h(a) \neq 0$. Differentiating both sides of (7.5.6), we obtain

$$f'(z) = \frac{(z-a)h'(z) - Nh(z)}{(z-a)^{N+1}}.$$

This expresses the fact that $z = a$ is pole of order $N + 1$ of f', since the function

$$g(z) = (z-a)h'(z) - Nh(z)$$

is analytic in a neighborhood of $z = a$, and is such that $g(a) = -Nh(a) \neq 0$. □

Solution of Exercise 7.2.14. Without loss of generality we will assume that $z_0 = 0$. There is $M \in \mathbb{N}$ and g analytic in some neighborhood V of 0 such that $g(0) \neq 0$ and is finite, and

$$f(z) = \frac{g(z)}{z^M}.$$

We note that M is uniquely defined by the condition $g(0) \neq 0$ and finite. Furthermore,

$$f'(z) = \frac{zg'(z) - Mg(z)}{z^{M+1}}, \quad z \in V \setminus \{0\}.$$

Assume now by contradiction that 0 is not an essential singularity of the function e^f. Then, there exist a unique $N \in \mathbb{Z}$ and h analytic in some neighborhood of 0, which may be assumed equal to V, such that $h(0)$ is finite and $h(0) \neq 0$ and

$$e^{f(z)} = z^N h(z), \quad z \in V. \tag{7.5.7}$$

Differentiating both sides of this equation we get

$$f'(z)e^{f(z)} = z^N h'(z) + Nz^{N-1}h(z),$$

and hence

$$e^{f(z)} = z^{M+N} \frac{zh'(z) + Nh(z)}{-Mg(z) + zg'(z)}$$

in a possibly smaller neighborhood of 0. The function

$$w(z) = \frac{zh'(z) + Nh(z)}{-Mg(z) + zg'(z)}$$

is analytic in a neighborhood of the origin, and

$$w(0) = -\frac{Nh(0)}{Mg(0)} \neq 0.$$

This contradicts the uniqueness of the power of z in (7.5.7). □

Remark 7.5.2. In a possibly quicker way one can use the logarithmic derivative (see formula $(4.2.3)$) to get from $(7.5.7)$

$$\frac{f'(z)}{f(z)} = \frac{h'(z)}{h(z)} + \frac{N}{z},$$

and so

$$\frac{g'(z)}{z^M} - \frac{Mg(z)}{z^{M+1}} = \frac{h'(z)}{h(z)} + \frac{N}{z},$$

leading to a contradiction since $M \geq 1$.

Solution of Exercise 7.2.15. Since f is odd there are only odd powers in the Laurent expansion at the origin, and we have

$$f(z) = \frac{\alpha}{z^2} + \sum_{n=0}^{\infty} \alpha_n z^{2n} = \frac{\alpha}{z^2}\left(1 + \sum_{n=0}^{\infty} \frac{\alpha_n}{\alpha} z^{2n+2}\right),$$

where $\alpha \neq 0$. The above expression is valid for $0 < |z| < 1$. Differentiating both sides we obtain

$$f'(z) = \frac{-2\alpha}{z^3} + \sum_{n=1}^{\infty} 2n\alpha_n z^{2n-1} = -\frac{2\alpha}{z^3}\left(1 - \sum_{n=1}^{\infty} \frac{n\alpha_n}{\alpha} z^{2n+2}\right).$$

Moreover,

$$(f(z) - \alpha_0)^3 = \frac{\alpha^3}{z^6}\left(1 + 3\frac{\alpha_1}{\alpha}z^4 + 3\frac{\alpha_2}{\alpha}z^6 + g(z)\right)$$

$$= \frac{\alpha^3}{z^6} + \frac{3\alpha^2\alpha_1}{z^2} + 3\alpha^2\alpha_2 + \frac{\alpha^3 g(z)}{z^6},$$

where the function $g(z)$ is a convergent power series with powers greater than or equal to 8, and

$$(f'(z))^2 = \frac{4\alpha^2}{z^6}\left(1 - 2\frac{\alpha_1}{\alpha}z^4 - 2\frac{2\alpha_2}{\alpha}z^6 + h(z)\right)$$

$$= \frac{4\alpha^2}{z^6} - \frac{8\alpha\alpha_1}{z^2} - 16\alpha\alpha_2 + \frac{4\alpha^2 h(z)}{z^6},$$

where the function $h(z)$ is a convergent power series with powers greater than or equal to 8. Therefore,

$$(f'(z))^2 - \frac{4}{\alpha}(f(z) - \alpha_0)^3 = -\frac{20\alpha\alpha_1}{z^2} - 28\alpha\alpha_2 + k(z),$$

where the function $k(z)$ is a convergent power series with powers greater than or equal to 2. Thus, the function

$$(f'(z))^2 - \frac{4}{\alpha}(f(z) - \alpha_0)^3 + 20\alpha_1(f(z) - \alpha_0) + 28\alpha\alpha_2$$

has a removable singularity at $z = 0$, which moreover is a zero. In view of the condition (7.2.1), this function has no pole at the points $m + ni$ and therefore is entire. Since it is biperiodic, it is a constant (see Exercise 6.8.8) and thus vanishes identically. (7.2.2) follows by developing

$$(f(z) - \alpha_0)^3 = f^3(z) - 3\alpha_0 f^2(z) + 3\alpha_0^2 f(z) - \alpha_0^3.$$ □

The Weierstrass function (see Exercise 6.8.8) is an example of a function with the properties of the preceding exercise.

Solution of Exercise 7.2.16. f has a power series expansion at the origin

$$f(z) = f_0 + f_1 z + \cdots$$

with radius of convergence equal to infinity. The point 0 is a pole of

$$f(1/z) = f_0 + \frac{f_1}{z} + \cdots,$$ (7.5.8)

and so there is only a finite number of coefficients in the Taylor series which are not equal to 0. Thus f is a polynomial.   □

We note that the converse statement in the previous result also holds. Every non-constant polynomial is an entire function with a pole at infinity.

Solution of Exercise 7.2.17. $z = 0$ is a removable singularity of the function $z/(e^z - 1)$ since

$$\lim_{z \to 0} \frac{z}{e^z - 1} = \frac{1}{\lim_{z \to 0} \frac{e^z - 1}{z}} = \frac{1}{1} = 1.$$

The zeros of $e^z - 1$ different from the origin and of smallest modulus are $z = \pm 2\pi i$. Thus, the Taylor expansion of f around $z = 0$ has radius of convergence $R = 2\pi$. Since, for $z \neq 0$,

$$\frac{z}{e^z - 1} = \frac{1}{1 + \frac{z}{2!} + \frac{z^2}{3!} + \cdots},$$

we have, for $|z| < 2\pi$,

$$\left(\sum_{n=0}^{\infty} \frac{z^n}{(n+1)!} \right) \left(\sum_{n=0}^{\infty} \frac{B_n}{n!} z^n \right) = 1.$$

We now compare the coefficients in z^n on both sides of the above equality for $n = 0, 1, \ldots$. We have $B_0 = 1$. Using the convolution formula for the coefficients of the product of two power series (see Exercise 4.4.7), we obtain that the nth coefficient in the power series expansion of the above product is equal to

$$\sum_{j+k=n} \frac{B_j}{j!(k+1)!} = \sum_{j=0}^{n} \frac{B_j}{j!(n+1-j)!} = \frac{1}{(n+1)!} \sum_{j=0}^{n} \binom{n+1}{j} B_j.$$

Thus, for $n \geq 1$,

$$\sum_{j=0}^{n} \binom{n+1}{j} B_j = 0.$$

The above equation leads in particular to

$$B_1 + \frac{B_0}{2} = 0, \quad n = 1,$$

$$\frac{B_2}{2} + \frac{B_1}{2} + \frac{B_0}{6} = 0, \quad n = 2,$$

and in particular we have

$$B_1 = -\frac{1}{2}, \quad B_2 = \frac{1}{6}.$$

Finally we have that $B_3 = B_5 = \cdots = 0$ since the function

$$\frac{z}{e^z - 1} - \left(1 - \frac{z}{2}\right) = \frac{z \, e^z + 1}{2 \, e^z - 1} - 1$$

is even.   □

Solution of Exercise 7.2.19. The function $\frac{z}{e^z - 1}$ has a removable singularity at the origin, and therefore there are functions $B_0(t), B_1(t), \ldots$ such that (7.2.5) holds in $|z| < 2\pi$. Using Cauchy's formula and Weierstrass' theorem (see Theorem 14.4.1) we have

$$\frac{B_n(t)}{n!} = \frac{1}{2\pi i} \int_{|z|=1} \frac{\left(\dfrac{z e^{tz}}{e^z - 1}\right)}{z^{n+1}} dz$$

$$= \sum_{p=0}^{\infty} \frac{t^p}{p!} \frac{1}{2\pi i} \int_{|z|=1} \frac{z^{p-n}}{e^z - 1} dz$$

$$= \sum_{p=0}^{n} \frac{t^p}{p!} \frac{B_{n-p}}{(n-p)!},$$

since, by definition of the Bernoulli numbers and in view of Cauchy's formula (for $p < n$) and Cauchy's theorem (for $p \geq n$),

$$\frac{1}{2\pi i} \int_{|z|=1} \frac{z^{p-(n+1)}}{e^z - 1} dz = \begin{cases} \frac{B_{n-p}}{(n-p)!}, & \text{if } p < n+1, \\ 0, & \text{if } p \geq n+1. \end{cases}$$

Equation (7.2.6) follows. The proof of (7.2.7) goes as follows:

$$\frac{B_n(t+1)}{n!} = \frac{1}{2\pi i} \int_{|z|=1} \frac{\left(\dfrac{z e^{(t+1)z}}{e^z - 1}\right)}{z^{n+1}} dz$$

$$= \frac{1}{2\pi i} \int_{|z|=1} \frac{\left(\dfrac{ze^{tz}(e^z - 1 + 1)}{e^z - 1}\right)}{z^{n+1}} dz$$

$$= \frac{1}{2\pi i} \int_{|z|=1} \frac{\left(\dfrac{ze^{tz}}{e^z - 1} + ze^{tz}\right)}{z^{n+1}} dz$$

$$= \frac{1}{2\pi i} \int_{|z|=1} \frac{\left(\dfrac{ze^{tz}}{e^z - 1}\right)}{z^{n+1}} dz + \frac{1}{2\pi i} \int_{|z|=1} \frac{ze^{tz}}{z^{n+1}} dz$$

$$= \frac{B_n(t)}{n!} + \frac{t^{n-1}}{(n-1)!},$$

and hence the result.

The proof of (7.2.8) involves the interchange of derivation and integral in the integral

$$\frac{1}{2\pi i} \int_{|z|=1} \frac{\left(\dfrac{ze^{tz}}{e^z - 1}\right)}{z^{n+1}} dz.$$

By rewriting explicitly this integral as

$$\frac{1}{2\pi} \int_0^{2\pi} \frac{e^{iu} e^{te^{iu}}}{(e^{e^{iu}} - 1)e^{inu}} du$$

one sees that the conditions of Theorem 14.6.1 are in force, and one can write

$$\frac{B_n'(t)}{n!} = \frac{1}{2\pi i} \int_{|z|=1} \frac{\left(\dfrac{z^2 e^{tz}}{e^z - 1}\right)}{z^{n+1}} dz$$

$$= \frac{1}{2\pi i} \int_{|z|=1} \frac{\left(\dfrac{ze^{tz}}{e^z - 1}\right)}{z^n} dz$$

$$= \frac{B_{n-1}(t)}{(n-1)!},$$

and hence $B_n'(t) = nB_{n-1}(t)$ for $n = 1, 2, \ldots$. $\square$

Solution of Exercise 7.2.20. The function $\frac{f(s)}{1-s}$ is analytic in the open unit disk, and it has a power series expansion centered at the origin, and with radius of convergence at least 1. See Exercise 4.4.7. Furthermore, since

$$\frac{1}{1-s} = \sum_{n=0}^{\infty} 1 \cdot s^n, \quad s \in \mathbb{D},$$

formula (4.4.14) from that same exercise, or the formula in Exercise 4.4.9, leads to

$$\frac{f(s)}{1-s} = \sum_{n=0}^{\infty} \left(\sum_{j=0}^{n} a_j \right) s^n.$$

It follows that

$$\int_{[0,z]} \frac{f(s)}{1-s} ds = \sum_{n=0}^{\infty} \frac{\sum_{j=0}^{n} a_j}{n+1} z^{n+1}. \tag{7.5.9}$$

Therefore the point $z = 0$ is a removable singularity of the function $C(z)$, and its power series expansion in the open unit disk is given by (7.5.9). □

Solution of Exercise 7.2.21. Assume by contradiction that such a function exists. Then, $|f(z)|^2$, and hence $|f(z)|$ is bounded in a neighborhood of the origin. Hence, $z = 0$ is a removable singularity of f, and f can be extended to an analytic function in a neighborhood of the origin by $f(0) = 0$. Differentiating $f(z)^2 = z$ at the origin we obtain

$$2f(0)f'(0) = 1,$$

which leads to a contradiction. □

See also [31, p. 140, 2.4.4].

Remark 7.5.3. There is also no function in $1 < |z| < 2$ such that $f(z)^2 = z$. See Exercise 5.8.5.

For a related problem, see also Exercise 5.4.3.

Solution of Exercise 7.2.22. We have already seen in Exercise 3.3.2 that the series converges. From the proof there the convergence is uniform on compact sets, and hence the series defines an analytic function, with a simple pole with residue 1 at the points $z = 0, 1, 2, \ldots$. The idea is to check that the function

$$q(z) \stackrel{\text{def.}}{=} \frac{1}{z} + \sum_{n=1}^{\infty} \left(\frac{1}{z-n} + \frac{1}{n} \right) - \pi \cot(\pi z)$$

is entire and bounded. By Liouville's theorem, it is a constant, which is then easily computed. The singularities of q consists of $\mathbb{N}_0$. Using formula (7.3.2) for the residue, we have, for $n \in \mathbb{N}_0$,

$$\mathrm{Res}(\pi \cot(\pi z), n) = \mathrm{Res}\left(\pi \frac{\cos(\pi z)}{\sin(\pi z)}, n \right) = \frac{\pi \cos(\pi z)}{\pi \cos(\pi z)} = 1.$$

Therefore q has only removable singularities and extends to an entire function, which we still call q. To show that q is bounded it is enough, by the maximum modulus principle, to show that q is uniformly bounded on the boundary of the squares with nodes

$$(\pm 1 \pm i)\left(N + \frac{1}{2}\right).$$

It is then readily seen that

$$\lim_{z \to 0} q(z) = 0.$$

We will skip the details and refer to [91, p. 189] for more information.

The second formula follows from the first by differentiation.   □

Solution of Exercise 7.3.3. Write

$$\frac{e^z - 1}{\sin^2 z} = \frac{z(1 + z/2 + z^2/3! + \cdots)}{z^2(1 - z^2/3! + z^4/5! + \cdots)^2}$$
$$= \frac{g(z)}{z},$$

where

$$g(z) = \frac{1 + z/2 + z^2/3! + \cdots}{(1 - z^2/3! + z^4/5! + \cdots)^2}$$

is analytic in a neighborhood of the origin, and $g(0) = 1 \neq 0$. Thus the origin is a simple pole, and by the formula (7.3.2) to compute residues, the residue is $g(0) = 1$.

Another way amounts to rewriting

$$\frac{e^z - 1}{\sin^2 z} = \frac{\dfrac{e^z - 1}{z}}{\left(\dfrac{\sin z}{z}\right)^2 z},$$

and using formula (7.3.2) with

$$f(z) = \begin{cases} \dfrac{e^z - 1}{z}, & z \neq 0, \\ 1, & z = 0, \end{cases} \quad \text{and} \quad g(z) = \begin{cases} \left(\dfrac{\sin z}{z}\right)^2 z, & z \neq 0, \\ 0, & z = 0. \end{cases} \quad □$$

Solution of Exercise 7.3.4. The function

$$\frac{z^{n-1}}{z^n - 1}$$

has simple poles at the points

$$z_\ell = e^{\frac{2\pi i \ell}{n}}, \quad \ell = 0, 1, \ldots, n - 1,$$

(that is, at the roots of unity of order n) and takes value 0 at infinity. Therefore it can be written as

$$\frac{z^{n-1}}{z^n - 1} = \sum_{\ell=0}^{n-1} \frac{A_\ell}{z - z_\ell}.$$

By the formula (7.3.2) for computing the residue, we have

$$\text{Res}\left(\frac{nz^{n-1}}{z^n - 1}, z_\ell\right) = \frac{nz^{n-1}}{nz^{n-1}}\Big|_{z=z_\ell} = 1, \quad \ell = 0, 1, \ldots, n-1,$$

and this proves formula (3.3.1). Recall that the residue at infinity is given by

$$\text{Res}(f, \infty) = -\text{Res}\left(1/z^2 f(1/z), 0\right).$$

Since here

$$\frac{1}{z^2} f\left(\frac{1}{z}\right) = \frac{n}{z(1 - z^n)},$$

we have

$$\text{Res}\left(\frac{nz^{n-1}}{z^n - 1}, \infty\right) = -n,$$

and we have, in accordance with Exercise 7.3.9, that the sum of all the residues is equal to 0.   □

Solution of Exercise 7.3.5. Let α be a zero of f with multiplicity n. Then, in some neighborhood V of α we have

$$f(z) = (z - \alpha)^n h(z),$$

where h is analytic in V and is such that $h(\alpha) \neq 0$. Therefore, for $z \neq \alpha$, and in a possibly smaller neighborhood, and using the formula (4.2.3) for the logarithmic derivative if need be, we have

$$\frac{f'(z)}{f(z)} = \frac{n}{z - \alpha} + \frac{h'(z)}{h(z)}, \tag{7.5.10}$$

so that

$$\text{Res}\left(\frac{f'}{f}, \alpha\right) = n.$$

Similarly, let β be a pole of f of multiplicity m. Then, in some neighborhood W of β, we have

$$f(z) = \frac{g(z)}{(z - \beta)^m},$$

where g is analytic in W and such that $g(\beta) \neq 0$. Therefore, for $z \neq \beta$, and in a possibly smaller neighborhood, and here too possibly using the formula (4.2.3) for the logarithmic derivative, we have

$$\frac{f'(z)}{f(z)} = \frac{-m}{z - \beta} + \frac{g'(z)}{g(z)},$$

so that

$$\operatorname{Res}\left(\frac{f'}{f},\beta\right) = -m.$$

The result is obtained by summing on all zeros and poles of f in the interior of Γ. $\qquad\square$

Solution of Exercise 7.3.7. We compute the residues of $\frac{g(z)f'(z)}{f(z)}$ at a zero and a pole of f and use the notation of the solution of the previous exercise. In the case of a pole α of multiplicity n, and using (7.5.10) we have

$$\frac{f'(z)g(z)}{f(z)} = \frac{ng(z)}{z-\alpha} + \underbrace{\frac{g(z)h'(z)}{h(z)}}_{\text{analytic in a neigborhood of }\alpha}, \qquad (7.5.11)$$

so that, in view of formula (7.3.2) we have

$$\operatorname{Res}\left(\frac{f'(z)g(z)}{f(z)},\alpha\right) = ng(\alpha).$$

Similarly, in the case of a pole β of multiplicity m, we have

$$\operatorname{Res}\left(\frac{f'(z)g(z)}{f(z)},\beta\right) = -mg(\beta).$$

It follows that the integral (7.3.11) is equal to

$$\sum_{j=1}^{N} n_j g(z_j) - \sum_{j=1}^{M} m_j g(w_j),$$

where we have denoted by $z_1,\ldots,z_N$ the zeros of f inside Γ, and by $n_1,\ldots,n_N$ their respective multiplicities, and by $w_1,\ldots,w_M$ the poles of f inside Γ, and by $m_1,\ldots,m_M$ their respective multiplicities. $\qquad\square$

Solution of Exercise 7.3.8. By assumption, there exist $r_0 > 0$ and a function h analytic and not vanishing in $|z-z_0| < r_0$ such that $f(z) = (z-z_0)^M h(z)$. In particular

$$\frac{f'(z)}{f(z)} = \frac{M}{z-z_0} + \frac{h'(z)}{h(z)}, \qquad 0 < |z-z_0| < r_0.$$

Thus for $r \in (0,r_0)$,

$$\int_{|z-z_0|=r} \frac{f'(z)}{f(z)} dz = \int_{|z-z_0|=r} \frac{M}{z-z_0} dz + \int_{|z-z_0|=r} \frac{h'(z)}{h(z)} dz$$
$$= 2\pi M$$

and so, by Exercise 7.3.5, the equation $f(z) = f(z_1)$ has exactly M solutions (counting multiplicity) for every $z_1 \in B(z_0, r)$ since the function

$$k \mapsto \frac{1}{2\pi i} \int_{|z-z_0|=r} \frac{f'(z)}{f(z) - k} dz$$

is continuous and integer-valued in a neighborhood of the origin   □

Solution of Exercise 7.3.9. (a) Multiply both sides of (7.3.12) by z, and let $z \to \infty$. Since $\deg P \geq 2$ we obtain the desired relation.

(b) The function

$$f(z) = \frac{1}{(z^{1000} + 1)(z - 3)}$$

has 1000 poles, say $z_1, \ldots, z_{1000}$, inside the circle $|z| = 2$, and one pole outside. By (7.3.13) we have

$$\sum_{n=1}^{1000} \mathrm{Res}(f, z_n) = -\mathrm{Res}(f, 3) = -\frac{1}{3^{1000} + 1}.$$

Thus, using the residue theorem and this last expression, we obtain

$$\int_{|z|=2} \frac{1}{(z^{1000} + 1)(z - 3)} dz = 2\pi i \sum_{n=1}^{1000} \mathrm{Res}(f, z_n) = -2\pi i \,\mathrm{Res}(f, 3) = -\frac{2\pi i}{3^{1000} + 1}.$$

□

Solution of Exercise 7.3.10. We write f as a sum of the form (7.3.12) and of a polynomial p. For any $R > 0$ we have (for instance because p has a primitive and the integration is over a closed path)

$$\int_{|z|=R} p(z)dz = 0$$

and so for R large enough (that is, strictly larger that $\max_{j=1,\ldots,k} |z_j|$) we have

$$\frac{1}{2\pi i} \int_{|z|=R} f(z)dz = \sum_{j=1}^{k} A_j.$$

To conclude we let $R \to \infty$ and use formula (7.3.9).   □

Solution of Exercise 7.3.11. We first note that (7.3.14) can be rewritten as

$$\frac{1}{2\pi} \int_0^{2\pi} \frac{e^{it} + z}{e^{it} - z} e^{int} dt = \frac{1}{2\pi i} \int_{|s|=1} \frac{s + z}{s - z} s^{n-1} ds.$$

For $n \geq 1$, Cauchy's formula applied to the function $s \mapsto (s+z)s^{n-1}$ gives the value $2z^n$ for the integral. For $n \leq -1$ the exactity relation (see Exercise 7.3.10) and the residue theorem give the value 0 for the integral since the residue at infinity is 0. For $n = 0$ we have

$$\frac{s+z}{s-z}\frac{1}{s} = -\frac{1}{s} + \frac{2}{s-z},$$

and the residue theorem gives the value 1 for the integral. Thus

$$\frac{1}{2\pi} \int_0^{2\pi} \frac{e^{it}+z}{e^{it}-z} e^{int} \, dt = \begin{cases} 2z^n, & n \geq 1, \\ 1, & n = 0, \\ 0, & n \leq -1. \end{cases}$$

If follows that, for a polynomial $p(z) = a_0 + \cdots + a_N z^N$ we have:

$$\frac{1}{2\pi} \int_0^{2\pi} \frac{e^{it}+z}{e^{it}-z} \left(\frac{p(e^{it}) + \overline{p(e^{it})}}{2} \right) dt$$

$$= \frac{1}{2\pi} \int_0^{2\pi} \frac{e^{it}+z}{e^{it}-z} \left(\frac{a_0 + \overline{a_0} + a_1 e^{it} + \cdots a_N e^{iNt} + \overline{a_1} e^{-it} + \cdots + \overline{a_N} e^{-iNt}}{2} \right) dt$$

$$= \frac{a_0 + \overline{a_0}}{2} + \sum_{n=1}^{N} a_n z^n$$

$$= p(z) - i \operatorname{Im} p(0). \qquad \square$$

Solution of Exercise 7.3.12. From the equality

$$(1+z)^n = \sum_{k=0}^{n} \binom{n}{k} z^k$$

we see that

$$\frac{(1+z)^n}{z^{\ell+1}} = \sum_{k=0}^{n} \frac{\binom{n}{k}}{z^{\ell+1-k}}.$$

Recall now that

$$\int_{|z|=1} \frac{dz}{z^{\ell+1-k}} = 0$$

unless $\ell + 1 - k = 1$, that is, unless $k = \ell$. Then the integral is equal to $2\pi i$. Thus,

$$\frac{1}{2\pi i} \int_{|z|=1} \frac{(1+z)^n}{z^{\ell+1}} dz = \sum_{k=0}^{n} \binom{n}{k} \frac{1}{2\pi i} \int_{|z|=1} \frac{dz}{z^{\ell-k+1}} = \binom{n}{\ell}. \qquad (7.5.12)$$

For z on the unit circle we have

$$\left| \frac{(1+z)^2}{7z} \right| < \frac{4}{7} < 1.$$

Thus:

$$\sum_{n=0}^{\infty} \frac{\binom{2n}{n}}{7^n} = \frac{1}{2\pi i} \sum_{n=0}^{\infty} \frac{1}{7^n} \int_{|z|=1} \frac{(1+z)^{2n}}{z^{n+1}} dz$$

$$= \frac{1}{2\pi i} \int_{|z|=1} \left(\sum_{n=0}^{\infty} \left(\frac{(1+z)^2}{7z} \right)^n \right) \frac{dz}{z} \quad \text{(by Weierstrass' theorem)}$$

$$= \frac{1}{2\pi i} \int_{|z|=1} \frac{1}{1 - \frac{(1+z)^2}{7z}} \frac{dz}{z}$$

$$= \frac{7}{2\pi i} \int_{|z|=1} \frac{dz}{7z - (1+z)^2}$$

$$= \frac{7}{2\pi i} \int_{|z|=1} \frac{dz}{5z - 1 - z^2}.$$

We apply the residue theorem to compute this last integral. The zeroes of the equation $5z - 1 - z^2 = 0$ are $z_\pm = (5 \pm \sqrt{21})/2$. Thus

$$\frac{1}{2\pi i} \int_{|z|=1} \frac{dz}{5z - 1 - z^2} = -\frac{1}{2\pi i} \int_{|z|=1} \frac{dz}{(z - z_-)(z - z_+)}$$

$$= -\operatorname{Res}\left(\frac{1}{(z - z_-)(z - z_+)}, z_- \right)$$

$$= -\frac{1}{z_- - z_+} = \frac{1}{\sqrt{21}},$$

and so the sum is equal to $7/\sqrt{21}$:

$$\sum_{n=0}^{\infty} \frac{\binom{2n}{n}}{7^n} = \frac{7}{\sqrt{21}} = \sqrt{\frac{7}{3}}.$$

Note that we could also just apply Cauchy's formula to the function $\frac{1}{z - z_+}$ to compute the last integral. □

Solution of Exercise 7.3.13. We follow the solution of the previous exercise. Take $|\zeta| > 4$. Then, $\frac{(1+z)^2}{\zeta} \in \mathbb{D}$ for $z \in \mathbb{T}$. We have (where we use Weierstrass' theorem to go from the first line to the second line):

$$\sum_{n=0}^{\infty} \frac{\binom{2n}{n}}{\zeta^n} = \frac{1}{2\pi i} \sum_{n=0}^{\infty} \frac{1}{\zeta^n} \int_{|z|=1} \frac{(1+z)^{2n}}{z^{n+1}} dz$$

$$= \frac{1}{2\pi i} \int_{|z|=1} \sum_{n=0}^{\infty} \left(\frac{(1+z)^2}{\zeta z} \right)^n \frac{dz}{z}$$

$$= \frac{1}{2\pi i} \int_{|z|=1} \frac{1}{1 - \frac{(1+z)^2}{\zeta z}} \frac{dz}{z}$$

$$= \frac{\zeta}{2\pi i} \int_{|z|=1} \frac{dz}{\zeta z - (1+z)^2}$$

$$= \frac{\zeta}{2\pi i} \int_{|z|=1} \frac{dz}{(\zeta - 2)z - 1 - z^2}.$$

Consider the equation

$$z^2 + z(2 - \zeta) + 1 = 0.$$

Its zeros are

$$z = \frac{\zeta - 2 \pm \sqrt{\zeta^2 - 4\zeta}}{2}.$$

We now wish to give an analytic meaning to this expression. More precisely, we wish to show that there exists an analytic square root to the function $\zeta^2 - 4\zeta$ in $|\zeta| > 4$, equal to $\sqrt{x^2 - 4x}$ for x real of modulus bigger than 4. Writing

$$\zeta^2 - 4\zeta = \zeta^2 \left(1 - \frac{4}{\zeta} \right),$$

we see that the function $\zeta f_{1/2}(\zeta)$, where the function $f_{1/2}$ has been defined in (4.4.9), answers the question. Thus, with

$$\zeta_\pm = \frac{\zeta - 2 \pm \zeta f_{1/2}(4/\zeta)}{2}$$

we have

$$\sum_{n=0}^{\infty} \frac{\binom{2n}{n}}{\zeta^n} = \frac{\zeta}{2\pi i} \int_{|z|=1} \frac{dz}{(\zeta - 2)z - 1 - z^2}$$

$$= \frac{\zeta}{2\pi i} \int_{|z|=1} \frac{dz}{-(z - \zeta_+)(z - \zeta_-)}.$$

We have $\zeta_- \in \mathbb{D}$ and therefore the above integral is equal to

$$\zeta \operatorname{Res} \left(\frac{1}{-(z - \zeta_+)(z - \zeta_-)}, \zeta_- \right) = \frac{\zeta}{\zeta_+ - \zeta_-} = \frac{1}{f_{1/2}(4/\zeta)}. \qquad \Box$$

We note that we retrieve for $\zeta = 7$ and $\zeta = 5$ the result in Exercise 7.3.12 and formula (7.3.16) respectively.

Solution of Exercise 7.3.14. From (7.5.12) we have

$$\binom{3n}{n} = \frac{1}{2\pi i} \int_{|z|=1} \frac{(1+z)^{3n}}{z^{n+1}} dz.$$

Hence, and using Weierstrass' theorem to justify the interchange of sum and integration,

$$\sum_{n=0}^{\infty} \left(\frac{2}{27}\right)^n \binom{3n}{n} = \sum_{n=0}^{\infty} \left(\frac{2}{27}\right)^n \left(\frac{1}{2\pi i} \int_{|z|=1} \frac{(1+z)^{3n}}{z^{n+1}} dz\right)$$

$$= \frac{1}{2\pi i} \int_{|z|=1} \sum_{n=0}^{\infty} \left(\frac{2(1+z)^3}{27z}\right)^n \frac{dz}{z}$$

$$= \frac{1}{2\pi i} \int_{|z|=1} \frac{1}{1 - \dfrac{2(1+z)^3}{27z}} \frac{dz}{z}$$

$$= \frac{1}{2\pi i} \int_{|z|=1} \frac{27 dz}{27z - 2(1+z)^3}.$$

The polynomial $p(z) = 27z - 2(1+z)^3$ vanishes at $z = 2$. To find its other two roots, one can divide it by $z - 2$ directly; one may also proceed as follows: Write

$$27z - 2(1+z)^3 = 27(z-2) - 2((1+z)^3 - 3^3)$$

$$= 27(z-2) - 2(1+z-3)((1+z)^2 + (1+z)3 + 3^2)$$

$$= (z-2)(27 - 2(1+z)^2 - 6(1+z) - 18)$$

$$= (z-2)(-2z^2 - 10z + 1).$$

In any event the other two zeros of $p(z)$ are

$$z_{\pm} = \frac{5 \pm 3\sqrt{3}}{-2}.$$

The only zero inside the unit circle is

$$z_- = \frac{3\sqrt{3} - 5}{2}$$

and the residue of $27/p(z)$ at this point is equal to

$$\mathrm{Res}\left(\frac{27}{p(z)}, z_-\right) = \frac{27}{(z_- - 2)(-4z_- - 10)}$$

$$= \frac{27}{\dfrac{3\sqrt{3} - 9}{2}(-6\sqrt{3})}$$

$$= \frac{1}{\sqrt{3} - 1} = \frac{\sqrt{3} + 1}{2},$$

and this ends the proof.   □

Solution of Exercise 7.4.1. Since $|f(z)| > |g(z)|$ on γ, we note that the functions f and $f + g$ do not vanish on γ, and the integrals in the hint given after the exercise make sense. Following this hint, we write

$$\int_\gamma \left(\frac{f'(z)}{f(z)} - \frac{(f+g)'(z)}{f(z)+g(z)} \right) dz = \int_\gamma \frac{f'(z)g(z) - f(z)g'(z)}{f(z)(f(z)+g(z))} dz$$

$$= -\int_\gamma \frac{\left(\frac{g}{f} \right)'(z)}{1 + \frac{g}{f}(z)} dz$$

where we have divided both numerator and denominator by f^2 on the right side. Thus, using Weierstrass' theorem,

$$\int_\gamma \left(\frac{f'(z)}{f(z)} - \frac{f'(z)+g'(z)}{f(z)+g(z)} \right) dz = -\int_\gamma \left(\frac{g}{f} \right)'(z) \left(\sum_{n=0}^\infty (-1)^n \left(\frac{g(z)}{f(z)} \right)^n \right) dz$$

$$= -\int_\gamma \left(\sum_{n=0}^\infty (-1)^n \left(\frac{g}{f} \right)'(z) \left(\frac{g(z)}{f(z)} \right)^n \right) dz$$

$$= -\sum_{n=0}^\infty (-1)^n \int_\gamma \left(\frac{g}{f} \right)'(z) \left(\frac{g(z)}{f(z)} \right)^n dz.$$

Each of the functions

$$\left(\frac{g}{f} \right)' \left(\frac{g}{f} \right)^n, \quad n = 0, 1, 2, \ldots$$

has a primitive. Hence, by Theorem 5.2.1, each of the integrals

$$\int_\gamma \left(\frac{g}{f} \right)'(z) \left(\frac{g(z)}{f(z)} \right)^n dz = 0, \quad n = 0, 1, 2, \ldots.$$

Therefore

$$\int_\gamma \left(\frac{f'}{f} - \frac{(f+g)'}{f+g} \right) dz = 0,$$

and f and $f + g$ have the same number of zeros inside γ by Exercise 7.3.5 since they have no poles there. □

Solution of Exercise 7.4.2. On the unit circle we have

$$|-f(z)| < 1 = |z^n|.$$

Rouché's theorem asserts then that the functions z^n and $z^n - f(z)$ have the same numbers of zeros in the open unit disk. This concludes the proof since z^n has a zero of order n at the origin. □

Solution of Exercise 7.4.3. For $|z| = 3/2$ we have

$$|f(z)| \leq \frac{3^3}{2^3} + 1 = \frac{35}{8} < \frac{81}{16} = |z^4| = |F(z)|.$$

Rouché's theorem asserts then that the functions F and $F + f$ have the same number of zeros in $|z| < 3/2$. Since F has a zero of order 4 there, the function $F(z) + f(z) = z^4 + z^3 + 1$ has four roots in $|z| < 3/2$. By the fundamental theorem of algebra it has altogether four roots in $\mathbb{C}$, and hence all its roots are in $|z| < 3/2$.   □

Solution of Exercise 7.4.4. With f and F as in the hint we have, for $|z| = 1$,

$$|f(z) + g(z)| = |F(z) - z| < 2 = |f(z)|,$$

and hence, by Rouché's theorem, f and $f + g$ have the same number of zeroes in the open unit disk.   □

Solution of Exercise 7.4.5. Take $f(z) = 1 - 3z$ and $g(z) = z^4$. We have

$$|f(z)| \geq |3z| - 1 = 2,$$

so that

$$|g(z)| = 1 < 2 \leq |f(z)|.$$

Thus, $f(z) + g(z) = z^4 - 3z + 1$ has exactly one zero in $|z| \leq 1$.   □

Solution of Exercise 7.4.6. Let $g(z) = z^4 + z^3$ and $f(z) = -4z + 1$. For $|z| = 1$ we have

$$|g(z)| \leq |z|^4 + |z|^3 = 2 < 3 = |4z| - 1 \leq |-4z + 1| = |f(z)|,$$

and so the equation $z^4 + z^3 - 4z + 1 = 0$ has one solution in the open unit disk. On the other hand, for $|z| = 3$ we have, with $F(z) = z^4$ and $G(z) = z^3 - 4z + 1$,

$$|G(z)| \leq |z|^3 + 4|z| + 1 = 3^3 + 4 \cdot 3 + 1 = 40 < 81 = |F(z)|.$$

Thus the equation $F(z) + G(z) = 0$ has four roots in $|z| < 3$, and hence three roots in the ring $1 < |z| < 3$.   □

Solution of Exercise 7.4.7. We follow the solution of Exercise 6.8.13 up to (b). We then consider $f(z) = 3z^2$ and $g(z) = z^4 + z + 1$. Then, on $|z| = 1.1$ we have

$$|g(z)| < |f(z)|,$$

and so p has two roots inside $|z| < 1.1$. But under (b), it has four roots there (on the unit circle), and we obtain a contradiction.   □

Solution of Exercise 7.4.8. It suffices to apply Rouché's theorem for $f(z) = ze^\lambda$ and $g(z) = e^z$. Since $\lambda > 1$ we have, for $|z| = 1$,

$$|g(z)| = |e^z| \le e^{|z|} = e < e^\lambda = |f(z)|.$$

Here we have used that

$$|e^z| = \left| \sum_{n=0}^{\infty} \frac{z^n}{n!} \right| \le \sum_{n=0}^{\infty} \frac{|z|^n}{n!} = e^{|z|}. \qquad \square$$

Solution of Exercise 7.4.9. To ease the notation we consider the case $z_0 = 0$. By definition of a zero of order N, there exist $r_0 > 0$ and h analytic in $B(0, r_0)$ and not vanishing there, such that

$$f(z) = z^N h(z), \quad z \in B(0, r_0).$$

Let $r \in (0, r_0)$ and let

$$\epsilon = \min_{|z|=r} |z^N h(z)|.$$

We note that $\epsilon > 0$. Let $a \in B(0, \epsilon)$. Rouché's theorem applied to $f(z) = z^N h(z)$ and $g(z) = -a$ insures that the equation $f(z) = a$ has exactly N solutions in $B(0, r)$ since the only zero of f there is $z = 0$, and it has order N.

The second item is a direct consequence of the first item. $\qquad \square$

Chapter 8

Computations of Definite Integrals Using the Residue Theorem

We have seen in Chapter 5 how the fundamental theorem of calculus for line integrals, or Cauchy's theorem, allow us to compute (in general real) definite integrals such as the Fresnel integrals. In that chapter no residues are computed. The approach in the present chapter is different. The main player is the residue theorem. There are numerous kinds of definite integrals which one can compute using this theorem, and in the present chapter we do not try to be exhaustive.

8.1 Integrals on the real line of rational functions

For real values of a and b, item (1) of the first exercise, and the second exercise, are taken from [75], which is a mine of problems. For item (2) of Exercise 8.1.1, see also Exercise 3.1.13. Recall that we have denoted by $\mathbb{C}_r$ the open right half-plane; see (1.1.43).

Exercise 8.1.1.

(1) *Let* $a, b \in \mathbb{C}_r$. *Compute*

$$\int_{\mathbb{R}} \frac{dx}{(x^2 + a^2)(x^2 + \overline{b^2})},$$

first for $a \neq \overline{b}$ *and then for* $a = \overline{b}$.

(2) *Show that for every choice of N and points $a_1, \ldots, a_N \in \mathbb{C}_r$, the $N \times N$ Hermitian matrix with (ℓ, j)-entry equal to*

$$\frac{1}{a_\ell + \overline{a_j}}$$

is non-negative (see Definition 16.3.1 for the latter).

The integral $\int_{\mathbb{R}} \frac{dx}{x^2+1}$ is easy to compute. Its generalizations

$$\int_{\mathbb{R}} \frac{dx}{x^{2n}+1} \quad \text{and} \quad \int_{\mathbb{R}} \frac{dx}{(x^2+1)^n}, \quad n = 1, 2, \ldots,$$

are a bit more difficult.

Exercise 8.1.2. *Compute*

$$\int_{\mathbb{R}} \frac{dx}{x^{2n}+1} \quad n = 1, 2, \ldots, \tag{8.1.1}$$

and

$$\int_{\mathbb{R}} \frac{x^2 dx}{x^{2n}+1} \quad n = 2, 3, \ldots.$$

We note the formula

$$\int_{\mathbb{R}} \frac{x^{2p}}{x^{2n}+1} dx = \frac{\frac{\pi}{n}}{\sin \frac{(2p+1)\pi}{2n}} \tag{8.1.2}$$

for $p < n$. See [153, p. 313], [200, Exercise 17, p. 188], [211, p. 267] for instance. These integrals diverge for $n \le p$. For $n = p$, the right side of (8.1.2) is negative, while the left side is equal to $+\infty$.

For computing the integral (8.1.1) using another method see [184, p. 236] and [175, p. 110]. The method in these two books shows in fact that the formula

$$\int_0^\infty \frac{dx}{x^p+1} = \frac{\pi}{p \sin \frac{\pi}{p}} \tag{8.1.3}$$

is valid for any integer $p \ge 2$ and not only for even p. See also Exercise 8.3.1 in the next section. The proof given there works for all integers greater than or equal to 2 and requires us to compute only one residue. Formula (8.1.3) is even true for any real $p > 1$; see [80, equation (9), p. 165], and Exercise 8.3.2 below. We give another computation of (8.1.3) in Exercise 8.3.3, valid for $p \in \{2, 3, 4, \ldots\}$, and computing p residues.

Question 8.1.3. *Interpret (8.1.3) in terms of analytic continuation.*

Exercise 8.1.4. *Compute*

$$\int_{\mathbb{R}} \frac{dx}{(x^{2n}+1)^2}, \quad n=1,2,\ldots.$$

More generally:

Exercise 8.1.5. *Using the monotone convergence theorem applied to the series*

$$\sum_{p=1}^{\infty} \frac{t^p}{(x^{2n}+1)^p}, \quad t \in (0,1)$$

compute the integrals

$$\int_{\mathbb{R}} \frac{dx}{(x^{2n}+1)^p}, \quad p=3,4,\ldots.$$

Exercise 8.1.6. *Compute*

$$\int_{\mathbb{R}} \frac{dx}{(x^2+1)^n}, \quad n=1,2,\ldots.$$

Exercise 8.1.7. *Show that*

$$\int_{\mathbb{R}} \frac{x\,dx}{(x^2+x+1)^2} = -\frac{2\pi}{3\sqrt{3}}.$$

Exercise 8.1.8. *Compute*

$$\int_{\mathbb{R}} \frac{dx}{(x^2+1)(x-2i)^2(x-3i)^3(x-4i)^4}.$$

Remark 8.1.9. In fact it is not needed to use the residue theorem to compute integrals of the form $\int_{\mathbb{R}} \frac{p(x)}{q(x)} dx$, where p and q are polynomials such that $\deg q \geq \deg p + 2$ and assuming moreover that q has no zeros on the real line. It is enough to use the partial fraction expansion of p/q; see Remarks 7.1.10 for the latter. More precisely, let $\alpha_1,\ldots,\alpha_N$ be the distinct zeros of q and, for $u=1,\ldots,N$, let a_u be the coefficient of the factor $\frac{1}{x-\alpha_u}$ in the partial fraction expansion of p/q (of course a_u is the residue of p/q at α_u). In [211, p. 266] on can find the formula (proved using only the partial fraction expansion)

$$\int_{\mathbb{R}} \frac{p(x)}{q(x)} dx = i\pi \sum_{u=1}^{N} e(\alpha_u)a_u,$$

where $e(\alpha_u)$ is the sign of the imaginary part of α_u. One gets the same formula as using the residue theorem, as is verified using the exactity relation (see (7.3.13) and Exercise 7.3.10).

8.2   Integrals on the real line of rational functions multiplied by a trigonometric function

We now compute integrals of the form

$$\int_{\mathbb{R}} \frac{p(x)}{q(x)} \sin ax\, dx \quad \text{or} \quad \int_{\mathbb{R}} \frac{p(x)}{q(x)} \cos ax\, dx,$$

where $a \in \mathbb{R}$ and where p and q are now polynomials with real coefficients, and still with deg $q \geq \deg p + 2$.

We begin with a simple computation, which has an important consequence. For the notion of positive definite function, see Definition 16.3.11. The function $e^{-|t-s|}$ appearing in the statement is the covariance function of a Gaussian process called the Ornstein–Uhlenbeck process.

Exercise 8.2.1.

(1) *Compute*

$$\int_{\mathbb{R}} \frac{e^{itx}}{x^2 + 1}\, dx, \quad t \in \mathbb{R}.$$

(2) *Show that the kernel*

$$f(t - s) = e^{-|t-s|}, \quad t, s \in \mathbb{R},$$

is positive definite.

Exercise 8.2.2. *Let C_R denote the closed upper half-circle with radius R. Show that*

$$\lim_{R\to\infty} \int_{C_R} \frac{\cos z}{z^2 + 1}\, dz$$

exists, and compute its value.

Exercise 8.2.3. *Compute*

$$\int_{\mathbb{R}} \frac{\cos ax}{x^4 + 1}\, dx, \quad a \geq 0.$$

The following exercise is taken from [75, Exercise 13.28, p. 146].

Exercise 8.2.4. *Show that the function f defined by*

$$f(z) = \int_{\mathbb{R}} \frac{\cos u}{(u^2 + 1)(u - z)^2}\, du \tag{8.2.1}$$

is analytic in $\mathbb{C}_+ = \operatorname{Im} z > 0$ and has an analytic extension to $\operatorname{Im} z > -1$.

Rational functions p/q which appeared in the previous exercises were such that $\deg q \geq \deg p + 2$, and without poles on the real line. We consider now the case when simple poles are allowed on the real line. Then, the integrals of the preceding type

$$\int_{\mathbb{R}} \frac{p(x)}{q(x)} e^{imx} dx,$$

where $\deg q \geq \deg p + 2$, but where now q is allowed to have simple poles on the real line, can be computed as above, with the principal value taken at the real poles. Let $b \in \mathbb{R}$ be a simple pole of q. Let γ_ϵ be the contour

$$\gamma_\epsilon(t) = b + \epsilon e^{i(\pi - t)}, \quad t \in [0, \pi].$$

Then we recall that

$$\lim_{\epsilon \to 0} \int_{\gamma_\epsilon} \frac{p(z)}{q(z)} e^{izm} dz = -\pi i \frac{p(b)}{q(b)} e^{imb}.$$

Indeed, write $q(z) = (z - b) h(z)$, where $h(b) \neq 0$. We have

$$\int_{\gamma_\epsilon} \frac{p(z)}{q(z)} e^{izm} dz = \int_0^\pi \frac{p(b + \epsilon e^{i(\pi - t)})}{h(b + \epsilon e^{i(\pi - t)}) \epsilon e^{i(\pi - t)}} e^{i(mb + m\epsilon e^{i(\pi - t)})} (-i) \epsilon e^{i(\pi - t)} dt$$

$$= -i \int_0^\pi \frac{p(b + \epsilon e^{i(\pi - t)})}{h(b + \epsilon e^{i(\pi - t)})} e^{i(mb + \epsilon e^{i(\pi - t)})} dt$$

$$\to -i\pi \frac{p(b)}{h(b)} e^{imb} \quad \text{as} \quad \epsilon \to 0.$$

This last expression can also be rewritten as

$$-\pi i \operatorname{Res}\left(\frac{p(z)}{q(z)} e^{imz}, b \right). \tag{8.2.2}$$

Exercise 8.2.5. *Compute*

$$\int_{\mathbb{R}} \frac{dx}{(x - a)(x^2 + 1)}, \quad a \in \mathbb{R}. \tag{8.2.3}$$

Not exactly of the form discussed above, but in the same vein, we have:

Exercise 8.2.6. *Let* $p, q \in \mathbb{N}_0$. *Compute*

$$\int_{\mathbb{R}} \frac{\cos(px) - \cos(qx)}{x^2} dx.$$

8.3   Integrals of rational functions on a half-line

The integrals alluded to in the title are of the form

$$\int_0^\infty r(x)dx$$

where the function $r(x)$ is rational, and the degree of its denominator is larger than the degree of the numerator by at least 2. We begin with a simple case, which is a continuation of Exercise 8.1.2.

Exercise 8.3.1. *Compute for $n = 2, 3, \dots$ the integral*

$$\int_0^\infty \frac{dx}{x^n + 1}.$$ (8.3.1)

In fact, it is not more difficult to compute the previous integral when n is replaced by any real $p > 1$.

Exercise 8.3.2. *Prove that, for any real $p > 1$,*

$$\int_0^\infty \frac{dx}{x^p + 1} = \frac{\pi}{p} \frac{1}{\sin(\frac{\pi}{p})}.$$

The computations in the solutions of the previous two exercises were *ad hoc*, and did not indicate a general method to compute integrals on a half-line. We now describe a general method to tackle such integrals. One considers contours of the following form: The contour is composed of four parts:

(i) $\gamma_{1,r,R,\epsilon}$ is the interval

$$\gamma_{1,r,R,\epsilon}(t) = i\epsilon + t, \quad t \in [r, R].$$

(ii) $\gamma_{2,R,\epsilon}$ is the part of the circle of radius $\sqrt{R^2 + \epsilon^2}$ and center 0, which starts at $i\epsilon + R$ and ends at $-i\epsilon + R$.

(iii) $\gamma_{3,r,R,\epsilon}$ is the interval

$$\gamma_{1,r,R,\epsilon}(t) = -i\epsilon + (R + r - t), \quad t \in [r, R].$$

(iv) $\gamma_{4,r,\epsilon}$ is the part of the circle of radius $\sqrt{\epsilon^2 + r^2}$ which connects the points $i\epsilon + r$ and $-i\epsilon + r$ with parametrization

$$\gamma_{4,r,R,\epsilon}(t) = \sqrt{\epsilon^2 + r^2}e^{it}, \quad t \in [\theta, 2\pi - \theta],$$

where $\theta \in [0, \pi/2]$ is such that

$$\tan \theta = \frac{\epsilon}{r}.$$

One then obtains the following formulas, which appear for instance in [81, p. 178] (and in most chapters on computations in integrals in complex analysis books). We leave the proofs as an exercise to the reader.

$$\int_0^\infty r(x)dx = -\sum_{z\neq 0} \operatorname{Res}(r(s)\ln s, z), \tag{8.3.2}$$

$$\int_0^\infty x^\alpha r(x)dx = \frac{2\pi i}{1-e^{2i\pi\alpha}} \sum_{z\neq 0} \operatorname{Res}(s^\alpha r(s), z). \tag{8.3.3}$$

In these expressions, for

$$s = \rho e^{i\theta}, \quad \text{with} \quad 0 < \theta < 2\pi,$$

we set:

$$s^\alpha = \exp\{\alpha(\ln\rho + i\theta)\} \quad \text{and} \quad \ln s = \ln\rho + i\theta.$$

As a first application we give yet another computation of the integral (8.3.1), where $n \in \{2, 3, 4, \ldots\}$.

Exercise 8.3.3. *Compute* (8.3.1) *using formula* (8.3.2).

We now use formula (8.3.2) to compute the integral appearing in the solution of Exercise 4.1.8; see (4.5.2) there.

Exercise 8.3.4. *Compute, for* $\theta \in (0, 2\pi)$,

$$\int_0^\infty \frac{du}{u^2 - 2u\cos\theta + 1}.$$

Integrals on $(-\infty, 0]$ are treated in a similar way, and the reader will easily adapt formulas (8.3.2)–(8.3.3). One now takes the determination of the logarithm with $\theta \in (-\pi, \pi)$, and we get

$$\int_{-\infty}^0 r(x)dx = \sum_{z\neq 0} \operatorname{Res}(r(s)\ln s, z),$$

$$\int_{-\infty}^0 |x|^\alpha r(x)dx = \frac{2\pi i}{e^{-i\pi\alpha} - e^{i\pi\alpha}} \sum_{z\neq 0} \operatorname{Res}(s^\alpha r(s), z), \tag{8.3.4}$$

where

$$s^\alpha = e^{\alpha(\ln|s|+i\theta)} \quad \text{for} \quad s \in \mathbb{C} \setminus (-\infty, 0].$$

One can also make a real change of variable $x \mapsto -x$ to go from one set of formulas to the other.

As an application we have the following:

Exercise 8.3.5. *Let $\alpha \in (0,1)$. Prove that for $z \in \mathbb{C} \setminus (-\infty, 0]$,*

$$z^{\alpha} = \frac{\sin \pi \alpha}{\pi} \int_{-\infty}^{0} |x|^{\alpha} \left\{ \frac{1}{x-z} - \frac{x}{x^2+1} \right\} dx + \cos \frac{\pi \alpha}{2}. \tag{8.3.5}$$

We remark the following: Let $z = \rho e^{i\theta}$, where $\theta \in (-\pi, \pi)$. For $\alpha \in (0,1)$ the function

$$z^{\alpha} = \rho^{\alpha}(\cos \alpha \theta + i \sin \alpha \theta)$$

is analytic in $\mathbb{C} \setminus (-\infty, 0]$, and has a positive imaginary part in the open upper half-plane; when $\alpha \in (-1, 0)$, it is the function $-z^{\alpha}$ which has a positive imaginary part in the open upper half-plane. These functions belong therefore to the Pick class, and admit a representation of the form (5.5.21), multiplied by the constant i if $\alpha \in (0,1)$ and $-i$ if $\alpha \in (-1, 0)$. This was verified directly in the preceding exercise. For $\alpha = \frac{1}{2}$ we get back the formula in [67, p. 27] (note that $\sqrt{\lambda}$ should be $\sqrt{|\lambda|}$ in the formula there).

The following two examples are taken from [116, p. 418].

Question 8.3.6. *Let $r \in (0,1)$. Show that*

$$\int_{0}^{\infty} \frac{t^r}{1+t^2} dt = \frac{\pi}{2 \cos \left(\frac{\pi r}{2} \right)}, \tag{8.3.6}$$

$$\int_{0}^{\infty} \frac{t^r}{(1+t)^2} dt = \frac{\pi r}{\sin(\pi r)}. \tag{8.3.7}$$

Remark 8.3.7. It follows from (8.3.6) and (8.3.7) that the functions

$$K_1(r, s) = \frac{\pi}{2 \cos \left(\frac{\pi(s+r)}{2} \right)}, \tag{8.3.8}$$

$$K_2(r, s) = \frac{\pi(r + s)}{\sin(\pi(r + s))} \tag{8.3.9}$$

are positive definite in $[0, 1/2)$; see Definition 16.3.11. The first one is moreover a *complete Nevanlinna–Pick kernel*, meaning that $1/K_1$ has one positive square. This can be seen from

$$\frac{1}{K_1(t, s)} = \frac{2}{\pi} \left(\cos \left(\frac{r}{2} \right) \cos \left(\frac{s}{2} \right) - \sin \left(\frac{r}{2} \right) \sin \left(\frac{s}{2} \right) \right).$$

See [CAPB2, p. 61 and p. 366] for a discussion of complete Nevanlinna–Pick kernels, and connections with a theorem of Kaluza (see [133] and [114, Theorem 22, p. 68] for the latter).

Exercise 8.3.8. *Compute, for $z \in \mathbb{C} \setminus (-\infty, 0]$,*

$$g(z) = \int_{-\infty}^{0} \left\{ \frac{1}{x-z} - \frac{x}{x^2+1} \right\} dx. \tag{8.3.10}$$

A quick solution would be as follows: We have that

$$g'(z) = \int_{-\infty}^{0} \frac{dx}{(x-z)^2},$$

which in turn is equal to

$$g'(z) = \frac{-1}{x-z}\Big|_{-\infty}^{0} = \frac{1}{z},$$

and hence $g(z) = \ln z$, up to an additive constant. This constant is shown to be equal to 0 by computing $g(i)$. Indeed,

$$g(i) = \int_{-\infty}^{0} \frac{xi+1}{(x-i)(x^2+1)}dx = \int_{-\infty}^{0} \frac{i}{x^2+1}dx = i\frac{\pi}{2} = \ln i.$$

We leave it to the reader to check this formula using the residue theorem.

We conclude this section with two integrals on $[0,\infty)$ which require a different, and simpler approach. To prove them it suffices to differentiate both sides and see that equality holds. See [32, p. 76]. In the second formula, Γ denotes the Gamma function.

Exercise 8.3.9. *Show that, for* $\operatorname{Re} z > 0$*, the formulas*

$$\ln(1+z) = \int_{0}^{\infty} (1 - e^{-zu})\frac{e^{-u}}{u}du, \tag{8.3.11}$$

$$z^\alpha = \frac{\alpha}{\Gamma(1-\alpha)} \int_{0}^{\infty} (1 - e^{-zu})\frac{du}{u^{\alpha+1}}, \tag{8.3.12}$$

where $\alpha \in (0,1)$*, hold.*

8.4 Integrals of rational expressions of the trigonometric functions

Exercise 8.4.1. *Show that, for* $a \in (-1,1)$*,*

$$\int_{0}^{2\pi} \frac{dt}{1 + a\cos t} = \frac{2\pi}{\sqrt{1-a^2}}. \tag{8.4.1}$$

In relation with the preceding exercise, see also Exercises 6.3.5 and 5.5.7.

More precisely, an application of Weierstrass' theorem to (8.4.1) leads to

$$\frac{2\pi}{\sqrt{1-a^2}} = \int_0^{2\pi} \frac{dt}{1 + a\cos t}$$

$$= \sum_{n=0}^{\infty} (-a)^n \int_0^{2\pi} \cos^n t\, dt$$

$$= \sum_{p=0}^{\infty} a^{2p} \int_0^{2\pi} \cos^{2p} t\, dt,$$

since the integrals corresponding to odd values of n vanish. Hence

$$\frac{2\pi}{\sqrt{1-a^2}} = \sum_{p=0}^{\infty} a^{2p} \int_0^{2\pi} \cos^{2p} t\, dt.$$

On the other hand it is well known that (see, e.g., [204, p. 135])

$$\frac{1}{\sqrt{1-a^2}} = 1 + \frac{1}{2}a^2 + \frac{1\cdot 3}{2\cdot 4}a^4 + \frac{1\cdot 3\cdot 5}{2\cdot 4\cdot 6}a^6 + \cdots.$$

Hence we return to the formula (3.9.6)

$$\int_0^{2\pi} \cos^{2p} t\, dt = 2\pi \frac{\binom{2p}{p}}{2^{2p}},$$

since, as is shown by an easy induction,

$$\frac{\binom{2p}{p}}{2^{2p}} = \frac{1\cdot 3\cdots (2p-1)}{2\cdot 4\cdots (2p)}.$$

Exercise 8.4.2. *Show that*

$$\int_0^{2\pi} \frac{dt}{1 + 8\cos^2 t} = \frac{2\pi}{3}.$$

(See [184, p. 217]).

8.5   Other examples

The residue theorem can also be used in numerous other instances to derive formulas. For example, it can be used to sum series of the form

$$\sum_{\mathbb{Z}} f(n) \quad \text{and} \quad \sum_{\mathbb{Z}} (-1)^n f(n)$$

where f is meromorphic in $\mathbb{C}$, and subject to some supplementary hypothesis. We present here an example, and prove, without resorting to Poisson's summation formula, the formula (see [40, (10), p. 357])

$$2a \sum_{n \in \mathbb{Z}} \frac{1}{a^2 + (\theta - 2\pi n)^2} = \frac{1 - r^2}{1 + r^2 - 2r \cos \theta}, \quad \theta \in \mathbb{R}, \tag{8.5.1}$$

where a and r are real and related by $r = e^{-a}$. We use the formula (see [36, (7.3), p. 129])

$$\sum_{n \in \mathbb{Z}} f(n) = - \sum_{z_k, \text{ pole of } f} \text{Res}(\pi f(z) \cot(\pi z), z_k), \tag{8.5.2}$$

where f is a function meromorphic in $\mathbb{C}$, whose poles are not in $\mathbb{Z}$, and such that the sum $\sum_{n \in \mathbb{Z}} f(n)$ converges. We take

$$f(z) = \frac{1}{a^2 + (\theta - 2\pi z)^2},$$

which has two poles, namely $z_+ = \frac{\theta + ia}{2\pi}$ and $z_- = \frac{\theta - ia}{2\pi}$. Using (8.5.2) we see that the sum is equal to

$$-\left\{ \text{Res}\left(\pi \frac{\cot(\pi z)}{a^2 + (\theta - 2\pi z)^2}, z_+ \right) + \text{Res}\left(\pi \frac{\cot(\pi z)}{a^2 + (\theta - 2\pi z)^2}, z_- \right) \right\}$$

$$= -\pi \left\{ \left(\frac{\cot(\pi z)}{4\pi(2\pi z - \theta)} \right)_{z = z_+} + \left(\frac{\cot(\pi z)}{4\pi(2\pi z - \theta)} \right)_{z = z_-} \right\}$$

$$= -\frac{1}{4ia} \left\{ \cot\left(\frac{\theta + ia}{2} \right) - \cot\left(\frac{\theta - ia}{2} \right) \right\}$$

$$= -\frac{1}{4ia} \frac{\cos(\frac{\theta + ia}{2}) \sin(\frac{\theta - ia}{2}) - \cos(\frac{\theta - ia}{2}) \sin(\frac{\theta + ia}{2})}{\sin(\frac{\theta - ia}{2}) \sin(\frac{\theta + ia}{2})}.$$

We now remark that

$$\cos\left(\frac{\theta + ia}{2} \right) \sin\left(\frac{\theta - ia}{2} \right) - \cos\left(\frac{\theta - ia}{2} \right) \sin\left(\frac{\theta + ia}{2} \right) = -\sin(ia).$$

To check this equality, it suffices to note that the derivative of the function on the left (with respect to θ) is identically 0, and hence the function is equal to its value at the origin. This formula, together with trigonometric equalities and the connections between trigonometric and hyperbolic functions (see (1.2.14) and

(1.2.15)) leads to

$$-\frac{1}{4ia}\frac{\cos(\frac{\theta+ia}{2})\sin(\frac{\theta-ia}{2}) - \cos(\frac{\theta-ia}{2})\sin(\frac{\theta+ia}{2})}{\sin(\frac{\theta-ia}{2})\sin(\frac{\theta+ia}{2})}$$

$$= \frac{1}{4ia}\frac{\sin ia}{(\sin(\frac{\theta}{2}))^2(\cos(\frac{ia}{2}))^2 - (\cos(\frac{\theta}{2}))^2(\sin(\frac{ia}{2}))^2}$$

$$= \frac{\sinh a}{4a\left((\sin(\frac{\theta}{2}))^2(\cosh(\frac{a}{2}))^2 - (\cos(\frac{\theta}{2}))^2(\sinh(\frac{a}{2}))^2\right)}$$

$$= \frac{\sinh a}{4a(\cosh^2(a/2) - \cos^2(\theta/2))}$$

$$= \frac{\sinh a}{2a(\cosh a - \cos\theta)}.$$

To conclude we note that

$$\sinh a = \frac{1/r - r}{2} = \frac{1 - r^2}{2r} \quad \text{and} \quad \cosh a = \frac{1/r + r}{2} = \frac{1 + r^2}{2r}.$$

Inserting these formulas in the last equation above we obtain (8.5.1). We refer to (3.9.15) for a related formula. Using the second formula in (3.9.15) we thus obtain, when $r < 1$,

$$\sum_{n\in\mathbb{Z}} \frac{2a}{a^2 + (\theta - 2\pi n)^2} = (1 - r^2)\sum_{n=1}^{\infty} r^{n-1}\frac{\sin(n\theta)}{\sin\theta}.$$

We now turn to the computation of the Gaussian integral, promised before Exercise 5.2.7. We follow [91, Exercise 17, p. 193], [27, p. 15]. The latter quotes [187] who himself quotes [137]. For another proof using the residue theorem, see [42, p. 381]. Admittedly we do not provide the motivation for the choice of the given contour and function.

Exercise 8.5.1. *For $R > 0$, consider the parallelogram Γ_R with vertices $-R$, R, $R + (1+i)\sqrt{\frac{\pi}{2}}$ and $-R + (1+i)\sqrt{\frac{\pi}{2}}$. Compute the Gaussian integral using the residue theorem for the function*

$$g(z) = \frac{e^{-z^2}}{1 + e^{-z(1+i)\sqrt{2\pi}}} \tag{8.5.3}$$

along this contour.

Let f be a function in $\mathbf{L}_\infty(\mathbb{R})$ (or, continuous and bounded on the real line if you want to avoid measurable functions). Then, the function

$$F(z) = \frac{1}{2\pi i}\int_{\mathbb{R}} \frac{f(u)du}{(u - z)^2} \tag{8.5.4}$$

is analytic in $\mathbb{C}\setminus\mathbb{R}$. We now want to inquire when is F analytic across the real axis.

Exercise 8.5.2. *Assume that f is entire, and that f is bounded in the closed upper half-plane. Then, the function F_+ defined by (8.5.4) in $\mathbb{C}_+$ has an extension to the whole of $\mathbb{C}$. Is this extension equal to the restriction F_- of F to the open lower half-plane $\mathbb{C}_-$?*

Exercise 8.5.3. *Let $\epsilon > 0$ and let f be analytic in the open half-plane $\operatorname{Im} z > -\epsilon$, and assume that*

$$\lim_{R \to \infty} \frac{\int_0^\pi |f(Re^{it})|dt}{R} = 0, \tag{8.5.5}$$

and

$$\int_{\mathbb{R}} \frac{|\operatorname{Re} f(x)|}{x^2 + 1} dx < \infty. \tag{8.5.6}$$

(These conditions hold in particular when f is bounded in the closed upper half-plane.)

(a) *Show that for all $z \in \mathbb{C}_+$ it holds that*

$$f(z) = i \operatorname{Im} f(i) + \frac{1}{\pi i} \int_{\mathbb{R}} \left\{ \frac{1}{x - z} - \frac{x}{x^2 + 1} \right\} \operatorname{Re} f(x)dx. \tag{8.5.7}$$

(b) *Assume that $\operatorname{Re} f(x) \geq 0$ on the real line. Show that*

$$\operatorname{Re} f(z) \geq 0, \quad z \in \mathbb{C}_+,$$

and show that the inequality is strict, unless $f \equiv 0$.

Exercise 8.5.4. *Compute for $z \in \mathbb{C} \setminus \mathbb{R}$,*

$$g(z) = \frac{1}{2\pi i} \int_{\mathbb{R}} \left\{ \frac{1}{x - z} - \frac{x}{x^2 + 1} \right\} dx.$$

8.6 Solutions

Solution of Exercise 8.1.1. (1) We note that ia and $i\bar{b}$ are in the open upper half-plane $\mathbb{C}_+$. Let

$$f(z) = \frac{1}{(z^2 + a^2)(z^2 + \bar{b}^2)}.$$

We first suppose $a \neq \bar{b}$. Then

$$\int_{\mathbb{R}} \frac{dx}{(x^2 + a^2)(x^2 + \bar{b}^2)} = 2\pi i \left\{ \operatorname{Res}(f, ia) + \operatorname{Res}(f, i\bar{b}) \right\}$$

$$= 2\pi i \left\{ \frac{1}{2ia(-a^2 + \bar{b}^2)} + \frac{1}{2i\bar{b}(a^2 - \bar{b}^2)} \right\}$$

$$= \frac{\pi}{a\bar{b}(a + \bar{b})}.$$

Now assume $a = \bar{b}$. Then

$$f(z) = \frac{1}{(z^2 + a^2)^2} = \frac{1}{(z + ia)^2 (z - ia)^2},$$

and we have

$$\operatorname{Res}(f, ia) = \left(\frac{1}{(z + ia)^2}\right)' \Big|_{z=ia} = -\frac{2}{(z + ia)^3}\Big|_{z=ia} = -\frac{2}{(2ia)^3} = \frac{-i}{4a^3}.$$

Thus,

$$\int_{\mathbb{R}} \frac{dx}{(x^2 + a^2)^2} = 2\pi i \operatorname{Res}(f, ia) = 2\pi i \frac{-i}{4a^3} = \frac{\pi}{2a^3}, \tag{8.6.1}$$

which coincides with the formula

$$\frac{\pi}{a\bar{b}(a + \bar{b})}$$

for $a = \bar{b}$.

(2) Set $f_a(x) = \frac{a}{x^2 + a^2}$. We have, for $a, b \in \mathbb{C}_r$,

$$\frac{1}{a + \bar{b}} = \int_{\mathbb{R}} f_a(x)\overline{f_b(x)}dx. \tag{8.6.2}$$

Therefore, for $N \in \mathbb{N}$, $a_1, \ldots, a_N \in \mathbb{C}_r$ and $c_1, \ldots, c_N \in \mathbb{C}$ we have

$$\sum_{\ell,j=1}^{N} \frac{\overline{c_\ell} c_j}{a_\ell + \overline{a_j}} = \int_{\mathbb{R}} \left(\sum_{\ell=1}^{N} \overline{c_\ell} f_{a_\ell}(x)\right) \overline{\left(\sum_{j=1}^{N} \overline{c_j} f_{a_j}(x)\right)} dx \geq 0. \tag{8.6.3}$$

$\square$

We remark that (8.6.2) expresses $\frac{1}{a+\bar{b}}$ as an inner product, and so we can conclude already from this equation that $\frac{1}{a+\bar{b}}$ is a positive definite kernel. We leave it to the student to check that the inequality in (8.6.3) is strict if all the a_ℓ are different and if at least one of the $c_\ell \neq 0$. Furthermore, the choice $a_\ell = \ell - \frac{1}{2}$ leads to: For every N the matrix

$$\begin{pmatrix} 1 & \frac{1}{2} & \frac{1}{3} & \cdots & \frac{1}{1+N} \\ \frac{1}{2} & \frac{1}{3} & \frac{1}{4} & \cdots & \frac{1}{2+N} \\ & & & & \\ \frac{1}{N+1} & \frac{1}{N+2} & & \cdots & \frac{1}{2N+1} \end{pmatrix} > 0, \quad \text{(and not only } \geq 0\text{)}.$$

Such a matrix, constant on the anti-diagonals, is an example of a Hankel matrix.

Solution of Exercise 8.1.2. We will show that the integral is equal to

$$\int_{\mathbb{R}} \frac{dx}{x^{2n} + 1} = \frac{\frac{\pi}{n}}{\sin \frac{\pi}{2n}}. \tag{8.6.4}$$

(See [45, Exercice 23, p. 117].) The zeroes of $z^{2n} + 1$ are

$$z_\ell = \exp i \left(\frac{\pi}{2n} + \frac{2\ell\pi}{2n} \right), \quad \ell = 0, 1, \ldots, 2n - 1.$$

The choices $\ell = 0, \ldots, n - 1$ correspond to the zeroes in the upper half-plane and so, by the residue theorem,

$$\int_{\mathbb{R}} \frac{dx}{x^{2n} + 1} = 2\pi i \left\{ \sum_{\ell=0}^{n-1} \text{Res} \left(\frac{1}{z^{2n} + 1}, z_\ell \right) \right\}$$

$$= 2\pi i \left\{ \sum_{\ell=0}^{n-1} \frac{1}{2n z_\ell^{2n-1}} \right\}$$

$$= -2\pi i \left\{ \sum_{\ell=0}^{n-1} \frac{z_\ell}{2n} \right\},$$

where we multiplied the denominator and numerator of the ℓth term by z_ℓ and used $z_\ell^{2n} = -1$ to go from the penultimate line to the last line.

We have

$$\sum_{\ell=0}^{n-1} z_\ell = e^{(i\pi)/2n} \sum_{\ell=0}^{n-1} (e^{(i\pi)/n})^\ell$$

$$= e^{(i\pi)/2n} \frac{1 - e^{(i\pi n)/n}}{1 - e^{(i\pi)/n}}$$

$$= \frac{2}{-2i \sin(\frac{\pi}{2n})},$$

and hence the result.

The computation of the second integral is done in much the same way: By the residue theorem,

$$\int_{\mathbb{R}} \frac{x^2 dx}{x^{2n} + 1} = 2\pi i \left\{ \sum_{\ell=0}^{n-1} \text{Res} \left(\frac{z^2}{z^{2n} + 1}, z_\ell \right) \right\}$$

$$= 2\pi i \left\{ \sum_{\ell=0}^{n-1} \frac{z_\ell^2}{2n z_\ell^{2n-1}} \right\}$$

$$= -2\pi i \left\{ \sum_{\ell=0}^{n-1} \frac{z_\ell^3}{2n} \right\}.$$

Furthermore, we have

$$\sum_{\ell=0}^{n-1} z_\ell^3 = e^{(3i\pi)/2n} \sum_{\ell=0}^{n-1} (e^{(3i\pi)/n})^\ell$$

$$= e^{(3i\pi)/2n} \frac{1 - e^{(3i\pi n)/n}}{1 - e^{(3i\pi)/n}}$$

$$= \frac{2}{-2i \sin(\frac{3\pi}{2n})}$$

and hence

$$\int_{\mathbb{R}} \frac{x^2 dx}{x^{2n} + 1} = \frac{\frac{\pi}{n}}{\sin \frac{3\pi}{2n}}. \tag{8.6.5}$$
□

We note the following: Letting $n \to \infty$ the integral (8.6.4) tends to 2. On the other hand the functions

$$f_n(x) = \frac{1}{x^{2n} + 1}$$

tend pointwise to the function

$$f(x) = \begin{cases} 1, & x \in (-1, 1), \\ \frac{1}{2}, & x = \pm 1, \\ 0, & |x| > 1, \end{cases}$$

and

$$f_n(x) \le f(x).$$

The dominated convergence theorem then allows the interchange of limit and integral in

$$\lim_{n\to\infty} \int_{\mathbb{R}} \frac{dx}{x^{2n} + 1} = \int_{\mathbb{R}} \left(\lim_{n\to\infty} \frac{1}{x^{2n} + 1} \right) dx = \int_{-1}^{1} dx = 2.$$

Similarly, (8.6.5) tends to $2/3$ as $n \to \infty$. This is consistent with

$$\lim_{n\to\infty} \int_{\mathbb{R}} \frac{x^2 dx}{x^{2n} + 1} = \int_{\mathbb{R}} \left(\lim_{n\to\infty} \frac{x^2}{x^{2n} + 1} \right) dx = \int_{-1}^{1} x^2 dx = \frac{2}{3},$$

where here too, the dominated convergence theorem allows us to interchange limit and integral.

Solution of Exercise 8.1.4. We use the notation of the solution of the previous exercises. As above we have

$$\int_{\mathbb{R}} \frac{dx}{(x^{2n} + 1)^2} = 2\pi i \left\{ \sum_{\ell=0}^{n-1} \text{Res} \left(\frac{1}{(z^{2n} + 1)^2}, z_\ell \right) \right\},$$

but now the points z_ℓ are zeros of multiplicity of the denominator. We set $g(z) = (z^{2n} + 1)^2$ and use formula (7.3.6) to compute the residue. Setting $f = 1$ in (7.3.6) we have:

$$\text{Res}\left(\frac{1}{(x^{2n}+1)^2}, z_\ell\right) = -\frac{\dfrac{g^{(3)}(z_\ell)}{3!}}{\left(\dfrac{g^{(2)}(z_\ell)}{2!}\right)^2}, \quad \ell \in \{0, \ldots, n-1\}. \tag{8.6.6}$$

$$g^{(2)} = 2u^{(2)}u + 2(u^{(1)})^2$$
$$g^{(3)} = 2u^{(3)}u + 6u^{(2)}u.$$

Since $u(z_\ell) = 0$ we obtain

$$g^{(2)}(z_\ell) = 2\left(2nz_\ell^{2n-1}\right)^2$$
$$= 8n^2 z_\ell^{-2}$$

and

$$g^{(3)}(z_\ell) = 6\left(2n(2n-1)z_\ell^{2n-2}\right) \cdot 2nz_\ell^{2n-1}$$
$$= 24n^2(2n-1)z_\ell^{-3}.$$

Thus,

$$\text{Re}\left(\frac{1}{(x^{2n}+1)^2}, z_\ell\right) = -\frac{\dfrac{4n^2(2n-1)}{z_\ell^{-3}}}{16n^4 z_\ell^{-4}}$$
$$= -\frac{2n-1}{2n}\frac{1}{2n}z_\ell.$$

Since $\sum_{\ell=0}^{n-1} z_\ell = -\frac{1}{i\sin(\frac{\pi}{2n})}$ we obtain

$$\int_{\mathbb{R}} \frac{1}{(x^{2n}+1)^2}dx = \frac{2n-1}{2n}\frac{\pi}{n\sin(\frac{\pi}{2n})} = \frac{2n-1}{2n}\int_{\mathbb{R}}\frac{1}{(x^{2n}+1)}dx. \tag{8.6.7}$$

$\square$

Solution of Exercise 8.1.5. Set $I_n = \int_{\mathbb{R}} \frac{1}{x^{2n}+1}dx$. For $t \in (0,1)$ we have

$$\sum_{p=1}^{\infty} \frac{t^p}{(x^{2n}+1)^p} = \frac{t}{x^{2n}+1-t}.$$

By Exercise 8.1.2 we have (with the change of variable $x \mapsto \sqrt[2n]{1-tu}$)

$$\int_{\mathbb{R}} \frac{t}{x^{2n}+1-t}dx = t \cdot \sqrt[2n]{1-t}^{1-2n}I_n, \tag{8.6.8}$$

and the monotone convergence theorem for series of positive functions (see Theorem 17.5.3) gives

$$\sum_{p=1}^{\infty} t^p \int_{\mathbb{R}} \frac{1}{(x^{2n}+1)^p} dx = \int_{\mathbb{R}} \left(\sum_{p=1}^{\infty} t^p \frac{1}{(x^{2n}+1)^p} \right) dx$$

$$= \int_{\mathbb{R}} \frac{t}{x^{2n}+1-t} dx$$

$$= t \cdot \sqrt[2n]{1-t}\, I_n.$$

Writing

$$t \cdot \sqrt[2n]{1-t} = t \left(1 + t\frac{2n-1}{2n} + \frac{t^2}{2} \frac{2n-1}{2n} \left(1 - \frac{2n-1}{2n} \right) + \cdots \right)$$

$$= \sum_{p=1}^{\infty} t^p c_{p,n},$$

where the coefficients $c_{p,n}$ are computed as in (4.4.9), we have

$$\int_{\mathbb{R}} \frac{1}{(x^{2n}+1)^p} dx = c_{p,n} \int_{\mathbb{R}} \frac{1}{x^{2n}+1} dx, \quad p=1,2,\dots.$$

The special case $p=2$ is proved in the previous exercise; see (8.6.8).   □

Solution of Exercise 8.1.6. Let

$$f(z) = \frac{1}{(z^2+1)^n}.$$

We have

$$\int_{\mathbb{R}} \frac{dx}{(x^2+1)^n} = 2\pi i \operatorname{Res}(f,i).$$

Let $g(z) = (z+i)^{-n}$. We have, for $n \geq 2$,

$$g^{(n-1)}(z) = (-1)^{(n-1)} \left(\prod_{\ell=0}^{n-2} (n+\ell) \right) (z+i)^{-2n+1},$$

and so

$$\operatorname{Res}(f,i) = \frac{g^{(n-1)}(i)}{(n-1)!}$$

$$= \frac{(-1)^{(n-1)} \prod_{\ell=0}^{n-2}(n+\ell)}{(2i)^{2n-1}(n-1)!}$$

$$= -i \frac{\prod_{\ell=0}^{n-2}(n+\ell)}{2^{2n-1}(n-1)!}.$$

Thus

$$\int_{\mathbb{R}} \frac{dx}{(x^2+1)^n} = \frac{2\pi \prod_{\ell=0}^{n-2}(n+\ell)}{2^{2n-1}(n-1)!} = \frac{2\pi(2n-2)!}{2^{2n-1}((n-1)!)^2}. \qquad \square$$

As a check, we note that $n = 2$ in the above formula leads to the formula (8.6.1) when $a = 1$.

Solution of Exercise 8.1.7. Let

$$f(z) = \frac{z}{(z^2+z+1)^2} = \frac{z}{(z-z_-)^2(z-z_+)^2},$$

with

$$z_{\pm} = \frac{-1 \pm i\sqrt{3}}{2}.$$

The point $z_+ = (-1 + i\sqrt{3})/2$ is in the open upper half-plane, and we have that

$$\int_{\mathbb{R}} \frac{x\,dx}{(x^2+x+1)^2} = 2\pi i \operatorname{Res}\ (f, z_+)$$

$$= 2\pi i \left(\frac{z}{(z-z_-)^2} \right)' \bigg|_{z=z_+}$$

$$= 2\pi i \left(\frac{1}{(z-z_-)^2} - \frac{2z}{(z-z_-)^3} \right) \bigg|_{z=z_+}$$

$$= 2\pi i \left(\frac{1}{(z_+-z_-)^2} - \frac{2z_+}{(z_+-z_-)^3} \right)$$

$$= 2\pi i \left(\frac{1}{(i\sqrt{3})^2} - \frac{(-1+i\sqrt{3})}{(-i3\sqrt{3})} \right) = -\frac{2\pi}{3\sqrt{3}}. \qquad \square$$

Solution of Exercise 8.1.8. Let

$$f(z) = \frac{1}{(z^2+1)(z-2i)^2(z-3i)^3(z-4i)^4}.$$

We have

$$\int_{\mathbb{R}} \frac{dx}{(x^2+1)(x-2i)^2(x-3i)^3(x-4i)^4} = 2\pi i \{\operatorname{Res}(f,i) + \operatorname{Res}\ (f,2i)$$

$$+ \operatorname{Res}\ (f,3i) + \operatorname{Res}\ (f,4i)\}.$$

We know by Exercise 7.3.9, that the sum of all the residues of f is equal to 0. Thus

$$\operatorname{Res}(f,i) + \operatorname{Res}\ (f,2i) + \operatorname{Res}\ (f,3i) + \operatorname{Res}\ (f,4i) = -\operatorname{Res}\ (f,-i)$$

$$= -\frac{1}{(-2i)(-3i)^2(-4i)^3(-5i)^4}$$

$$= \frac{1}{2 \cdot 9 \cdot 4^3 \cdot 5^4},$$

and hence

$$\int_{\mathbb{R}} \frac{dx}{(x^2+1)(x-2i)^2(x-3i)^3(x-4i)^4} = \frac{2\pi i}{2 \cdot 9 \cdot 4^3 \cdot 5^4}. \qquad \Box$$

The reader should check directly that the above integral is indeed purely imaginary without computing it explicitly.

Solution of Exercise 8.2.1. (1) The case $t = 0$ is clear and can also be obtained from Exercise 8.1.2 if need be. We now assume $t > 0$, and define

$$f(z) = \frac{e^{itz}}{z^2+1}.$$

The function $z \mapsto e^{itz}$ is bounded in the closed upper half-plane, and the residue theorem gives us

$$\int_{\mathbb{R}} f(x)dx = 2\pi i \operatorname{Res}(f, i) = 2\pi i \frac{e^{-t}}{2i} = \pi e^{-t}.$$

By symmetry it follows that

$$\int_{\mathbb{R}} \frac{e^{itx}}{x^2+1} dx = \pi e^{-|t|}, \quad t \in \mathbb{R}.$$

(2) We have, for $t, s \in \mathbb{R}$,

$$e^{-|t-s|} = \frac{1}{\pi} \int_{\mathbb{R}} \frac{e^{ix(t-s)}}{x^2+1} dx. \qquad (8.6.9)$$

Thus, for $N \in \mathbb{N}, t_1, \ldots, t_N \in \mathbb{R}$ and $c_1, \ldots, c_N \in \mathbb{C}$ we have

$$\sum_{\ell,j=1}^{N} \overline{c_\ell} c_j e^{-|t_\ell - t_j|} = \frac{1}{\pi} \int_{\mathbb{R}} \frac{\left(\sum_{\ell=1}^{N} \overline{c_\ell} e^{ixt_\ell}\right)\overline{\left(\sum_{j=1}^{N} \overline{c_j} e^{ixt_j}\right)}}{x^2+1} dx$$

$$\geq 0. \qquad \Box$$

As in the case of Exercise 8.1.1, we remark that (8.6.9) expresses $e^{-|t-s|}$ as an inner product, and so we can conclude already from this equation that $e^{-|t-s|}$ is a positive definite kernel, that is $e^{-|t|}$ is positive definite. Hence, the formula

$$e^{-|t|} = \frac{1}{\pi} \int_{\mathbb{R}} \frac{e^{itx}}{x^2+1} dx \qquad (8.6.10)$$

is an illustration of Bochner's theorem. See Theorem 16.3.14 for the latter. It is also an easy computation of a Fourier transform (see Exercise 13.5.1 for the definition of the latter): The Fourier transform of the function $\frac{1}{x^2+1}$ is equal to $\pi e^{-|t|}$.

Solution of Exercise 8.2.2. On the one hand,

$$\int_{\mathbb{R}} \frac{\cos x}{x^2 + 1} dx = \operatorname{Re} \int_{\mathbb{R}} \frac{e^{ix}}{x^2 + 1} dx$$

$$= \operatorname{Re} \left\{ 2\pi i \operatorname{Res} \left(\frac{e^{iz}}{z^2 + 1}, i \right) \right\}$$

$$= \operatorname{Re} \left\{ 2\pi i \frac{e^{-1}}{2i} \right\}$$

$$= \frac{\pi}{e}.$$

On the other hand, the residue theorem applied to the function $\frac{\cos z}{z^2+1}$ in the closed upper half-circle of radius $R > 1$ leads to

$$\int_{-R}^{R} \frac{\cos x}{x^2 + 1} dx + \int_{C_R} \frac{\cos z}{z^2 + 1} dz = 2\pi i \operatorname{Res} \left(\frac{\cos z}{z^2 + 1}, i \right) = 2\pi i \frac{\cosh 1}{2i} = \pi \frac{e + e^{-1}}{2}.$$

Thus

$$\lim_{R \to \infty} \int_{C_R} \frac{\cos z}{z^2 + 1} dz = \pi \left\{ \frac{e + e^{-1}}{2} - e^{-1} \right\} = \pi \sinh 1. \qquad \square$$

Solution of Exercise 8.2.3. Let γ_R denote the contour made of the interval $[-R, R]$ and of the half-circle of radius R and centered at the origin and which is in the upper half-plane. Consider the function $f(z) = e^{iaz}/(z^4 + 1)$. We have

$$\int_{\mathbb{R}} f(x) dx = \lim_{R \to \infty} \int_{\gamma_R} f(z) dz = 2\pi i \left\{ \operatorname{Res} \left(f, e^{\frac{i\pi}{4}} \right) + \operatorname{Res} \left(f, e^{\frac{3i\pi}{4}} \right) \right\}.$$

But for z_0 a zero of $z^4 + 1$ we have

$$\operatorname{Res}(f, z_0) = \frac{e^{iaz_0}}{4z_0^3} = -\frac{z_0 e^{iaz_0}}{4},$$

and so the integral is equal to (the real part of)

$$2\pi i \left(\frac{-e^{\frac{i\pi}{4}} e^{iae^{\frac{i\pi}{4}}}}{4} + \frac{-e^{\frac{3i\pi}{4}} e^{iae^{\frac{3i\pi}{4}}}}{4} \right) = \frac{\pi e^{-\frac{a}{\sqrt{2}}}}{\sqrt{2}} \left(\cos(a/\sqrt{2}) + \sin(a/\sqrt{2}) \right). \qquad \square$$

When $a = 0$, the previous result corresponds to the case $n = 2$ in Exercise 8.6.4.

Solution of Exercise 8.2.4. We first prove that f is indeed analytic for $\operatorname{Im} z > 0$. This requires the interchange of an infinite sum and of an integral. The integral is not on a compact interval, and we cannot use Weierstrass' theorem (Theorem 14.4.1), and we will use the dominated convergence theorem. Let $z_0 \in \mathbb{C}_+$, and let

$$d = \operatorname{dist}(z_0, \mathbb{R}) = \operatorname{Im} z_0.$$

For $z \in \mathbb{C}_+$ such that $|z - z_0| < d$ we have

$$\left| \frac{z - z_0}{u - z_0} \right| < 1, \quad \forall u \in \mathbb{R}.$$

Furthermore,

$$f(z) = \int_{\mathbb{R}} \frac{\cos u}{(u^2 + 1)(u - z_0 - (z - z_0))^2} du$$

$$= \int_{\mathbb{R}} \frac{\cos u}{(u^2 + 1)(u - z_0)^2} \left\{ \sum_{n=1}^{\infty} n \left(\frac{z - z_0}{u - z_0} \right)^{n-1} \right\} du$$

$$= \sum_{n=1}^{\infty} a_n (z - z_0)^{n-1}$$

where

$$a_n = n \int_{\mathbb{R}} \frac{\cos u}{(u^2 + 1)(u - z_0)^{n+1}} du,$$

and where the interchange of sum and integral is done using the dominated convergence theorem. The function f admits a power series development around every point in $\mathbb{C}_+$, and is therefore analytic there.

We note that formula (8.2.1) makes sense for $\operatorname{Im} z > 0$ but does not make sense for z on the real line. The fact that there is an analytic extension to the asserted set requires us to find another formula for f. We now compute directly f using the residue theorem. We have

$$f(z) = \frac{1}{2} \left\{ \int_{\mathbb{R}} \frac{e^{iu}}{(u^2 + 1)(u - z)^2} du + \int_{\mathbb{R}} \frac{e^{-iu}}{(u^2 + 1)(u - z)^2} du \right\}. \qquad (8.6.11)$$

Therefore, for z in the open upper half-plane,

$$\int_{\mathbb{R}} \frac{e^{iu}}{(u^2 + 1)(u - z)^2} du = 2\pi i \left\{ \operatorname{Res} \left(\frac{e^{iu}}{(u^2 + 1)(u - z)^2}, i \right) \right.$$

$$\left. + \operatorname{Res} \left(\frac{e^{iu}}{(u^2 + 1)(u - z)^2}, z \right) \right\}$$

$$= 2\pi i \left\{ \frac{e^{-1}}{2i(i - z)^2} + \frac{i e^{iz}(z^2 + 1) - 2z e^{iz}}{(z^2 + 1)^2} \right\}$$

and

$$\int_{\mathbb{R}} \frac{e^{-iu}}{(u^2 + 1)(u - z)^2} du = -2\pi i \operatorname{Res} \left(\frac{e^{-iu}}{(u^2 + 1)(u - z)^2}, -i \right) = -2\pi i \frac{e^{-1}}{-2i(z + i)^2}.$$

From these two formulas we see that the right side of (8.6.11) is analytic in the band

$$-1 < \operatorname{Im} z < 1. \qquad (8.6.12)$$

It follows that the function defined by the right side of (8.2.1) in the upper half-plane has an analytic extension to $\operatorname{Im} z > -1$ given by the formula

$$\pi i \left\{ \frac{e^{-1}}{2i(i-z)^2} + \frac{ie^{iz}(z^2+1) - 2ze^{iz}}{(z^2+1)^2} + \frac{e^{-1}}{2i(z+i)^2} \right\}$$

in the band (8.6.12). $\quad\square$

Solution of Exercise 8.2.5. Using formula (8.2.2), the integral (8.2.3) is equal, in the sense of the principal value, to

$$\int_{\mathbb{R}} \frac{dx}{(x-a)(x^2+1)} = \pi i \operatorname{Re}\left(\frac{1}{(z-a)(z^2+1)}, a\right) + 2\pi i \operatorname{Re}\left(\frac{1}{(z-a)(z^2+1)}, i\right)$$

$$= \pi i \frac{1}{a^2+1} + 2\pi i \frac{1}{(i-a)(2i)}$$

$$= -\frac{\pi a}{a^2+1}.$$

$\quad\square$

Solution of Exercise 8.2.6. We take the function

$$f(z) = \frac{e^{ipz} - e^{iqz}}{z^2}.$$

We integrate f on the contour appearing in Exercise 5.3.2. We obtain

$$\int_{\mathbb{R}} f(x)dx - \pi i \operatorname{Res}(f, 0) = 0,$$

but $\operatorname{Res}(f, 0) = i(p - q)$, and hence

$$\int_{\mathbb{R}} f(x)dx = \pi i i(p - q) = \pi(q - p).$$

$\quad\square$

As a verification, let $p = 0$ and $q = 2$. We obtain

$$\int_{\mathbb{R}} \frac{1 - \cos 2x}{x^2}dx = 2\pi,$$

in accordance with Exercise 5.3.3 and (5.9.15). Similarly, the choice $p = 0$ and $q = 1$ gives

$$\int_{\mathbb{R}} \frac{1 - \cos x}{x^2}dx = \pi.$$

Solution of Exercise 8.3.1. We consider the positively oriented contour Γ_R made of the three following arcs:
(a) The closed interval $[0, R]$:

$$\gamma_{1,R}(x) = x, \quad x \in [0, R].$$

(b) The arc of circle of radius R and angle varying from 0 to $\frac{2\pi i}{n}$:

$$\gamma_{2,R}(t) = Re^{it}, \quad t \in \left[0, \frac{2\pi}{n}\right].$$

(c) The interval joining the point $Re^{\frac{2\pi i}{n}}$ to the origin:

$$\gamma_{3,R}(x) = (R-x)e^{\frac{2\pi i}{n}}, \quad x \in [0, R].$$

For $R > 1$ there is only one zero of $z^n + 1$ inside the contour, namely $z = \exp\frac{i\pi}{n}$. Thus the residue theorem applied to $f(z) = \frac{1}{z^n+1}$ and the contour γ_R leads to

$$\int_0^R \frac{dx}{x^n+1} + \int_0^{\frac{2\pi}{n}} \frac{Rie^{it}dt}{R^ne^{int}+1} - \int_0^R \frac{e^{\frac{2\pi i}{n}}dx}{(R-x)^nx^n+1} = 2\pi i\,\mathrm{Res}\left(\frac{1}{z^n+1}, \frac{i\pi}{n}\right)$$

$$= \frac{2\pi i}{ne^{\frac{(n-1)i\pi}{n}}}$$

$$= -\frac{2\pi i e^{\frac{i\pi}{n}}}{n}.$$

The change of variable $t \mapsto R - t$ and letting $R \to \infty$ lead to

$$\left(1 - e^{\frac{2\pi i}{n}}\right)\int_0^\infty \frac{dx}{x^n+1} = -\frac{2\pi i e^{\frac{i\pi}{n}}}{n}.$$

Thus

$$\int_0^\infty \frac{dx}{x^n+1} = \frac{-2\pi i e^{\frac{i\pi}{n}}}{n(1 - e^{\frac{2\pi i}{n}})} = \frac{\pi}{n\sin\frac{\pi}{n}}. \qquad (8.6.13)$$

□

The dominated convergence theorem shows that

$$\lim_{n\to\infty}\int_0^\infty \frac{dx}{x^n+1} = \int_0^\infty \left(\lim_{n\to\infty}\frac{1}{x^n+1}\right)dx = 1.$$

This is also, of course,

$$\lim_{n\to\infty}\frac{\pi}{n\sin\frac{\pi}{n}}.$$

Solution of Exercise 8.3.2. The solution of the preceding exercise has to be appropriately adapted. Let $\ln z$ be the function

$$\ln z = \ln r + i\theta,$$

with $z = re^{i\theta} \in \mathbb{C} \setminus (-\infty, 0]$ and $\theta \in (-\pi, \pi)$. The function $\frac{1}{1+x^p}$ is the restriction to $(0, +\infty)$ of the function

$$f(z) = \frac{1}{1 + \exp(p\ln z)},$$

which is analytic in $\mathbb{C} \setminus (-\infty, 0]$. The contour in the preceding exercise has to be changed, in order to exclude the origin. For $0 < \epsilon < R < \infty$ we consider the positively oriented contour $\Gamma_{\epsilon,R}$ made of the four following arcs:

(a) The closed interval $[\epsilon, R]$:

$$\gamma_{1,\epsilon,R}(x) = x, \quad x \in [\epsilon, R].$$

(b) The arc of circle of radius R and angle varying from 0 to $\frac{2\pi i}{p}$:

$$\gamma_{2,R}(t) = Re^{it}, \quad t \in \left[0, \frac{2\pi}{p}\right].$$

(c) The interval joining the points $Re^{\frac{2\pi i}{p}}$ and $\epsilon e^{\frac{2\pi i}{p}}$:

$$\gamma_{3,\epsilon,R}(x) = (R - x)e^{\frac{2\pi i}{p}}, \quad x \in [0, R - \epsilon].$$

(d) The arc of circle of radius ϵ and angle varying from $\frac{2\pi i}{p}$ to 0:

$$\gamma_{4,\epsilon}(t) = \epsilon e^{it}, \quad t \in \left[\frac{2\pi}{p}, 0\right].$$

For $R > 1$ there is only one zero of $f(z)$ inside the contour, namely $z_p = e^{\frac{i\pi}{p}}$ since, with $z = \rho e^{i\theta}$,

$$f(z) = 0 \iff \exp(p \ln z) = -1$$
$$\iff ip\theta = i(2k + 1)\pi, \quad k \in \mathbb{Z} \quad \text{and} \quad \exp(p \ln \rho) = 1.$$

The residue of f at z_p is

$$\mathrm{Res}(f, z_p) = \frac{1}{(\exp(p \ln z))'|_{z=z_p}} = \frac{1}{(p/z \exp(p \ln z))|_{z=z_p}}.$$

But $\ln z_p = \frac{i\pi}{p}$ and so

$$\mathrm{Res}(f, z_p) = -\frac{e^{\frac{i\pi}{p}}}{p}.$$

The residue theorem applied to the contour $\Gamma_{\epsilon,R}$, and letting $R \to \infty$ and $\epsilon \to 0$ lead then to the result. $\qquad\square$

Solution of Exercise 8.3.3. By formula (8.3.2), we have:

$$\int_0^\infty \frac{dx}{x^n + 1} = -\sum_{k=0}^{n-1} \mathrm{Res}\left(\frac{\ln(z)}{z^n + 1}, e^{\frac{i\pi}{n} + \frac{2ki\pi}{n}}\right)$$

$$= -\sum_{k=0}^{n-1} \frac{\left(\frac{i\pi}{n} + \frac{2ki\pi}{n}\right)}{ne^{(n-1)\left(\frac{i\pi}{n} + \frac{2ki\pi}{n}\right)}} \qquad (8.6.14)$$

$$= \frac{e^{\frac{i\pi}{n}}}{n^2} \sum_{k=0}^{n-1} (i\pi + 2ki\pi)e^{\frac{2ki\pi}{n}},$$

where the sum begins in fact at $k = 1$, and where we used formula (7.3.2) to compute the residue. But

$$\sum_{k=0}^{n-1} e^{\frac{2ki\pi}{n}} = 0,$$

and (see (3.4.14) with $z = e^{\frac{2i\pi}{n}}$)

$$\sum_{k=0}^{n-1} k e^{\frac{2ki\pi}{n}} = e^{\frac{2i\pi}{n}} \cdot \frac{(n-1)e^{\frac{2ni\pi}{n}} - ne^{\frac{2(n-1)i\pi}{n}} + 1}{(1 - e^{\frac{2i\pi}{n}})^2}$$

$$= e^{\frac{2i\pi}{n}} \cdot \frac{n - 1 - ne^{-\frac{2i\pi}{n}} + 1}{(1 - e^{\frac{2i\pi}{n}})^2}$$

$$= \frac{n}{e^{\frac{2i\pi}{n}} - 1}.$$

Plugging this expression in the last line in (8.6.14) we obtain:

$$\int_0^\infty \frac{dx}{x^n + 1} = \frac{2\pi i e^{\frac{i\pi}{n}}}{n^2} \frac{n}{e^{\frac{2i\pi}{n}} - 1} = \frac{\pi}{n} \frac{1}{\sin(\frac{\pi}{n})}. \qquad \square$$

Solution of Exercise 8.3.4. We use formula (8.3.2). Then

$$r(x) = x^2 - 2x \cos\theta + 1,$$

with zeroes equal to

$$z_\pm = e^{\pm i\theta}.$$

See (1.6.24) with $z_0 = e^{i\theta}$ if need be.

Assume first that $\theta \neq \pi$. Then

$$\mathrm{Res}\left(\frac{\ln(z)}{r(z)}, z_+\right) = \frac{i\theta}{2i \sin\theta} = \frac{\theta}{2 \sin\theta} \quad \text{and} \quad \mathrm{Res}\left(\ln(z)r(z), z_-\right) = -\frac{2\pi - \theta}{2 \sin\theta},$$

and hence the integral is equal to

$$\int_0^\infty \frac{du}{u^2 - 2u \cos\theta + 1} = \frac{\pi - \theta}{\sin\theta}. \qquad (8.6.15)$$

The case $\theta = \pi$ is treated in the same way; r has now a zero of multiplicity 2, and

$$\mathrm{Res}\left(\frac{\ln z}{r(z)}, -1\right) = -1.$$

One sees also that formula

$$\frac{\pi - \theta}{\sin\theta}$$

is extended continuously for $\theta = \pi$ by taking its value to be equal to 1 at this point. $\qquad \square$

As a check for (8.6.15), let us put $\theta = \pi/2$. Then, the right side of (8.6.15) is equal to $\pi/2$. On the other hand, now $\cos\theta = 0$, and the left side of (8.6.15) is equal to

$$\int_0^\infty \frac{du}{u^2+1} = \frac{\pi}{2},$$

as can be seen from formula (8.6.4) with $n = 1$ (this formula gives twice the integral $\int_0^\infty \frac{du}{u^2+1}$).

Solution of Exercise 8.3.5. Let

$$g(s) = s^\alpha \frac{sz+1}{(s-z)(s^2+1)}, \qquad s \in \mathbb{C} \setminus (-\infty, 0],$$

where z, viewed as a parameter, also belongs to $\mathbb{C} \setminus (-\infty, 0]$. We have

$$\int_{-\infty}^0 |x|^\alpha \left\{ \frac{1}{x-z} - \frac{x}{x^2+1} \right\} dx = \int_{-\infty}^0 |x|^\alpha \frac{xz+1}{(x-z)(x^2+1)} dx$$

$$= \frac{2\pi i}{e^{i\pi\alpha} - e^{-i\pi\alpha}} \{\text{Res}(g, z)$$

$$+ \text{Res}(g, i) + \text{Res}\,(g, -i)\}.$$

By definition of s^α, we have

$$i^\alpha = e^{\frac{i\pi\alpha}{2}} \quad \text{and} \quad (-i)^\alpha = e^{-\frac{i\pi\alpha}{2}},$$

so that the residues are equal to

$$\text{Res}(g, z) = z^\alpha \frac{z^2+1}{z^2+1} \qquad\qquad = z^\alpha,$$

$$\text{Res}(g, i) = e^{\frac{i\pi\alpha}{2}} \frac{iz+1}{(i-z)(2i)} \qquad = -\frac{e^{\frac{i\pi\alpha}{2}}}{2},$$

$$\text{Res}(g, -i) = e^{-\frac{i\pi\alpha}{2}} \frac{-iz+1}{(-i-z)(-2i)} = -\frac{e^{-\frac{i\pi\alpha}{2}}}{2},$$

and the right side of the last line above becomes

$$\frac{2\pi i}{e^{i\pi\alpha} - e^{-i\pi\alpha}} \{\text{Res}(g, z) + \text{Res}(g, i) + \text{Res}\,(g, -i)\}$$

$$= \frac{2\pi i}{e^{i\pi\alpha} - e^{-i\pi\alpha}} \left\{ z^\alpha - \frac{e^{\frac{i\pi\alpha}{2}}}{2} - \frac{e^{-\frac{i\pi\alpha}{2}}}{2} \right\}$$

$$= \frac{2\pi i}{e^{i\pi\alpha} - e^{-i\pi\alpha}} \left\{ z^\alpha - \cos\frac{\pi\alpha}{2} \right\}$$

$$= \frac{\pi}{\sin\pi\alpha} \left\{ z^\alpha - \cos\frac{\pi\alpha}{2} \right\},$$

and formula (8.3.5) follows. $\qquad\qquad\qquad\qquad\qquad\qquad\qquad\qquad$ □

Solution of Exercise 8.3.8. We take the function

$$g(s) = \ln s \frac{sz + 1}{(s - z)(s^2 + 1)},$$

where $\ln s$ is the principal branch of the logarithm in $\mathbb{C} \setminus (-\infty, 0]$, that is

$$\ln s = \ln |s| + i\theta, \quad \text{where} \quad s = |s| e^{i\theta}, \quad \theta \in (-\pi, \pi).$$

We take the following contour, composed of four parts, where $0 < r < R$ and $\epsilon > 0$:

(i) $\gamma_{1,r,R,\epsilon}$ is the interval

$$\gamma_{1,r,R,\epsilon}(t) = i\epsilon + t, \quad t \in [-R, -r].$$

(ii) $\gamma_{2,r,R,\epsilon}$ is the part of the circle of radius $\sqrt{\epsilon^2 + r^2}$ which connects the points $i\epsilon - r$ and $-i\epsilon - r$ with parametrization

$$\gamma_{2,r,R,\epsilon}(t) = \sqrt{\epsilon^2 + r^2} e^{it}, \quad t \in [-\theta, \pi + \theta],$$

where $\theta \in [0, \pi/2]$ is such that

$$\tan \theta = \frac{\epsilon}{r}.$$

(iii) $\gamma_{3,r,R,\epsilon}$ is the interval

$$\gamma_{3,r,R,\epsilon}(t) = -i\epsilon + (-r - t), \quad t \in [r, R - r].$$

(iv) $\gamma_{4,R,\epsilon}$ is the part of the circle of radius $\sqrt{R^2 + \epsilon^2}$ and center 0, which starts at $i\epsilon - R$ and ends at $-i\epsilon - R$.

By the residue theorem, and letting ϵ and r go to 0 and R go to infinity we obtain for $z \neq \pm i$:

$$\int_{(-\infty,0]} (i\pi) \frac{sz + 1}{(s - z)(s^2 + 1)} ds - \int_{(-\infty,0]} (-i\pi) \frac{sz + 1}{(s - z)(s^2 + 1)} ds$$
$$= 2\pi i \left\{ \operatorname{Res}(g(s), z) + \operatorname{Res}(g(s), -i) + \operatorname{Res}(g(s), i) \right\}.$$

But

$$\operatorname{Res}(g(s), -i) = -\frac{i\pi}{2} \frac{-iz + 1}{(-i - z)(-2i)} = i\frac{\pi}{4},$$

and similarly for the residue at i. Hence

$$\int_{(-\infty,0]} \frac{sz + 1}{(s - z)(s^2 + 1)} ds = \ln z + i\frac{\pi}{4} - i\frac{\pi}{4} = \ln z,$$

first for $z \neq \pm i$, and then to all of $\mathbb{C} \setminus (-\infty, 0]$ by analytic continuation. $\square$

Solution of Exercise 8.3.9. We have

$$\left(\int_0^\infty (1 - e^{-zu}) \frac{e^{-u}}{u} du\right)' = \int_0^\infty e^{-(z+1)u} du$$

$$= \frac{e^{-(z+1)u}}{-(z+1)} \Big|_{u=0}^{u=\infty}$$

$$= \frac{1}{1+z},$$

and hence the result. For the second example, we proceed as follows. On the one hand,

$$(z^\alpha)' = (e^{\alpha \ln z})'$$

$$= \alpha \frac{1}{z} e^{\alpha \ln z}$$

$$= \alpha z^{\alpha-1}.$$

On the other hand,

$$\left(\frac{\alpha}{\Gamma(1-\alpha)} \int_0^\infty (1 - e^{-zu}) \frac{du}{u^{\alpha+1}}\right)' = \frac{\alpha}{\Gamma(1-\alpha)} \int_0^\infty \frac{e^{-zu}}{u^\alpha} du.$$

To compute this last integral, we first consider the case where $z = x > 0$. Then, the change of variable $xu = v$ leads to

$$\frac{\alpha}{\Gamma(1-\alpha)} \int_0^\infty \frac{e^{-xu}}{u^\alpha} du = \frac{\alpha x^{\alpha-1}}{\Gamma(1-\alpha)} \int_0^\infty e^{-v} v^{-\alpha} dv = \alpha x^{\alpha-1}.$$

By analytic continuation we have, for all z in the open right half-plane,

$$\frac{\alpha}{\Gamma(1-\alpha)} \int_0^\infty \frac{e^{-zu}}{u^\alpha} du = \alpha z^{\alpha-1}.$$

Therefore both sides of (8.3.12) have the same derivative. To check that both sides coincide in $\operatorname{Re} z > 0$, we check that there is equality for $z = 1$. The left side is trivially equal to 1. Using integration by parts we see that the right side is equal to

$$\frac{\alpha}{\Gamma(1-\alpha)} \int_0^\infty (1 - e^{-u}) \frac{du}{u^{\alpha+1}} = \frac{\alpha}{\Gamma(1-\alpha)} \left\{ \left(\frac{(1 - e^{-u})u^{-\alpha}}{-\alpha}\right)_{u=0}^{u=\infty} \right.$$

$$\left. + \frac{1}{\alpha} \int_0^\infty e^{-u} u^{-\alpha} du \right\}$$

$$= \frac{\alpha}{\Gamma(1-\alpha)} \frac{\Gamma(1-\alpha)}{\alpha}$$

$$= 1. \qquad \square$$

Solution of Exercise 8.4.1. If $a = 0$ the result is trivial and we assume now $a \neq 0$ (so, $a \in (-1, 0) \cup (0, 1)$). We have

$$\int_0^{2\pi} \frac{dt}{1 + a \cos t} = \int_{\mathbb{T}} \frac{dz}{iz \left(1 + \frac{a}{2} \left(z + \frac{1}{z}\right)\right)}$$

$$= \frac{1}{i} \int_{\mathbb{T}} \frac{2 \, dz}{az^2 + 2z + a}$$

$$= \frac{1}{i} \int_{\mathbb{T}} \frac{2 \, dz}{a(z - z_1)(z - z_2)},$$

where

$$z_1 = \frac{-1 - \sqrt{1 - a^2}}{a} \quad \text{and} \quad z_2 = \frac{-1 + \sqrt{1 - a^2}}{a}.$$

The point z_1 is outside the closed unit disk while the point z_2 is inside the open unit disk; thus the residue theorem (or Cauchy's formula) leads to

$$\int_0^{2\pi} \frac{dt}{1 + a \cos t} = \frac{1}{i} \int_{\mathbb{T}} \frac{2 \, dz}{a(z - z_1)(z - z_2)}$$

$$= 2\pi i \frac{1}{i} \operatorname{Res} \left(\frac{2}{a(z - z_1)(z - z_2)}, z_2\right)$$

$$= 2\pi \frac{2}{a \dfrac{2\sqrt{1 - a^2}}{a}}$$

$$= \frac{2\pi}{\sqrt{1 - a^2}}. \qquad \square$$

Solution of Exercise 8.4.2. Since $\cos t = (z + 1/z)/2$ for $z = e^{it}$ we have

$$\int_0^{2\pi} \frac{dt}{1 + 8 \cos^2 t} = \int_{|z|=1} \frac{1}{1 + 2 \left(z^2 + 2 + \frac{1}{z^2}\right)} \frac{dz}{iz}$$

$$= \int_{|z|=1} \frac{z \, dz}{i(2z^4 + 5z^2 + 2)}$$

$$= \int_{|z|=1} \frac{z \, dz}{2i(z^2 + 2)(z + i/\sqrt{2})(z - i/\sqrt{2})}$$

$$= 2\pi i \left\{ \operatorname{Res} \left(\frac{z}{2i(z^2 + 2)(z + i/\sqrt{2})(z - i/\sqrt{2})}, \frac{i}{\sqrt{2}}\right) \right.$$

$$\left. + \operatorname{Res} \left(\frac{z}{2i(z^2 + 2)(z + i/\sqrt{2})(z - i/\sqrt{2})}, -\frac{i}{\sqrt{2}}\right) \right\}$$

$$= 2\pi i \left\{ \frac{z}{2i(z^2+2)(z+i/\sqrt{2})} \bigg|_{z=\frac{i}{\sqrt{2}}} \right.$$

$$\left. + \frac{z}{2i(z^2+2)(z-i/\sqrt{2})} \bigg|_{z=-\frac{i}{\sqrt{2}}} \right\}$$

$$= \frac{2\pi}{3}. \qquad \qquad \square$$

Solution of Exercise 8.5.1. By the residue theorem

$$\int_{\Gamma_R} g(z)dz = 2\pi i \operatorname{Res}(g(z), z_0)$$

with $z_0 = \frac{1+i}{2}\sqrt{\frac{\pi}{2}}$. We compute the residue using formula (7.3.2) and get:

$$\operatorname{Res}(g(z), z_0) = \frac{e^{-z^2}}{-(1+i)\sqrt{2\pi}e^{-(1+i)\sqrt{2\pi}z}} \bigg|_{z=z_0}$$

$$= \frac{e^{-i\frac{\pi}{4}}}{(1+i)\sqrt{2\pi}}$$

$$= \frac{1-i}{(1+i)\sqrt{2}} \frac{1}{\sqrt{2\pi}}$$

$$= -\frac{i}{2\sqrt{\pi}}.$$

Thus, $\int_{\Gamma_R} g(z)dz = \sqrt{\pi}$. Remarking that

$$g(z) - g(z + 2z_0) = e^{-z^2}$$

we have

$$\int_{[-R,R]} g(z)dz + \int_{[R+2z_0, -R+2z_0]} g(z)dz = \int_{[-R,R]} e^{-z^2}dz.$$

To conclude it remains to show that

$$\lim_{R\to\infty} \int_{R, R+2z_0} g(z)dz = \lim_{R\to\infty} \int_{-R, -R+2z_0} g(z)dz = 0.$$

We consider only the first integral and take the parametrization

$$\gamma(t) = R + 2z_0 t, \quad t \in [0, 1].$$

Then we have

$$|e^{-\gamma(t)^2}| \le e^{-R^2} e^{2|z_0|R + 4|z_0|^2}, \quad t \in [0, 1]$$

and

$$\left|e^{-\gamma(t)(1+i)\sqrt{2\pi}}\right| = e^{-\sqrt{2\pi}R} \cdot e^{4|z_0|t}.$$

Thus for R large enough we have $\left|e^{-\gamma(t)(1+i)\sqrt{2\pi}}\right| \le \frac{1}{2}$, and

$$|g(\gamma(t))| \le 2e^{-R^2} e^{2|z_0|R + 4|z_0|^2}.$$

The claim follows then from inequality (5.1.3).   □

Solution of Exercise 8.5.2. By the residue theorem we have, for $z \in \mathbb{C}_+$,

$$F_+(z) = f'(z).$$

Thus the function f' is an analytic extension to $\mathbb{C}$ of the function F_+.

For z is the lower half-plane, the function $u \mapsto \frac{f(u)}{(u-z)^2}$ is analytic in $\mathbb{C}_+$, and the residue theorem will lead to $F_-(z) \equiv 0$ in $\mathbb{C}_-$. Thus the analytic extension of F_+ to the lower half-plane will coincide with F_- if and only if f is a constant.   □

Solution of Exercise 8.5.3. (a) We consider the contour C_R consisting of the interval $[-R, R]$ (in view of (8.5.6), the definite integral can be computed as a limit of the integral on the symmetric interval $[-R, R]$) and of the semi-circle of center 0 and radius R which lies in the upper half-plane $\mathbb{C}_+$. Let

$$g(s) = f(s)\left\{\frac{1}{s-z} - \frac{s}{s^2+1}\right\} = f(s)\frac{sz+1}{(s-z)(s-i)(s+i)}.$$

By the residue theorem we have, for $z \in \mathbb{C}_+$,

$$\int_{C_R} f(s)\frac{sz+1}{(s-z)(s-i)(s+i)}\,ds = 2\pi i\,\{\mathrm{Res}(g,z) + \mathrm{Res}(g,i)\}$$

$$= 2\pi i\left(f(z) + f(i)\frac{iz+1}{(i-z)(2i)}\right)$$

$$= 2\pi i\left(f(z) - \frac{f(i)}{2}\right).$$

Still for $z \in \mathbb{C}_+$ and with the same contour we have

$$\int_{C_R} f(s)\frac{s\bar{z}+1}{(s-\bar{z})(s-i)(s+i)}\,ds = 2\pi i f(i)\frac{i\bar{z}+1}{(i-\bar{z})(2i)} = 2\pi i\left(-\frac{f(i)}{2}\right).$$

In view of (8.5.5), the integral on the half-circle goes to 0 as $R \to \infty$, and we get

$$\int_\mathbb{R} f(x)\frac{xz+1}{(x-z)(x^2+1)}\,dx = 2\pi i\left(f(z) - \frac{f(i)}{2}\right),$$

$$\int_\mathbb{R} f(x)\frac{x\bar{z}+1}{(x-\bar{z})(x^2+1)}\,dx = 2\pi i\left(-\frac{f(i)}{2}\right).$$

Taking the conjugate of the second equation, and adding to the first equation side by side we get

$$\int_{\mathbb{R}} \frac{xz + 1}{(x - z)(x^2 + 1)} (2 \operatorname{Re} f(x)) dx = 2\pi i (f(z) - i \operatorname{Im} f(i)),$$

and hence we get (8.5.7)

$$f(z) = i \operatorname{Im} f(i) + \frac{1}{\pi i} \int_{\mathbb{R}} \frac{xz + 1}{(x - z)(x^2 + 1)} (\operatorname{Re} f(x)) dx$$

$$= i \operatorname{Im} f(i) + \frac{1}{\pi i} \int_{\mathbb{R}} \left\{ \frac{1}{x - z} - \frac{x}{x^2 + 1} \right\} (\operatorname{Re} f(x)) dx, \quad z \in \mathbb{C}_+.$$

(b) From the above expression we get

$$\operatorname{Re} f(z) = \frac{z - \bar{z}}{2\pi i} \int_{\mathbb{R}} \frac{\operatorname{Re} f(x)}{|x - z|^2} dx, \tag{8.6.16}$$

and hence $\operatorname{Re} f(z) \geq 0$ for $z \in \mathbb{C}_+$. Assume now that for some $z \in \mathbb{C}_+$, it holds that $\operatorname{Re} f(z) = 0$. Then, (8.6.16) implies that $\operatorname{Re} f(x) \equiv 0$ on the real line. Hence f is a unitary constant, as is seen from (8.5.7). □

More generally we have, for z and w off the real line,

$$f(z) + \overline{f(w)} = \frac{z - \bar{w}}{2\pi i} \int_{\mathbb{R}} \frac{\operatorname{Re} f(x)}{(x - z)(x - \bar{w})} dx.$$

Therefore the kernel

$$\frac{f(z) + \overline{f(w)}}{-i(z - \bar{w})} = \frac{1}{2\pi} \int_{\mathbb{R}} \frac{\operatorname{Re} f(x)}{(x - z)(x - \bar{w})} dx$$

is positive definite in $\Omega = \mathbb{C} \setminus \mathbb{R}$.

Solution of Exercise 8.5.4. Rewrite

$$\frac{1}{x - z} - \frac{x}{x^2 + 1} = \frac{xz + 1}{(x - z)(x + i)(x - i)}.$$

For $z \in \mathbb{C}_+$ we have by the residue theorem

$$g(z) = \operatorname{Res} \left(\frac{xz + 1}{(x - z)(x + i)(x - i)}, x = z \right)$$

$$+ \operatorname{Res} \left(\frac{xz + 1}{(x - z)(x + i)(x - i)}, x = i \right)$$

$$= 1 - \frac{1}{2} = \frac{1}{2}.$$

The function g satisfies the symmetry

$$g(\bar{z}) = -\overline{g(z)}, \quad z \in \mathbb{C} \setminus \mathbb{R},$$

and is therefore identically equal to $-1/2$ in the open lower half-plane. □

Part III

Applications and
More Advanced Topics

Part II

Applications and
More Advanced Topics

Chapter 9

Harmonic Functions

The notion of a harmonic function arises from real analysis and appeared before the notion of an analytic function. In some sense, analytic functions were defined to study in an efficient way problems related to harmonic functions.

9.1   Harmonic functions

Let $f(z) = u(x, y) + iv(x, y)$ be a function analytic in some open set Ω. The functions u and v admit partial derivatives of all orders, and thus the Cauchy–Riemann equations

$$\frac{\partial u}{\partial x}(x, y) = \frac{\partial v}{\partial y}(x, y),$$

$$\frac{\partial u}{\partial y}(x, y) = -\frac{\partial v}{\partial x}(x, y), \quad x + iy \in \Omega,$$

imply

$$\frac{\partial^2 u}{\partial x^2}(x, y) = \frac{\partial^2 v}{\partial x \partial y}(x, y),$$

$$\frac{\partial^2 u}{\partial y^2}(x, y) = -\frac{\partial v}{\partial y \partial x}(x, y), \quad x + iy \in \Omega.$$

The smoothness of v implies that the mixed derivatives are equal; adding these two equations we get

$$\Delta u \overset{\text{def.}}{=} \frac{\partial^2 u}{\partial x^2} + \frac{\partial^2 u}{\partial y^2} = 0.$$

Definition 9.1.1. A real-valued function defined in an open subset of $\mathbb{R}^2$ is *harmonic* if it has continuous partial derivatives of order 2 and if it solves the Laplace equation in Ω:

$$\Delta u = 0.$$

Exercise 9.1.2. *Show that the functions*

$$u(x, y) = e^x \cos y, \qquad (x, y) \in \mathbb{R}^2,$$
$$u(x, y) = \cos x \cosh y, \quad (x, y) \in \mathbb{R}^2,$$
$$u(x, y) = \ln(x^2 + y^2), \quad (x, y) \in \mathbb{R}^2 \setminus \{(0, 0)\}$$

are harmonic.

Exercise 9.1.3. *Show that a harmonic function u is locally the real part of an analytic function, and in particular has partial derivatives of all orders.*

Exercise 9.1.4. *Let $u_1(x, y)$ and $u_2(x, y)$ be two real-valued function harmonic in the open sets Ω_1 and Ω_2 respectively, and assume that the range of the function*

$$(x, y) \mapsto \left(\frac{\partial u_2}{\partial x}, -\frac{\partial u_2}{\partial y} \right)$$

is inside Ω_1. Show that the function

$$U(x, y) = u_1 \left(\frac{\partial u_2}{\partial x}, -\frac{\partial u_2}{\partial y} \right) \tag{9.1.1}$$

is harmonic in Ω_2.

Exercise 9.1.5. *Given f analytic and not vanishing in an open set Ω. Show that $\ln |f(z)|$ is harmonic in Ω.*

From Exercise 1.2.8 it is clear that the function $u(x, y)$ defined by (9.1.2) below is harmonic. The point in Exercise 9.1.6 is to provide a direct and different proof.

Exercise 9.1.6. *Show that*

$$u(x, y) = \frac{\sin x}{\cos x + \cosh y} \tag{9.1.2}$$

is harmonic in

$$\Omega = \{z = x + iy, \ -\pi < x < \pi \ \text{and} \ y \in \mathbb{R}\}.$$

Remark 9.1.7. If $u(x, y)$ is harmonic in a neighborhood of the origin, then $u(x, y)$ is the real part of the function

$$F(z) = 2u(z/2, z/2i) - u(0, 0). \tag{9.1.3}$$

This is a well-known formula; see for instance [45]. A good exercise (see Exercise 9.1.9) is to prove this formula for polynomials. The formula does not hold if $u(x, y)$ is not defined in a neighborhood of the origin, as one can see with the function $u(x, y) = \ln(x^2 + y^2)$.

Exercise 9.1.8. *Apply formula (9.1.3) to $u(x, y) = \dfrac{\sin x}{\cos x + \cosh y}$.*

Exercise 9.1.9. *Prove formula (9.1.3) for a polynomial.*

Exercise 9.1.10. *Let $u(x,y)$ and $v(x,y)$ admit partial derivatives which are continuous in an open set Ω, and assume that u and v satisfy the Cauchy–Riemann equations in Ω. Assume moreover that $u^2 + v^2 \neq 0$ in Ω. Show that the function*

$$\frac{u\dfrac{\partial u}{\partial x} + \dfrac{\partial v}{\partial x}v}{u^2 + v^2}$$

is harmonic in Ω.

Zeroes of harmonic functions are not isolated (see for instance the sets where the functions $u(x,y) = x^2 - y^2$ and $\ln(x^2 + y^2)$ vanish), but we have:

Exercise 9.1.11. *Let u be harmonic in the connected open set Ω and assume that u vanishes in a set $B(z_0, r) \subset \Omega$. Show that u is identically equal to 0 in Ω.*

9.2 Harmonic conjugate

Let Ω be an open subset of $\mathbb{R}^2$, and let u be harmonic in Ω. The function $v(x,y)$ is called a *conjugate harmonic* of u if the function $f(z) = u(x,y) + iv(x,y)$ is analytic in Ω.

Exercise 9.2.1. *Assume that v is a conjugate harmonic of the harmonic function u in an open set Ω. Let a and b be two real numbers. Show that $V(x,y) = bu(x,y) + av(x,y)$ is a harmonic conjugate of the function $U(x,y) = au(x,y) - bv(x,y)$ in Ω.*

Exercise 9.2.2. *Assume that u and v are harmonic in an open set Ω and that v is the harmonic conjugate of u in Ω. Define functions U and V by*

$$U(x,y) = e^{u(x,y)^2 - v(x,y)^2} \cos 2u(x,y)v(x,y),$$
$$V(x,y) = e^{u(x,y)^2 - v(x,y)^2} \sin 2u(x,y)v(x,y).$$

Show that U and V are harmonic in Ω and that V is a harmonic conjugate to U there.

A real-valued function u harmonic in the open set Ω need not have a harmonic conjugate (see Exercise 9.2.5 below), but we have the following key fact (see also Exercise 9.1.3):

Exercise 9.2.3. *Let u be real-valued and harmonic in the open set Ω. Then $-\frac{\partial u}{\partial y}$ is a harmonic conjugate of $\frac{\partial u}{\partial x}$. In other words, the function*

$$\frac{\partial u}{\partial x} - i\frac{\partial u}{\partial y} \tag{9.2.1}$$

is analytic in Ω

Remark 9.2.4. One should not confuse the contexts of equations (4.2.6) and (9.2.1). The first equation gives the derivative of an analytic function. The second formula defines an analytic function from a given harmonic function; this analytic function need not have a primitive in Ω.

It was proved in Exercise 9.1.3 that a harmonic conjugate always exists *locally*, that is, in a neighborhood of a point, but need not exist globally. If it exists, it is harmonic.

Exercise 9.2.5.

(a) *Show that the function* $\ln(x^2+y^2)$ *has no harmonic conjugate in* $\mathbb{R}^2\setminus\{(0,0)\}$.

(b) *Show that it has a harmonic conjugate in* $\mathbb{R}^2 \setminus \{(x,0)\,;\, x\leq 0\}$.

We have now the penultimate characterization of simply-connected sets to be given in these notes.

Definition 9.2.6. A connected open set $\Omega \subset \mathbb{R}^2$ is *simply-connected* if every function harmonic in Ω admits a harmonic conjugate in Ω.

Strictly speaking, the definition of a simply-connected set does not require the set to be connected. The existence of a harmonic conjugate has then to be checked in every connected component of the given set.

Theorem 9.2.7. *Let Ω be an open connected subset of $\mathbb{C}$ and let u be harmonic in Ω. Then, u has a harmonic conjugate in Ω if and only if*

$$\int_C -\frac{\partial u}{\partial y}dx + \frac{\partial u}{\partial x}dy = 0 \qquad\qquad (9.2.2)$$

for all closed piecewise smooth paths C in Ω.

Proof. Assume that u admits a harmonic conjugate, say v, in Ω. Then, the function $f(z) = u(x,y)+iv(x,y)$ is analytic in Ω and its derivative is equal to the function

$$g(z) = \frac{\partial u}{\partial x} - i\frac{\partial u}{\partial y}. \qquad\qquad (9.2.3)$$

So, by Theorem 5.2.1,

$$\int_C g(z)dz = \int_C f'(z)dz = 0.$$

Conversely, if (9.2.2) is in force for every closed curve C inside Ω, it holds that

$$\int_C g(z)dz = \int_C \frac{\partial u}{\partial x}dx + \frac{\partial u}{\partial y}dy + i\int_C -\frac{\partial u}{\partial y}dx + \frac{\partial u}{\partial x}dy = 0,$$

where we have also used (5.1.12). Hence g has a primitive $F(z) = U(x,y)+iV(x,y)$ in Ω. Since

$$F'(z) = \frac{\partial U}{\partial x} - i\frac{\partial U}{\partial y},$$

we see that, up to a real constant, $U = u$ and hence V is a harmonic conjugate to u. $\square$

In the following exercise, the set Ω is *not* required to be connected.

Exercise 9.2.8. *Let $u(x,y)$ be harmonic in an open set Ω. Let $n, m \in \mathbb{N}_0$ such that $n + m > 0$. Show that the function*

$$U(x,y) = \frac{\partial^{n+m} u}{\partial x^n \partial y^m}(x,y)$$

is harmonic in Ω; prove that U has a harmonic conjugate, and find all functions f analytic in Ω and with real part U.

Exercise 9.2.9. *Let f be an entire function and assume that $\operatorname{Re} f'(z) = x^2 - y^2 + 6xy$. The value $f(0)$ is known. Find f.*

Exercise 9.2.10. *Prove that the function*

$$u(x,y) = y \cos y \sinh x + x \sin y \cosh x$$

is harmonic in $\mathbb{R}^2$, and find its harmonic conjugate.

Exercise 9.2.11. *Prove formula (9.5.2), which is given in the solution of Exercise 9.2.10.*

More generally, one has the following formula for the harmonic conjugate:

Exercise 9.2.12. *Let Ω be a simply-connected set, and let u be harmonic in Ω.*

(a) *Prove that*

$$\int_C -\frac{\partial u}{\partial y} dx + \frac{\partial u}{\partial x} dy = 0$$

for every closed contour C.

(b) *Fix a point $(x_0, y_0) \in \Omega$. Prove that the function*

$$v(x,y) = \int_{C_{x,y}} -\frac{\partial u}{\partial y} dx + \frac{\partial u}{\partial x} dy \qquad (9.2.4)$$

is the harmonic conjugate of u (up to an additive constant), where $C_{x,y}$ is any path connecting (x_0, y_0) to (x,y).

We recall that the Cauchy–Riemann equations in polar coordinates are given by

$$\begin{aligned} \frac{\partial u}{\partial \rho} &= \frac{1}{\rho}\frac{\partial v}{\partial \theta}, \\ \frac{\partial v}{\partial \rho} &= -\frac{1}{\rho}\frac{\partial u}{\partial \theta}, \end{aligned} \qquad (9.2.5)$$

and that the Laplacian in polar coordinates is given by

$$\Delta u(\rho, \theta) = \frac{\partial^2 u}{\partial \rho^2} + \frac{1}{\rho}\frac{\partial u}{\partial \rho} + \frac{1}{\rho^2}\frac{\partial^2 u}{\partial \theta^2}. \tag{9.2.6}$$

The functions

$$u(\rho, \theta) = \ln \rho \quad \text{and} \quad v(\rho, \theta) = \theta$$

($\theta \in (-\pi, \pi)$ for instance, or more generally in an open interval of length 2π) are harmonic since

$$\ln z = \ln \rho + i\theta.$$

Exercise 9.2.13. *Find all functions harmonic in $\mathbb{R}^2 \setminus \{(x,0)\,;\, x \leq 0\}$ of the form $f(\rho)$ and of the form $f(\theta)$, where f is a real-valued function of class C_2.*

Exercise 9.2.14. *Show that the function*

$$u(\rho, \theta) = \rho\theta \cos\theta + \rho \sin\theta \ln \rho$$

is harmonic in $\Omega = \mathbb{R}^2 \setminus \{(x,0)\,;\, x \leq 0\}$, and compute its harmonic conjugate.

9.3 Various

Exercise 9.3.1. *Let u be a harmonic function such that u^2 is also harmonic. Show that u is constant.*

Exercise 9.3.2. *Let u and v be harmonic in $\mathbb{R}^2$ and assume that v is the harmonic conjugate of u. Assume that*

$$u^3 - 3uv^2 \geq 0 \quad \text{in} \quad \mathbb{R}^2.$$

Find u and v.

The preceding exercise relies on the following fact: If u is a function harmonic in the whole plane and is bounded from above or from below, then u is a constant.

Exercise 9.3.3. *Prove the previous claim.*

Exercise 9.3.4. *Find all functions u harmonic in $\mathbb{R}^2$ and satisfying*

$$\frac{\partial u}{\partial x} \leq 0.$$

The next exercise is almost identical to the previous one.

Exercise 9.3.5. *Find all functions u harmonic in $\mathbb{R}^2$ and such that $\frac{\partial u}{\partial x} \geq 5$.*

Let f be analytic in Ω. We saw in Exercise 9.1.5 that $\ln |f|$ is harmonic in Ω when it does not vanish there. On the other hand, the modulus function $|f|$ is not harmonic, but subharmonic

$$\Delta|f| \geq 0,$$

as illustrated by the examples

$$\Delta|z|^2 = 4,$$

and

$$\Delta r = \frac{1}{r}, \quad r = |z|. \tag{9.3.1}$$

Exercise 9.3.6. *Let f be analytic and non-vanishing in the open set Ω. Show that*

$$\Delta|f| = \frac{|f'|^2}{|f|}.$$

The case $f(z) = z^n$, $n \in \mathbb{Z}$, leads to

$$\Delta|z|^n = n^2|z|^{n-2}. \tag{9.3.2}$$

When replacing f by f^2 in the preceding exercise, one can remove the hypothesis that f does not vanish, and obtain the formula

$$\Delta|f^2| = 4|f'|^2. \tag{9.3.3}$$

More generally, one has the following formulas (see [75, Exercises 8.41 and 8.42]): Let F be a function of class C_2 on $\mathbb{R}$, and let f be analytic in some domain Ω and not vanishing there. Then

$$\Delta F(|f(z)|) = |f'(z)|^2 \left\{ F^{(2)}(|f(z)|) + \frac{F^{(1)}(|f(z)|)}{|f(z)|} \right\}. \tag{9.3.4}$$

The special choice $F(t) = t^n$ and $f(z) = z$ leads to (9.3.2). Furthermore, choosing $F(t) = \ln(1 + t^2)$ we have

$$\Delta(\ln(1 + |f(z)|^2)) = \frac{|f'(z)|^2}{(1 + |f(z)|^2)^2},$$

$$\Delta(e^{p|f(z)|}) = p^2 e^{p|f(z)|} \left(p + \frac{1}{|f(z)|} \right) |f'(z)|^2, \quad p \in \mathbb{R}. \tag{9.3.5}$$

Exercise 9.3.7. *Let u be a subharmonic function defined in $B(z_0, R)$.*

(1) *Show that for every $r \in (0, R)$ it holds that*

$$u(x_0, y_0) \leq \frac{1}{2\pi} \int_0^{2\pi} u(x_0 + r\cos t, y_0 + r\sin t)dt. \tag{9.3.6}$$

(2) *Prove the strict maximum principle for subharmonic functions.*

As a corollary of (9.3.6) we have the following result, which expresses the fact that the Bergman space of the open unit disk is a reproducing kernel Hilbert space.

Exercise 9.3.8. *Let f be a function analytic in the open unit disk and such that*

$$\|f\|_{\mathcal{B}}^2 \stackrel{\text{def.}}{=} \iint_{\mathbb{D}} |f(z)|^2 \, dx dy < \infty.$$

Show that for every $z_0 \in \mathbb{D}$ there exists a number $K_{z_0} > 0$ such that

$$|f(z_0)| \le K_{z_0} \cdot \|f\|_{\mathcal{B}}. \tag{9.3.7}$$

Exercise 9.3.9.

(a) *Find all harmonic functions of the form $u(x,y) = \varphi(x^2 - y^2)$, where φ is of class C_2 on all of $\mathbb{R}$.*

(b) *Find all harmonic functions of the form $u(x,y) = \varphi(y/x)$.*

Exercise 9.3.10. *Let Ω be an open connected set, and let*

$$u(x,y) = \operatorname{Re}(\overline{z} f(z) + g(z)),$$

where f and g are analytic in Ω. Show that

$$\Delta^2 u = 0. \tag{9.3.8}$$

Functions which satisfy (9.3.8) are called biharmonic, and appear in mathematical physics, in particular in elasticity. See [33, p. 246], [226].

9.4   The Dirichlet problem

We conclude this chapter with a discussion of the Dirichlet problem.

Definition 9.4.1. Let f be a real-valued piecewise continuous function on the unit circle $\mathbb{T}$. The *Dirichlet problem* consists in finding a function u harmonic in the open unit disk, and equal to f on $\mathbb{T}$, in the sense that

$$\lim_{r \to 1} u(re^{i\theta}) = f(e^{i\theta}),$$

when f is continuous at the point $e^{i\theta}$.

As is well known, see for instance [101, Theorem 9.14, p. 180], the problem is solvable, its solution is unique and is given by the formula

$$u(z) = \operatorname{Re} \frac{1}{2\pi} \int_0^{2\pi} \frac{e^{it} + z}{e^{it} - z} f(e^{it}) dt, \quad z \in \mathbb{D}. \tag{9.4.1}$$

As an application, solve the following exercise (see [148, pp. 355–356]).

Exercise 9.4.2. *Compute the solution of the Dirichlet problem when*

$$f(e^{it}) = \begin{cases} 1, & t \in [0, \pi), \\ -1, & t \in [\pi, 2\pi). \end{cases}$$

More challenging is the case when one considers the punctured unit disk $\mathbb{D}\backslash\{0\}$. Indeed, $\mathbb{D}\backslash\{0\}$ is not simply-connected, as is seen for instance by computing

$$\int_{|z|=1/2} \frac{dz}{z}.$$

Therefore not every function harmonic in $\mathbb{D}\backslash\{0\}$ has a harmonic conjugate there. These points are illustrated in the following exercise, taken from [148, pp. 356–357].

Exercise 9.4.3. *Find all functions u harmonic in $0 < |z| < 1$, continuous on $0 < |z| \leq 1$, and vanishing on the unit circle.*

9.5 Solutions

Solution of Exercise 9.1.2. The first two examples are clear since

$$e^x \cos y = \operatorname{Re} e^z \quad \text{and} \quad \cos x \cosh y = \operatorname{Re} \cos z.$$

See (1.2.17) if needed for the latter. The third exercise will follow from the general claim in Exercise 9.1.5, but it is also a good exercise to check directly that $\ln(x^2 + y^2)$ is harmonic in $\mathbb{R}^2 \backslash \{(0,0)\}$. See in that respect the discussion after the proof of Exercise 9.1.5. $\qquad\square$

Solution of Exercise 9.1.3. Consider the function

$$g(z) = \frac{\partial u}{\partial x}(x, y) - i\frac{\partial u}{\partial y}(x, y),$$

so that

$$\operatorname{Re} g(z) \overset{\text{def.}}{=} U(x, y) = \frac{\partial u}{\partial x}(x, y) \quad \text{and} \quad \operatorname{Im} g(z) \overset{\text{def.}}{=} V(x, y) = -\frac{\partial u}{\partial y}(x, y).$$

We have

$$\frac{\partial U}{\partial x} = \frac{\partial V}{\partial y}$$

since u is assumed harmonic. Since u has continuous partial derivatives of order 2, we have

$$\frac{\partial^2 u}{\partial x \partial y} = \frac{\partial^2 u}{\partial y \partial x},$$

and in particular

$$\frac{\partial U}{\partial y} = -\frac{\partial V}{\partial x} = \frac{\partial^2 u}{\partial x \partial y}.$$

Therefore U and V satisfy the Cauchy–Riemann equations in Ω. They are moreover differentiable in Ω (as functions of (x, y)) since u has continuous partial derivatives of order 2. Therefore, the Cauchy–Riemann equations imply that g is $\mathbb{C}$-differentiable in Ω. Since Ω is convex, g has a primitive in Ω, say $f(z) = U_1(x, y) + iV_1(x, y)$, and

$$f'(z) = \frac{\partial U_1}{\partial x}(x, y) - i\frac{\partial U_1}{\partial y}(x, y).$$

Therefore, U_1 is equal to u, up to a constant. Choosing this constant to be 0, the corresponding f is the real part of u in Ω.   □

Solution of Exercise 9.1.4. Since u_2 is harmonic in Ω_2, the function

$$g(z) = \frac{\partial u_2}{\partial x}(x, y) - i\frac{\partial u_2}{\partial y}(x, y)$$

is analytic there. Let $(x_0, y_0) \in \Omega_2$, let $z_0 = x_0 + iy_0$, and let W be a convex open neighborhood of $g(z_0)$ in Ω_1. There exists an open neighborhood of (x_0, y_0) in Ω_2 such that $g(V) \subset W$. In W, the function u_1 is the real part of an analytic function, say f. So the function $f \circ g$ is analytic in V, and its real part

$$U(x, y) = \mathrm{Re}(f(g(z)) = u_1 \left(\frac{\partial u_2}{\partial x}(x, y), -\frac{\partial u_2}{\partial y}(x, y) \right)$$

is harmonic in V. Since all such V cover all of Ω_2, the function (9.1.1) is harmonic in Ω_2.   □

Solution of Exercise 9.1.5. In any convex open subset U of Ω the function f has an analytic logarithm: There exists a function g analytic in U and such that

$$f(z) = e^{g(z)}, \quad z \in U.$$

Taking absolute value on both sides of the above equality we get

$$|f(z)| = e^{\mathrm{Re}\, g(z)}.$$

Thus the function

$$\ln |f(z)| = \mathrm{Re}\, g(z)$$

is harmonic in U since the real part of g is harmonic there. This ends the proof since every point of Ω has a convex open neighborhood in Ω.   □

Note that the claim in the preceding exercise can also be proved directly: Let $f = u + iv$ be analytic in Ω, and let $T(x, y) = \ln(u^2 + v^2)$. We have

$$T_x = 2\frac{uu_x + vv_x}{u^2 + v^2},$$

and so

$$
\begin{aligned}
T_{xx} &= 2\frac{u_x^2 + uu_{xx} + v_x^2 + vv_{xx}}{u^2 + v^2} - 2\frac{(uu_x + vv_x)^2}{(u^2 + v^2)^2} \\
&= 2\frac{u_x^2 + uu_{xx} + v_x^2 + vv_{xx}}{u^2 + v^2} - 2\frac{u^2u_x^2 + v^2v_x^2 + 2uvu_xv_x}{(u^2 + v^2)^2}.
\end{aligned}
$$

Similarly,

$$
\begin{aligned}
T_{yy} &= 2\frac{u_y^2 + uu_{yy} + v_y^2 + vv_{yy}}{u^2 + v^2} - 2\frac{(uu_y + vv_y)^2}{(u^2 + v^2)^2} \\
&= 2\frac{u_y^2 + uu_{yy} + v_y^2 + vv_{yy}}{u^2 + v^2} - 2\frac{u^2u_y^2 + v^2v_y^2 + 2uvu_yv_y}{(u^2 + v^2)^2}.
\end{aligned}
$$

Taking into account that $\Delta u = \Delta v = 0$ and that, thanks to the Cauchy–Riemann equations,

$$u_x v_x + u_y v_y = 0,$$

we obtain that $\Delta T = 0$.

Solution of Exercise 9.1.6. A natural way is to first check that $\Delta u = 0$ and then to solve the Cauchy–Riemann equations. We will proceed in a different way. Recall that

$$\cosh(iz) = \frac{\exp(iz) + \exp(-iz)}{2} = \cos z.$$

Thus

$$
\begin{aligned}
u(x, y) &= \frac{\sin(\frac{z + \bar{z}}{2})}{\cos(\frac{z + \bar{z}}{2}) + \cosh(\frac{z - \bar{z}}{2i})} \\
&= \frac{\sin(\frac{z + \bar{z}}{2})}{\cos(\frac{z + \bar{z}}{2}) + \cos(\frac{z - \bar{z}}{2})} \\
&= \frac{\sin\frac{z}{2}\cos\frac{\bar{z}}{2} + \sin\frac{\bar{z}}{2}\cos\frac{z}{2}}{\cos\frac{z}{2}\cos\frac{\bar{z}}{2} - \sin\frac{z}{2}\sin\frac{\bar{z}}{2} + \cos\frac{z}{2}\cos\frac{\bar{z}}{2} + \sin\frac{z}{2}\sin\frac{\bar{z}}{2}} \\
&= \frac{\sin\frac{z}{2}\cos\frac{\bar{z}}{2} + \sin\frac{\bar{z}}{2}\cos\frac{z}{2}}{2\cos\frac{z}{2}\cos\frac{\bar{z}}{2}} \\
&= \frac{\sin\frac{z}{2}}{2\cos\frac{z}{2}} + \frac{\sin\frac{\bar{z}}{2}}{2\cos\frac{\bar{z}}{2}} \\
&= \operatorname{Re}\tan\left(\frac{z}{2}\right).
\end{aligned}
$$

Here we used that $\overline{\sin(z)} = \sin(\overline{z})$ and similarly for $\cos z$ since the coefficients of the Maclaurin series of $\cos z$ and $\sin z$ are real, and thus

$$\tan(\overline{z}) = \overline{\tan(z)}. \qquad \square$$

Solution of Exercise 9.1.8. Using (9.1.3), we have

$$u(x, y) = \frac{\sin x}{\cos x + \cosh y}$$

$$= 2 \operatorname{Re} \frac{\sin(z/2)}{\cos(z/2) + \cosh(z/2i)}$$

$$= 2 \operatorname{Re} \frac{\sin(z/2)}{\cos(z/2) + \cos(z/2)}$$

$$= \operatorname{Re} \tan(z/2). \qquad \square$$

Solution of Exercise 9.1.9. It suffices to check the formula for $F(z) = z^n$. Since

$$z^n = (x + iy)^n$$

$$= \sum_{\substack{k, \\ 2k \leq n}} (-1)^k \binom{n}{2k} x^{2k} y^{n-2k} + i \sum_{\substack{k, \\ 2k+1 \leq n}} (-1)^k \binom{n}{2k+1} x^{2k+1} y^{n-2k-1},$$

we have to prove that

$$z^n = 2 \sum_{\substack{k, \\ 2k \leq n}} (-1)^k \binom{n}{2k} (z/2)^{2k} (z/2i)^{n-2k},$$

that is,

$$z^n = 2 \sum_{\substack{k, \\ 2k \leq n}} (-1)^k \binom{n}{2k} \frac{z^n}{2^n (-1)^k},$$

that is, we have to prove that

$$2^{n-1} = \sum_{\substack{k, \\ 2k \leq n}} \binom{n}{2k}. \tag{9.5.1}$$

But from

$$0^n = (1 - 1)^n$$

$$= \left(\sum_{\substack{k, \\ 2k \leq n}} 1^{n-2k} (-1)^{2k} \binom{n}{2k} \right) + \left(\sum_{\substack{k, \\ 2k+1 \leq n}} 1^{n-2k-1} (-1)^{2k-1} \binom{n}{2k+1} \right)$$

we obtain

$$\sum_{\substack{k, \\ 2k \le n}} \binom{n}{2k} = \sum_{\substack{k, \\ 2k+1 \le n}} \binom{n}{2k+1}.$$

On the other hand,

$$2^n = (1+1)^n = \sum_{\substack{k, \\ 2k \le n}} \binom{n}{2k} + \sum_{\substack{k, \\ 2k+1 \le n}} \binom{n}{2k+1}.$$

Hence,

$$\sum_{\substack{k, \\ 2k \le n}} \binom{n}{2k} = \sum_{\substack{k, \\ 2k+1 \le n}} \binom{n}{2k+1} = \frac{2^n}{2} = 2^{n-1}.$$

Thus, (9.5.1) holds and this concludes the proof. □

Solution of Exercise 9.1.10. From the hypothesis on u and v these functions are differentiable in Ω, and so the function $F(z) = u(x,y) + iv(x,y)$ is analytic in Ω. Its derivative

$$F'(z) = \frac{\partial u}{\partial x}(x,y) - i\frac{\partial u}{\partial y}(x,y) = \frac{\partial u}{\partial x}(x,y) + i\frac{\partial v}{\partial x}(x,y)$$

is analytic. The function F does not vanish in Ω since $u^2 + v^2 \ne 0$ there, and hence the function $\dfrac{F'}{F}$ is analytic in Ω and so its real part is harmonic. We have

$$\mathrm{Re}\,\frac{F'}{F} = \frac{u\dfrac{\partial u}{\partial x} + \dfrac{\partial v}{\partial x}v}{u^2 + v^2},$$

and this ends the proof. □

One can build a number of exercises based on the same principle. Take $F = u + iv$ analytic in Ω and compute (for instance)

$$\mathrm{Re}\,\frac{F''}{F}, \quad \mathrm{Re}\,\frac{F}{F'}, \quad \mathrm{Im}\,\frac{F'}{F''}, \ldots.$$

In the computations, note that

$$F''(z) = \frac{\partial^2 u}{\partial x^2} - i\frac{\partial^2 u}{\partial x \partial y},$$

as follows from formula (4.2.6), since the real part of F' is $\frac{\partial u}{\partial x}$.

Solution of Exercise 9.1.11. The function

$$g(z) = \frac{\partial u}{\partial x}(x, y) - i\frac{\partial u}{\partial y}(x, y), \quad (x, y) \in \Omega,$$

is analytic in Ω and vanishes identically in $B(z_0, r)$. Therefore it vanishes identically in Ω. It follows that

$$\frac{\partial u}{\partial x}(x, y) = \frac{\partial u}{\partial y}(x, y) = 0, \quad \forall (x, y) \in \Omega.$$

Thus u is constant in Ω, and therefore is identically equal to 0 there since it vanishes at z_0. $\qquad\square$

Solution of Exercise 9.2.1. We show that $U + iV$ is analytic in Ω. We have

$$
\begin{aligned}
U(x, y) + iV(x, y) &= (au(x, y) - bv(x, y)) + i(bu(x, y) + av(x, y)) \\
&= (a + ib)u(x, y) + (ai - b)v(x, y) \\
&= (a + ib)u(x, y) + i(a + ib)v(x, y) \\
&= (a + ib)(u(x, y) + iv(x, y)),
\end{aligned}
$$

which ends the proof since by assumption the function $f(z) = u(x, y) + iv(x, y)$ is analytic in Ω. $\qquad\square$

Solution of Exercise 9.2.2. It is enough to check that the function $F(z) = U(x, y) + iV(x, y)$ is analytic in Ω. By hypothesis, the function $f(z) = u(x, y) + iv(x, y)$ is analytic in Ω. One has

$$
\begin{aligned}
U(x, y) + iV(x, y) &= e^{u(x,y)^2 - v(x,y)^2} \cos 2u(x, y)v(x, y) \\
&\quad + ie^{u(x,y)^2 - v(x,y)^2} \sin 2u(x, y)v(x, y) \\
&= e^{u(x,y)^2 - v(x,y)^2} (\cos 2u(x, y)v(x, y) + i \sin 2u(x, y)v(x, y)) \\
&= e^{u(x,y)^2 - v(x,y)^2} e^{2iu(x,y)v(x,y)} \\
&= e^{u(x,y)^2 - v(x,y)^2 + 2iu(x,y)v(x,y)} \\
&= e^{(u(x,y) + iv(x,y))^2} = e^{f(z)^2},
\end{aligned}
$$

and so F is analytic in Ω. $\qquad\square$

Another (and much longer) proof is by direct computations.

Solution of Exercise 9.2.3. It is enough to verify that the pair of functions

$$\frac{\partial u}{\partial x} \quad \text{and} \quad -\frac{\partial u}{\partial y}$$

satisfy the Cauchy–Riemann equations. The first Cauchy–Riemann equation follows from $\Delta u = 0$, and the second from the possibility to interchange partial derivatives with respect to x and y in view of the assumed smoothness of u. $\qquad\square$

Solution of Exercise 9.2.5. (a) Assume by contradiction that there is a harmonic conjugate v. Then, the function

$$f(z) = \ln(x^2 + y^2) + iv(x, y)$$

is analytic in $\mathbb{C} \setminus \{0\}$. Using formula (4.2.6), we see that its derivative is equal to

$$f'(z) = \frac{2x}{x^2 + y^2} - i\frac{2y}{x^2 + y^2} = \frac{2}{z}.$$

Thus

$$0 = \int_{|\zeta|=1} f'(\zeta)d\zeta = \int_{|\zeta|=1} \frac{2d\zeta}{\zeta} = 4\pi i,$$

which cannot be.

(b) Let $\theta \in (-\pi, \pi)$ be defined by (1.1.19). Then, $v = 2\theta$ is such that $f(z) = 2\ln z$. $\square$

Solution of Exercise 9.2.8. We know that u has partial derivatives of all orders, and so U makes sense. That U is harmonic follows from

$$\Delta U = \Delta\left(\frac{\partial^{n+m} u}{\partial x^n \partial y^m}\right) = \frac{\partial^{n+m}(\Delta u)}{\partial x^n \partial y^m} = 0.$$

We note also the following. If m is even: $m = 2p$ for some $p \in \mathbb{N}$, then the harmonicity of u (see (9.5.3) below if needed) implies that

$$U(x, y) = (-1)^p \frac{\partial^{n+2p} u}{\partial x^{n+2p}}.$$

Similarly, if m is odd, $m = 2p + 1$, then

$$U(x, y) = (-1)^p \frac{\partial^{n+2p+1} u}{\partial x^{n+2p} \partial y}.$$

In the first case,

$$U(x, y) = \mathrm{Re}(-1)^p f^{(n+2p)}(z),$$

while in the second case,

$$U(x, y) = \mathrm{Re}\, i(-1)^p f^{(n+2p+1)}(z),$$

since, in view of formula (4.2.6),

$$f^{(\ell)}(z) = \frac{\partial^{\ell} u}{\partial x^{\ell}}(x, y) - i\frac{\partial^{\ell} u}{\partial x^{\ell-1} \partial y}(x, y), \quad \ell \in \mathbb{N}.$$

When the set Ω is connected, $V = \mathrm{Im}\, f^{(n+2p)}(z)$ or $V = \mathrm{Im}\, i(-1)^p f^{(n+2p+1)}(z)$ (depending on whether m is even or odd) is, up to a real constant, the harmonic conjugate of U. When Ω is not connected, a different constant has to be added to V on every connected component of Ω. $\square$

Solution of Exercise 9.2.9. We know that

$$\operatorname{Re} z^2 = x^2 - y^2 \quad \text{and} \quad \operatorname{Re} -iz^2 = 2xy,$$

and hence

$$\operatorname{Re} f'(z) = \operatorname{Re}(1 - 3i)z^2.$$

Thus

$$f(z) = \frac{1 - 3i}{3} z^3 + f(0). \qquad \qquad \square$$

Solution of Exercise 9.2.10. We will find directly an analytic function f with real part u. This will prove directly that u is indeed harmonic. This method is fine only if you know ahead that the given function u is harmonic!

We replace the trigonometric and hyperbolic functions by their values in terms of the exponential functions and get to

$$u(x, y) = y \left(\frac{e^{iy} + e^{-iy}}{2} \right) \left(\frac{e^x - e^{-x}}{2} \right) + x \left(\frac{e^{iy} - e^{-iy}}{2i} \right) \left(\frac{e^x + e^{-x}}{2} \right)$$

$$= \frac{y}{4} \left(e^{x+iy} + e^{x-iy} - e^{iy-x} - e^{-x-iy} \right)$$

$$\quad + \frac{x}{4i} \left(e^{x+iy} + e^{iy-x} - e^{-iy+x} - e^{-iy-x} \right)$$

$$= \frac{y}{4} \left(e^z + e^{\bar{z}} - e^{-\bar{z}} - e^{-z} \right) + \frac{x}{4i} \left(e^z + e^{-\bar{z}} - e^{\bar{z}} - e^{-z} \right)$$

$$= e^z \frac{x + iy}{4i} + e^{\bar{z}} \frac{iy - x}{4i} - e^{-z} \frac{x + iy}{4i} + e^{-\bar{z}} \frac{iy - x}{4i}$$

$$= \frac{1}{4i} \left((ze^z - ze^{-z}) - \overline{(ze^z - ze^{-z})} \right).$$

Hence,

$$u(x, y) = \operatorname{Re} \frac{1}{2i} (ze^z - ze^{-z}) = \operatorname{Re} \frac{1}{i} (z \sinh z).$$

The harmonic conjugate of u is given by

$$v(x, y) = \operatorname{Im} \frac{1}{i} (z \sinh z).$$

Another way is to use the formula for the harmonic conjugate

$$v(x, y) = \int_0^y \frac{\partial u}{\partial x}(x, t)dt - \int_0^x \frac{\partial u}{\partial y}(s, 0)ds. \qquad (9.5.2)$$

$$\square$$

Solution of Exercise 9.2.11. We compute $\dfrac{\partial v}{\partial x}$ and show that the second Cauchy–Riemann equation holds. We have:

$$\frac{\partial v}{\partial x}(x,y) = \int_0^y \frac{\partial^2 u}{\partial x^2}(x,t)dt - \frac{\partial u}{\partial y}(x,0)$$

$$= -\int_0^y \frac{\partial^2 u}{\partial y^2}(x,t)dt - \frac{\partial u}{\partial y}(x,0)$$

$$= \frac{\partial u}{\partial y}(x,0) - \frac{\partial u}{\partial y}(x,y) - \frac{\partial u}{\partial y}(x,0)$$

$$= -\frac{\partial u}{\partial y}(x,y).$$

To go from the first line to the second, we used that u is harmonic and hence

$$\frac{\partial^2 u}{\partial x^2}(x,y) = -\frac{\partial^2 u}{\partial y^2}(x,y). \tag{9.5.3}$$

□

Solution of Exercise 9.2.12. The function (9.2.3) is analytic in the simply-con-nected set Ω. Therefore the integral of g around any closed path in Ω vanishes. Let $C_{x,y}$ be a path connecting (x_0, y_0) to (x, y), with parametrization $\gamma(t) = x(t) + iy(t)$, $t \in [a, b]$. The integral

$$f(z) = \int_{C_{x,y}} g(\zeta)d\zeta$$

is independent of the specific choice of $C_{x,y}$ and depends only on (x_0, y_0) and (x, y). Note that f is analytic in Ω. Furthermore

$$f(z) - f(z_0) = \int_a^b \left(\frac{\partial u}{\partial x}(x(t), y(t)) - i\frac{\partial u}{\partial y}(x(t), y(t)) \right) (x'(t) + iy'(t))dt.$$

Taking real and imaginary part we obtain

$$\operatorname{Re} f(x,y) - \operatorname{Re} f(x_0, y_0) = \int_a^b \left(\frac{\partial u}{\partial x}(x(t), y(t))x'(t) + \frac{\partial u}{\partial y}(x(t), y(t))y'(t) \right) dt,$$

$$\operatorname{Im} f(x,y) - \operatorname{Im} f(x_0, y_0) = \int_a^b \left(\frac{\partial u}{\partial x}(x(t), y(t))y'(t) - \frac{\partial u}{\partial y}(x(t), y(t))x'(t) \right) dt.$$

Since

$$\frac{du(x(t), y(t))}{dt} = \frac{\partial u}{\partial x}(x(t), y(t))x'(t) + \frac{\partial u}{\partial y}(x(t), y(t))y'(t),$$

the first equation leads to $\operatorname{Re} f(x, y) = u(x, y)$, up to a constant. It follows that $\operatorname{Im} f$ is a harmonic conjugate of u, and the above formula for $\operatorname{Im} f$ is just (9.3.3).

□

Solution of Exercise 9.2.13. From (9.2.6) we get

$$f^{(2)}(\rho) + \frac{1}{\rho} f^{(1)}(\rho) = 0,$$

which readily leads to

$$f(\rho) = A \ln \rho + B,$$

while in the second case, we have

$$f(\theta) = A\theta + B,$$

where, in both cases, A and B are arbitrary constants. □

Solution of Exercise 9.2.14. The first Cauchy–Riemann equation in polar coordinates gives (see (9.2.5))

$$\theta \cos \theta + (\ln r + 1) \sin \theta = \frac{1}{r} \frac{\partial v}{\partial \theta},$$

and hence

$$v(r, \theta) = r(\theta \sin \theta + \cos \theta - (\ln r + 1) \cos \theta) + k(r).$$

The unknown $k(r)$ is obtained using the second Cauchy–Riemann equation:

$$\theta \sin \theta + \cos \theta - (2 + \ln r) \cos \theta + k'(r) = \theta \sin \theta - \cos \theta - \ln r \cos \theta.$$

It follows that $k(r)$ is a constant, say k.

We note that

$$u + iv = r\theta e^{i\theta} - ir \ln r e^{i\theta} = -iz \ln z.$$ □

Solution of Exercise 9.3.1. We have the formulas

$$(u^2)_x = 2u_x u,$$
$$(u^2)_{xx} = 2(u_x)^2 + 2u_{xx} u,$$
$$(u^2)_y = 2u_y u,$$
$$(u^2)_{yy} = 2(u_y)^2 + 2u_{yy} u.$$

Thus

$$\Delta(u^2) = 2\left(u_x^2 + u_y^2\right) + 2u\Delta u.$$

From $\Delta u = \Delta(u^2) = 0$ we obtain

$$u_x^2 + u_y^2 = 0.$$

Hence $u_x = u_y = 0$ and u is constant (on the connected components of its domain of definition). □

In the preceding exercise, one can replace u^2 by $\varphi(u)$ where $\varphi(t)$ is any function of class C_2 (that is, with continuous second derivative), and whose second derivative is non-zero, at the possible exception of a finite number of points (this latter can also be relaxed). Indeed,

$$(\varphi(u))_x = u_x \varphi'(u),$$
$$(\varphi(u))_{xx} = (u_x)^2 \varphi''(u) + u_{xx} \varphi'(u),$$
$$(\varphi(u))_y = u_y \varphi'(u),$$
$$(\varphi(u))_{yy} = (u_y)^2 \varphi''(u) + u_{yy} \varphi'(u).$$

Hence

$$\Delta\varphi(u) = \varphi'(u)\Delta u + \left(u_x^2 + u_y^2\right)\varphi''(u).$$

If $\varphi(u)$ and u are both harmonic, we get that

$$\left(u_x^2 + u_y^2\right)\varphi''(u) = 0,$$

and hence the result if φ'' vanishes at most at a finite number of points. □

Solution of Exercise 9.3.2. The function $f(z) = u(x, y) + iv(x, y)$ is entire and so is the function $F(z) = f(z)^3$. We have

$$\operatorname{Re} F(z) = u(x, y)^3 - 3u(x, y)v(x, y)^2.$$

Consider the entire function $G(z) = e^{-F(z)}$. Its modulus is

$$|G(z)| = e^{-\operatorname{Re} F(z)} \le 1 \quad \text{since} \quad \operatorname{Re} F(z) \ge 0.$$

By Liouville's theorem G is constant. Thus

$$G'(z) = -3f'(z)f^2(z)G(z) \equiv 0.$$

Thus $f'(z)f(z)^2 \equiv 0$. Since the zeros of F are isolated, we get that $f'(z) = 0$ first at those points where $f(z) \ne 0$ and then on all of $\mathbb{C}$ by continuity of f'. Hence $f(z)$ is a constant and so are u and v. □

Solution of Exercise 9.3.3. We assume that u is bounded from above. The case where u is bounded from below is treated in the same way after replacing u by $-u$.

Since u is harmonic in the whole plane, it has a harmonic conjugate v on $\mathbb{R}^2$, given for instance by formula (9.5.2). Assume that there is M such that

$$u(x, y) \le M, \quad \forall (x, y) \in \mathbb{R}^2.$$

The function

$$H(z) = \exp(u(x, y) + iv(x, y) - M)$$

is entire. It is bounded in the complex plane since

$$|H(z)| = e^{u(x,y)-M} \le 1.$$

By Liouville's theorem, H is constant and it follows that its modulus is constant. Taking the logarithm of $H(z)$ we obtain that u is constant. $\qquad\square$

Solution of Exercise 9.3.4. The function $G(z) = 1 - \frac{\partial u}{\partial x} + i\frac{\partial u}{\partial y}$ is entire and satisfies

$$\operatorname{Re} G(z) \ge 1.$$

The function $1/G$ is thus entire and bounded by 1 in modulus since

$$\frac{1}{|G(z)|^2} = \frac{1}{(1 - \frac{\partial u}{\partial x})^2 + (\frac{\partial u}{\partial y})^2}.$$

By Liouville's theorem, the function $1/G$ is constant and so is G. It follows that $\frac{\partial u}{\partial x}$ is equal to a constant, say A. Thus

$$u(x,y) = Ax + B(y).$$

Since $B(y)$ is harmonic, we have $B(y) = By + C$ for some real constants B and C. Thus

$$u(x,y) = Ax + By + C.$$

Moreover, $A \le 0$ since $\frac{\partial u}{\partial x} \le 0$. $\qquad\square$

Solution of Exercise 9.3.5. The function $\frac{\partial u}{\partial x}$ is also harmonic in the plane and thus has a harmonic conjugate, say $t(x,y)$, such that

$$f(z) = \frac{\partial u}{\partial x} + it(x,y)$$

is entire. We have that $|f(z)| \ge 5$ since $\frac{\partial u}{\partial x} \ge 5$. So the function f does not vanish in the complex plane, and $1/f$ is bounded in modulus by $1/5$. By Liouville's theorem, $1/f$, and hence f, is constant. Thus $\frac{\partial u}{\partial x}$ is a constant, and we have

$$u(x,y) = Ax + B(y),$$

with $A \ge 5$. Since $\Delta u = 0$, it follows that $B(y) = By + C$ for some arbitrary real constants B and C. $\qquad\square$

Solution of Exercise 9.3.6. Let $|f| = \sqrt{u^2 + v^2}$ be the modulus of f. Then a direct computation shows that

$$\frac{\partial |f|}{\partial x} = \frac{uu_x + vv_x}{|f|},$$

$$\frac{\partial^2 |f|}{\partial x^2} = \frac{u_x^2 + v_x^2 + uu_{xx} + vv_{xx}}{|f|} - \frac{(uu_x + vv_x)^2}{|f|^3},$$

and similarly for the second derivative of $|f|$ with respect to y,

$$\frac{\partial^2 |f|}{\partial y^2} = \frac{u_y^2 + v_y^2 + uu_{yy} + vv_{yy}}{|f|} - \frac{(uu_y + vv_y)^2}{|f|^3}. \tag{9.5.4}$$

Since $\Delta u = \Delta v = 0$, adding (9.5.4) and the equation preceding (9.5.4) leads to:

$$\Delta |f| = \frac{|f|^2 (u_x^2 + u_y^2 + v_x^2 + v_y^2) - (uu_x + vv_x)^2 - (uu_y + vv_y)^2}{|f|^3}.$$

Furthermore, the formula for the derivative and the Cauchy–Riemann equations lead to

$$u_x^2 + u_y^2 + v_x^2 + v_y^2 = 2|f'|^2 \quad \text{and} \quad |f\overline{f'}|^2 = (uu_x + vv_x)^2 + (uu_y + vv_y)^2,$$

and hence the formula for $\Delta |f|$. □

Note that $f(z) = z$ leads to (9.3.1).

Solution of Exercise 9.3.7. (1) We define for $r \in (0, R)$ the function

$$F(r) = \frac{1}{2\pi} \int_0^{2\pi} u(x_0 + r\cos t, y_0 + r\sin t)dt.$$

We have

$$F'(r) = \frac{1}{2\pi} \int_0^{2\pi} \left(\cos t \frac{\partial u}{\partial x}(x_0 + r\cos t, y_0 + r\sin t) \right.$$

$$\left. + \sin t \frac{\partial u}{\partial y}(x_0 + r\cos t, y_0 + r\sin t) \right) dt$$

$$= \int_{\mathbb{T}} P(x, y)dx + Q(x, y)dy,$$

with

$$P(x, y) = -\frac{1}{2\pi} \frac{\partial u}{\partial y}(x_0 + rx, y_0 + ry) \quad \text{and} \quad Q(x, y) = \frac{1}{2\pi} \frac{\partial u}{\partial x}(x_0 + rx, y_0 + ry).$$

Using Green's theorem, we have that

$$F'(r) = \frac{1}{2\pi} \int_{\mathbb{D}} (\Delta u)(x_0 + rx, y_0 + ry)dxdy \geq 0.$$

Thus

$$u(x_0, y_0) = F(0) \leq F(r) = \frac{1}{2\pi} \int_0^{2\pi} u(x_0 + r\cos t, y_0 + r\sin t)dt.$$

We note that this last equation can be rewritten as

$$\int_0^{2\pi} (u(x_0, y_0) - u(x_0 + r\cos t, y_0 + r\sin t))dt \le 0. \qquad (9.5.5)$$

(2) Let u be a subharmonic function in an open connected set Ω, and let (x_0, y_0) be a local maximum of u. Thus, there exists $R > 0$ such that, for all $r \in (0, R)$ and all $t \in [0, 2\pi]$,

$$u(x_0, y_0) \ge u(x_0 + r\cos t, y_0 + r\sin t).$$

Together with (9.5.5), this leads to

$$u(x_0, y_0) = u(x_0 + r\cos t, y_0 + r\sin t), \quad \forall r \in (0, R) \quad \text{and} \quad t \in [0, 2\pi].$$

If now u is bounded in Ω, the preceding argument shows that the set of points where the maximum is attained is open. But the set where it is not attained is also open; since Ω is connected and the first set is not empty, u is a constant. $\square$

Solution of Exercise 9.3.8. Consider r such that $B(z_0, R) \subset \mathbb{D}$. Since $|f(z)^2|$ is subharmonic (see formula (9.3.3)), it follows from the preceding exercise that

$$|f(z_0)^2| \le \frac{1}{2\pi} \int_0^{2\pi} |f(x_0 + r\cos t, y_0 + r\sin t)^2|dt.$$

Therefore, for $r_0 < R$,

$$\frac{r_0^2}{2}|f(z_0)^2| = \int_0^{r_0} |f(z_0)^2| r dr \le \frac{1}{2\pi} \int_0^{r_0} \left(\int_0^{2\pi} |f(x_0 + r\cos t, y_0 + r\sin t)^2|dt \right) r dr$$

$$= \frac{1}{2\pi} \iint_{B(z_0, r_0)} |f(z)|^2 dx dy$$

$$\le \|f\|_B^2,$$

where we have used the formula for the change of variable in the double integral to go from the first to the second line. $\square$

Remark 9.5.1. The preceding proof holds for more general domains. In the case of the disk, another proof would go as follows: Let $f(z) = \sum_{n=0}^{\infty} a_n z^n$. Then

$$\|f\|_B^2 = \sum_{n=0}^{\infty} \frac{|a_n|^2}{n+1},$$

and therefore

$$|f(z_0)| \leq \sum_{n=0}^{\infty} |a_n||z_0|^n$$

$$= \sum_{n=0}^{\infty} \frac{|a_n|}{\sqrt{n+1}} \sqrt{n+1}|z_0|^n$$

$$\leq \left(\sum_{n=0}^{\infty} \frac{|a_n|^2}{n+1}\right)^{1/2} \left(\sum_{n=0}^{\infty} (n+1)|z_0|^{2n}\right)^{1/2}$$

$$= \frac{1}{1-|z_0|^2}\|f\|_{\mathcal{B}},$$

where we have used the Cauchy–Schwarz inequality and the equality

$$\sum_{n=0}^{\infty} (n+1)|z_0|^{2n} = \frac{1}{(1-|z_0|^2)^2}.$$

Definition 9.5.2. The space of functions analytic in the open unit disk for which $\|f\|_{\mathcal{B}}$ is finite is a Hilbert space called the *Bergman space*.

The inequality (9.3.7) expresses that the Bergman space is a reproducing kernel Hilbert space; its reproducing kernel is $\frac{1}{(1-z\overline{w})^2}$.

Solution of Exercise 9.3.9. In case (a) we have

$$u_x(x, y) = 2x\varphi'(x^2 - y^2),$$
$$u_{xx}(x, y) = 2\varphi'(x^2 - y^2) + 4x^2\varphi''(x^2 - y^2),$$
$$u_y(x, y) = -2y\varphi'(x^2 - y^2),$$
$$u_{yy}(x, y) = -2\varphi'(x^2 - y^2) + 4y^2\varphi''(x^2 - y^2).$$

Thus

$$\Delta u = 4(x^2 + y^2)\varphi''(x^2 - y^2),$$

and so $\varphi(t) = At + B$ for some real constants A and B and so

$$u(x, y) = A(x^2 - y^2) + B.$$

In case (b),

$$u_x(x, y) = -\frac{y}{x^2}\varphi'(y/x),$$
$$u_{xx}(x, y) = \frac{2y}{x^3}\varphi'(y/x) + \frac{y^2}{x^4}\varphi''(y/x),$$
$$u_y(x, y) = \frac{1}{x}\varphi'(y/x),$$
$$u_{yy}(x, y) = \frac{1}{x^2}\varphi''(y/x).$$

Thus

$$\Delta u = 0 \quad \Longleftrightarrow \quad \frac{2y}{x}\varphi'(y/x) + ((y/x)^2 + 1)\varphi''(y/x) = 0.$$

Thus

$$2t\varphi'(t) + (t^2 + 1)\varphi''(t) = 0,$$

that is

$$\varphi(t) = A\arctan t + B. \qquad \square$$

Solution of Exercise 9.3.10. Recall that $\Delta = 4\partial_z\partial_{\bar{z}}$, where the operators ∂_z and $\partial_{\bar{z}}$ have been defined in (4.2.15). We have

$$\partial_{\bar{z}}(\bar{z}f) = f + \bar{z}\partial_{\bar{z}}f,$$

and therefore

$$\partial_{\bar{z}^2}^2(\bar{z}f) = 2\partial_{\bar{z}}f + \bar{z}\partial_{\bar{z}^2}^2 f.$$

So, taking into account that

$$\Delta f = 4\partial_z\partial_{\bar{z}}f = 0,$$

we have

$$\partial_z(\partial_{\bar{z}^2}^2(\bar{z}f)) = 0,$$

and in particular $\Delta^2 f = 0$, and so also $\Delta^2\operatorname{Re} f = 0.$ $\qquad \square$

Solution of Exercise 9.4.2. Using formula (9.4.1) we have

$$u(z) = \operatorname{Re}\left(\frac{1}{2\pi}\int_0^\pi \frac{e^{it} + z}{e^{it} - z}dt - \frac{1}{2\pi}\int_\pi^{2\pi} \frac{e^{it} + z}{e^{it} - z}dt\right)$$

$$= \operatorname{Re}\left(\frac{1}{2\pi}\int_0^\pi \left(1 + 2\sum_{n=1}^\infty z^n e^{-int}\right)dt \right.$$

$$\left. -\frac{1}{2\pi}\int_\pi^{2\pi}\left(1 + 2\sum_{n=1}^\infty z^n e^{-int}\right)dt\right)$$

$$= \frac{1}{\pi}\operatorname{Re}\left(\sum_{n=1}^\infty z^n\left(\int_0^\pi e^{-int}dt - \int_\pi^{2\pi} e^{-int}dt\right)\right)$$

$$= \frac{2}{\pi}\operatorname{Re}\sum_{n=1}^\infty z^n\frac{(-1)^n - 1}{-in}$$

$$= \sum_{p=0}^\infty \frac{4r^{2p+1}\sin(2p+1)\theta}{(2p+1)\pi}. \qquad \square$$

Solution of Exercise 9.4.3. Let u be a solution. It will have a harmonic conjugate if and only if equality (9.2.2)

$$\int_C -\frac{\partial u}{\partial y}dx + \frac{\partial u}{\partial x}dy = 0$$

holds for all closed curve C in Ω. The integral (9.2.2) depends only on the homology class of the curve, and needs to be checked only for the circle of radius $1/2$. Set

$$H = \int_{|z|=1/2} -\frac{\partial u}{\partial y}dx + \frac{\partial u}{\partial x}dy = 0,$$

where by $|z| = 1/2$ we mean the curve

$$\gamma(t) = \frac{e^{it}}{2}, \quad t \in [0, 2\pi].$$

Since

$$\int_{|z|=1/2} -\frac{\partial \ln(x^2+y^2)}{\partial y}dx + \frac{\partial \ln(x^2+y^2)}{\partial x}dy = \int_{|z|=1/2} -\frac{2y}{x^2+y^2}dx + \frac{2x}{x^2+y^2}dy$$

$$= 4\pi,$$

Theorem 9.2.7 insures that the function

$$U(x,y) = u(x,y) - \frac{H}{4\pi}\ln(x^2+y^2) \tag{9.5.6}$$

has a harmonic conjugate in $\mathbb{D} \setminus \{0\}$. The function u satisfies the boundary condition on the unit circle if and only if U does, and so the problem is reduced to the case where a harmonic conjugate exists. Let V be a harmonic conjugate of U. The function $F(z) = U(x,y) + iV(x,y)$ is analytic in $0 < |z| < 1$ and therefore has a Laurent expansion of the form

$$F(z) = \sum_{n=0}^{\infty} a_n z^n + \sum_{n=1}^{\infty} \frac{b_n}{z^n},$$

where the series

$$\sum_{n=0}^{\infty} a_n z^n$$

converges in $|z| < 1$, and the series

$$\sum_{n=1}^{\infty} \frac{b_n}{z^n}$$

converges in $\mathbb{C}\setminus\{0\}$, and is in particular continuous on the unit circle. The function U is continuous in $0 < |z| \leq 1$ if and only if

$$\operatorname{Re} \sum_{n=0}^{\infty} a_n z^n$$

is continuous in $0 < |z| \leq 1$. Writing $U(e^{it}) \equiv 0$ leads to

$$\operatorname{Re}\left\{\sum_{n=0}^{\infty} a_n e^{int} + \sum_{n=1}^{\infty} b_n e^{-int}\right\} \equiv 0.$$

It follows that $\operatorname{Re} a_0 = 0$ and

$$a_n = -\overline{b_n}, \quad n = 1, 2, \ldots.$$

Solutions $u(x, y)$ are therefore of the form

$$\frac{H}{4\pi} \ln(x^2 + y^2) + \operatorname{Re}\left\{h(z) - \overline{h(1/\bar{z})}\right\},$$

where h is an arbitrary entire function, and H is an arbitrary real number. $\square$

Chapter 10

Conformal Mappings

Riemann's mapping theorem asserts that a simply-connected domain different from $\mathbb{C}$ is conformally equivalent to the open unit disk: There exists an analytic bijection from Ω onto $\mathbb{D}$ (that the inverse is itself analytic is automatic; see Exercise 10.2.4). In this chapter we closely follow Chapters 5 and 6 of [45] and present some related exercises. The chapter is smaller than the previous ones, but is certainly of key importance in the theory of analytic functions. To quote [195, p. 1], Riemann's theorem *is one of those results one would like to present in a one-semester intro-ductory course in complex variables, but often does not for lack of sufficient time. The proof requires also some topology, which is not always known by students of a first complex variable course.*

10.1 Uniform convergence on compact sets

The proof of Riemann's theorem is not constructive, and uses deep properties of the topology of the space of functions analytic in an open set. We review here some of these properties. The solutions of the following two questions will not be given here.

Question 10.1.1. *Let Ω be an open connected subset of $\mathbb{C}$. Show that there exists an increasing sequence $(K_n)_{n\in\mathbb{N}}$ of compact subsets of Ω with the following property: Given any compact subset K of Ω, there exists $N \in \mathbb{N}$ such that*

$$K \subset \bigcup_{n=1}^{N} K_n.$$

Question 10.1.2. *Let Ω and $(K_n)_{n\in\mathbb{N}}$ be as in the previous exercise. Show that (see*

[45, (3.3), p. 149])

$$d(f,g) = \sum_{n=1}^{\infty} \frac{1}{2^n} \min(1, \max_{z \in K_n} |f(z) - g(z)|) \tag{10.1.1}$$

defines a metric on the space $A(\Omega)$ of functions analytic in Ω.

Convergence of a sequence in the metric (10.1.1) is equivalent to uniform convergence on every compact subspace of Ω. The space $A(\Omega)$ endowed with this metric has a key property: A subset of $A(\Omega)$ is compact if and only if it is both closed and bounded. Locally convex Hausdorff barreled topological vector spaces for which this property holds are called Montel spaces. See, e.g., [214, Definition 34.2, p. 356]. We also refer to [CAPB2], where some of these definitions and concepts are reviewed. Bounded here does not mean boundedness with respect to the metric, but boundedness in a topological vector space. Recall:

Definition 10.1.3. Let V denote a topological vector space on the complex numbers or on the real numbers. The set $U \subset V$ is called *bounded* if for every neighborhood W of the origin there exists $\lambda > 0$ such that

$$U \subset \lambda W.$$

The above characterization of compact sets is the key in the proof of Riemann's theorem. We refer to [109] for a thorough study of the metric spaces where (sequential) compactness is equivalent to being closed and bounded.

10.2   One-to-oneness

It is an important fact that an analytic function is one-to-one in a neighborhood of a point where its derivative does not vanish. For the following exercise, see [148, p. 372].

Exercise 10.2.1. *Let f be analytic in a convex open set Ω and assume that $\operatorname{Re} f'(z) > 0$ in Ω. Show that f is one-to-one in Ω.*

Note that an analytic function which is one-to-one on an open set Ω is said to be univalent in that set. As a corollary of this exercise we get the following very important result. For the converse statement, namely that when $f'(z_0) = 0$ there is no neighborhood of z_0 in which the function is one-to-one, see Exercise 7.3.8.

Theorem 10.2.2. *An analytic function is univalent in a neighborhood of any point where its derivative does not vanish.*

Indeed, if $f'(z_0) \neq 0$, then at least one of the numbers $\operatorname{Re} f'(z_0)$ and $\operatorname{Im} f'(z_0)$ is not zero. Without loss of generality we may assume that $\operatorname{Re} f'(z_0) > 0$ (otherwise replace f by $-f$ or $\pm if$ depending on the case). By continuity, $\operatorname{Re} f'(z) > 0$ in an open disk around z_0. We can then apply the precedent result since a disk is in particular convex.

Exercise 10.2.3. *Give a solution of Exercise 5.2.9 using Exercise 10.2.1.*

We note the following:

$$f'(z) = 1 + \sum_{n=2}^{\infty} n a_n z^{n-1},$$

and in particular

$$|f'(z)| \geq 1 - \left| \sum_{n=2}^{\infty} n a_n z^{n-1} \right|$$

$$\geq 1 - \sum_{n=2}^{\infty} n |a_n| |z|^{n-1}$$

$$\geq 1 - \sum_{n=2}^{\infty} n |a_n| > 0, \quad \forall z \in \mathbb{D}.$$

Thus, by Theorem 10.2.2, f is one-to-one in a neighborhood of every point in $\mathbb{D}$. This is a local result. We want a direct solution of a global result: f is one-to-one in $\mathbb{D}$.

Exercise 10.2.4. *Assume that the analytic function f is one-to-one in Ω. Show that the formula (see, e.g., [42, p. 180])*

$$g(z) = \frac{1}{2\pi i} \int_{\gamma} \frac{s f'(s)}{f(s) - z} ds, \tag{10.2.1}$$

where γ is a closed simple contour, defines the inverse of f inside γ.

Formula (10.2.1) shows in particular that f^{-1} is analytic.

We now consider the case where the derivative vanishes at a given point. It is no loss of generality to assume that the function itself also vanishes at that point.

Exercise 10.2.5. *Let f be analytic in the open subset Ω and assume that $z_0 \in \Omega$ is a zero of order N of f. Show that there is a function g which is analytic and one-to-one in some open neighborhood $U \subset \Omega$ of z_0 and such that*

$$f(z) = g(z)^N, \quad z \in U. \tag{10.2.2}$$

With the preceding exercises at hand we can state the following key result, called the *open mapping theorem* (see also Exercise 7.4.9).

Theorem 10.2.6. *Let Ω be an open subset of $\mathbb{C}$ and let f be analytic in Ω. Then, $f(\Omega)$ is an open subset of $\mathbb{C}$.*

Proof. Take first a point $\omega \in \Omega$ where $f'(\omega) \neq 0$. By Exercise 10.2.4 the function f is one-to-one in an open neighborhood U of ω, with analytic inverse. The inverse h of f,

$$h:\ f(U) \longrightarrow U,$$

is in particular continuous, and therefore

$$f(U) = h^{-1}(U) \subset f(\Omega)$$

is an open neighborhood of $f(\omega)$ which lies inside $f(\Omega)$. If $f'(\omega) = 0$, we first remark that for any $N \in \mathbb{N}$ the map $z \mapsto z^N$ maps open balls into open balls, and therefore open sets into open sets. Write f in the form (10.2.2). There is an open neighborhood U of ω where $g(z)$ is one-to-one. By the above argument, $g(U)$ is open, and so is $f(U) = g(U)^N$. $\square$

Exercise 10.2.7. *In the notation and hypothesis of Exercise 6.1.9, show that the set Ω_0 contains uncountably many points.*

We conclude this section with an important fact on univalent functions, which comes into play in the proof of Riemann's mapping theorem. See [45, Proposition 2.2, p. 147, p. 191].

Exercise 10.2.8. *Let Ω be open and connected, and let $(s_n)_{n \in \mathbb{N}}$ be a sequence of functions univalent in Ω, which converge uniformly on compact subsets of Ω. The limit is then either a constant or univalent.*

10.3   Conformal mappings

Simply-connected sets have already been characterized in a number of ways. Geometrically, Riemann's mapping theorem expresses the following characterization:

Definition 10.3.1. A connected open subset Ω of $\mathbb{C}$ which is different from $\mathbb{C}$ is *simply-connected* if it is conformally equivalent to the open unit disk.

Question 10.3.2. *Show that any open disk is conformally equivalent to any open half-plane.*

We recall that the Blaschke factors (1.1.44), possibly multiplied by a constant of modulus 1,

$$\varphi(z) = c\frac{z - a}{1 - \bar{a}z}$$

are the only conformal mappings from the open unit disk onto itself. Taking into account this fact allows to solve the following exercise.

Exercise 10.3.3.

(1) *Show that the conformal maps from the open upper half-plane $\mathbb{C}_+$ onto itself are exactly the Moebius maps which can be written in the form*

$$\varphi(z) = \frac{az+b}{cz+d},$$

where a, b, c, d are real and such that $ad - bc = 1$.

(2) *Show that any two points in $\mathbb{C}_+$ can be related by such a conformal map.*

The proof of Riemann's mapping theorem (see for instance H. Cartan's [45]) uses the fact that a connected open subset Ω of the complex plane is simply connected if and only if every non-vanishing function analytic in Ω admits an analytic logarithm. The proof can be divided into three steps (and here, we follow [45]):

(a) Reduce to the case where $\Omega \subset \mathbb{D}$ and $0 \in \Omega$.

(b) Show that the existence of a conformal map is equivalent to the solution of a maximum problem.

(c) Show that the maximum problem has a solution.

Steps (a) and (b) use, each once only once, the assumed existence of an analytic logarithm. Step (c) uses topology tools which are somewhat beyond the scope of the present book. The content of the following question is Step (b).

Question 10.3.4. *Let Ω be an open subset of $\mathbb{D}$, containing the origin, and with the property that every non vanishing function analytic in Ω has an analytic logarithm. Let M denote the set of univalent functions from Ω into $\mathbb{D}$ such that $f(0) = 0$. Show that the range of f is $\mathbb{D}$ if and only if*

$$|f'(0)| = \max_{g \in M} |g'(0)|.$$

Hints: One direction is relatively easy, and uses the Schwarz lemma. For the other direction, proceed by contradiction, and use Theorem 5.7.6 (see [45]).

Question 10.3.5. *Show that $\tan z$ is a conformal map from the strip*

$$L_1 = \{(x, y) \, ; \, x \in (-\pi/4, \pi/4) \quad \text{and} \quad y \in \mathbb{R}\}$$

onto the open unit disk.

Exercise 10.3.6. *Find a conformal map between the open right half-plane and the quarter-plane*

$$\{(x, y) \, ; \, 0 < x < |y|\} \, .$$

Exercise 10.3.7. *Let $\mathbb{D}$ denote the open unit disk and $\mathbb{C}_+$ denote the open upper half-plane. Show that the map*

$$\varphi(z) = \frac{z - i(z^2 + 1)}{z + i(z^2 + 1)}$$

is a conformal mapping from $\mathbb{D}_+ = \mathbb{D} \cap \mathbb{C}_+$ onto $\mathbb{D}$. What happens on the boundary?

Exercise 10.3.8 (see [75, Exercice 35.33, p. 329]). *Let $\alpha \in (0, \pi/2)$ and define*

$$D_\alpha = \left\{ z \in \mathbb{C};\, |z \pm i \cot \alpha| < \frac{1}{\sin \alpha} \right\}.$$

Show that the map

$$c(z) = \frac{\left(\frac{1+z}{1-z}\right)^{\frac{\pi}{2\alpha}} - 1}{\left(\frac{1+z}{1-z}\right)^{\frac{\pi}{2\alpha}} + 1}$$

is conformal from D_α onto $\mathbb{D}$, and that its inverse is given by

$$c^{-1}(z) = \frac{\left(\frac{1+z}{1-z}\right)^{\frac{2\alpha}{\pi}} - 1}{\left(\frac{1+z}{1-z}\right)^{\frac{2\alpha}{\pi}} + 1}.$$

The following exercise can be found for instance in [53, p. 203], [168, Exercise 2, p. 196], and [18, § 10.4.4, pp. 308–311]. We follow the solution of that latter reference. In the statement the function $\sqrt{1 - s^4}$ is defined via (4.4.9).

Exercise 10.3.9. *Show that the map*

$$z \mapsto c(z) = \int_{[0,z]} \frac{ds}{\sqrt{1 - s^4}} \tag{10.3.1}$$

is conformal from $\mathbb{D}$ onto a square.

Hint. Following [18, § 10.4.4, pp. 308–311] we suggest to solve the exercise along the steps below:

Step 1: Show that the map c extends continuously to the closed unit disk, and that (see [18, p. 310])

$$c(e^{i\theta}) = M + e^{i\frac{3\pi}{4}} \int_0^\theta \frac{du}{\sqrt{2\sin(2u)}}, \quad \theta \in \left[0, \frac{\pi}{4}\right]. \tag{10.3.2}$$

for some constant $M > 0$.

Step 2: Show that the image of the unit circle is the boundary of a square. Exercise 3.5.7 plays an important role in this step. It is also useful to note that

$$c(iz) = ic(z), \quad z \in \mathbb{D}. \tag{10.3.3}$$

Step 3: Compute $\frac{1}{2\pi i} \int_{|z|=1} \frac{c'(z)}{c(z)} dz$.

10.4 Solutions

Solution of Exercise 10.2.1. Let z_1 and z_2 be in Ω. Since Ω is convex, the closed interval

$$[z_1, z_2] = \{z_1 + t(z_2 - z_1) \,;\, t \in [0,1]\} \subset \Omega.$$

By the fundamental theorem of calculus for analytic functions,

$$\begin{aligned}
f(z_2) - f(z_1) &= \int_{[z_1, z_2]} f'(z) dz \\
&= (z_2 - z_1) \int_0^1 f'(z_1 + t(z_2 - z_1)) dt \\
&= (z_2 - z_1) \left\{ \mathrm{Re} \left(\int_0^1 f'(z_1 + t(z_2 - z_1)) dt \right) \right. \\
&\qquad\qquad \left. + i \, \mathrm{Im} \left(\int_0^1 f'(z_1 + t(z_2 - z_1)) dt \right) \right\}.
\end{aligned}$$

(10.4.1)

Since $\mathrm{Re}\, f'(z) > 0$ in Ω we have

$$\mathrm{Re} \left(\int_0^1 f'(z_1 + t(z_2 - z_1)) dt \right) > 0. \qquad (10.4.2)$$

It follows from (10.4.2) that

$$\begin{aligned}
f(z_2) - f(z_1) &= (z_2 - z_1) \left\{ \mathrm{Re} \left(\int_0^1 f'(z_1 + t(z_2 - z_1)) dt \right) \right. \\
&\qquad\qquad \left. + i \, \mathrm{Im} \left(\int_0^1 f'(z_1 + t(z_2 - z_1)) dt \right) \right\}
\end{aligned}$$

that $f(z_1) \neq f(z_2)$ if $z_1 \neq z_2$. □

Solution of Exercise 10.2.3. We have, for $z \in \mathbb{D}$,

$$\mathrm{Re}\, f'(z) = 1 - \mathrm{Re} \sum_{n=2}^\infty n a_n z^{n-1} \geq 1 - \left| \sum_{n=2}^\infty n a_n z^{n-1} \right| \geq 1 - \sum_{n=2}^\infty n |a_n| > 0.$$

It suffices then to apply the previous exercise. □

Solution of Exercise 10.2.4. We have, for z_0 inside γ,

$$\begin{aligned}
g(f(z_0)) &= \frac{1}{2\pi i} \int_\gamma \frac{\dfrac{s f'(s)}{f(s) - f(z_0)}}{s - z_0} ds \\
&= \left(\frac{\dfrac{s f'(s)}{f(s) - f(z_0)}}{s - z_0} \right)_{s = z_0} \\
&= z_0.
\end{aligned}$$
□

Solution of Exercise 10.2.5. By definition of a zero of order N we can write in some neighborhood $W \subset \Omega$ of z_0,

$$f(z) = (z - z_0)^N h(z),$$

where h is analytic in W and does not vanish there. We can always assume W to be convex (for instance, W may be chosen to be an open disk with center z_0 and small enough radius). Then, the function h has an analytic logarithm in W, and therefore also an analytic root of order N: There is a function h_0 analytic in W and such that

$$h(z) = h_0(z)^N, \quad z \in W.$$

We therefore have $f(z) = ((z - z_0)h_0(z))^N, \quad z \in W$. The function $g(z) = (z - z_0)h_0(z)$ is analytic in W. It is one-to-one in a neighborhood $U \subset W$ of z_0 since

$$g'(z)|_{z=z_0} = ((z - z_0)h_0'(z) + h_0(z))|_{z=z_0} = h_0(z_0) \neq 0. \qquad \square$$

The following solution is taken from [10, pp. 4–5].

Solution of Exercise 10.2.7. We use the notation of Exercises 4.1.13 and 6.1.9. We know from Exercise 6.1.9 that there is a point $\mu \in \Omega$ such that

$$|a(\mu)| = |b(\mu)| \neq 0.$$

The map

$$\sigma(z) = \frac{b(z)}{a(z)}$$

is analytic in the open set $\Omega \setminus \mathcal{Z}(a)$. The image $\sigma(\Omega \setminus \mathcal{Z}(a))$ is an open set, and therefore there exists an $r > 0$ such that

$$B(\sigma(\mu), r) \subset \sigma(\Omega \setminus \mathcal{Z}(a)).$$

The image $\sigma(\Omega \setminus \mathcal{Z}(a))$ contains in particular an arc of a circle, and the claim follows. $\qquad \square$

Solution of Exercise 10.2.8. We first remark that the limit function s is indeed analytic, since the convergence is uniform on compact subsets of Ω. Assume that s is not a constant, but that there are two points a_1 and a_2 in Ω such that

$$s(a_1) = s(a_2) \overset{\text{def.}}{=} c.$$

The function $s(z) - c$ has isolated zeroes (since it is not a constant), and therefore we can find two closed neighborhoods

$$B_c(a_1, \rho_1) = \{z \in \Omega; |z - a_1| \leq \rho_1\}$$

and

$$B_c(a_2, \rho_2) = \{z \in \Omega; |z - a_2| \leq \rho_1\},$$

with ρ_1 and ρ_2 strictly positive, such that

$$B_c(a_1, \rho_1) \cap B_c(a_2, \rho_2) = \emptyset,$$

and such that

$$s(z) - c \neq 0,$$

both in $B_c(a_1, \rho_1) \setminus \{a_1\}$ and in $B_c(a_2, \rho_2) \setminus \{a_2\}$. Set

$$m_\ell = \min_{|a_\ell - z| = \rho_\ell} |s(z) - c|, \quad \ell = 1, 2.$$

We have that $m_1 > 0$ and $m_2 > 0$. Furthermore, since the neighborhoods $B_c(a_1, \rho_1)$ and $B_c(a_2, \rho_2)$ are compact, there exists $N \in \mathbb{N}$ such that

$$n \geq N \implies \forall z \in B_c(a_1, \rho_1) \cup B_c(a_2, \rho_2), \quad |s_n(z) - s(z)| < m_\ell, \quad \ell = 1, 2.$$

Thus, for all $z \in B_c(a_\ell, \rho_\ell)$, $\ell = 1, 2$, we have

$$|s_n(z) - s(z)| < m_\ell \leq |s(z) - c|.$$

From Rouché's theorem (see Exercise 7.4.1), we have that $s_n(z) - c$ vanishes in $B_c(a_\ell, \rho_\ell)$ for $\ell = 1, 2$, contradicting the fact that the s_n are univalent since $B_c(a_1, \rho_1) \cap B_c(a_2, \rho_2) = \emptyset$. $\square$

Solution of Exercise 10.3.3.
(1) The map $\varphi(z) = \frac{1+iz}{1-iz}$ sends conformally $\mathbb{C}_+$ onto $\mathbb{D}$. It follows that the conformal maps of $\mathbb{C}_+$ onto itself are, in terms of matrices, of the form

$$\begin{pmatrix} i & 1 \\ -i & 1 \end{pmatrix}^{-1} \begin{pmatrix} k & ku \\ \overline{u} & 1 \end{pmatrix} \begin{pmatrix} i & 1 \\ -i & 1 \end{pmatrix} = \frac{1}{2i} \begin{pmatrix} i(k(1-u)+(1-\overline{u})) & k(1+u)-(1+\overline{u}) \\ -k(1-u)+(1-\overline{u}) & i(k(1+u)+(1+\overline{u})). \end{pmatrix}$$

with $k \in \mathbb{T}$ and $u \in \mathbb{D}$. Let $k = e^{i\theta}$ with $\theta \in \mathbb{R}$. Dividing the entries of the above matrix by $e^{i\frac{\theta}{2}}\sqrt{1-|u|^2}$ we obtain the matrix

$$\frac{1}{\sqrt{1-|u|^2}} \begin{pmatrix} \mathrm{Re}\, e^{i\frac{\theta}{2}}(1-u) & \mathrm{Im}(e^{i\frac{\theta}{2}}(1+u)) \\ -\mathrm{Im}(e^{i\frac{\theta}{2}}(1-u)) & \mathrm{Re}(e^{i\frac{\theta}{2}}(1+u)) \end{pmatrix}, \tag{10.4.3}$$

which is of the required form. Conversely for any $\varphi(z) = \frac{az+b}{cz+d}$ where a, b, c, d are real and such that $ad - bc = 1$ we have

$$\mathrm{Im}\, \varphi(z) = \frac{\mathrm{Im}\, z}{|cz+d|^2}$$

and so φ sends $\mathbb{C}_+$ onto itself.

(2) The result is a direct consequence of Exercise 2.3.5. $\square$

Remark 10.4.1. When $u = 0$ the matrix (10.4.3) becomes

$$\begin{pmatrix} \cos(\frac{\theta}{2}) & \sin(\frac{\theta}{2}) \\ -\sin(\frac{\theta}{2}) & \cos(\frac{\theta}{2}) \end{pmatrix}.$$

Solution of Exercise 10.3.6. The open right half-plane $\mathbb{C}_r$ consists of the complex numbers $z = re^{it}$ with $r > 0$ and $t \in (-\frac{\pi}{2}, \frac{\pi}{2})$. The map $\sqrt{z} = \sqrt{r}e^{i\frac{t}{2}}$ is a conformal map from $\mathbb{C}_r$ onto the quarter-plane, with inverse map z^2. $\square$

Solution of Exercise 10.3.7. The map $G(z) = \dfrac{1-z}{1+z}$ is conformal from the open right half-plane $\mathbb{C}_r = \{z = x + iy \in \mathbb{C} \; ; \; x > 0\}$ onto $\mathbb{D}$. It is therefore enough to check that $G^{-1} \circ \varphi$ is conformal from $\mathbb{D}_+$ onto the right half-plane. But $G^{-1}(z) = \dfrac{1-z}{1+z}$ and so

$$G^{-1} \circ \varphi(z) = i\frac{z^2 + 1}{z}.$$

Let us write

$$\psi(z) = i\frac{z^2 + 1}{z} = i\left(z + \frac{1}{z}\right). \tag{10.4.4}$$

We now proceed in a number of steps.

Step 1: ψ *is one-to-one from* $\mathbb{D}_+$ *onto its range.*

Indeed, assume that $\psi(z_1) = \psi(z_2)$. Then, in view of (10.4.4),

$$z_1 - z_2 + \frac{1}{z_1} - \frac{1}{z_2} = 0,$$

that is

$$(z_1 - z_2)(1 - \frac{1}{z_1 z_2}) = 0.$$

Thus $z_1 = z_2$ or $z_1 = \dfrac{1}{z_2}$. Since we assume that both z_1 and z_2 belong to $\mathbb{D}_+$ we have $z_1 = z_2$.

Step 2: *The range of* ψ *is inside* $\mathbb{C}_r$.

Indeed, with $z = x + iy$,

$$\psi(z) = i\left((x + iy) + \frac{x - iy}{x^2 + y^2} \right)$$

$$= y\left(\frac{1}{x^2 + y^2} - 1 \right) + i\left(\frac{x}{x^2 + y^2} + x \right).$$

But for $z \in \mathbb{D}_+$ we have

$$y > 0 \quad \text{and} \quad \frac{1}{x^2 + y^2} > 1,$$

and so

$$y\left(\frac{1}{x^2+y^2}-1\right)>0,$$

that is $\operatorname{Re}\psi(z)>0$.

Step 3: ψ *is onto* $\mathbb{C}_r$.

Indeed, for w such that $\operatorname{Re}w>0$ consider the equation $\psi(z)=w$. We have

$$z^2+izw+1=0,$$

and thus the product of the two roots of this second degree equation is equal to 1. Since, in view of the previous step,

$$0<\operatorname{Re}w=y\left(\frac{1}{x^2+y^2}-1\right),$$

we see that one of them is in $\mathbb{D}_+$. □

Solution of Exercise 10.3.8. We proceed in a number of steps:

(1) *The map*

$$z\mapsto\psi(z)=\frac{1+z}{1-z}$$

is conformal from $\mathbb{D}$ *onto the open right half-plane, with inverse* $\psi^{-1}(z)=\frac{z-1}{z+1}$.

This follows from

$$\frac{1+z}{1-z}+\frac{1+\bar{z}}{1-\bar{z}}=2\frac{1-|z|^2}{|1-z|^2}.$$

(2) *Let* $z=re^{it}$ *with* $r>0$ *and* $t\in(-\pi,\pi)$. *The map*

$$z\mapsto p_\alpha(z)=z^{\frac{\pi}{2\alpha}}=r^{\frac{\pi}{2\alpha}}e^{i\frac{\pi}{2\alpha}t} \tag{10.4.5}$$

is conformal from the domain

$$C_{r,\alpha}=\{z\in\mathbb{C};0<x<(\tan\alpha)|y|\}$$

onto the open right half-plane.

This is because $z\in C_{r,\alpha}$ if and only if it is of the form $z=re^{i\theta}$, where $\theta\in(-\alpha,\alpha)$. Under the map (10.4.5) the angle has now range $(-\frac{\pi}{2},\frac{\pi}{2})$.

(3) *The map* $\psi^{-1}(z)=\frac{z-1}{z+1}$ *is conformal from* $C_{r,\alpha}$ *onto* D_α.

We note that the boundary of $C_{r,\alpha}$ consists of the two rays $re^{\pm i\alpha}$, with $r\in[0,\infty)$. We first check that this boundary is sent onto the boundary of D_α. We consider the ray $re^{i\alpha}$. The other one is treated in the same way. Let therefore

$$x+iy=\frac{re^{i\alpha}-1}{re^{i\alpha}+1}=\frac{r^2-1}{r^2+1+2r\cos\alpha}+i\frac{2r\sin\alpha}{r^2+1+2r\cos\alpha}$$

be in the image of this ray under ψ^{-1}. We have

$$y + \cot\alpha = \frac{1}{\sin\alpha}\frac{2r + (1+r^2)\cos\alpha}{r^2 + 1 + 2r\cos\alpha}.$$

Thus

$$
\begin{aligned}
|x + iy + i\cot\alpha|^2 &= \left(\frac{r^2 - 1}{r^2 + 1 + 2r\cos\alpha}\right)^2 + \frac{1}{\sin^2\alpha}\left(\frac{2r + (1+r^2)\cos\alpha}{r^2 + 1 + 2r\cos\alpha}\right)^2 \\
&= \frac{(r^2 - 1)^2\sin^2\alpha + (r^2+1)^2\cos^2\alpha + 4r^2 + 4r(r^2+1)\cos\alpha}{(\sin^2\alpha)(r^2 + 1 + 2r\cos\alpha)^2} \\
&= \frac{1}{\sin^2\alpha},
\end{aligned}
$$

and similarly when α is replaced by $-\alpha$. Since the image of $z = 1$ under ψ^{-1} is $z = 0$ we conclude that ψ^{-1} is conformal from $C_{r,\alpha}$ onto D_α. The claim on the inverse of c follows from the fact that $c = \psi^{-1}\circ p_\alpha\circ\psi$ (where p_α is defined by (10.4.5)). □

Solution of Exercise 10.3.9.

Step 1: Let $\alpha_0, \alpha_1, \alpha_2, \dots$ be defined by

$$\frac{1}{\sqrt{1-z}} = \sum_{n=0}^{\infty} \alpha_n z^n, \quad z \in \mathbb{D}.$$

For $z \in \mathbb{D}$ we have

$$
\begin{aligned}
c(z) &= \int_{[0,z]}\frac{ds}{\sqrt{1 - s^4}} = \int_0^1 \frac{z}{\sqrt{1 - z^4 t^4}}\,dt \\
&= \sum_{n=0}^{\infty}\alpha_n z^{4n+1}\int_0^1 t^{4n}\,dt
\end{aligned}
$$

(where one can use, for instance, the dominated convergence theorem to interchange the sum and the integral)

$$= \sum_{n=0}^{\infty}\frac{\alpha_n z^{4n+1}}{n+1}, \quad z \in \mathbb{D}.$$

The coefficients $\alpha_0, \alpha_1, \dots$ satisfy (3.5.9), and so this last expression defines a function analytic in $\mathbb{D}$ (namely, $c(z)$) and continuous in the closed unit disk $\overline{\mathbb{D}}$. By Exercise 3.5.7, we have for $\theta \in [0, 2\pi]$

$$c(e^{i\theta}) = M + i\int_0^\theta\left(\sum_{n=0}^{\infty}\alpha_n e^{i(4n+1)u}\right)du, \quad \text{where} \quad M = \sum_{n=0}^{\infty}\frac{\alpha_n}{n+1}.$$

To conclude the first step we show that

$$i \sum_{n=0}^{\infty} a_n e^{i(4n+1)u} = \frac{e^{i\frac{3\pi}{4}}}{\sqrt{2\sin(2u)}}, \quad u \in \left(0, \frac{\pi}{4}\right). \tag{10.4.6}$$

To that purpose, let $t \in (0, 1)$. By Theorem 3.5.1 the sum $\sum_{n=0}^{\infty} a_n t^n e^{4inu}$ converges for $u \in (0, \frac{\pi}{4})$. By Theorem 3.5.4, and for such u, we have:

$$ie^{iu} \lim_{\substack{t \to 1 \\ t \in (0,1)}} \sum_{n=0}^{\infty} a_n t^n e^{4inu} = ie^{iu} \sum_{n=0}^{\infty} a_n e^{4inu}.$$

On the other hand,

$$i \sum_{n=0}^{\infty} a_n t^n e^{i(4n+1)u} = \frac{ie^{iu}}{\sqrt{1 - t^4 e^{4iu}}}.$$

Consider the polar decomposition

$$\frac{ie^{iu}}{\sqrt{1 - t^4 e^{4iu}}} = \rho_t(u) e^{i\theta_t(u)},$$

with $\theta_t(u) \in (0, \frac{\pi}{4})$. We have

$$\rho_t(u) = \frac{1}{|\sqrt{1 - t^4 e^{4iu}}|}$$

$$= \frac{1}{\sqrt[4]{1 + t^8 - 2t^4 \cos(4u)}}$$

$$\longrightarrow \frac{1}{\sqrt[4]{2 - 2\cos(4u)}} = \frac{1}{\sqrt{2\sin(2u)}},$$

as $t \to 1$. Moreover,

$$\rho_t(u)^2 e^{2i\theta_t(u)} = \frac{-e^{2iu}}{1 - t^4 e^{4iu}} \to \frac{-i}{2\sin(2u)},$$

as $t \to 1$, and so $\lim_{t \to 1} 2\theta_t(u) = \frac{3\pi}{2}$.

Step 2: It follows from (10.3.2) that c maps $[0, \frac{\pi}{4}]$ into a closed interval. On the other hand, the formula (10.3.3)

$$c(iz) = \int_{[0,iz]} \frac{ds}{\sqrt{1 - s^4}} = \int_0^1 \frac{izdt}{\sqrt{1 - (iz)^4}} = ic(z), \quad z \in \mathbb{D},$$

still holds on the boundary using radial limits since $\lim_{\substack{r \to 1 \\ r \in (0,1)}} c(re^{i\theta})$ exists for $\theta \in [0, 2\pi] \setminus \{0, \frac{\pi}{2}, \frac{3\pi}{2}, 2\pi\}$, and shows that the image of $[\frac{\pi}{4}, \frac{\pi}{2}]$ is an interval of the same length, rotated by $\pi/2$ in the trigonometric sense. The same holds for the other two quadrants, and the image of the unit circle is a square.

Step 3: Let $w \in \mathbb{D}$ and let $r \in (|w|, 1)$. By Exercise 7.3.5

$$\frac{1}{2\pi i} \int_{|z|=r} \frac{c'(z)}{c(z) - w} dz$$

is equal to the number of solutions of the equation $c(z) = w$ in $B(0, r)$. The function $w \mapsto \frac{1}{2\pi i} \int_{|z|=r} \frac{c'(z)}{c(z)-w} dz$ takes integer values and is continuous. It is constant on open connected sets, and so equal to its value at $w = 0$. On the other hand, by the dominated convergence theorem

$$\lim_{r \to 1} \int_{|z|=r} \frac{c'(z)}{c(z)} dz = \int_{|z|=1} \frac{c'(z)}{c(z)} dz.$$

To conclude, note that, by definition of the winding number,

$$\frac{1}{2\pi i} \int_{|z|=1} \frac{c'(z)}{c(z)} dz = 1. \qquad \square$$

Remark 10.4.2. A variation of the preceding arguments will show that the application

$$c(z) = \int_{[0,z]} \frac{ds}{(1 - s^n)^{\frac{2}{n}}}$$

maps conformally the open unit disk onto the interior of a regular polygon with n sides, the length of the side being equal to

$$\frac{2\pi}{n} \frac{\Gamma\left(1 - \frac{2}{n}\right)}{\left(\Gamma\left(1 - \frac{1}{n}\right)\right)^2} = \frac{1}{n} 2^{1-\frac{4}{n}} \frac{\left(\Gamma\left(\frac{1}{2} - \frac{1}{n}\right)\right)^2}{\Gamma\left(1 - \frac{2}{n}\right)}. \tag{10.4.7}$$

See [168, Exercise 4, p. 196], [195, Example 5.1, p. 48].

Using Legendre's duplication formula (see, e.g., [53, p. 212], [146, (1.2.3) p. 3])

$$\sqrt{\pi}\Gamma(2z) = 2^{2z-1}\Gamma(z)\Gamma\left(z + \frac{1}{2}\right) \tag{10.4.8}$$

it is readily seen that both expressions in (10.4.7) coincide. Indeed, it is equivalent to prove that

$$\sqrt{\pi}\Gamma\left(1 - \frac{2}{n}\right) = 2^{-\frac{2}{n}}\Gamma\left(\frac{1}{2} - \frac{1}{n}\right)\Gamma\left(1 - \frac{1}{n}\right), \tag{10.4.9}$$

which is (10.4.8) with $z = \frac{1}{2} - \frac{1}{n}$.

Remark 10.4.3. We will not discuss here the Schwarz–Christoffel formula (see, e.g., [168, Chapter 5, §6, p. 189], [195, p. 42]), which allows to build conformal maps onto certain polygons.

Chapter 11

A Taste of Linear System Theory and Signal Processing

In the present chapter, we briefly discuss some links between the theory of analytic functions and the theory of linear systems. We refer to the books [89], [117], [170], [171], [178] for more information. The reader should be aware that more recent advances in linear system theory, in the setting of several complex variables, non-commuting variables, or stochastic setting, to name a few, require much more involved tools. Still it is necessary to master the elementary setting outlined here before going to these more advanced areas.

We recall that we denote by $\mathbf{L}_2(\mathbb{R})$ and $\mathbf{L}_2(-F, F)$ the Lebesgue spaces of functions measurable and square summable with respect to the Lebesgue measure, on $\mathbb{R}$ and on $(-F, F)$ respectively.

11.1 Continuous signals

A continuous signal of finite energy is modeled by a continuous complex-valued function f defined on the real line, and its energy will be by definition

$$\int_{\mathbb{R}} |f(t)|^2 dt.$$

The integral is a Riemann integral, but the fact that we consider f with this norm forces us to consider measurable functions and the Lebesgue space $\mathbf{L}_2(\mathbb{R})$. See Chapter 17 for a brief review of these notions.

The spectrum of the signal f is by definition its inverse Fourier transform (13.5.3):

$$\check{f}(u) = \frac{1}{2\pi} \int_{\mathbb{R}} e^{iut} f(t) dt,$$

so that

$$f(t) = \int_{\mathbb{R}} e^{-itu} \check{f}(u)\,du.$$

The above expression is the decomposition of f along frequencies (technically, it would be better to have $2\pi u$ rather than u for frequencies, but we will stick to the present definition of the Fourier transform). We are interested in signals which have spectrum with finite support. It then follows that the signal itself is the restriction on the real line of an entire function. If the spectrum has support in the closed interval $[-F, F]$, the signal can be written as

$$f(t) = \frac{1}{2F} \int_{[-F,F]} e^{-itu} m(u)\,du, \tag{11.1.1}$$

where $m \in \mathbf{L}_2(-F, F)$ denotes the spectrum. The representation (11.1.1) expresses that the signal f is built from frequencies in a bounded domain (that is, f is a *band limited signal*). This is a characteristic of physical systems. The factor $\frac{1}{2F}$ is a normalization to have nicer formulas in the sequel. We recognize with (11.1.1) a function similar to the ones appearing in Exercises 3.4.13 and 4.2.14. In particular, f is the restriction to the real line of the entire function

$$f(z) = \frac{1}{2F} \int_{[-F,F]} e^{-izu} m(u)\,du, \quad z \in \mathbb{C}.$$

Besides being entire, this function has a special property:

Exercise 11.1.1. *Show that there exists $K > 0$ such that*

$$|f(z)| \le Ke^{F|z|}, \quad \forall z \in \mathbb{C}. \tag{11.1.2}$$

Entire functions which admit a bound of the form (11.1.2) are called of exponential type, and the smallest F in (11.1.2) is the exponential type of the function. That every entire function which admits a bound of the form (11.1.2) can be written as (11.1.1) with $m \in \mathbf{L}_1(-F, F)$ is a deep result, called the Paley–Wiener theorem. See for instance [71, §3.3, p. 158], [72, §2.2, p. 28]. Here we restrict

$$m \in \mathbf{L}_2(-F, F) \subset \mathbf{L}_1(-F, F)$$

because we want an underlying Hilbert space structure.

To summarize, physical considerations in modeling signals (having a band limited spectrum) make it natural to consider a very special class of entire functions (entire functions of exponential type).

11.2   Sampling

Since the function f in (11.1.1) has an analytic extension to the whole complex plane, one can ask the question of reconstructing f from a discrete set of values. From an engineering point of view this is an important issue. The surprising answer to this question is a result called the sampling theorem, which we present in this section; see Theorem 11.2.1. The sampling theorem has a long history, and we refer to the paper [159] for a historical account. We mention that a version of the sampling theorem already appears in the 1915 paper [220] of E.T. Whittaker. We refer to the papers of Claude Shannon [196], [197]. This last paper refers in particular to the 1935 book [221, Ch. IV] of J.M. Whittaker for an earlier version of the sampling theorem. See also [26, p. 258].

We note that there is no need of analytic functions to prove the sampling theorem. On the other hand, the result is somewhat of a mystery to students who have no background in analytic functions.

We consider $\mathbf{L}_2(-F, F)$ with the normalized inner product (17.7.3)

$$\langle m, n \rangle = \frac{1}{2F} \int_{(-F,F)} m(t)\overline{n(t)}dt.$$

Theorem 11.2.1. *Let $m \in \mathbf{L}_2(-F, F)$, with $F \in (0, \infty)$, and let f be defined by* (11.1.1)

$$f(t) = \frac{1}{2F} \int_{(-F,F)} e^{-itu} m(u)du.$$

Then

$$f(t) = \sum_{n \in \mathbb{Z}} f\left(\frac{\pi n}{F}\right) \frac{\sin(Ft - n\pi)}{Ft - n\pi}, \qquad (11.2.1)$$

where the limit is pointwise, and uniformly on compact subsets of $\mathbb{C}$ (with $t \in \mathbb{C}$). Finally

$$\int_{\mathbb{R}} |f(t)|^2 dt = \frac{\pi}{F} \sum_{n \in \mathbb{Z}} \left|f\left(\frac{\pi n}{F}\right)\right|^2. \qquad (11.2.2)$$

For instance consider the choice $F = 2$ and

$$m(u) = \begin{cases} 1, & u \in [-1, 1], \\ 0, & u \in [-2, 2] \setminus [-1, 1]. \end{cases}$$

Then

$$f(t) = \frac{\sin t}{2t}$$

and (11.2.2) becomes

$$\frac{1}{4} \int_{\mathbb{R}} \left(\frac{\sin t}{t}\right)^2 dt = \frac{\pi}{2} \sum_{n \in \mathbb{Z}} \left|f\left(\frac{\pi n}{2}\right)\right|^2$$

$$= \frac{\pi}{2}\left(\frac{1}{4} + 2\sum_{k \in \mathbb{N}_0} \left|\frac{1}{2} \frac{\sin(\frac{(2k+1)\pi}{2})}{\frac{(2k+1)\pi}{2}}\right|^2\right)$$

$$= \frac{\pi}{2}\left(\frac{1}{4} + 2\sum_{k \in \mathbb{N}_0} \frac{1}{(2k+1)^2\pi^2}\right).$$

Using Exercise 5.3.3, this leads to (6.7.1)

$$\sum_{k=0}^{\infty} \frac{1}{(2k+1)^2} = \frac{\pi^2}{8}.$$

Exercise 11.2.2. *Give a direct proof of* (6.7.1) *taking into account* (1.3.14).

To prove Theorem 11.2.1 we use the expansion of an $\mathbf{L}_2(-F, F)$ function along an orthogonal basis. To characterize functions which admit a representation (11.1.1) is a more delicate matter, and uses the Phragmén–Lindelöf principle (we will not recall its definition here). See [72, p. 28] for more information.

Exercise 11.2.3. *The space $\mathcal{H}_F$ of functions of the form*

$$f(z) = \frac{1}{2F} \int_{(-F,F)} m(u)e^{-izu} du, \quad m \in \mathbf{L}_2(-F, F), \tag{11.2.3}$$

with norm

$$\|f\|_{\mathcal{H}_F} = \|m\|_{\mathbf{L}_2(-F,F)} \tag{11.2.4}$$

is the reproducing kernel Hilbert space of entire functions with reproducing kernel

$$K_F(z, w) = \frac{\sin(Fz - F\overline{w})}{Fz - F\overline{w}}, \quad z, w \in \mathbb{C}.$$

In view of the isometry property (13.5.2) of the Fourier transform, we see that the space $\mathcal{H}_F$ defined in the preceding exercise is in fact, up to a unitary constant, isometrically included in $\mathbf{L}_2(\mathbb{R}, dx)$. More precisely, we have

$$f(t) = \frac{\widehat{m}(t)}{2F},$$

and so

$$\|f\|^2_{\mathbf{L}_2(\mathbb{R},dx)} = \frac{2\pi}{2F} \|m\|^2_{\mathbf{L}_2(-F,F)}.$$

Thus

$$\|f\|^2_{\mathcal{H}_F} = \|m\|^2_{\mathbf{L}_2(-F,F)}$$
$$= \frac{2F}{2\pi} \|f\|^2_{\mathbf{L}_2(\mathbb{R},dx)}.$$

Exercise 11.2.4. *Prove formula* (11.2.1).

11.3 Time-invariant causal linear systems

A linear continuous operator T,

$$u \in \mathbf{L}_2(\mathbb{R}) \mapsto Tu \in \mathbf{L}_2(\mathbb{R})$$

from $\mathbf{L}_2(\mathbb{R})$ into itself, is called a *linear system* when one views the elements of $\mathbf{L}_2(\mathbb{R})$ as signals with finite energy. The function u is then called the *input signal* and the function Tu is called the *output signal*. The (linear) system is called *dissipative* if the norm of the operator is less than or equal to 1:

$$\forall u \in \mathbf{L}_2(\mathbb{R}), \quad \|Tu\|_{\mathbf{L}_2(\mathbb{R})} \le \|u\|_{\mathbf{L}_2(\mathbb{R})}.$$

It will be called causal if the following property holds for every $t \in \mathbb{R}$: If the input function u has support $(-\infty, t)$, then the output function has also support in $(-\infty, t)$.

We are in particular interested in operators which have a *kernel representation* in the form

$$Tf(t) = \int_{\mathbb{R}} k(t, s) f(s) ds, \tag{11.3.1}$$

or as convolution operators

$$Tf(t) = \int_{\mathbb{R}} k(t - s) f(s) ds, \tag{11.3.2}$$

when the kernel $k(t, s)$ is required to depend only on the difference $t - s$.

Not every continuous linear operator from $\mathbf{L}_2(\mathbb{R})$ admits such a representation. To ensure such a representation for *every* continuous operator, one has to restrict the domain to a set of test functions and extend the range to the dual space of distributions. Continuity is then understood with respect to the topology of the Schwartz space and of its dual, and Schwartz' kernel theorem insures then a counterpart of (11.3.1) with a distribution $k(t, s)$. This is a fascinating line of research (see [109], [110] for instance for the background of the kernel theorem, and Zemanian's book [227] for applications to the theory of linear systems). Here we are interested in a simpler kind of linear systems, namely systems $y = Tu$ given by

$$(L(y))(z) = h(z)(L(u))(z)$$

where L denotes the Laplace transform. Such systems are time-invariant and characterized by a convolution in continuous time.

Exercise 11.3.1. *Let* $(A, B, C, D) \in \mathbb{C}^{N \times N} \times \mathbb{C}^{N \times p} \times \mathbb{C}^{q \times N} \times \mathbb{C}^{q \times p}$, *and consider the equations*

$$\begin{aligned} x'(t) &= Ax(t) + Bu(t), \\ y(t) &= Cx(t) + Du(t), \quad t \ge 0 \end{aligned} \tag{11.3.3}$$

where the functions x, u *and* y *are respectively* $\mathbb{C}^N$-*valued,* $\mathbb{C}^p$-*valued and* $\mathbb{C}^q$-*valued. Assume that* $x(0) = 0$ *and that the Laplace transform* $L(u)$ *has a positive axis of convergence. Show that the function* $L(y)$ *has a positive axis of convergence and that*

$$(L(y))(z) = h(z)(L(u))(z), \tag{11.3.4}$$

where

$$h(z) = D + C(zI_N - A)^{-1}B. \tag{11.3.5}$$

The equations (11.3.4) are called *state space equations*, and the vector $x(t)$ is called the state at time t. The expression (11.3.5) is called a *realization* of the rational matrix-valued function h. See Chapter 11 for more on this notion.

11.4   Discrete signals and systems

A discrete signal will be a sequence $(u_n)_{n \in \mathbb{N}_0}$ of complex numbers, indexed by $\mathbb{N}_0$ (or sometimes by $\mathbb{Z}$). Its z-transform is the power series

$$u(z) = \sum_{n=0}^{\infty} u_n z^n.$$

The energy of the signal is its ℓ_2 norm

$$\|u\|_{\ell_2} = \sqrt{\left(\sum_{n=0}^{\infty} |u_n|^2 \right)},$$

and we see that the space of signals of finite energy is nothing else than the Hardy space $\mathbf{H}_2(\mathbb{D})$. See Definition 5.6.11 for the latter. It is therefore reasonable to think that function theory in $\mathbf{H}_2(\mathbb{D})$ should have implications, and applications, in signal theory.

A bounded linear system will be a linear bounded operator from ℓ_2 into itself. It translates into a linear bounded operator T from $\mathbf{H}_2(\mathbb{D})$ into itself. The linear system will be called dissipative if it is moreover a contraction

$$\|Tu\|_{\mathbf{H}_2(\mathbb{D})} \leq \|u\|_{\mathbf{H}_2(\mathbb{D})}, \quad \forall u \in \mathbf{H}_2(\mathbb{D}).$$

An important class of linear systems is defined by multiplication operators: The input sequence $(u_n)_{n \in \mathbb{N}_0}$ and the output sequence $(y_n)_{n \in \mathbb{N}_0}$ are related by

$$y(z) = h(z)u(z), \tag{11.4.1}$$

where $h(z) = \sum_{n=0}^{\infty} h_n z^n$ is convergent in $\mathbb{D}$. Therefore, $(y_n)_{n \in \mathbb{N}_0}$ is the convolution of $(h_n)_{n \in \mathbb{N}_0}$ and $(u_n)_{n \in \mathbb{N}_0}$. See (4.4.14) for the latter. The function h is called the *transfer function* of the system, and its Taylor coefficients at the origin are called the *impulse response*.

Not every h will lead to a bounded operator. We have:

Theorem 11.4.1. *The relation* (11.4.1) *defines a bounded linear operator from* $\mathbf{H}_2(\mathbb{D})$ *into itself if and only if h is analytic and bounded in the open unit disk. It defines a dissipative linear operator from* $\mathbf{H}_2(\mathbb{D})$ *into itself if and only if h is analytic and contractive in the open unit disk.*

The proof of Theorem 11.4.1 relies on the characterization (5.6.7) of the space $\mathbf{H}_2(\mathbb{D})$. If s is analytic and contractive in the open unit disk, then for every $f \in \mathbf{H}_2(\mathbb{D})$ and every $r \in (0,1)$

$$|s(re^{it})f(re^{it})|^2 \le |f(re^{it})|^2,$$

and thus

$$\int_0^{2\pi} |s(re^{it})f(re^{it})|^2 dt \le \int_0^{2\pi} |f(re^{it})|^2 dt.$$

It follows that $\|sf\|_{\mathbf{H}_2(\mathbb{D})} \le \|f\|_{\mathbf{H}_2(\mathbb{D})}$. We refer for instance to [6] for a proof of the converse statement.

Functions analytic and contractive (in modulus) in the open unit disk played an important role in Section 6.4 and were called there Schur functions.

The preceding discussion focused on scalar-valued signals and systems, but one can also consider the matrix-valued case. Then for a sequence $(u_n)_{n \in \mathbb{N}_0}$ of $\mathbb{C}^N$ vectors, the series

$$\sum_{n=0}^{\infty} u_n z^n = \sum_{n=0}^{\infty} \begin{pmatrix} u_{n1} \\ u_{n2} \\ \vdots \\ u_{nN} \end{pmatrix} z^n$$

with

$$u_n = \begin{pmatrix} u_{n1} \\ u_{n2} \\ \vdots \\ u_{nN} \end{pmatrix}$$

is a column vector with each entry being a scalar power series. The radius of convergence of this series is by definition the smallest of the radiuses of convergence of the N power series

$$\sum_{n=0}^{\infty} u_{nj} z^n, \quad j = 1, \dots, N.$$

See also Exercise 12.2.4.

11.5 The Schur algorithm

In Section 6.5 we have first met the recursion (6.5.7)

$$f_0(z) = f(z),$$

$$f_{n+1}(z) = \begin{cases} \dfrac{f_n(z) - f_n(0)}{z(1 - \overline{f_n(0)}f_n(z))}, & z \in \mathbb{D} \setminus \{0\}, \\ f_n'(0), & z = 0, \end{cases} \qquad n = 0, 1, \dots$$

where f is analytic and contractive in the open unit disk. The coefficients $\rho_n = f_n(0)$ are called the Schur coefficients, or reflection coefficients of the function f.

The Schur algorithm allows to solve in an iterative way classical interpolation problems such as:

Problem 11.5.1 (The Carathéodory–Fejér interpolation problem). *Given numbers $a_0, \ldots, a_N$, find all (if any) Schur functions f such that*

$$\frac{f^{(n)}(0)}{n!} = a_n, \quad n = 0, \ldots, N.$$

Problem 11.5.2 (The Nevanlinna–Pick interpolation problem). *Given N pairs of numbers $(z_1, w_1), \ldots, (z_N, w_N)$ in $\mathbb{D}^2$, find all (if any) Schur functions f such that*

$$f(z_n) = w_n, \quad n = 1, \ldots, N.$$

Exercise 11.5.3. *Let $f \in \mathcal{S}$. Then, show that the Schur algorithm applied to f ends after a finite number of times ($N \geq 0$) if and only if f is a finite Blaschke product, or a unitary constant (this being the case when $N = 0$).*

For instance, if

$$f(z) = \frac{z-a}{1-z\bar{a}} \frac{z-b}{1-z\bar{b}},$$

then

$$f_1(z) = \frac{z-c}{1-z\bar{c}},$$

where c is given by (1.1.47),

$$c = \frac{(1-|a|^2)b + (1-|b|^2)a}{1-|ab|^2},$$

and

$$f_2(z) \equiv 1.$$

Indeed, we have for $z \neq 0$,

$$
\begin{aligned}
f_1(z) &= \frac{1}{z} \frac{\dfrac{z-a}{1-z\bar{a}}\dfrac{z-b}{1-z\bar{b}} - ab}{1 - \bar{a}\bar{b}\dfrac{z-a}{1-z\bar{a}}\dfrac{z-b}{1-z\bar{b}}} \\
&= \frac{1}{z} \frac{(z-a)(z-b) - ab(1-z\bar{a})(1-z\bar{b})}{(1-z\bar{a})(1-z\bar{b}) - \bar{a}\bar{b}(z-a)(z-b)} \\
&= \frac{1}{z} \frac{z^2(1-|ab|^2) - z(a+b-ab(\bar{a}+\bar{b}))}{1-|ab|^2 - z(\bar{a}+\bar{b}) + \bar{a}\bar{b}(z(a+b)+ab)} \\
&= \frac{z-c}{1-\bar{c}z},
\end{aligned}
$$

and

$$f_2(z) = \frac{1}{z} \frac{\dfrac{z-c}{1-z\bar{c}} + c}{1 + \bar{c}\dfrac{z-c}{1-z\bar{c}}} = \frac{1}{z} \frac{z(1-|c|^2)}{1-|c|^2} \equiv 1.$$

Theorem 11.5.4. *Let $f(z) = \sum_{n=0}^{\infty} f_n z^n$ be a power series converging in a neighborhood of the origin. Then, f is analytic and contractive in the open unit disk if and only if either:*

(a) *Applying the Schur algorithm to f, we have*

$$|f_n(0)| < 1, \quad \forall n \in \mathbb{N}_0,$$

or

(b) *$f(0)$ has modulus 1 (and then f is a unitary constant), or the numbers $f_n(0)$ are strictly contractive up to a finite rank, say N_0, and $f_{N_0+1}(z)$ is a unitary constant.*

In view of the following question, we recall the notation (2.3.4)

$$T_M(z) = \frac{az+b}{cz+d},$$

where

$$M = \begin{pmatrix} a & b \\ c & d \end{pmatrix}.$$

Question 11.5.5. *Let us assume that the Schur function f in the recursion (6.5.7) is such that*

$$|f_n(0)| < 1, \quad n = 0, 1, \ldots, N.$$

Then, setting

$$\rho_n = f_n(0), \quad n = 0, 1, \ldots, N,$$

and using the notation (2.3.4) show that

$$f(z) = T_{M_N(z)}(f_{N+1}(z)) \tag{11.5.1}$$

where

$$M_N(z) = \prod_{n=0}^{\overset{\frown}{N}} \begin{pmatrix} 1 & \rho_n \\ \bar{\rho}_n & 1 \end{pmatrix} \begin{pmatrix} z & 0 \\ 0 & 1 \end{pmatrix}. \tag{11.5.2}$$

Assume that $|\rho_n| < 1$, $n = 0, 1, \ldots$. The infinite product $\lim_{N \to \infty} M_N(z)$ diverges for every point z, with the possible exception of $z = 1$. A related infinite product, which plays a key role in the theory, converges on the unit circle:

Exercise 11.5.6. *Assume that* $|\rho_n| < 1$, $n = 0, 1, \ldots$, *and that, moreover*

$$\sum_{n=0}^{\infty} |\rho_n| < \infty.$$

Then, for every z *of modulus 1, the limit*

$$\lim_{N \to \infty} M_N(z) \begin{pmatrix} z^{-N-1} & 0 \\ 0 & 1 \end{pmatrix}$$

exists.

The following result gives four equivalent characterizations of Schur functions. The first one is on the level of a first complex variable course, while the second, third and fourth characterizations require (easy) functional analysis tools. These last three characterizations are much more conducive to defining counterparts of Schur functions for the extensions of linear system theory mentioned in the introduction of the chapter.

Theorem 11.5.7. *Let* f *be a function defined in the open unit disk. The following are equivalent:*

(1) f *is analytic and contractive in the open unit disk.*

(2) *The kernel*

$$k_f(z, w) = \frac{1 - f(z)\overline{f(w)}}{1 - z\overline{w}}$$

   is positive definite in the open unit disk.

(3) *There exist a Hilbert space* $\mathcal{H}$ *and a coisometric operator matrix*

$$\begin{pmatrix} A & B \\ C & D \end{pmatrix} : \mathcal{H} \oplus \mathbb{C} \longrightarrow \mathcal{H} \oplus \mathbb{C},$$

   such that

$$f(z) = D + zC(I_{\mathcal{H}} - zA)^{-1}B.$$

(4) *The Taylor coefficients of* f *are of the form*

$$f_n = \begin{cases} D, & n = 0, \\ CA^{n-1}B, & n = 1, 2, \ldots, \end{cases}$$

   where A, B, C, D *are as in* (3).

11.6 Solutions

Solution of Exercise 11.1.1. Let

$$f(z) = \frac{1}{2F} \int_{[-F,F]} e^{-izu} m(u) du.$$

Using (1.2.5) we have

$$|e^{izt}| \leq e^{|z|F}.$$

Therefore, using the Cauchy–Schwarz inequality (16.1.5), we have

$$|f(z)| \leq \frac{e^{|z|F}}{2F} \int_{(-F,F)} |m(u)| du$$

$$\leq \frac{e^{|z|F}}{2F} \left(\int_{(-F,F)} |m(u)|^2 du \right)^{1/2} \left(\int_{(-F,F)} 1 du \right)^{1/2}$$

$$= K e^{|z|F}$$

with

$$K = \frac{\int_{(-F,F)} |m(u)|^2 du)^{1/2}}{\sqrt{2F}} < \infty,$$

since $m \in \mathbf{L}_2(-F, F)$. □

Solution of Exercise 11.2.2. We have

$$\sum_{n=1}^{\infty} \frac{1}{n^2} = \sum_{k=1}^{\infty} \frac{1}{(2k)^2} + \sum_{k=0}^{\infty} \frac{1}{(2k+1)^2}$$

$$= \frac{1}{4} \sum_{n=1}^{\infty} \frac{1}{n^2} + \sum_{k=0}^{\infty} \frac{1}{(2k+1)^2}.$$

Taking into account (1.3.14) we have

$$\sum_{k=0}^{\infty} \frac{1}{(2k+1)^2} = \frac{\pi^2}{6} - \frac{1}{4}\frac{\pi^2}{6} = \frac{\pi^2}{8}.$$ □

Solution of Exercise 11.2.3. From the estimate in the previous exercise we see that the integral (11.2.3) is well defined for every $z \in \mathbb{C}$. The function is entire. For continuous m this follows from the same arguments as for Exercise 4.4.19. As explained after the proof of that exercise for the interval $(0, 1)$, the statement is still true for functions $m \in \mathbf{L}_2(-F, F)$.

Let now $f \in \mathcal{H}_F$ be such that $f(z) \equiv 0$. Then, the choice $z = \frac{\pi n}{F}$ gives

$$\int_{(-F,F)} m(u) e^{\frac{-\pi i n u}{F}} du = 0, \quad \forall n \in \mathbb{Z}.$$

But the functions

$$f_n(u) = e^{\frac{\pi i n u}{F}}, \quad n \in \mathbb{Z}, \tag{11.6.1}$$

form an orthonormal basis of $\mathbf{L}_2(-F, F)$ (see Exercise 17.7.5). It follows that $m \equiv 0$ (as an element of $\mathbf{L}_2(-F, F)$). Therefore (11.2.4) indeed defines a norm, and $\mathcal{H}_F$ is a Hilbert space since $\mathbf{L}_2(-F, F)$ is a Hilbert space. Let for $z, w \in \mathbb{C}$,

$$K_F(z, w) = \frac{1}{2F} \int_{(-F,F)} e^{-i\overline{w}u} e^{izu} du = \frac{\sin(Fz - F\overline{w})}{Fz - F\overline{w}}.$$

Then for $f \in \mathcal{H}_F$ and $w \in \mathbb{C}$ we have that

$$f(w) = \frac{1}{2F} \int_{(-F,F)} m(u) e^{-iwt} dt = \langle f(\cdot), K_F(\cdot, w) \rangle_{\mathcal{H}_F}. \qquad \square$$

Solution of Exercise 11.2.4. Take $m \in \mathbf{L}_2(-F, F)$. Then

$$m(u) = \sum_{n \in \mathbb{Z}} \left(\frac{1}{2F} \int_{(-F,F)} m(s) e^{-\frac{i\pi s n}{F}} ds \right) e^{\frac{i\pi u n}{F}}$$

where the limit is in the norm of $\mathbf{L}_2(-F, F)$. By Parseval's equality, this sum becomes

$$f(\cdot) = \sum_{n \in \mathbb{Z}} f\left(\frac{\pi n}{F}\right) K_F\left(\cdot, \frac{\pi n}{F}\right), \tag{11.6.2}$$

where the equality is in the norm of $\mathcal{H}_F$. Let $z \in \mathbb{C}$ and $e_z(u) = e^{izu}$. Using the continuity of the inner product or Parseval equality we have with f_n as in (11.6.1)

$$\langle m, e_z \rangle_{\mathbf{L}_2(-F,F)} = \sum_{n \in \mathbb{Z}} \langle m, f_n \rangle_{\mathbf{L}_2(-F,F)} \langle f_n, e_z \rangle_{\mathbf{L}_2(-F,F)}.$$

In other words

$$f(z) = \sum_{n \in \mathbb{Z}} f\left(\frac{\pi n}{F}\right) K_F\left(z, \frac{\pi n}{F}\right), \quad z \in \mathbb{C}. \tag{11.6.3}$$

Here the convergence is pointwise, and uniform on bounded sets since the kernel is bounded on bounded sets. $\qquad \square$

Equation (11.6.3) can also be obtained directly from (11.6.2) since convergence in norm implies pointwise convergence in a reproducing kernel Hilbert space (see Exercise 16.3.13).

Solution of Exercise 11.3.3. It suffices to apply the Laplace transform on both sides of the state space equations. $\qquad \square$

Note that the transfer function is analytic at infinity. In the discrete case, the transfer function is analytic at the origin. See Exercise 12.2.4.

Solution of Exercise 11.5.3. Suppose that f is not a unitary constant and that the Schur algorithm ends after a finite number of steps. Then, there is an $N \in \mathbb{N}_0$ such that

$$f_n(0) \in \mathbb{D}, \quad n = 0, 1, 2, \ldots, N,$$

and $f_{N+1}(z)$ is a unitary constant. Formula (11.5.2) leads to

$$f(z) = T_{M_N(z)}(f_{N+1}).$$

For $|z| = 1$, the Moebius transform with matrix

$$\begin{pmatrix} 1 & \rho_n \\ \overline{\rho_n} & 1 \end{pmatrix} \begin{pmatrix} z & 0 \\ 0 & 1 \end{pmatrix}$$

sends the unit circle onto itself, and so does $T_{M_n(z)}$. Therefore, the function f is unitary on the unit circle. It follows from Exercise 6.3.4 that f is a finite Blaschke product, that is

$$f(z) = cz^L \prod_{n=1}^{M} b_{w_n}(z), \tag{11.6.4}$$

where $|c| = 1$, $L, M \in \mathbb{N}_0$ and the factors b_{w_n} are defined by (1.1.44), with $w_n \neq 0$.

Conversely, assume that f is a finite Blaschke product. We show that applying the Schur algorithm to f we obtain a finite Blaschke product with one less factor. If $L > 0$ in (11.6.4) this is clear. Assume now $L = 0$, and set

$$p(z) = c \prod_{n=1}^{M} (z - w_n) \quad \text{and} \quad q(z) = \prod_{n=1}^{M} (1 - \overline{w_n} z). \tag{11.6.5}$$

We have

$$f_1(z) = \frac{\overline{q(0)}}{q(0)} \frac{\left(c \dfrac{p(z)q(0) - p(0)q(z)}{z} \right)}{q(z)\overline{q(0)} - p(z)\overline{p(0)}} = c \frac{\left(\dfrac{p(z) - p(0)q(z)}{z} \right)}{q(z) - p(z)\overline{p(0)}}.$$

The coefficient of the power z^M in the polynomial $p(z) - p(0)q(z)$ is equal to

$$c \left(1 - \prod_{n=1}^{M} (-w_n) \prod_{n=1}^{M} (-\overline{w_n}) \right) = c \left(1 - \prod_{n=1}^{M} |w_n|^2 \right) \neq 0.$$

Thus the polynomial $p(z) - p(0)q(z)$ has degree M. It vanishes at the origin, and so the function

$$\frac{p(z) - p(0)q(z)}{z}, \quad z \neq 0,$$

defines a polynomial of degree $M - 1$ (with value at the origin equal to $p'(0) - p(0)q'(0)$). The coefficient of the power z^M in the polynomial $q(z)$ is equal to $\prod_{n=1}^{M}(-\overline{w_n})$. Therefore, the coefficient of the power z^M in the polynomial

$$q(z) - p(z)\overline{p(0)}$$

is equal to

$$\prod_{n=1}^{M}(-\overline{w_n}) - c\overline{p(0)} = 0.$$

Therefore,

$$\deg(q(z) - p(z)\overline{p(0)}) \leq M - 1.$$

We want to show that $\deg f_1 = M - 1$. Since f_1 is unitary on the unit circle, it will then follow that f_1 is also a finite Blaschke product (see Exercise 6.3.4), but with one less factor.

To check that $\deg f_1 = M - 1$, we will show that the polynomials

$$\frac{p(z) - p(0)q(z)}{z} \quad \text{and} \quad q(z) - p(z)\overline{p(0)}$$

have no common zeros. Since $q(z) - p(z)\overline{p(0)}$ has value $1 - |p(0)|^2 > 0$ at the origin, it is enough to check that the polynomials

$$p(z) - p(0)q(z) \quad \text{and} \quad q(z) - p(z)\overline{p(0)}$$

have no common zeros. If $z_0 \in \mathbb{C}$ is such that

$$p(z_0) = p(0)q(z_0) \quad \text{and} \quad q(z_0) = p(z_0)\overline{p(0)}, \qquad (11.6.6)$$

we obtain

$$p(z_0)(1 - |p(0)|^2) = 0,$$

and hence $p(z_0) = 0$, and hence, by (11.6.6), we also have $q(z_0) = 0$. But this is not possible since, by (11.6.5), p and q have no common zero. It follows that $\deg f_1 = M - 1$. $\qquad\square$

Solution of Exercise 11.5.6. Set

$$S_n(z) = \begin{pmatrix} z^n & 0 \\ 0 & 1 \end{pmatrix} \begin{pmatrix} 1 & \rho_n \\ \overline{\rho_n} & 1 \end{pmatrix} \begin{pmatrix} z^{-n} & 0 \\ 0 & 1 \end{pmatrix}.$$

We have

$$\begin{pmatrix} z & 0 \\ 0 & 1 \end{pmatrix} M_N(z) \begin{pmatrix} z^{-N-1} & 0 \\ 0 & 1 \end{pmatrix} = \prod_{n=0}^{\widehat{N}} S_n(z). \qquad (11.6.7)$$

Furthermore,

$$S_n(z) = \begin{pmatrix} 1 & 0 \\ 0 & 1 \end{pmatrix} + \begin{pmatrix} 0 & z^n \rho_n \\ z^{-n} \overline{\rho_n} & 0 \end{pmatrix} = I_2 + A_n(z),$$

with

$$A_n(z) = \begin{pmatrix} 0 & z^n \rho_n \\ z^{-n} \overline{\rho_n} & 0 \end{pmatrix}.$$

With

$$\|A\|_\infty = \max_{i,j=1,2} |a_{ij}|$$

we have, for $|z| = 1$,

$$\|A_n(z)\|_\infty = \rho_n.$$

Therefore

$$\sum_{n=0}^{\infty} \|A_n(z)\|_\infty$$

converges for every point on the unit circle. Since all norms are equivalent in $\mathbb{C}^{2\times 2}$ (see (16.1.2) for the definition of equivalent norms), we have that the infinite product (11.6.7) also converges in view of Theorem 3.7.3,   □

Chapter 12

Rational Functions

Complex-valued rational functions are by definition functions which are meromorphic on the Riemann sphere, or equivalently, which are quotient of polynomials. They form thus a class of *a priori* very simple objects, where the notions of degree, zeros, poles, and factorization are quite obvious. An important place where rational functions appear besides pure mathematics is linear system theory. They are then transfer functions of certain classes of linear systems. Even in the scalar case, some problems for rational functions are far from obvious, as is illustrated by the following multipoint interpolation problem (see [12] and Section 12.3):

Given complex numbers $w_1, \ldots, w_N, a_1, \ldots, a_N$ and b, describe the set of all rational functions $r(z)$ with no poles at the points $w_1, \ldots, w_N$ and such that

$$\sum_{n=1}^{N} a_n r(w_n) = b. \tag{12.0.1}$$

A $\mathbb{C}^{p \times q}$-valued function R will be called rational if each of its entries is rational, or, equivalently, if it can be written in one of the forms

$$\frac{P(z)}{p(z)}, \quad P_1(z)^{-1} P_2(z), \quad \text{or} \quad P_3(z) P_4(z)^{-1},$$

where P, P_2 and P_3 are $\mathbb{C}^{p \times q}$-valued polynomials, p is a scalar non-identically vanishing polynomial, and P_1 and P_4 are respectively $\mathbb{C}^{p \times p}$-valued and $\mathbb{C}^{q \times q}$-valued polynomials with non-identically vanishing determinant. The exercises presented in this chapter pertain mainly to the theory of matrix-valued rational functions, and in particular to the above questions. The situation is much more involved than in the scalar case. For instance, what are the correct definitions of degree, zero, pole, and factorization? The literature is vast, and we refer in particular to [130] for a thorough survey of the main definitions, properties, and applications of matrix-valued rational functions.

12.1   First properties

Recall that a function f is said to be meromorphic in an open set Ω if its singular points in Ω, if any, are poles. Equivalently, f is a quotient of two functions analytic in Ω (the proof is trivial when there are a finite number of poles, and otherwise involves an infinite product to factor out the poles of f). We will say that f is meromorphic in the extended complex plane $\mathbb{P}$ if moreover, the point ∞ is also a pole or a removable singularity of f, that is, if $z = 0$ is a pole or a removable singularity of the function $f(1/z)$.

Exercise 12.1.1. *Let φ be a (non-trivial) Moebius map. Then, f is rational if and only if $f \circ \varphi$ is rational.*

Exercise 12.1.2. *A function f is meromorphic in $\mathbb{P}$ if and only if it is rational, that is, if and only if it is a quotient of two polynomials.*

The following theorem gathers various equivalent characterizations of a matrix-valued rational function with no pole at the origin. Some of the various equivalences are proved as exercises in the sequel. In the statement, R_0 denotes the backward-shift operator; see (4.2.26).

Theorem 12.1.3. *Let r be $\mathbb{C}^{p \times q}$-valued function, analytic in a neighborhood of the origin, and with power series expansion $r(z) = a_0 + a_1 z + \cdots$ there. Then the following are equivalent:*

(1) *r is a rational function (or more precisely, r is the restriction to the given neighborhood of a rational function).*

(2) *r can be written in the form*

$$r(z) = D + zC(I - zA)^{-1}B, \qquad (12.1.1)$$

where $D = r(0)$ and A, B, C are matrices of appropriate dimensions.

(3) *There exist matrices A, B, C of appropriate dimensions such that*

$$a_n = CA^{n-1}B, \quad n = 1, 2, \ldots.$$

(4) *The linear span of the functions $R_0 rc, R_0^2 rc, \ldots$ is finite-dimensional, when c runs through $\mathbb{C}^q$.*

Equation (12.1.1) is called a realization of the rational function r, and already appeared in the previous chapter; see (11.3.5) there. One also denotes a realization by the block matrix

$$\begin{pmatrix} A & B \\ C & D \end{pmatrix}. \qquad (12.1.2)$$

Such realizations have numerous applications in fields such as the theory of linear systems and optimal control. See [30], [68], [132].

We note that an expression of the form (12.1.1) is highly non-unique. If (12.1.2) is a realization of the (say, $\mathbb{C}^{p \times q}$-valued) rational function r, so is

$$\begin{pmatrix} T & 0 \\ 0 & I_p \end{pmatrix} \begin{pmatrix} A & B \\ C & D \end{pmatrix} \begin{pmatrix} T^{-1} & 0 \\ 0 & I_q \end{pmatrix}$$

for any invertible matrix T. When the size of A is minimal, this is the only degree of freedom.

In Exercise 12.1.5, which gives a characterization of a rational function with no pole at the origin, we do not assume *a priori* that the power series $a_0 + a_1 z + \cdots$ has a strictly positive radius of convergence. One part of the proof uses item (3) in the previous theorem and the Cayley–Hamilton theorem, see for instance [143, Theorem 3.1, p. 561] for the case of matrices with entries in a commutative ring. We recall this theorem just before the statement of the exercise. The other part of the proof involves elementary results in the theory of difference equations, and we refer the reader to the book [73] for more information on the subject.

Theorem 12.1.4 (Cayley–Hamilton). *Let $A \in \mathbb{C}^{M \times M}$, and let $p(z) = \det(zI_M - A)$. Then p is a monic polynomial of degree M (that is, with coefficient of z^M equal to 1) which satisfies $p(A) = 0$.*

The polynomial p in the Cayley–Hamilton theorem is called the characteristic polynomial.

Exercise 12.1.5. *Let $a_0, a_1, \ldots$ be a sequence of matrices in $\mathbb{C}^{p \times q}$. Then there exists a rational function with power series expansion*

$$r(z) = a_0 + a_1 z + \cdots$$

at the origin if and only there exist $M \in \mathbb{N}$ and complex numbers $c_0, \ldots, c_{M-1}$ such that

$$a_{M+n} + c_{M-1} a_{M+n-1} + \cdots + c_0 a_n = 0, \quad n = 0, 1, \ldots. \tag{12.1.3}$$

In the previous exercise, one can ask what happens if one is given only a finite sequence of matrices, say $a_0, \ldots, a_N$. There is always a rational function which answers the question, namely the matrix-polynomial $p(z) = a_0 + a_1 z + \cdots + a_N z^N$, but one may ask for the set of all solutions under additional constraints, for instance being contractive in the open unit disk, or having a real positive part there.

Rational functions form a field, which is denoted by $\mathbb{C}(X)$. The next "simplest" field of analytic functions is the field of elliptic functions; see Section 13.1. For the time being we just recall:

Theorem 12.1.6. *Let f be a meromorphic function such that*

$$f(z+1) = f(z+i), \quad \forall z \in \mathbb{C} \quad \text{where } f \text{ is defined.}$$

Let $\wp$ be the Weierstrass function (defined below by (13.1.3)). Then, there are two rational functions r_1 and r_2 such that

$$f(z) = r_1(\wp(z)) + \wp'(z)r_2(\wp(z)). \tag{12.1.4}$$

12.2 Realizations of rational functions

As a corollary of the proof of Exercise 12.1.2 we have the partial fraction representation of a rational function (see for instance [168, p. 116]): A function is rational if and only if it is of the form

$$f(z) = \sum_{n=1}^{N} \sum_{j=1}^{k_n} \frac{A_j}{(z - z_n)^j} + p(z), \tag{12.2.5}$$

where p is a polynomial (corresponding to the pole at infinity) and where the A_j are complex numbers, with $A_{k_n} \neq 0$ when there are finite poles. The order of the pole z_n is by definition k_n. From this result one gets (after the change of variable $z \mapsto 1/z$) the second item, in the scalar case, of Theorem 12.1.3. We note that (12.1.1) is called a realization centered at the origin, while (12.2.6) is called a realization centered at infinity.

Exercise 12.2.1. *Show that a function r analytic at ∞ is rational if and only if it can be written as*

$$r(z) = D + C(zI_N - A)^{-1}B, \tag{12.2.6}$$

where $D = r(\infty)$, $N \in \mathbb{N}$, C is a row vector with N components, B is a column vector with N components and A is an $N \times N$ matrix..

Hint for Exercise 12.2.1. First prove that, for any $w \in \mathbb{C}$, the function $\frac{1}{z-w}$ admits a realization. Next, for r_1 and r_2 two rational functions analytic at infinity, with realizations

$$r_j(z) = D_j + C_j(zI_{N_j} - A_j)^{-1}B_j, \quad j = 1, 2,$$

prove the following two formulas:

$$r_1(z)r_2(z) = D + C(zI_N - A)^{-1}B, \tag{12.2.7}$$

where $N = N_1 + N_2$, $D = D_1 D_2$ and

$$C = \begin{pmatrix} C_1 & D_1 C_2 \end{pmatrix}, \quad B = \begin{pmatrix} B_1 D_2 \\ B_2 \end{pmatrix} \quad \text{and} \quad A = \begin{pmatrix} A_1 & B_1 C_2 \\ 0 & A_2 \end{pmatrix},$$

and

$$r_1(z) + r_2(z) = D + C(zI_N - A)^{-1}B, \tag{12.2.8}$$

where $N = N_1 + N_2$, $D = D_1 + D_2$ and

$$C = \begin{pmatrix} C_1 & C_2 \end{pmatrix}, \quad B = \begin{pmatrix} B_1 \\ B_2 \end{pmatrix} \quad \text{and} \quad A = \begin{pmatrix} A_1 & 0 \\ 0 & A_2 \end{pmatrix}.$$

Conclude by using the partial fraction expansion of r.

A maybe easier proof of the realization theorem, still using (12.2.5) is based on the formula given in Exercise 12.2.2 below. We leave to the student to work out the details of the modified proof. Before stating Exercise 12.2.2 we recall that a Jordan cell is an $n \times n$ matrix of the form

$$
J(w) = \begin{pmatrix}
w & 1 & 0 & 0 & 0 & \cdots \\
0 & w & 1 & 0 & 0 & \cdots \\
0 & 0 & w & 1 & 0 & \cdots \\
\vdots & & & & & \vdots \\
\vdots & & & & & \vdots \\
0 & \cdots & 0 & 0 & w & 1 \\
0 & 0 & \cdots & 0 & 0 & w
\end{pmatrix},
\tag{12.2.9}
$$

where $w \in \mathbb{C}$ (for $N = 1$ one sets $J(w) = w$). For completeness, we also recall that any $N \times N$ matrix with complex entries is similar to a block matrix with Jordan cells as block entries.

Exercise 12.2.2. *Let $w \in \mathbb{C}$. Show that*

$$
\frac{1}{(z-w)^n} = C(zI_n - J(w))^{-1}B,
\tag{12.2.10}
$$

where $J(w)$ is given by (12.2.9) and

$$
C = \begin{pmatrix} 1 & 0 & \cdots & 0 & 0 \end{pmatrix}, \quad and \quad B = \begin{pmatrix} 0 \\ 0 \\ \vdots \\ 0 \\ 1 \end{pmatrix}.
$$

We now present another realization of a rational function, called the backward shift realization, and which plays an important role in linear system theory. See for instance [94].

Exercise 12.2.3. *Let r be a $\mathbb{C}^{p \times q}$-valued rational function, and let α be a point of analyticity of r (meaning that r is analytic in some open neighborhood of α).*

(1) *Show that the linear span $\mathcal{M}$ of the functions*

$$
z \mapsto R_\alpha^n r c, \quad n = 1, 2, \ldots, \quad c \in \mathbb{C}^q,
$$

is finite-dimensional, say of dimension N.

(2) *Define operators*

$$
\begin{pmatrix} A & B \\ C & D \end{pmatrix} : \begin{pmatrix} \mathcal{M} \\ \mathbb{C}^q \end{pmatrix} \longrightarrow \begin{pmatrix} \mathcal{M} \\ \mathbb{C}^p \end{pmatrix}
$$

by
$$Af = R_\alpha f, \quad Bc = R_\alpha rc, \quad Cf = f(\alpha), \quad Dc = r(\alpha)c.$$

Show that

$$r(z) = r(\alpha) + (z - \alpha)C(I_M - (z - \alpha)A)^{-1}B. \qquad (12.2.11)$$

When the rational function r is analytic in a neighborhood of the origin, one can put $\alpha = 0$ in (12.2.11), and one then looks at realization of the form (12.1.1), that is,

$$r(z) = D + zC(I_N - zA)^{-1}B. \qquad (12.2.12)$$

When A is invertible, one can rewrite (12.2.12) as

$$r(z) = D - CA^{-1}B - CA^{-1}(zI_N - A^{-1})^{-1}A^{-1}B,$$

which is of the form (12.2.6).

Formula (12.2.12) gives links with the theory of linear systems (see Chapter 11 for more information on these) and the notion of state space equations. The equations in the next exercise are called state space equations. They define a special, but very important, class of not necessarily bounded linear systems. The vector x_n in the equation (12.2.13) is called the state of the system at time n.

Exercise 12.2.4. *Let $(A, B, C, D) \in \mathbb{C}^{N \times N} \times \mathbb{C}^{N \times p} \times \mathbb{C}^{q \times N} \times \mathbb{C}^{q \times p}$, and consider the equations*

$$\begin{aligned} x_{n+1} &= Ax_n + Bu_n, \\ y_n &= Cx_n + Du_n, \quad n = 0, 1, \ldots \end{aligned} \qquad (12.2.13)$$

where, for $n = 0, 1, \ldots$,

$$x_n \in \mathbb{C}^N, \quad u_n \in \mathbb{C}^p \quad and \quad y_n \in \mathbb{C}^q.$$

Assume that $x_0 = 0$. Assume that the series $u(z)$ has a positive radius of convergence. Show that the series $y(z)$ has a positive radius of convergence and that

$$y(z) = h(z)u(z), \qquad (12.2.14)$$

where

$$h(z) = D + zC(I_N - zA)^{-1}B.$$

When are the entries of h in $\mathbf{H}_2(\mathbb{D})$?

Exercise 12.2.5. *Let $w \in \mathbb{D}$ and let b_w be the associated Blaschke factor (1.1.44):*

$$b_w(z) = \frac{z - w}{1 - z\overline{w}}.$$

Find a realization of b_w of the form (12.2.12). Find a realization of $b_{w_1} b_{w_2} \cdots b_{w_N}$ where $w_1, w_2, \ldots, w_N \in \mathbb{D}$.

More generally, and as suggested by our colleague Prof. Izchak Lewkowicz:

Exercise 12.2.6. *Let $N \in \mathbb{N}$, $w \in \mathbb{D}$ and let b_w be the associated Blaschke factor* (1.1.44). *Find a realization of $b_w(z^N)$,*

$$b_w(z^N) = \frac{z^N - w}{1 - z^N \overline{w}}.$$

Exercise 12.2.7. *Let $r(z) = D + C(zI_N - A)^{-1}B$ be a realization of the $\mathbb{C}^{p \times p}$-valued rational function r, assumed analytic and invertible at infinity. Show that*

$$r(z)^{-1} = D^{-1} - D^{-1}C(zI_N - A^{\times})^{-1}BD^{-1}$$

is a realization of r^{-1} with

$$A^{\times} = A - BD^{-1}C.$$

Motivated by electrical engineering applications, it is of interest to relate the properties of the matrix

$$\begin{pmatrix} A & B \\ C & D \end{pmatrix}$$

and of the associated rational function r. See for instance the discussion after the proof of Exercise 12.2.5. We also mention the *positive real lemma*, which has numerous applications in electrical engineering (see for instance [76]):

Theorem 12.2.8 (The positive real lemma). *Let r be a rational $\mathbb{C}^{p \times p}$-valued function analytic in a neighborhood of infinity, and let $r(z) = D + C(zI_N - A)^{-1}B$ be a minimal realization of r. Then, the following are equivalent:*

(a) $$\operatorname{Re} r(iy) \geq 0, \quad \forall y \in \mathbb{R} \text{ such that } r(iy) \text{ exists,}$$

and

(b) *There exists an invertible Hermitian matrix H such that*

$$\begin{pmatrix} H & 0 \\ 0 & I_p \end{pmatrix}^{*} \begin{pmatrix} A & B \\ C & D \end{pmatrix} + \begin{pmatrix} A & B \\ C & D \end{pmatrix}^{*} \begin{pmatrix} H & 0 \\ 0 & I_p \end{pmatrix} \geq 0.$$

12.3 Multipoint interpolation

The purpose of this section is to present a decomposition theorems for rational functions, and the implication to a family of interpolation problems, called multipoint interpolation problems. The section is based on [12].

Exercise 12.3.1. *Let p be a polynomial of degree N, and let r be a rational function analytic in neighborhoods of the zeros of p. Show that there exist uniquely determined rational functions $r_1, \ldots, r_N$, analytic in a neighborhood of the origin and such that*

$$r(z) = \sum_{j=1}^{N} z^{j-1} r_j(p(z)). \tag{12.3.1}$$

Hint. Assume first that the zeros of p are simple. The solution involves then a Vandermonde determinant. In the case of non simple zeroes one uses a generalized Vandermonde determinant. The hypothesis that the zeros of p are not poles of r can be removed, but makes some arguments easier.

Exercise 12.3.2. *Given N pairwise different points $w_1, \ldots, w_N$ and numbers $a_1, \ldots, a_N$ not all equal to 0, and $c \in \mathbb{C}$, find all rational functions r analytic in neighborhoods of $w_1, \ldots, w_N$ and such that*

$$\sum_{n=1}^{N} a_n r(w_n) = c. \tag{12.3.2}$$

Hint. Use the decomposition (12.3.1) with $p(z) = \prod_{n=1}^{N}(z - w_n)$.

Problem 12.3.3. *Let p be a polynomial of degree N. Let α and β be such that the roots $w_1(\alpha), \ldots, w_N(\alpha)$ and $w_1(\beta), \ldots, w_N(\beta)$ of the equations $p(z) = \alpha$ and $p(z) = \beta$ are all distinct ($w_u(\alpha) \neq w_v(\beta)$ for $u, v = 1, \ldots, N$), and define*

$$\left(R_\alpha^{(p)} f\right)(z) = \frac{f(z)}{p(z) - \alpha} - \sum_{u=1}^{N} \frac{f(w_u(\alpha))}{p'(w_u(\alpha))(z - w_u(\alpha))}, \tag{12.3.3}$$

for a function analytic in neighborhood of the points $w_1(\alpha), \ldots, w_N(\alpha)$, and similarly for $R_\beta^{(p)}$. Then the resolvent equation

$$R_\alpha^{(p)} - R_\beta^{(p)} = (\alpha - \beta) R_\alpha^{(p)} R_\beta^{(p)}$$

holds.

Exercise 12.3.4. *Let $f(z) = \frac{1}{p(z) - \lambda}$, and let $\alpha \neq \lambda$. Then,*

$$(R_\alpha^{(p)} f)(z) = -\frac{1}{\alpha - \lambda} f(z). \tag{12.3.4}$$

12.4   Solutions

Solution of Exercise 12.1.1. The result holds because the composition of a polynomial with a Moebius map is a rational function, and since the inverse of an invertible Moebius map is a Moebius map. □

Solution of Exercise 12.1.2. In the proof of Exercise 7.1.9 (Weierstrass' theorem), the entire functions $f_1, \ldots, f_N$ are now polynomials since the points z_j are poles of f. The function f_0 is also a polynomial since it is the principal part of $f(1/z)$ at the origin, and $f(1/z)$ has (at most) a pole at the origin. It follows that f is rational. The converse statement is clear. □

Solution of Exercise 12.1.5. Since r is rational it admits a realization and there are matrices A, B, C such that $a_n = CA^{n-1}B$ for $n = 1, 2, \ldots$. Let $p(z) = z^M + c_{M-1}z^{M-1} + \cdots + c_1 A + c_0$ denote the characteristic polynomial of the matrix A. We have

$$A^M + c_{M-1}A^{M-1} + \cdots + c_1 A + c_0 I_M = 0,$$

and thus, for $n = 1, 2, \ldots$

$$A^{M+n-1} + c_{M-1}A^{M+n-2} + \cdots + c_1 A + c_0 A^{n-1} = 0. \tag{12.4.5}$$

Multiplying this equation on the left by C and on the right by B we obtain (12.1.3).

Conversely, any solution of the difference equation (12.1.3) corresponds to a series $f(z) = \sum_{n=0}^{\infty} a_n z^n$ with a positive radius of convergence. Equation (12.1.3) leads to

$$c_0 \frac{f(z) - a_0}{z} + c_1 \frac{f(z) - a_0 - a_1 z}{z^2} + \cdots +$$
$$+ c_{M-1} \frac{f(z) - a_0 - \cdots - a_{M-1}z^{M-1}}{z^M} + \frac{f(z) - a_0 - \cdots - a_M z^M}{z^{M+1}} = 0,$$

from which we get that f is rational. $\qquad\square$

The proofs of the following exercises involve matrix computations, which are elementary, but to which most second (or even third year students) have not been exposed.

Solution of Exercise 12.2.1. We follow the steps given in the hint, and proceed in a number of steps.

Step 1: *The function* $\frac{1}{z-w}$ *admits a realization centered at infinity.*

Indeed, it suffices to take $N = 1$ and

$$A = w, \quad C = B = 1, \quad \text{and} \quad D = 0.$$

Step 2: *We prove* (12.2.7):

With the notation after the statement of the exercise we have

$$r_1(z)r_2(z) = (D_1 + C_1(zI_{N_1} - A_1)^{-1}B_1)(D_2 + C_2(zI_{N_2} - A_2)^{-1}B_2)$$
$$= D_1 D_2 + C_1(zI_{N_1} - A_1)^{-1}B_1 D_2$$
$$+ D_1 C_2(zI_{N_2} - A_2)^{-1}B_2$$
$$+ C_1(zI_{N_1} - A_1)^{-1}B_1 C_2(zI_{N_2} - A_2)^{-1}B_2$$
$$= D_1 D_2 + \begin{pmatrix} C_1 & D_1 C_2 \end{pmatrix} X \begin{pmatrix} B_1 D_2 \\ B_2 \end{pmatrix},$$

where

$$X = \begin{pmatrix} (zI_{N_1} - A_1)^{-1} & (zI_{N_1} - A_1)^{-1}B_1C_2(zI_{N_2} - A_2)^{-1} \\ 0 & (zI_{N_2} - A_2)^{-1} \end{pmatrix}.$$

Computing

$$\begin{pmatrix} zI_{N_1} - A_1 & -B_1C_2 \\ 0 & zI_{N_2} - A_2 \end{pmatrix}$$

$$\times \begin{pmatrix} (zI_{N_1} - A_1)^{-1} & (zI_{N_1} - A_1)^{-1}B_1C_2(zI_{N_2} - A_2)^{-1} \\ 0 & (zI_{N_2} - A_2)^{-1} \end{pmatrix}$$

$$= I_{N_1+N_2},$$

we conclude the proof. Note that the proof is for matrix-valued functions.

Step 3: *We prove* (12.2.8):

Since (12.2.7) has been proved for matrix-valued functions, (12.2.8) is seen to be a special case of (12.2.7) by writing

$$r_1(z) + r_2(z) = \begin{pmatrix} r_1(z) & I_p \end{pmatrix} \begin{pmatrix} I_q \\ r_2(z) \end{pmatrix}$$

where r_1 and r_2 are $\mathbb{C}^{p \times q}$-valued.

Step 4: *The result follows then from the partial fraction expansion* (12.2.5).

Indeed, there is no polynomial term $p(z)$ in (12.2.5) since we assume analyticity at infinity. It suffices to apply the preceding steps to obtain a realization for each term of the form $\frac{1}{(z-z_n)^j}$ and hence, to obtain a realization for the sum (12.2.5).   □

Solution of Exercise 12.2.2. Let J denote the matrix

$$J = J(0) = \begin{pmatrix} 0 & 1 & 0 & 0 & 0 & \cdots \\ 0 & 0 & 1 & 0 & 0 & \cdots \\ 0 & 0 & 0 & 1 & 0 & \cdots \\ \vdots & & & & & \\ \vdots & & & & & \vdots \\ 0 & \cdots & 0 & 0 & 0 & 1 \\ 0 & 0 & \cdots & 0 & 0 & 0 \end{pmatrix}.$$

Thus

$$J(w) = wI_n + J.$$

Note that J is nilpotent: $J^n = 0_{n \times n}$. Therefore

$$(zI_n - J(w))^{-1} = ((z - w)I_n - J)^{-1}$$

$$= (z - w)^{-1} \left(I_n - \frac{J}{z - w} \right)^{-1}$$

$$= (z - w)^{-1} \left(I_n + \frac{J}{z - w} + \frac{J^2}{(z - w)^2} + \cdots + \frac{J^{n-1}}{(z - w)^{n-1}} \right).$$

The entry in the right upper corner of this matrix is $\frac{1}{(z-w)^n}$, and the realization (12.2.10) follows. $\qquad\square$

It follows from the previous exercise that every term of the form $\frac{1}{(z-w)^n}$ admits a realization. Using (12.2.5) and (12.2.8) we obtain the realization theorem for all rational functions analytic at infinity.

Solution of Exercise 12.2.3.
(1) The first claim follows from the partial fraction decomposition and from the formula[7]

$$\left(R_\alpha^n \left(\frac{1}{(\cdot - w)^u} \right) \right)(z) = - \sum_{v=0}^{u-1} \frac{(\alpha - w)^{u-1-v}}{(z - w)^{u-v}}. \qquad (12.4.6)$$

(2) By construction the operator A sends $\mathcal{M}$ into itself and the operator B sends $\mathbb{C}^q$ into $\mathcal{M}$. Let now $h \in \mathcal{M}$ and let $z \in \mathbb{C}$ such that $(I_\mathcal{M} - (z - \alpha)A)$ is invertible. Here we view z as a parameter, and we denote by λ the variable on which depend the functions in $\mathcal{M}$. Set

$$g = (I_\mathcal{M} - (z - \alpha)A)^{-1}h.$$

Thus

$$h = (I_\mathcal{M} - (z - \alpha)A)g,$$

that is, pointwise,

$$h(\lambda) = g(\lambda) - (z - \alpha)\frac{g(\lambda) - g(\alpha)}{\lambda - \alpha}.$$

Thus

$$h(z) = g(z) - (z - \alpha)\frac{g(z) - g(\alpha)}{z - \alpha}$$

$$= g(\alpha)$$

$$= C(I_\mathcal{M} - (z - \alpha)A)^{-1}g.$$

We obtain the realization by applying the above to $h = Bc$ $\qquad\square$

---

[7]of course it is much quicker to rewrite the left side of (12.4.6) as $R_\alpha^n \frac{1}{(z-w)^u}$.

484 Chapter 12. Rational Functions

Remark 12.4.1. The point evaluation formula

$$g(\alpha) = C(I_M - (z - \alpha)A)^{-1}g \qquad (12.4.7)$$

plays an important role in the theory of reproducing kernel spaces.

Solution of Exercise 12.2.4. We divide the proof into two steps.

    Step 1: *The (vector-valued) power series*

$$x(z) = \sum_{n=0}^{\infty} x_n z^n$$

has a strictly positive radius of convergence.

A priori $x(z)$ may converge only at the origin. We have

$$\frac{x(z) - x(0)}{z} = \sum_{n=0}^{\infty} x_{n+1} z^n.$$

It therefore follows from the first equation in (12.2.13) that

$$\frac{x(z) - x(0)}{z} = Ax(z) + Bu(z),$$

and so

$$x(z) = x(0) + z(I_N - zA)^{-1}Bu(z). \qquad (12.4.8)$$

This formula for $x(z)$ shows that it has a strictly positive radius of convergence since $u(z)$ has a strictly positive radius of convergence.

    Step 2: *Assume* $x(0) = 0$. *The series* $y(z)$ *is given by* (12.2.14):

Indeed, the second equation in (12.2.13) leads to

$$\sum_{n=0}^{\infty} y_n z^n = C \sum_{n=0}^{\infty} x_n z^n + D \sum_{n=0}^{\infty} u_n z^n.$$

Taking into account (12.4.8) we obtain

$$y(z) = zC(I_N - zA)^{-1}Bu(z) + Du(z) = h(z)u(z),$$

with $h(z) = D + zC(I_N - zA)^{-1}B$. This matrix-valued function will have its entries in $\mathbf{H}_2(\mathbb{D})$ if and only if all its entries have no poles in the closed unit disk. A *sufficient*, but in general not *necessary*, condition will be that the spectrum of A is in the open unit disk. □

Solution of Exercise 12.2.5. We write

$$b_w(z) = \frac{z - w}{1 - z\overline{w}}$$

$$= -w + \frac{z - w}{1 - z\overline{w}} + w$$

$$= -w + \frac{z(1 - |w|^2)}{1 - z\overline{w}}.$$

So, a realization of b_w is given by

$$\begin{pmatrix} A & B \\ C & D \end{pmatrix} = \begin{pmatrix} \overline{w} & \sqrt{1 - |w|^2} \\ \sqrt{1 - |w|^2} & -w \end{pmatrix}. \tag{12.4.9}$$

For two points, formula (12.2.7) gives the realization

$$\begin{pmatrix} A & B \\ C & D \end{pmatrix} = \left(\begin{array}{ccc|c} \overline{w_1} & \sqrt{1 - |w_1|^2}\sqrt{1 - |w_2|^2} & & -w_2\sqrt{1 - |w_1|^2} \\ 0 & \overline{w_2} & & \sqrt{1 - |w_2|^2} \\ \hline \sqrt{1 - |w_1|^2} & -w_1\sqrt{1 - |w_2|^2} & & w_1 w_2 \end{array} \right).$$

$$\tag{12.4.10}$$

We leave it to the interested student to develop the formula for $N > 2$. □

We note that the realizations (12.4.9) and (12.4.10) are unitary. This is no coincidence. Any rational function analytic at the origin, without poles in the open unit disk and unitary on the unit circle admits a unitary realization centered at the origin. See [100], and see also [11] for the case of poles inside the disk.

Solution of Exercise 12.2.6. We first assume $w \neq 0$. Then, $b_w(z^N)$ is analytic at infinity, and $b_w(\infty) = -1/\overline{w}$. We consider the realization of the type (12.2.6). Let $z_0, \ldots, z_{N-1}$ be the roots of order N of $1/\overline{w}$. The z_j are simple poles of $b_w(z^N)$, with residues computed by formula (7.3.2),

$$\operatorname{Res}(b_w(z^N), z_j) = \frac{z_j^N - w}{-N z_j^{N-1}\overline{w}}$$

$$= \frac{\frac{1}{\overline{w}} - w}{-N} z_j \quad (\text{since } z_j^N = 1/\overline{w})$$

$$= -\frac{1 - |w|^2}{N\overline{w}} z_j, \quad j = 0, \ldots, N - 1$$

and thus

$$b_w(z^N) = -\frac{1}{\overline{w}} - \frac{1 - |w|^2}{N} \sum_{j=0}^{N-1} \frac{z_j}{z - z_j}$$

$$= D + C(zI_N - A)^{-1}B,$$

with $D = -\frac{1}{\overline{w}}$, and as possible choice of A, B and C the matrices

$$A = \begin{pmatrix} z_0 & 0 & 0 & \cdots & 0 \\ 0 & z_1 & 0 & \cdots & 0 \\ \vdots & \vdots & & & \vdots \\ 0 & & \cdots & 0 & 0 \\ 0 & 0 & \cdots & 0 & z_{N-1} \end{pmatrix} \in \mathbb{C}^{N \times N}, \quad B = \begin{pmatrix} 1 \\ 1 \\ \vdots \\ 1 \end{pmatrix} \in \mathbb{C}^{N \times 1},$$

and

$$C = -\frac{1 - |w|^2}{N\overline{w}} \begin{pmatrix} z_0 & z_1 & \cdots & z_{N-1} \end{pmatrix} \in \mathbb{C}^{N \times N}.$$

Since the $z_j \neq 0$, we can look for a realization of the form (12.2.12). We obtain

$$b_w(z^N) = -\frac{1}{\overline{w}} - \frac{1 - |w|^2}{N\overline{w}} \sum_{j=0}^{N-1} \frac{z_j}{z - z_j}$$

$$= -\frac{1}{\overline{w}} + \frac{1 - |w|^2}{N\overline{w}} \sum_{j=0}^{N-1} \frac{1}{1 - \frac{z}{z_j}}$$

$$= -\frac{1}{\overline{w}} + \frac{1 - |w|^2}{\overline{w}} + \frac{1 - |w|^2}{N\overline{w}} \sum_{j=0}^{N-1} \left(\frac{1}{1 - \frac{z}{z_j}} - 1 \right)$$

$$= -w + z \frac{1 - |w|^2}{N\overline{w}} \sum_{j=0}^{N-1} \frac{1}{z_j} \frac{1}{1 - \frac{z}{z_j}},$$

which is of the form (12.2.12), with now $D = -w$, and as possible choice of A, B and C the matrices

$$A = \begin{pmatrix} z_0^{-1} & 0 & 0 & \cdots & 0 \\ 0 & z_1^{-1} & 0 & \cdots & 0 \\ \vdots & \vdots & & & \vdots \\ 0 & & \cdots & 0 & 0 \\ 0 & 0 & \cdots & 0 & z_{N-1}^{-1} \end{pmatrix} \in \mathbb{C}^{N \times N}, \quad B = \begin{pmatrix} 1 \\ 1 \\ \vdots \\ 1 \end{pmatrix} \in \mathbb{C}^{N \times 1},$$

and

$$C = \frac{1 - |w|^2}{N\overline{w}} \begin{pmatrix} z_0^{-1} & z_1^{-1} & \cdots & z_{N-1}^{-1} \end{pmatrix} \in \mathbb{C}^{N \times N}.$$

Finally, the case $w = 0$ is obtained by setting $w = 0$ and replacing z by $1/z$ in (12.2.10).   □

Solution of Exercise 12.2.7. We assume that r is $\mathbb{C}^{p \times p}$-valued. We have

$$(D + C(zI_N - A)^{-1}B)(D^{-1} - D^{-1}C(zI_N - A^\times)^{-1}BD^{-1})$$
$$= I_p - C(zI_N - A^\times)^{-1}BD^{-1}$$
$$+ C(zI_N - A)^{-1}BD^{-1}$$
$$- C(zI_N - A)^{-1}BD^{-1}C(zI_N - A^\times)^{-1}BD^{-1}.$$

Taking into account that

$$BD^{-1}C = A - A^\times = (zI_N - A^\times) - (zI_N - A), \qquad (12.4.11)$$

we have

$$
\begin{aligned}
C(zI_N - A)^{-1}&BD^{-1}C(zI_N - A^\times)^{-1}BD^{-1}\\
&= C(zI_N - A)^{-1}\left\{(zI_N - A^\times) - (zI_N - A)\right\}(zI_N - A^\times)^{-1}BD^{-1}\\
&= C(zI_N - A)^{-1}BD^{-1} - C(zI_N - A^\times)^{-1}BD^{-1}.
\end{aligned}
$$

Substituting this formula in the formula above (12.4.11) we obtain the result. $\square$

The previous exercise can also be solved using the following well-known result:

Proposition 12.4.2. *Let U and V be two matrices, respectively in $\mathbb{C}^{p\times q}$ and $\mathbb{C}^{q\times p}$. Then, $I_p - UV$ is invertible if and only if $I_q - VU$ is invertible, and one has*

$$(I_p - UV)^{-1} = I_p + U(I_q - VU)^{-1}V. \qquad (12.4.12)$$

It suffices to write

$$r(z) = D(I_p + zD^{-1}C(I_N - zA)^{-1}B)$$

and apply formula (12.4.12) to

$$U = -zD^{-1}C \quad \text{and} \quad V = (I_N - zA)^{-1}B.$$

As for (12.4.12) it is proved in the following way: Assume first that $I_q - VU$ is invertible. We have

$$
\begin{aligned}
(I_p - UV)(I_p + U(I_q - VU)^{-1}V) &= I_p - UV + U(I_q - VU)^{-1}V\\
&\quad - UVU(I_q - VU)^{-1}V.
\end{aligned}
$$

But

$$-UVU(I_q - VU)^{-1}V = (U(I_q - VU) - U)(I_q - VU)^{-1}V = UV - U(I_q - VU)^{-1}V,$$

and hence

$$(I_p - UV)(I_p + U(I_q - VU)^{-1}V) = I_p.$$

This concludes the proof of (12.4.12) since the claim is symmetric in U and V.

Solution of Exercise 12.3.1. We will only consider the case where the zeros of p are simple, and write $p(z) = \prod_{k=1}^N (z - z_k)$, with $z_k \neq z_\ell$ for $k \neq \ell$. We first show that (12.3.1), if it exists, is unique. To that purpose, assume that we have

$$0 \equiv \sum_{j=1}^N z^{j-1} r_j(p(z)), \qquad (12.4.13)$$

and plug in this expression $z = z_k$ for $k = 1, \ldots, N$. We obtain

$$0 = \sum_{j=1}^{N} z_k^{j-1} r_j(0), \quad k = 1, \ldots, N,$$

that is

$$\begin{pmatrix} r_1(0) & r_2(0) & \cdots & r_N(0) \end{pmatrix} \begin{pmatrix} 1 & 1 & \cdots & 1 \\ z_1 & z_2 & \cdots & z_N \\ z_1^2 & z_2^2 & \cdots & z_N^2 \\ \vdots & \vdots & & \vdots \\ z_1^{N-1} & z_2^{N-1} & \cdots & z_N^{N-1} \end{pmatrix} = \begin{pmatrix} 0 & 0 & \cdots & 0 \end{pmatrix}.$$

The Vandermonde matrix

$$\begin{pmatrix} 1 & 1 & \cdots & 1 \\ z_1 & z_2 & \cdots & z_N \\ z_1^2 & z_2^2 & \cdots & z_N^2 \\ \vdots & \vdots & & \vdots \\ z_1^{N-1} & z_2^{N-1} & \cdots & z_N^{N-1} \end{pmatrix}$$

is invertible, and so $r_1(0) = r_2(0) = \cdots = r_N(0) = 0$. Writing $r_j(z) = z s_j(z)$ for $j = 1, \ldots$, we have from (12.4.13)

$$0 \equiv \sum_{j=1}^{N} z^{j-1} s_j(p(z)), \tag{12.4.14}$$

and the same argument shows that

$$s_1(0) = \cdots = s_N(0) = 0.$$

Iterating we get that the functions $r_1, r_2, \ldots, r_N$ vanish identically.

   To prove existence we do not use the analyticity hypothesis of the given function r at the zeros of p, and take advantage of the partial fraction decomposition (12.2.5) of a rational function, and prove that the decomposition holds for polynomials, for functions of the form $\frac{1}{z-a}$ ($a \in \mathbb{C}$), and that if two functions admit a decomposition (12.3.1) so do their sum and their product. More precisely, for the monomial z^j with $j < N$, the decomposition (12.3.1) is trivial: $r_j \equiv 1$ and $r_k \equiv 0$ for $k \in \{0, \ldots, N\} \setminus \{j\}$. For the monomial z^N, the division

$$z^N = a_N p(z) + a_{N-1} z^{N-1} + \cdots + a_0$$

leads to the decomposition (12.3.1) with

$$r_j(z) = a_j, \quad j = 2, \ldots, N \quad \text{and} \quad r_1(z) = a_N z.$$

For the rational function $\frac{1}{z-a}$ we write

$$\frac{1}{z-a} = \frac{1}{p(z)-p(a)} \frac{p(z)-p(a)}{z-a} = \sum_{j=0}^{N-1} c_{a,j} z^j \frac{1}{p(z)-p(a)},$$

where the numbers $c_{a,0}, \ldots, c_{a,N-1}$ are defined by

$$\frac{p(z)-p(a)}{z-a} = \sum_{j=0}^{N-1} c_{a,j} z^j.$$

Hence $r_j(z) = \frac{c_{a,j}}{z-p(a)}$ for $j = 1, \ldots, N$.

That the sum of two functions admitting a decomposition (12.3.1) also admits such a decomposition is trivial. For the case of the product it is enough to consider two functions of the form

$$r(z) = z^u t_u(p(z)) \quad \text{and} \quad s(z) = z^v s_v(p(z)),$$

where $u, v \in \{0, \ldots, N-1\}$ and t_u and s_v are rational. But,

$$r(z)s(z) = z^{u+v} t_u(p(z)) s_v(p(z)).$$

Let

$$z^{u+v} = \sum_{j=0}^{N-1} z^j r_j(p(z))$$

be the corresponding decomposition of z^{u+v} (which is very simple when $u+v < N$). We obtain

$$r(p(z))s(p(z)) = \sum_{j=0}^{N-1} z^j \left(r_j(p(z)) t_u(p(z)) s_v(p(z)) \right).$$

The proof for a general rational function analytic in neighborhoods of the zeros of p follows from combining these various results. $\qquad \square$

Solution of Exercise 12.3.2. Let r be a solution, which we write in the form (12.3.1) with $p(z) = \prod_{n=1}^N (z - w_n)$. Condition (12.3.2) gives

$$\sum_{n=1}^N a_n r(w_n) = \sum_{n=1}^N a_n \left(\sum_{j=1}^N w_n^{j-1} r_j(0) \right)$$

$$= \sum_{j=1}^N \left(\sum_{n=1}^N a_n w_n^{j-1} \right) r_j(0)$$

$$= \sum_{j=1}^N \overline{\xi_j} r_j(0),$$

where we have set

$$\overline{\xi_j} = \sum_{n=1}^{N} a_n w_n^{j-1}, \quad j = 1, \ldots, N.$$

The multipoint interpolation condition (12.3.1) reduces thus to a one *point*, but *tangential* interpolation condition

$$\xi^* R(0) = c, \quad \text{where} \quad \xi = \begin{pmatrix} \xi_1 \\ \vdots \\ \xi_N \end{pmatrix} \quad \text{and} \quad R(z) = \begin{pmatrix} r_1(z) \\ \vdots \\ r_N(z) \end{pmatrix}.$$

Let (assuming $\xi \neq 0$)

$$P = \frac{\xi \xi^*}{\xi^* \xi}.$$

We leave to the reader to check that a $\mathbb{C}^N$-valued rational function R satisfies $\xi^* R(0) = c$ if and only if it can be written as

$$R(z) = c\xi + (I_N - P + zP)\, G(z),$$

where G is an arbitrary $\mathbb{C}^N$-valued rational function without a pole at the origin. The description of the set of functions r follows. $\square$

Solution of Exercise 12.3.4. We have

$$(R_\alpha^{(p)} f)(z) = \frac{1}{(p(z) - \lambda)(p(z) - \alpha)} - \sum_{u=1}^{N} \frac{\dfrac{1}{p(w_u(\alpha)) - \lambda}}{p'(w_u(\alpha))(z - w_u(\alpha))}$$

$$= \frac{1}{\alpha - \lambda}\left(\frac{1}{p(z) - \alpha} - \frac{1}{p(z) - \lambda} \right) - \sum_{u=1}^{N} \frac{1}{\underbrace{(p(w_u(\alpha)) - \lambda)}_{=\,\alpha} p'(w_u(\alpha))(z - w_u(\alpha))}$$

$$= \frac{1}{\alpha - \lambda}\frac{1}{p(z) - \alpha} - \frac{1}{\alpha - \lambda}\frac{1}{p(z) - \lambda} - \frac{1}{\alpha - \lambda}\underbrace{\sum_{u=1}^{N} \frac{1}{p'(w_u(\alpha))(z - w_u(\alpha))}}_{=\,\frac{1}{p(z)-\alpha}}$$

$$= -\frac{1}{\alpha - \lambda} f(z).$$ $\square$

Chapter 13

Special Functions and Transforms

In this short chapter we present some exercises on elliptic functions and on the Mellin transform. We also briefly discuss some aspects of the Fourier transform pertaining to the Bargmann transform.

13.1 Elliptic functions

The first exercise is taken from the book of Choquet on topology [46, p. 315], [47, p. 299]. The purpose of the exercise is to build a *meromorphic* bi-periodic function on $\mathbb{C}$ (thus it has a lattice of periods). Such functions are called elliptic. For more on elliptic functions expressed as infinite products, see for instance [167, pp. 286–290]. See also Exercise 7.2.15.

Exercise 13.1.1. *Let $k \in \mathbb{C}$ with $|k| > 1$.*

(a) *Show that the infinite product*

$$P(z) = \prod_{\ell=1}^{\infty} \left(1 + \frac{z}{k^\ell}\right)$$

   converges for all $z \neq -k^\ell$, $\ell = 1, 2, \ldots$.

(b) *Show that*

$$P(kz) = (1 + z)P(z).$$

(c) *Set $S(z) = P(z)P(1/z)(1+z)$. Show that $S(kz) = kzS(z)$.*

(d) *Let $a_1, \ldots, a_n, b_1, \ldots, b_n$ be distinct points in $\mathbb{C}$ such that*

$$a_1 \cdots a_n = b_1 \cdots b_n, \qquad (13.1.1)$$

and let $M(z) = \dfrac{S(a_1 z) \cdots S(a_n z)}{S(b_1 z) \cdots S(b_n z)}$. Show that $M(kz) = M(z)$.

(e) *Set $G(z) = M(e^z)$. What can be said about G?*

Remark 13.1.2. An additive analog of (13.1.1) comes into play in Exercise 13.3.3. See equation (13.3.2) there.

Exercise 13.1.3. *Using Exercise 3.6.2, show that the function*

$$\wp(z) = \frac{1}{z^2} + \sum_{\substack{p,q \in \mathbb{Z} \\ (p,q) \neq (0,0)}} \frac{1}{(z - (p + iq))^2} - \frac{1}{(p + iq)^2}$$

is analytic in $\mathbb{C} \setminus \mathbb{Z} + i\mathbb{Z}$.

The function $\wp$ is called the Weierstrass function (associated to the lattice $\mathbb{C} \setminus \mathbb{Z} + i\mathbb{Z}$). It has only poles and satisfies

$$\wp(z + 1) = \wp(z + i) = \wp(z),$$

and hence is an elliptic function. It follows as a consequence of Exercise 7.2.15 that the function $\wp$ satisfies a differential equation of the form

$$(\wp')^2 = g_0 \wp^3 + g_1 \wp^2 + g_2 \wp + g_3$$

for complex numbers g_0, g_1, g_2 and g_3 such that $g_0 \neq 0$.

The function $\wp$ is closely related to the function ϑ appearing in Exercise 13.2.1. See [162, p. 25].

Question 13.1.4.

(1) *Find the decomposition (12.1.4) for $f(z) = \wp''(z)$.*
(2) *Compare the decompositions (12.1.4) for a general elliptic function and its derivative.*

In contrast with the case of rational functions we have:

Question 13.1.5. *Show that the composition of two (non-trivial) elliptic functions is not elliptic.*

13.2 The ϑ function

Exercise 13.2.1. *Let $\tau \in \mathbb{C}$ be such that $\operatorname{Im} \tau > 0$. Show that the function*

$$\vartheta(z, \tau) = \sum_{n \in \mathbb{Z}} e^{i\pi n^2 \tau + 2\pi i n z}$$

is entire (as a function of z), and that it satisfies

$$\vartheta(z+1,\tau) = \vartheta(z,\tau), \qquad\qquad (13.2.1)$$

$$\vartheta(z+\tau,\tau) = e^{-i\pi\tau-2\pi i z}\vartheta(z,\tau). \qquad\qquad (13.2.2)$$

Show that

$$\vartheta\left(\frac{1+\tau}{2},\tau\right) = 0. \qquad\qquad (13.2.3)$$

The function ϑ is called the theta function with characteristic τ. See [162] for a thorough study of these functions and of their applications.

In Exercise 13.2.2 we now show that $\frac{1+\tau}{2}$ is the only zero of ϑ modulo $\mathbb{Z}+\tau\mathbb{Z}$.

Exercise 13.2.2. *Show that the zeros of the function*

$$\vartheta(z,\tau) = \sum_{n\in\mathbb{Z}} e^{in^2\tau+2\pi i n z}$$

are

$$\frac{1+\tau}{2} + m + \tau n, \quad n, m \in \mathbb{Z}.$$

13.3 An application to periodic entire functions

Exercise 13.3.1. *Let f be an entire function and assume that*

$$f(z+1) = f(z).$$

Show that there is a function g analytic in $\mathbb{C}\setminus\{0\}$ such that

$$f(z) = g(e^{2\pi i z}).$$

Show that there exist complex numbers $c_n, n\in\mathbb{Z}$ such that

$$f(z) = \sum_{n\in\mathbb{Z}} c_n e^{2\pi i n z},$$

where the convergence is uniform on every closed strip inside every closed horizontal strip.

Exercise 13.3.2. *Let $\tau \in \mathbb{C}$ be such that $\operatorname{Im}\tau > 0$. Apply the previous result to find all entire functions f such that, for some pre-assigned complex numbers a and b,*

$$f(z+1) = f(z),$$

$$f(z+\tau) = e^{az+b}f(z). \qquad\qquad (13.3.1)$$

See [162, pp. 2–3].

Exercise 13.3.3. *Let f be a non-identically vanishing entire function satisfying the conditions (13.3.1), and let $a_1, \ldots, a_N, b_1, \ldots, b_N$ be complex numbers such that*

$$\sum_{n=1}^{N} a_n = \sum_{n=1}^{N} b_n. \tag{13.3.2}$$

Show that the function

$$q(z) = \prod_{n=1}^{N} \frac{f(z - a_n)}{f(z - b_n)}$$

is elliptic.

13.4 The Γ function and the Mellin transform

The Mellin transform is defined by the formula

$$(M(f))(z) = \int_0^{\infty} t^{z-1} f(t) dt \tag{13.4.1}$$

for appropriate functions f defined on $(0, \infty)$, and where for $t > 0$ and $z \in \mathbb{C}$ we set

$$t^z = e^{z \ln t}.$$

We refer to [50, Chapitre II] for more information. The case $f(t) = e^{-t}$ leads to the important Gamma function (see (3.1.11)

$$\Gamma(z) = \int_0^{\infty} t^{z-1} e^{-t} dt.$$

In the following exercise, the convergence of the integral (3.1.11) is studied. In Exercise 13.4.2 we will see that the function Γ defined in the following exercise is in fact analytic in $\operatorname{Re} z > 0$ (and in fact by analytic continuation, in $\mathbb{C} \setminus \{0, -1, -2, \ldots\}$.

Exercise 13.4.1. *Show that the integral (3.1.11) converges for every z such that $\operatorname{Re} z > 0$. Show that, for real $x > 0$, it holds that*

$$\Gamma(x + 1) = x\Gamma(x). \tag{13.4.2}$$

We now turn to a proof of the analyticity of the Gamma function (see (3.1.11) and the previous exercise).

Exercise 13.4.2. *Show that the Γ function*

$$\Gamma(z) = \int_0^{\infty} t^{z-1} e^{-t} dt$$

is analytic in $\operatorname{Re} z > 0$.

Hint. Consider compact sets of the form

$$K = \{(x,y)\,;\ m \le x \le M \text{ and } -R \le y \le R\},$$

with $m > 0$ and $R > 0$. Show that the series of functions

$$\Gamma_n(z) = \int_{1/n}^{n} t^{z-1}e^{-t}dt, \quad n = 1,2,\ldots,$$

converges uniformly on K to Γ.

Exercise 13.4.3. *Let Γ denote the Gamma function defined by (3.1.11). Show that*

$$\Gamma(z) = \lim_{n\to\infty} \frac{n!n^z}{z(z+1)\cdots(z+n)}, \quad \text{Re } z > 0. \tag{13.4.3}$$

Hint (See for instance [23, Exercise 2.6.2, p. 119].). Apply the dominated convergence theorem (see Theorem 17.5.2) to the series of functions

$$f_n(t) = 1_{[0,n]}(t)\left(1 - \frac{t}{n}\right)^n t^{z-1},$$

where we have denoted by $1_{[0,n]}(t)$ the indicator function of the interval $[0,n]$:

$$1_{[0,n]}(t) = \begin{cases} 1, & \text{if } t \in [0,n], \\ 0, & \text{otherwise.} \end{cases}$$

Exercise 13.4.4 (see [50, pp. 49–50]).

(a) *Show that the Mellin transform of e^{-t^2} is equal to $\frac{1}{2}\Gamma(z/2)$.*

(b) *Show that the Mellin transforms of $\cos t$ and $\sin t$ are respectively*

$$\Gamma(z)\cos\frac{\pi z}{2} \quad \text{and} \quad \Gamma(z)\sin\frac{\pi z}{2}, \quad \text{with Re } z \in (0,1).$$

In the following exercise implicit is the hypothesis that there exist real numbers c_1 and c_2 such that $\int_0^\infty u^{c_j-1}|f_j(u)|du < \infty$ for $j = 1,2$.

Exercise 13.4.5. *Let f_1 and f_2 be functions with Mellin transforms F_1 and F_2 respectively.*

(1) *Show that the Mellin transform of the function*

$$\int_0^\infty f_1(u)f_2(t/u)\frac{du}{u} \tag{13.4.4}$$

is F_1F_2.

(2) *compute (13.4.4) when $f_1(u) = f_2(u) = e^{-u}$.*

13.5   The Fourier transform

The Fourier transform is defined by

$$\widehat{f}(\lambda) = \int_{\mathbb{R}} e^{-i\lambda x} f(x) dx, \qquad (13.5.1)$$

first for functions in $\mathbf{L}_1(\mathbb{R}, dx)$. In general $\widehat{f}$ will not belong to $\mathbf{L}_2(\mathbb{R}, dx)$. The Fourier transform maps the Schwartz space of rapidly vanishing smooth functions onto itself in an isometric way up to a multiplicative constant, and extends, up to a multiplicative constant, to an isometry from $\mathbf{L}_2(\mathbb{R}, dx)$ onto itself:

$$\|f\|_{\mathbf{L}_2(\mathbb{R},dx)} = \frac{1}{\sqrt{2\pi}} \|\widehat{f}\|_{\mathbf{L}_2(\mathbb{R},dx)}. \qquad (13.5.2)$$

Note that $\widehat{f}$ is not, in general, a function but rather an equivalence class of functions. Furthermore, the Fourier transform of an arbitrary element $f \in \mathbf{L}_2(\mathbb{R}, dx)$ is not given directly by formula (13.5.1) (which will not make sense in general), but is defined in terms of limits. Its inverse is given by the formula

$$\check{f}(x) = \frac{1}{2\pi} \int_{\mathbb{R}} e^{i\lambda x} f(\lambda) d\lambda, \qquad (13.5.3)$$

and we have

$$\|f\|_{\mathbf{L}_2(\mathbb{R},dx)} = \frac{1}{\sqrt{2\pi}} \|\check{f}\|_{\mathbf{L}_2(\mathbb{R},dx)}. \qquad (13.5.4)$$

As an illustration of the preceding inversion formula, consider the function $g(x) = \frac{1}{x^2+1}$. Its Fourier transform was computed to be $h(\lambda) = \pi e^{-|\lambda|}$. See (8.6.10). Thus, from (13.5.3),

$$\check{h}(x) = \frac{1}{2\pi} \int_{\mathbb{R}} e^{i\lambda x} h(\lambda) d\lambda$$

$$= \frac{1}{2} \left\{ \int_0^\infty e^{-\lambda} e^{i\lambda x} d\lambda + \int_{-\infty}^0 e^{\lambda} e^{i\lambda x} d\lambda \right\}$$

$$= \frac{1}{2} \left\{ \frac{-1}{ix-1} + \frac{1}{ix+1} \right\}$$

$$= g(x).$$

We follow [206, pp. 42–43] for the next exercise.

Exercise 13.5.1. *For $R > 0$, consider the closed contour*

$$\gamma_R = \gamma_{1,R} + \gamma_{2,R} + \gamma_{3,R} + \gamma_{4,R},$$

defined as follows:

(i) $\gamma_{1,R}$ is the interval $[-R, R]$.

(ii) $\gamma_{2,R}$ is the interval $[R, R + iy]$.

(iii) $\gamma_{3,R}$ is the interval $[R + iy, -R + iy]$.

(iv) $\gamma_{4,R}$ is the interval $[-R + iy, -R]$.

(1) *By computing the integral of the function e^{-z^2} along this rectangle and using the value of the Gaussian integral (5.2.6), show that, for $y \in \mathbb{R}$,*

$$\int_{\mathbb{R}} e^{-t^2} e^{-2ity}\, dt = \sqrt{\pi} e^{-y^2}. \tag{13.5.5}$$

(2) *Using (13.5.5) compute the even moments (5.2.7).*

We now discuss some aspects of the theory of Hermite functions. More exercises and details can be found in [CAPB2]. By making the change of variables $z \mapsto \sqrt{2}z$ and $t \mapsto \sqrt{2}t$, and a normalization we first rewrite (5.6.4) as

$$e^{2tz-t^2} = \sum_{n=0}^{\infty} \frac{H_n(z)}{n!} t^n. \tag{13.5.6}$$

We have

$$H_n(z) = (-1)^n e^{z^2} \left(e^{-z^2} \right)^{(n)}, \tag{13.5.7}$$

as is seen by writing $e^{2tz-t^2} = e^{z^2} e^{-(t-z)^2}$ and considering the Taylor expansion centered at $t = 0$ of the function $t \mapsto e^{-(t-z)^2}$.

Question 13.5.2. *Prove that*

$$\int_{\mathbb{R}} e^{-u^2} H_n(u) H_m(u)\, du = \sqrt{\pi} 2^n n! \delta_{n,m}. \tag{13.5.8}$$

Hint. Denoting by α_{nm} the left side of (13.5.8) compute, using (13.5.6), the generating function

$$\sum_{n,m=0}^{\infty} \alpha_{nm} z^n w^n.$$

The functions $\eta_0, \eta_1, \ldots$ with

$$\eta_n(z) = \frac{e^{\frac{z^2}{2}}}{\sqrt[4]{\pi} 2^{n/2} \sqrt{n!}}, \quad n = 0, 1, \ldots \tag{13.5.9}$$

are called the Hermite functions. They belong to the Schwartz space, and form an orthonormal basis of $\mathbf{L}_2(\mathbb{R}, dx)$.

The map which to η_n associates the function $\frac{z^n}{\sqrt{n!}}$ extends to a unitary operator from $\mathbf{L}_2(\mathbb{R}, dx)$ onto the Fock space. It is called the Bargmann transform.

Question 13.5.3. *The Bargmann transform can be written as*

$$F(z) = \frac{1}{\sqrt[4]{\pi}} \int_{\mathbb{R}} e^{\{-\frac{1}{2}(z^2+u^2)+\sqrt{2}zu\}} f(u)\,du.$$

We conclude by mentioning that

$$\widehat{\eta_n} = (-i)^n \eta_n, \quad n = 0, 1, \ldots.$$

13.6 Solutions

Solution of Exercise 13.1.1. Since $|k| > 1$ the series with general term z/k^n is absolutely convergent for any $z \in \mathbb{C}$. Thus, by Theorem 3.7.1, the infinite product converges for every z not equal to $-k^n$, $n = 1, 2, \ldots$ (and the corresponding function, extended to be 0 at these points, is entire).

To prove (b) we write

$$P(kz) = \prod_{n=1}^{\infty}\left(1 + \frac{kz}{k^n}\right) = \prod_{n=1}^{\infty}\left(1 + \frac{z}{k^{n-1}}\right)$$

$$= \prod_{n=0}^{\infty}\left(1 + \frac{z}{k^n}\right)$$

$$= (1+z)P(z).$$

We now turn to (c). From (b) we have $P(k/z) = (1 + 1/z)P(1/z)$, and replacing z by kz in the above expression,

$$P(1/z) = \left(1 + \frac{1}{kz}\right)P(1/kz) \quad \text{and hence} \quad P(1/kz) = \frac{kz}{1+kz}P(1/z).$$

Thus

$$S(kz) = P(kz)P(1/kz)(1+kz)$$

$$= (1+z)P(z)P(1/z)\frac{kz}{1+kz}(1+kz)$$

$$= (1+z)P(z)P(1/z)kz$$

$$= kzS(z).$$

(d) Using (c) we have

$$M(kz) = \frac{S(a_1 kz)\cdots S(a_n kz)}{S(b_1 kz)\cdots S(b_n kz)}$$

$$= \frac{ka_1 z S(a_1 z)\cdots ka_n z S(a_n z)}{kb_1 z S(b_1 z)\cdots kb_n z S(b_n z)}$$

$$= M(z)\frac{a_1\cdots a_n}{b_1\cdots b_n} = M(z)$$

since we assumed $a_1\cdots a_n = b_1\cdots b_n$.

(e) Let $w \in \mathbb{C}$ be such that $k = \exp w$. Since $|k| > 1$ the numbers w and $2\pi i$ are linearly independent over $\mathbb{Z}$. We cannot find m and n such that $mw + 2\pi in = 0$. Indeed, if there are such m and n, then $e^{mw} = e^{-2\pi in} = 1$ and so $k^m = 1$ contradicting the assumption $|k| > 1$. Moreover, we have

$$G(z + mw + n2\pi i) = M(e^{z+mw+n2\pi i}) = M(e^{z+mw}) = M(k^m e^z) = M(e^z) = G(z)$$

where we used (d) with e^z in place of z. Thus, $G(z)$ is bi-periodic since w and 2π are linearly independent over $\mathbb{Z}$. $\square$

Solution of Exercise 13.1.3. It follows from the proof of Exercise 3.6.2 that the convergence of the family of functions is uniform on compact sets, and therefore the limit is analytic. $\square$

Solution of Exercise 13.2.1. Let $L > 0$. We have, with $z = x + iy$,

$$\left|e^{i\pi n^2 \tau + 2\pi inz}\right| = e^{-\pi n^2 \operatorname{Im}\tau} \cdot e^{-2\pi ny} \leq e^{-\pi n^2 \operatorname{Im}\tau} \cdot e^{2\pi |n| L}$$

for $|y| \leq L$. We now show that the series converge uniformly in every band of the form $|\operatorname{Im} z| \leq L$, $L > 0$. For L fixed, there exists $n_0 \in \mathbb{N}$ such that

$$|n| \geq n_0 \longrightarrow \left|\frac{2\pi L}{n}\right| \leq \frac{\pi \operatorname{Im}\tau}{2}.$$

Thus for $|n| \geq n_0$ we have

$$\left|e^{i\pi n^2 \tau + 2\pi inz}\right| = e^{-\pi n^2 \operatorname{Im}\tau} \cdot e^{-2\pi ny} \leq e^{-\frac{n^2 \pi \operatorname{Im}\tau}{2}}.$$

Therefore the series converge uniformly on each band of the asserted form, and ϑ is an entire function of z.

(13.2.1) follows from the periodicity of the exponentials $e^{2\pi inz}$. Equality (13.2.2) is proved as follows:

$$
\begin{aligned}
\vartheta(z + \tau, \tau) &= \sum_{n \in \mathbb{Z}} e^{i\pi n^2 \tau + 2i\pi n(z+\tau)} \\
&= \sum_{n \in \mathbb{Z}} e^{i\pi\tau(n^2 + 2n) + 2\pi inz}, \quad \text{and, completing the square,} \\
&= \sum_{n \in \mathbb{Z}} e^{i\pi\tau(n+1)^2 + 2\pi inz - i\pi\tau} \\
&= e^{-i\pi\tau - 2\pi iz} \cdot \sum_{n \in \mathbb{Z}} e^{i\pi\tau(n+1)^2 + 2\pi i(n+1)z} \\
&= e^{-i\pi\tau - 2\pi iz} \vartheta(z, \tau).
\end{aligned}
$$

We now prove (13.2.3). Using (13.2.2) with $z = \frac{1-\tau}{2}$ we obtain

$$\vartheta\left(\frac{1+\tau}{2}, \tau\right) = \vartheta\left(\frac{1-\tau}{2} + \tau, \tau\right)$$

$$= e^{-i\pi\tau - 2\pi i\frac{1-\tau}{2}} \vartheta\left(\frac{1-\tau}{2}, \tau\right)$$

$$= e^{-i\pi} \vartheta\left(\frac{1-\tau}{2}, \tau\right), \quad \text{and, using (13.2.1),}$$

$$= -\vartheta\left(\frac{1-\tau}{2} - 1, \tau\right)$$

$$= -\vartheta\left(\frac{1+\tau}{2}, \tau\right),$$

and hence the result since ϑ is an even function of z.   □

Solution of Exercise 13.2.2. We already know from Exercise 13.2.1 that ϑ vanishes at the point $\frac{1+\tau}{2}$, and hence, because of (13.2.1) and (13.2.2) at all the points

$$\frac{1+\tau}{2} + m + \tau n, \quad m, n \in \mathbb{Z}.$$

The entire function $\vartheta(z, \tau)$ may vanish *a priori* for some points on the parallelogram with nodes $0, 1, \tau$ and $1+\tau$. By making a small translation by a complex number a, we obtain a parallelogram P_a, with nodes $a, 1+a, \tau+a, 1+\tau+a$, which still contains $\frac{1+\tau}{2}$, but on which ϑ does not vanish. We have

$$\int_{P_a} \frac{\vartheta'(z, \tau)}{\vartheta(z, \tau)} dz = \int_{[a, 1+a]} \frac{\vartheta'(z, \tau)}{\vartheta(z, \tau)} dz + \int_{[1+a, 1+\tau+a]} \frac{\vartheta'(z, \tau)}{\vartheta(z, \tau)} dz$$

$$+ \int_{[1+a+\tau, a+\tau]} \frac{\vartheta'(z, \tau)}{\vartheta(z, \tau)} dz + \int_{[a+\tau, a]} \frac{\vartheta'(z, \tau)}{\vartheta(z, \tau)} dz \qquad (13.6.1)$$

since ϑ has period 1 with respect to z (see (13.2.1)), the function $\frac{\vartheta'}{\vartheta}$ is also periodic with period 1 with respect to z and we have

$$\int_{[a+\tau, a]} \frac{\vartheta'(z, \tau)}{\vartheta(z, \tau)} dz = \int_{[1+a, 1+\tau+a]} \frac{\vartheta'(z, \tau)}{\vartheta(z, \tau)} dz = -\int_{[1+a+\tau, 1+a]} \frac{\vartheta'(z, \tau)}{\vartheta(z, \tau)} dz.$$

Thus the second and fourth integrals on the right side of (13.6.1) cancel each other. We now compare the first and the third integral, taking into account (13.2.2). Using for instance the property (4.2.3) of the logarithmic derivative, (13.2.2) leads to

$$\frac{\vartheta'(z+\tau, \tau)}{\vartheta(z+\tau, \tau)} = \frac{\vartheta'(z, \tau)}{\vartheta(z, \tau)} - 2\pi i.$$

It follows that

$$\int_{[1+a+\tau,a+\tau]} \frac{\vartheta'(z,\tau)}{\vartheta(z,\tau)}dz = \int_{[1+a,a]} \frac{\vartheta'(z+\tau,\tau)}{\vartheta(z+\tau,\tau)}dz$$

$$= \int_{[1+a,a]} \left(\frac{\vartheta'(z,\tau)}{\vartheta(z,\tau)} - 2\pi i\right)dz$$

$$= -\int_{[a,1+a]} \frac{\vartheta'(z,\tau)}{\vartheta(z,\tau)}dz + 2\pi i.$$

Thus the first and the third integral in (13.6.1) sum up to $2\pi i$, and so

$$\frac{1}{2\pi i}\int_{Pa} \frac{\vartheta'(z,\tau)}{\vartheta(z,\tau)}dz = 1.$$

Since ϑ is entire, it follows from (7.3.5) that $\frac{1+\tau}{2}$ is the only zero of ϑ in P_a, and hence the result. □

Solution of Exercise 13.3.1. We define a function g in $\mathbb{C}\setminus(-\infty,0]$ by

$$g(\zeta) = f\left(\frac{\ln\rho + i\theta}{2\pi i}\right), \quad \text{with} \quad \zeta = \rho e^{i\theta}, \quad \theta \in (-\pi,\pi).$$

For $\zeta = e^{2\pi i z}$ and z in the strip $|x| < 1/2$ we have

$$g(e^{2\pi i z}) = f(z).$$

The function g is analytic in $\mathbb{C}\setminus(-\infty,0]$. Take $x < 0$ to be a point on the negative axis. We have

$$\lim_{\substack{\zeta\to x \\ \text{Im}\,\zeta>0}} g(\zeta) = f\left(\frac{\ln x + i\pi}{2\pi i}\right),$$

and

$$\lim_{\substack{\zeta\to x \\ \text{Im}\,\zeta<0}} g(\zeta) = f\left(\frac{\ln x - i\pi}{2\pi i}\right).$$

The fact that f is periodic with period 1 leads to the continuity of g on $(-\infty,0)$. Using Morera's theorem we conclude that g is analytic in $\mathbb{C}\setminus\{0\}$, and therefore has a Laurent expansion, which converges uniformly in every ring of the form $r < |\zeta| < R$ (r and R are strictly positive numbers such that $r < R$):

$$g(\zeta) = \sum_{n\in\mathbb{Z}} c_n\zeta^n.$$

Thus,

$$f(z) = g(e^{2\pi i z}) = \sum_{n\in\mathbb{Z}} c_n e^{2\pi i n z},$$

where by analytic continuation, z is arbitrary in $\mathbb{C}$, and where the convergence is uniform in every closed horizontal strip. □

For more on the subject, see for instance [42, Exercise 11.10, p. 365], [75, Exercice 34.10, p. 307], [193, pp. 106–107]. As an application of the previous exercise, prove the following result (see [193, (2.23-12) and (2.23-13), p. 108]):

$$\frac{1}{\tan \pi z} = \begin{cases} -i(1 + 2\sum_{n=1}^{\infty} e^{2\pi i n z}), & \text{Im } z > 0, \\ i(1 + 2\sum_{n=1}^{\infty} e^{-2\pi i n z}), & \text{Im } z < 0. \end{cases}$$

Solution of Exercise 13.3.2. We follow [162, pp. 3-4]. In view of Exercise 13.3.1 we look for f, not identically vanishing, and of the form

$$f(z) = \sum_{n \in \mathbb{Z}} c_n(\tau) e^{2\pi i n z}. \tag{13.6.2}$$

The condition

$$f(z + \tau) = e^{az+b} f(z)$$

leads to

$$\sum_{n \in \mathbb{Z}} c_n(\tau) e^{2\pi i n (z+\tau)} = e^{az+b} \sum_{n \in \mathbb{Z}} c_n(\tau) e^{2\pi i n z}.$$

Replacing z by $z + 1$ in this expression we obtain (since we assume $f \not\equiv 0$)

$$e^a = 1,$$

that is, $a = 2\pi i k_0$ for some $k_0 \in \mathbb{Z}$. Comparing the coefficient of $e^{2\pi i n z}$ we have

$$c_n(\tau) = c_{n-k_0}(\tau) e^{-2\pi i n \tau} e^b = c_{n-k_0}(\tau) e^{b+2\pi n \, \text{Im} \, \tau} e^{-2\pi i n \, \text{Re} \, \tau}.$$

When $k_0 > 0$, the coefficients $c_n(\tau)$ go exponentially fast in modulus to infinity, and the series (13.6.2) diverges. We leave it to the student to consider the cases $k_0 = 0$ and $k_0 < 0$. □

Solution of Exercise 13.3.3. The function q is meromorphic in the plane since it is the quotient of two entire functions. Since f has period 1, all the functions $f(z - a_n)$ and $f(z - b_n)$ have also period 1, and so has the function q. We now show, using the second equality in (13.3.1), that q has also period τ. We have

$$q(z + \tau) = \prod_{n=1}^{N} \frac{f(z + \tau - a_n)}{f(z + \tau - b_n)}$$

$$= \prod_{n=1}^{N} \frac{e^{a(z-a_n)+b} f(z - a_n)}{e^{a(z-b_n)+b} f(z - b_n)}$$

$$= \frac{e^{aNz - a(\sum_{n=1}^{N} a_n) + Nb}}{e^{aNz - a(\sum_{n=1}^{N} b_n) + Nb}} q(z)$$

$$= q(z),$$

in view of (13.3.2).   □

The student will recognize in (13.3.2) a condition similar to (13.1.1) in Exercise 13.1.1.

Solution of Exercise 13.4.1. Let $z = x + iy$. We have

$$|t^{z-1}| = |e^{\{(z-1)\ln t\}}| = e^{(x-1)\ln t} = t^{x-1}.$$

The integral $\int_0^1 t^{x-1} dt$ converges for $x > 0$, and so the integral $\int_0^1 t^{z-1} e^{-t} dt$ converges absolutely for $\operatorname{Re} z > 0$. As for the convergence at infinity of the integral (3.1.11)

$$\int_0^\infty t^{x-1} e^{-t} dt,$$

we proceed as follows (the same argument will be used later in the solution of Exercise 13.4.2): Write

$$t^{x-1} e^{-t} = e^{\{((x-1)\frac{\ln t}{t} - 1)t\}}.$$

For a given $x > 0$, there exists $M > 0$ such that

$$t \geq M \implies \left|(x-1)\frac{\ln t}{t}\right| \leq \frac{1}{2}.$$

Then,

$$(x-1)\frac{\ln t}{t} - 1 \leq \left|(x-1)\frac{\ln t}{t}\right| - 1 \leq -\frac{1}{2},$$

and we have

$$\int_M^\infty t^{x-1} e^{-t} dt \leq \int_M^\infty e^{-\frac{t}{2}} dt < \infty.$$

Finally, equation (13.4.2) is proved by integration by parts.   □

Solution of Exercise 13.4.2. We follow the method given in the hint after the exercise. By Theorem 6.2.3 each of the functions Γ_n is analytic in $\operatorname{Re} z > 1$. Furthermore, for $z \in K$ we have

$$\left|\int_n^\infty t^{z-1} e^{-t} dt\right| \leq \int_n^\infty e^{(M-1)\ln t - t} dt = \int_n^\infty e^{(\frac{(M-1)\ln t}{t} - 1)t} dt.$$

For a given M there exists n_0 such that

$$t \geq n_0 \implies 0 < \frac{(M-1)\ln t}{t} < \frac{1}{2},$$

and therefore, for $n \geq n_0$,

$$\left|\int_n^\infty t^{z-1} e^{-t} dt\right| \leq \int_n^\infty e^{-\frac{t}{2}} dt \to 0, \quad \text{as} \quad n \to \infty.$$

Similarly, still for $z = x + iy \in K$, we have

$$\left| \int_0^{1/n} t^{z-1} e^{-t} dt \right| \leq \int_0^{1/n} e^{(x-1)\ln t} dt$$

$$= \int_0^{1/n} t^{x-1} dt$$

$$= \frac{1}{xn^x} \leq \frac{1}{m \cdot n^m}.$$

It follows that, for $n \geq n_0$,

$$|\Gamma(z) - \Gamma_n(z)| \leq \int_n^\infty e^{-\frac{t}{2}} dt + \frac{1}{m \cdot n^m}$$

uniformly in K (and in fact uniformly in the band $m \leq x \leq M$), and so Γ is analytic as the uniform limit on compact sets of analytic functions.   □

Solution of Exercise 13.4.3. We follow [23, p. 119]. In view of (1.2.6), we have that, for every $t \in [0, \infty)$,

$$\lim_{n \to \infty} f_n(t) = e^{-t} t^{z-1}.$$

Moreover, in view of item (a) in Exercise 3.2.6,

$$|f_n(t)| \leq \left(1 - \frac{t}{n} \right)^n t^{x-1} \leq e^{-t} t^{x-1}.$$

The dominated convergence theorem (see Theorem 17.5.2) leads to

$$\lim_{n \to \infty} \int_0^\infty f_n(t) dt = \int_0^\infty (\lim_{n \to \infty} f_n(t)) dt$$

$$= \int_0^\infty e^{-t} t^{z-1} dt$$

$$= \Gamma(z).$$

It remains to show that

$$\int_0^n \left(1 - \frac{t}{n} \right)^n t^{z-1} dt = \frac{n! n^z}{z(z+1) \cdots (z+n)}.$$

As suggested in [23] this is done by repeated integration by parts. Indeed, we have

$$\int_0^n \left(1 - \frac{t}{n}\right)^n t^{z-1}\, dt = \frac{n}{n} \int_0^n \left(1 - \frac{t}{n}\right)^{n-1} \frac{t^z}{z}\, dt$$

$$= \frac{n(n-1)}{n^2} \int_0^n \left(1 - \frac{t}{n}\right)^{n-2} \frac{t^{z+1}}{z(z+1)}\, dt$$

$$\vdots$$

$$= \frac{n(n-1)\cdots 2}{n^{n-1}} \int_0^n \frac{t^{z+(n-1)}}{z(z+1)\cdots(z+n-1)}\, dt$$

$$= \frac{n!}{n^n} \frac{n^{z+n}}{z(z+1)\cdots(z+n-1)(z+n)}$$

$$= \frac{n!\, n^z}{z(z+1)\cdots(z+n)}. \qquad \square$$

Solution of Exercise 13.4.4. (a) The first equality follows directly from the change of variable $t = \sqrt{u}$. Indeed,

$$\int_0^\infty e^{-t^2} t^{z-1}\, dt = \int_0^\infty e^{-u} u^{\frac{z-1}{2}} \frac{du}{2\sqrt{u}} = \frac{\Gamma\left(\frac{z}{2}\right)}{2}.$$

(b) The other two integrals are computed using Cauchy's theorem as follows. Consider the function of the complex variable s defined by

$$f(s) = e^{is+(z-1)\ln s},$$

where $\ln s$ is the principal branch of the logarithm in $\mathbb{C} \setminus (-\infty, 0]$, that is

$$\ln s = \ln \rho + i\theta,$$

where $s = \rho e^{i\theta}$ with $\theta \in (-\pi, \pi)$. We consider the closed path consisting of the following four parts:

(i) The interval $[r, R]$, with $0 < r < R < \infty$.
(ii) The arc of circle C_R parametrized by

$$\gamma_R(u) = Re^{iu}, \quad u \in \left[0, \frac{\pi}{2}\right].$$

(iii) The interval $[iR, ir]$.
(iv) The arc of circle c_r parametrized by

$$\gamma_r(u) = re^{iu}, \quad u \in \left[\frac{\pi}{2}, 0\right].$$

By Cauchy's theorem, the integral of f on this closed path is equal to 0. On the other hand,

$$\int_{[r,R]} f(s)ds = \int_r^R e^{it}t^{z-1}dt \to \int_0^\infty e^{it}t^{z-1}dt,$$

as $r \to 0$ and $R \to \infty$, and, with the parametrization $\gamma(t) = it$, with $t \in [R,r]$,

$$\int_{[iR,ir]} f(s)ds = \int_R^r e^{-t+(z-1)(\ln t+i\frac{\pi}{2})}idt$$

$$= -e^{i(z-1)\frac{\pi}{2}} \int_r^R e^{-t}t^{z-1}idt$$

$$= -e^{-i\frac{\pi}{2}}e^{iz\frac{\pi}{2}} \int_r^R e^{-t}t^{z-1}idt$$

$$\to -e^{iz\frac{\pi}{2}}\Gamma(z)$$

as $r \to 0$ and $R \to \infty$. We now show that

$$\lim_{r\to 0}\int_{C_r} f(s)ds = 0 \quad \text{and} \quad \lim_{R\to\infty}\int_{C_R} f(s)ds = 0. \qquad (13.6.3)$$

The first of these limits is computed as follows:

$$\left|\int_{C_r} f(s)ds\right| = \left|-\int_0^{\pi/2} e^{ire^{iu}+(z-1)(\ln r+iu)}rie^{iu}du\right|$$

$$\le e^{|\operatorname{Im}z|\frac{\pi}{2}}\int_0^{\pi/2} e^{-r\sin u}r^x du \qquad (13.6.4)$$

$$\le \frac{\pi}{2}r^x e^{|\operatorname{Im}z|\frac{\pi}{2}}$$

$$\to 0,$$

as $r \to 0$. In the computation we have used that, with $z = x + iy$,

$$\left|e^{(z-1)(\ln r+iu)}\right| \cdot r = e^{(x-1)\ln r - yu} \cdot r \le r^x e^{|\operatorname{Im}z|\frac{\pi}{2}},$$

since $e^{-yu} \le e^{|y|u} \le e^{|\operatorname{Im}z|\frac{\pi}{2}}$. In computing the limit (13.6.4) we have used that $x > 0$. To show that the second limit goes to 0 we make use of the fact that $x < 1$. Making use of (5.9.5) and of (13.6.4) with R instead of r we have

$$\left|\int_{C_R} f(s)ds\right| \le e^{|\operatorname{Im}z|\frac{\pi}{2}} \cdot R^x \cdot \frac{\pi}{R} \longrightarrow 0,$$

as $R \to \infty$ since $x < 1$. Therefore we have

$$\int_0^\infty e^{it}t^{z-1}dt = e^{iz\frac{\pi}{2}}\Gamma(z).$$

Take first $z = x$ real. Comparing the real and imaginary parts of this equality we obtain the asserted formulas for $x > 0$. They extend to complex z with $x \in (0, 1)$ by analytic extension. □

Solution of Exercise 13.4.5.

(1) To compute the integral

$$\int_0^\infty t^{z-1} \left(\int_0^\infty f_1(u) f_2(t/u) \frac{du}{u} \right) dt$$

we make the change of variable $(u, t) \mapsto (u, uv)$. The Jacobian matrix (see (4.2.7)) is equal to

$$J(u, v) = \begin{pmatrix} 1 & 0 \\ v & u \end{pmatrix}.$$

and $\det J(u, v) = u$. Thus, by the theorem on change of variables for double integrals, we can write:

$$\int_0^\infty t^{z-1} \left(\int_0^\infty f_1(u) f_2(t/u) \frac{du}{u} \right) dt = \int_0^\infty \int_0^\infty u^{z-1} v^{z-1} u \frac{dudv}{u}$$

$$= \left(\int_0^\infty u^{z-1} f_1(u) du \right) \left(\int_0^\infty v^{z-1} f_2(v) dv \right),$$

where the various interchanges of integrals are done using the dominated convergence theorem.

(2) In the case $f_1(u) = f_2(u) = e^{-u}$ we have:

$$\int_0^\infty f_1(u) f_2(t/u) \frac{du}{u} = \int_0^\infty e^{-u - \frac{t}{u}} \frac{du}{u}$$

$$= \int_0^\infty e^{-\sqrt{t}(v + \frac{1}{v})} \frac{dv}{v} \quad \text{(with the change of variable } u = \sqrt{t}v)$$

$$= \int_{-\infty}^\infty e^{-2\sqrt{t} \cosh a} da \quad \text{(with the change of variable } v = e^a)$$

$$= 2K_0(2\sqrt{t}),$$

with

$$K_0(x) = \int_0^\infty e^{-x \cosh a} da. \tag{13.6.5}$$

□

Remark 13.6.1. The function K_0 defined in (13.6.5) is the K Bessel function of order 0. See, e.g., [50, p. 7 and p. 50]. We have

$$\int_0^\infty t^{z-1} 2K_0(2\sqrt{t}) dt = (\Gamma(z))^2. \tag{13.6.6}$$

Setting $z = n + 1$ in the previous expression gives

$$\int_0^\infty t^n 2K_0(2\sqrt{t})dt = (n!)^2. \tag{13.6.7}$$

This fact is used in [13, 16] to study (and in particular give a geometric character-ization of the elements of) the reproducing kernel Hilbert space with reproducing kernel

$$\sum_{n=0}^\infty \frac{z^n \overline{w}^n}{(n!)^2}.$$

Solution of Exercise 13.5.1.

(1) For $y = 0$, (13.5.5) is the value of the Gaussian integral, which we assume known. See the discussion after (5.2.6). The integral under consideration is an even function of y, and we take $y > 0$. We give to Γ_R the positive orientation. We then have the following parametrizations for the components of Γ_R (we do not stress the dependence on y in the notation):

$$\begin{aligned}
\gamma_{1,R}(t) &= t, & t &\in [-R, R], \\
\gamma_{2,R}(t) &= R + it, & t &\in [0, y], \\
\gamma_{3,R}(t) &= -t + iy, & t &\in [-R, R], \\
\gamma_{4,R}(t) &= -R + i(y - t), & t &\in [0, y].
\end{aligned}$$

Since e^{-z^2} is defined by a power series centered at the origin, and converging in all of $\mathbb{C}$, it has a primitive in $\mathbb{C}$ and we can write

$$\int_{\Gamma_R} e^{-z^2} dz = 0, \quad \forall R > 0,$$

that is,

$$\int_{\gamma_{1,R}} e^{-z^2} dz + \int_{\gamma_{2,R}} e^{-z^2} dz + \int_{\gamma_{3,R}} e^{-z^2} dz + \int_{\gamma_{4,R}} e^{-z^2} dz = 0, \quad \forall R > 0. \tag{13.6.8}$$

We have

$$\left| \int_{\gamma_{2,R}} e^{-z^2} dz \right| = \left| \int_0^y e^{-(R^2 + 2Rti - t^2)} i \, dt \right|$$

$$\leq \int_0^y e^{-R^2 + t^2} dt$$

$$= e^{-R^2} \int_0^y e^{t^2} dt \longrightarrow 0 \quad \text{as} \quad R \longrightarrow \infty.$$

Similarly,

$$\lim_{R \to \infty} \int_{\gamma_{4,R}} e^{-z^2} dz = 0.$$

Therefore letting $R \to \infty$ in (13.6.8) and using the value of the Gaussian integral we obtain

$$e^{-y^2} \int_{\mathbb{R}} e^{-t^2} dt = e^{-y^2} \sqrt{\pi} = \int_{\mathbb{R}} e^{-t^2} e^{-2ity} dt. \qquad (13.6.9)$$

See for instance [206, p. 43].

(2) Using the dominated convergence theorem and the power series expansion of e^{-2ity} we rewrite (13.6.9) as

$$\sqrt{\pi} \left(\sum_{u=0}^{\infty} (-1)^u \frac{y^{2u}}{u!} \right) = \sum_{n=0}^{\infty} \frac{(-2iy)^n}{n!} \left(\int_{\mathbb{R}} e^{-t^2} t^n dt \right).$$

The odd moments vanish. Setting $n = 2u$ in the equality above and comparing the coefficient of y^{2u} we obtain the even moments:

$$\sqrt{\pi} \frac{(-1)^u}{u!} = \frac{(-1)^u(-2)^{2u}}{(2u)!} \left(\int_{\mathbb{R}} e^{-t^2} t^{2u} dt \right), \quad u = 0, 1, \ldots$$

and hence

$$\int_{\mathbb{R}} e^{-t^2} t^{2u} dt = \sqrt{\pi} \frac{(2u)!}{u! 2^{2u}}. \qquad (13.6.10)$$

□

Remark 13.6.2. The right side of (13.6.10) can be rewritten as

$$\sqrt{\pi} \frac{(2u-1)!!}{2^u}$$

where $n!! = n(n-2)(n-3) \cdots$.

Part IV

Appendix

Chapter 14

Some Useful Theorems

In this chapter we collect a number of results from real analysis, which are useful to solve the exercises. The results presented are along one main theme: How to interchange two operations in analysis (for instance order of integration in a double integral, integration of a function depending on a parameter and derivation with respect to this parameter,...). Most, if not all, of the results, can be proved by elementary methods, but are also special cases of general theorems from the theory of integration (such as the dominated convergence theorem, Fubini's theorem,...). Some aspects of this theory are reviewed in Chapter 17. Finally, note that we consider complex-valued functions. The results are easily derived in the complex case from their real counterparts. In fact, they are sometimes still valid for functions and sequences with values in a Banach space or a Banach algebra, but a discussion of this latter point is far outside the framework of this book.

14.1   Differentiable functions of two real variables

We here recall the definition of a differentiable function of two real variables. The case of functions with domain and range inside Banach spaces is given in Section 16.1. See Definition 16.1.13.

Definition 14.1.1. A real-valued function $t(x, y)$ defined in a neighborhood of the point $(x_0, y_0) \in \mathbb{R}^2$ is said to be *differentiable* at (x_0, y_0) if there exist real numbers a and b such that

$$\lim_{\substack{x \to x_0, \\ y \to y_0}} \frac{t(x, y) - t(x_0, y_0) - a(x - x_0) - b(y - y_0)}{\sqrt{(x - x_0)^2 + (y - y_0)^2}} = 0. \qquad (14.1.1)$$

It is well known that a necessary (but in no way sufficient) condition for differentiability at the point (x_0, y_0) is that t has first-order partial derivatives at this point. The numbers a and b are unique and equal to

$$a = \frac{\partial t}{\partial x}(x_0, y_0) \quad \text{and} \quad b = \frac{\partial t}{\partial y}(x_0, y_0)$$

Differentiability can be written in an equivalent way as follows: The function t admits first-order partial derivatives at the point (x_0, y_0) and there exists a function $E(x, y)$ such that

$$t(x, y) = t(x_0, y_0) + (x - x_0)\frac{\partial t}{\partial x}(x_0, y_0) + (y - y_0)\frac{\partial t}{\partial y}(x_0, y_0)$$
$$+ \sqrt{(x - x_0)^2 + (y - y_0)^2}\, E(x, y) \tag{14.1.2}$$

and

$$\lim_{\substack{x \to x_0, \\ y \to y_0}} E(x, y) = 0.$$

The function $E(x, y)$ is uniquely defined, and is equal to

$$E(x, y) = \frac{t(x, y) - t(x_0, y_0) - (x - x_0)\frac{\partial t}{\partial x}(x_0, y_0) - (y - y_0)\frac{\partial t}{\partial y}(x_0, y_0)}{\sqrt{(x - x_0)^2 + (y - y_0)^2}}. \tag{14.1.3}$$

Condition (14.1.2) is often more convenient that (14.1.1) to work with.

The following classical counter-example shows that continuity of the function and existence of partial derivatives at a given point do not imply differentiability at that point.

Example 14.1.2. *Let*

$$t(x, y) = \begin{cases} \dfrac{xy}{\sqrt{x^2 + y^2}}, & \text{if } (x, y) \neq (0, 0), \\ 0, & \text{if } (x, y) = (0, 0). \end{cases}$$

Then, t is continuous at the point $(0, 0)$, but is not differentiable there.

Discussion. The continuity at the origin follows from the inequality

$$|t(x, y)| \leq \frac{\left(\dfrac{x^2 + y^2}{2}\right)}{\sqrt{x^2 + y^2}} = \frac{\sqrt{x^2 + y^2}}{2}, \quad (x, y) \neq (0, 0).$$

The partial derivatives at the origin exist and are equal to 0, as follows from

$$\frac{t(x, 0) - t(0, 0)}{x} \equiv 0 \quad \text{and} \quad \frac{t(0, y) - t(0, 0)}{y} \equiv 0.$$

On the other hand, t is not differentiable at the origin since

$$\frac{t(x,y) - t(0,0) - t_x(0,0)x - t_y(0,0)y}{\sqrt{x^2 + y^2}} = \frac{xy}{x^2 + y^2}.$$

This last expression vanishes for x or y equal to 0, and is equal to $1/2$ for $x = y$. Therefore, it has no limit as $(x,y) \to (0,0)$.

In Example 14.1.2 the partial derivatives are not continuous at the point $(0,0)$. A sufficient condition for differentiability is given in the next theorem:

Theorem 14.1.3. *Assume that the function t admits partial derivatives in a neighborhood of (x_0, y_0) and that they are continuous at the point (x_0, y_0). Then, t is differentiable at the point (x_0, y_0).*

See for instance [45, p. 67] for a discussion.

14.2   Cauchy's multiplication theorem

The following result is due to Cauchy. It is also called the *Cauchy multiplication theorem*. Equality (14.2.2) can be obtained under weaker notations; these are then results due to Mertens and Abel; see [112, p. 199]. First some notation: For a sequence $a = (a_n)_{n \in \mathbb{N}_0}$ we set

$$\|(a_n)\|_1 = \sum_{n=0}^{\infty} |a_n|.$$

Furthermore, if $b = (b_n)_{n \in \mathbb{N}_0}$ is another sequence, the convolution, or the Cauchy product, of the sequences a and b is the sequence defined by

$$(a * b)_n = \sum_{m=0}^{n} a_m b_{n-m}.$$

The convolution of two sequences has appeared a number of times in the book, in particular in the setting of discrete signals, see Section 11.4.

Theorem 14.2.1. *Let $a = (a_n)_{n \in \mathbb{N}_0}$ and $b = (b_n)_{n \in \mathbb{N}_0}$ be two sequences of numbers such that $\|a\|_1$ and $\|b\|_1$ are both finite. Then*

$$\|a * b\|_1 \leq \|a\|_1 \cdot \|b\|_1, \tag{14.2.1}$$

and

$$\sum_{n=0}^{\infty} \left(\sum_{p=0}^{n} a_p b_{n-p} \right) = \left(\sum_{n=0}^{\infty} a_n \right) \left(\sum_{n=0}^{\infty} b_n \right). \tag{14.2.2}$$

Proof. Inequality (14.2.1) follows from (14.2.2) since

$$|c_n| \le \sum_{j=0}^{n} |a_j| \cdot |b_{n-j}|,$$

and hence, for $N \in \mathbb{N}_0$,

$$\sum_{n=0}^{N} |c_n| \le \sum_{n=0}^{N} \sum_{j=0}^{n} |a_j| \cdot |b_{n-j}| = \sum_{j=0}^{N} |a_j| \left(\sum_{n=j}^{N} |b_{n-j}| \right)$$

$$\le \sum_{j=0}^{N} |a_j| \left(\sum_{n=j}^{\infty} |b_{n-j}| \right) \tag{14.2.3}$$

$$= \left(\sum_{j=0}^{N} |a_j| \right) \left(\sum_{n=0}^{\infty} |b_n| \right) \le \left(\sum_{j=0}^{\infty} |a_j| \right) \left(\sum_{n=0}^{\infty} |b_n| \right).$$

It follows in particular that the series $(c_n)_{n \in \mathbb{N}_0}$ converges absolutely. To compute its sum, we will first assume that $\sum_{n=0}^{\infty} a_n \ne 0$; by replacing a_n by

$$\frac{a_n}{\sum_{p=0}^{\infty} a_p},$$

we may consider the case where

$$\sum_{n=0}^{\infty} a_n = 1.$$

Set $c_n = \sum_{p=0}^{n} a_p b_{n-p}$. Then

$$c_0 + \cdots + c_n = \sum_{j=0}^{n} \sum_{p=0}^{j} a_p b_{j-p}$$

$$= \sum_{p=0}^{n} a_p \sum_{j=p}^{n} b_{j-p}$$

$$= \sum_{p=0}^{n} a_p \sum_{j=0}^{n-p} b_j \tag{14.2.4}$$

$$= \sum_{p=0}^{n} a_p B_{n-p}$$

$$= a_0 B_n + a_1 B_{n-1} + \cdots + a_n B_0,$$

where

$$B_p = \sum_{j=0}^{p} b_j, \quad p = 0, 1, \ldots, n.$$

We have

$$|B_p| \le \sum_{j=0}^{\infty} |b_j|, \quad \forall p \in \mathbb{N}_0.$$

Let $B = \sum_{j=0}^{\infty} b_j$, and fix a $n_0 \in \mathbb{N}$. We have, for $n \ge n_0$,

$$c_0 + \cdots + c_n - B = a_0 B_n + a_1 B_{n-1} + \cdots + a_n B_0 - \left(\sum_{m=0}^{\infty} a_m \right) B \qquad (14.2.5)$$

$$= \sum_{j=0}^{n_0} a_j (B_{n-j} - B) + \sum_{j=n_0+1}^{n} a_j B_{n-j} - B \left(\sum_{j=n_0+1}^{\infty} a_j \right).$$

Given $\epsilon > 0$ choose n_0 such that

$$\sum_{j=n_0+1}^{\infty} |a_j| < \epsilon$$

and

$$n \ge n_0 \implies |B_n - B| < \epsilon.$$

Let M be such that $|B_n| \le M$ for all $n \in \mathbb{N}_0$, and take $n \ge 2n_0$. Then $n - j \ge n_0$ in (14.2.5), and we obtain

$$|c_0 + \cdots + c_n - B| \le \epsilon \left(\sum_{j=0}^{\infty} |a_j| \right) + (M + |B|)\epsilon,$$

and hence the result.

Assume now that $\sum_{j=0}^{\infty} a_j = 0$. We want to show that

$$\sum_{j=0}^{\infty} c_j = 0.$$

We replace a_0 by $a_0 + \eta$ for some $\eta \ne 0$. We denote by c' the convolution of this modified sequence and of b. We have now

$$c_0' = (a_0 + \eta) b_0,$$
$$c_1' = (a_0 + \eta) b_1 + a_1 b_0,$$
$$\vdots$$

By (14.2.3) the series c, and hence the series c', converges absolutely. Furthermore,

$$\sum_{n=0}^{\infty} c_n' = \eta \left(\sum_{n=0}^{\infty} b_n \right) + \sum_{n=0}^{\infty} c_n.$$

From the first part of the proof we have

$$\sum_{n=0}^{\infty} c'_n = \eta B.$$

Therefore $\sum_{n=0}^{\infty} c_n = 0$, that is, (14.2.2) also holds in the present case.   □

14.3   Summable families

Let J denote some set. Recall that a family $(a_j)_{j \in J}$ of complex numbers indexed by $j \in J$ is called summable if there exists a complex number L such that for every $\epsilon > 0$ there exists a finite subset $J_0 \subset J$ with the following property: For any finite subset $J_1 \supset J_0$,

$$\left| L - \sum_{j \in J_1} a_j \right| < \epsilon.$$

In the present book, the concept of summable family is of importance in particular in the definition of the Weierstrass function $\wp$. See Exercises 3.6.2 and 13.1.3.

Typically, the set of indices is equal to $J = \mathbb{N}_0^2$. In this case we have:

Theorem 14.3.1. *Let $(a_{\ell,k})_{\ell,k \in \mathbb{N}_0^2}$ be a sequence indexed by $\mathbb{N}_0^2$, and assume that*

$$\sum_{\ell=0}^{\infty} \left(\sum_{k=0}^{\infty} |a_{\ell,k}| \right) < \infty.$$

Then

$$\sum_{k=0}^{\infty} \left(\sum_{\ell=0}^{\infty} |a_{\ell,k}| \right) < \infty,$$

and the family $(a_{\ell,k})_{(\ell,k) \in \mathbb{N}_0^2}$ is summable. Moreover, its limit can be computed using any ordering of the indices. In particular,

$$L = \sum_{\ell=0}^{\infty} \left(\sum_{k=0}^{\infty} a_{\ell,k} \right) = \sum_{k=0}^{\infty} \left(\sum_{\ell=0}^{\infty} a_{\ell,k} \right). \qquad (14.3.1)$$

This result can be used to prove that the function

$$e^z = \sum_{n=0}^{\infty} \frac{z^n}{n!}$$

satisfies (1.2.4):

$$e^{z_1 + z_2} = e^{z_1} e^{z_2}, \quad z_1, z_2 \in \mathbb{C}.$$

Indeed, define

$$a_{\ell,k} = \begin{cases} \dfrac{z_1^\ell}{\ell!} \dfrac{z_2^{k-\ell}}{(k-\ell)!}, & \text{if } k \geq \ell, \\ 0, & \text{otherwise.} \end{cases}$$

This family satisfies the condition of the theorem since

$$\sum_{\ell=0}^\infty \left\{ \sum_{k=0}^\infty |a_{\ell k}| \right\} = \sum_{\ell=0}^\infty \frac{|z_1|^\ell}{\ell!} \left\{ \sum_{k=\ell}^\infty \frac{|z_2|^{k-\ell}}{(k-\ell)!} \right\}$$

$$= \sum_{\ell=0}^\infty \frac{|z_1|^\ell}{\ell!} \left\{ \sum_{k=0}^\infty \frac{|z_2|^k}{k!} \right\}$$

$$= \sum_{\ell=0}^\infty \frac{|z_1|^\ell}{\ell!} e^{|z_2|}$$

$$= e^{|z_1|} e^{|z_2|} < \infty.$$

The same computation without the absolute values leads to

$$\sum_{\ell=0}^\infty \left\{ \sum_{k=0}^\infty a_{\ell k} \right\} = e^{z_1} e^{z_2}.$$

On the other hand,

$$\sum_{k=0}^\infty \left\{ \sum_{\ell=0}^\infty a_{\ell k} \right\} = \sum_{k=0}^\infty \left\{ \sum_{\ell=0}^k \frac{z_1^\ell}{\ell!} \frac{z_2^{k-\ell}}{(k-\ell)!} \right\}$$

$$= \sum_{k=0}^\infty \frac{1}{k!} \left\{ \sum_{\ell=0}^k z_1^\ell z_2^{k-\ell} \frac{k!}{\ell!(k-\ell)!} \right\}$$

$$= \sum_{k=0}^\infty \frac{(z_1 + z_2)^k}{k!}$$

$$= e^{z_1 + z_2},$$

and hence $e^{z_1} e^{z_2} = e^{z_1 + z_2}$.

The example (see [63, p. 97])

$$a_{\ell k} = \begin{cases} \dfrac{1}{\ell^2 - k^2}, & \text{if } \ell \neq k, \\ 0 & \text{if } \ell = k, \end{cases}$$

where ℓ and k belong to $\mathbb{N}_0$, illustrates what can happen when less stringent hypotheses are set on the family. We check that

$$\sum_{\ell=0}^\infty \left(\sum_{k=0}^\infty a_{\ell k} \right) = -\sum_{k=0}^\infty \left(\sum_{\ell=0}^\infty a_{\ell k} \right) \neq 0.$$

As proposed in [63, p. 97], we first show that

$$\sum_{k=1}^{\infty} a_{\ell k} = \frac{-3}{4\ell^2} \tag{14.3.2}$$

for $\ell \geq 1$. Indeed, writing

$$\frac{1}{\ell^2 - k^2} = \frac{1}{2\ell}\left\{\frac{1}{\ell+k} + \frac{1}{\ell-k}\right\},$$

for any integers N, ℓ such that $N > \ell$, we have[8]

$$\sum_{k=1}^{N} a_{\ell k} = \frac{1}{2\ell}\left\{-\frac{1}{\ell} - \frac{1}{2\ell} + \sum_{k=N-\ell+1}^{N+\ell} \frac{1}{k}\right\}. \tag{14.3.3}$$

We first consider the case $N = \ell + 1$; (14.3.3) becomes

$$\sum_{k=1}^{\ell+1} a_{\ell k} = \frac{1}{2\ell}\left\{-\frac{1}{\ell} - \frac{1}{2\ell} + \sum_{k=2}^{2\ell+1} \frac{1}{k}\right\}. \tag{14.3.4}$$

To prove (14.3.4) we write:

$$\sum_{k=1}^{\ell+1} a_{\ell k} = \sum_{k=1}^{\ell+1} \frac{1}{2\ell}\left(\frac{1}{\ell-k} + \frac{1}{\ell+k}\right)$$

$$= \frac{1}{2\ell}\left\{\frac{1}{\ell} - \frac{1}{\ell} + \frac{1}{\ell+1} + \frac{1}{\ell-1} + \frac{1}{\ell+2} + \frac{1}{\ell-2} + \cdots + \right.$$

$$+ \cdots + \frac{1}{2\ell-3} + \frac{1}{3} + \frac{1}{2\ell-2} + \frac{1}{2} +$$

$$\left. + \frac{1}{2\ell-1} + \frac{1}{1} + \frac{1}{2\ell} - \frac{1}{2\ell} + \frac{1}{2\ell+1} + \frac{1}{-1}\right\}$$

$$= \frac{1}{2\ell}\left\{-\frac{1}{\ell} - \frac{1}{2\ell} + \sum_{k=2}^{2\ell+1} \frac{1}{k}\right\},$$

where the canceling terms $\frac{1}{2\ell} - \frac{1}{2\ell}$ and $\frac{1}{\ell} - \frac{1}{\ell}$ are added since $a_{\ell\ell} = 0$ and since the sum begins at $k = 1$. Note also that the terms $\frac{1}{1}$ and $\frac{1}{-1}$ cancel each other.

---

[8]For instance, for $\ell = 5$ and $N = 8$ we have

$$\sum_{k=1}^{8} a_{5,k} = \frac{1}{10}\left\{\frac{1}{4} + \frac{1}{6} + \frac{1}{3} + \frac{1}{7} + \frac{1}{2} + \frac{1}{8}\right.$$

$$\left. +1 + \frac{1}{9} - 1 + \frac{1}{11} - \frac{1}{2} + \frac{1}{12} - \frac{1}{3} + \frac{1}{13}\right\},$$

and the sum in brackets is equal to: $-1/5 - 1/10 + \sum_{k=4}^{13} 1/k$.

For a given ℓ we now prove (14.3.3) for all $N > \ell$ by induction as follows. Assuming the formula true at rank N we have at rank $N+1$:

$$\frac{1}{2\ell}\left\{-\frac{1}{\ell}-\frac{1}{2\ell}+\sum_{k=N+1-\ell+1}^{N+1+\ell}\frac{1}{k}\right\}=\frac{1}{2\ell}\left\{-\frac{1}{\ell}-\frac{1}{2\ell}+\sum_{k=N-\ell+2}^{N+\ell+1}\frac{1}{k}\right\}$$

$$=\frac{1}{2\ell}\left\{-\frac{1}{\ell}-\frac{1}{2\ell}+\sum_{k=N-\ell+1}^{N+\ell}\frac{1}{k}\right\}$$

$$+\frac{1}{2\ell}\left\{-\frac{1}{N+1-\ell}+\frac{1}{N+1+\ell}\right\}.$$

Using the induction hypothesis at rank N this last sum is equal to:

$$\sum_{k=1}^{N}a_{\ell k}+\frac{1}{\ell^2-(N+1)^2}=\sum_{k=1}^{N+1}a_{\ell k},$$

which proves the induction hypothesis at rank $N+1$.

Letting $N \to \infty$ we obtain (14.3.2). Thus,

$$\sum_{\ell=0}^{\infty}\left(\sum_{k=0}^{\infty}a_{\ell k}\right)=\sum_{k=0}^{\infty}a_{0k}+\sum_{\ell=1}^{\infty}\left\{a_{\ell 0}+\sum_{k=1}^{\infty}a_{\ell k}\right\}$$

$$=-\frac{\pi^2}{6}+\sum_{\ell=1}^{\infty}\left(\frac{1}{\ell^2}-\frac{3}{4\ell^2}\right)$$

$$=-\frac{\pi^2}{6}+\frac{1}{4}\frac{\pi^2}{6}<0.$$

The result follows since $a_{k\ell}=-a_{\ell k}$.

Other counterexamples may be found in [115, pp. 124–127]. For instance (see [115, 7.25, p. 125]), the family

$$a_{\ell k}=\begin{cases}1, & \text{if } \ell=k,\\ -\dfrac{1}{2^{k-\ell}}, & \text{if } \ell<k,\\ 0, & \text{if } \ell>k,\end{cases}$$

where $\ell,p \in \mathbb{N}_0$, is such that the series

$$\sum_{k=0}^{\infty}a_{\ell k}$$

converges for every $\ell \in \mathbb{N}_0$, and the series

$$\sum_{\ell=0}^{\infty}a_{\ell k}$$

converges for every $k \in \mathbb{N}_0$. Furthermore, both the series

$$\sum_{\ell=0}^{\infty}\sum_{k=0}^{\infty} a_{\ell k} \quad \text{and} \quad \sum_{k=0}^{\infty}\sum_{\ell=0}^{\infty} a_{\ell k}$$

converge, but we have:

$$\sum_{\ell=0}^{\infty}\sum_{k=0}^{\infty} a_{\ell k} = 2 \quad \text{and} \quad \sum_{k=0}^{\infty}\sum_{\ell=0}^{\infty} a_{\ell k} = 0.$$

14.4   Weierstrass' theorem

In this section we present theorems on interchanging limit and integral. The first result is quite easy to prove. The second, albeit a particular case of the first, seems more useful and is called the *Weierstrass theorem*.

Theorem 14.4.1. *Let* $(f_n)_{n \in \mathbb{N}_0}$ *be a sequence of functions continuous on a finite closed interval* $[a, b]$ *and converging uniformly to a function* f. *Then*

$$\lim_{n \to \infty} \int_a^b f_n(x)dx = \int_a^b (\lim_{n \to \infty} f_n(x))dx.$$

We note that, in view of the uniform convergence, $\lim_{n \to \infty} f_n$ is continuous on $[a, b]$ and so the integral on the right side exists. Since a uniform limit of Riemann integrable functions is still Riemann integrable, continuity may be weakened to Riemann integrability in the above theorem. A counterpart of Theorem 14.4.1 without uniform convergence is not possible since the pointwise limit of continuous functions need not be Riemann integrable (but see Theorem 17.1.3 below and the related discussion there). One has then to resort to Lebesgue integration. See Chapter 17. See also the discussion after the following theorem.

Theorem 14.4.2. *Let* $[a, b] \subset \mathbb{R}$, *and let* $(f_n)_{n \in \mathbb{N}_0}$ *be a sequence of continuous functions from* $[a, b]$ *into* $\mathbb{C}$. *Assume that there exists a sequence of numbers* $(M_n)_{n \in \mathbb{N}_0}$ *such that*

$$\max_{[a,b]} |f_n(x)| \le M_n, \quad and \quad \sum_{n=0}^{\infty} M_n < \infty. \tag{14.4.1}$$

Then

$$\int_a^b (\sum_{n=0}^{\infty} f_n(x))dx = \sum_{n=0}^{\infty} \int_a^b f_n(x)dx. \tag{14.4.2}$$

The hypothesis insures that the function

$$x \mapsto \sum_{n=0}^{\infty} f_n(x) \tag{14.4.3}$$

is continuous, as a uniform limit of functions continuous on the interval $[a, b]$. In particular the integral on the left-hand side of (14.4.2) makes sense. As in Theorem 14.4.1 the functions f_n may be assumed Riemann integrable. The function $x \mapsto \sum_{n=0}^{\infty} f_n(x)$ is Riemann integrable, and Theorem 14.4.2 still holds. The function (14.4.3) is in fact Lebesgue integrable. If one leaves the realm of the Riemann integral and goes to the setting of measurable functions and of the Lebesgue integral, much more general theorems hold; in particular Theorems 14.4.1 and 14.4.2 are special cases of the Lebesgue dominated convergence theorem; see Theorems 17.5.2 and 17.5.4 respectively.

14.5   Weak forms of Fubini's theorem

Fubini's theorem, that is, interchanging order of integration in double integrals, appears in particular when one proves analyticity using Morera's theorem. The general result involves measure theory. We here give two versions of this theorem, both set in the framework of continuous (rather than measurable) functions.

Theorem 14.5.1. *Let $f(t, s)$ be a complex-valued function continuous for $t, s \in [a, b] \times [c, d]$. Then*

$$\int_a^b \left(\int_c^d f(t, s) ds \right) dt = \int_c^d \left(\int_a^b f(t, s) dt \right) ds.$$

Either integral then coincides with the double integral

$$\iint_{[a,b] \times [c,d]} f(t, s) dt ds,$$

see any course on advanced calculus for a definition of the latter.

As a corollary we have the following result, which is used in particular in the proof of Theorem 6.2.3.

Theorem 14.5.2. *Let Ω be an open connected set, and let $F(z, s)$ be a function continuous in $(z, s) \in \Omega \times [c, d]$. Let $\gamma : [a, b] \mapsto \Omega$ be a path in Ω. Then*

$$\int_\gamma \left(\int_c^d F(z, s) ds \right) dz = \int_c^d \left(\int_\gamma F(z, s) dz \right) ds. \qquad (14.5.1)$$

Theorem 14.5.1 cannot be used to study for instance the Gamma function (3.1.11). For such cases, we will need the following result (see also the Majorant criterion in [188, p. 48], which can be easily obtained from Theorem 14.5.3)):

Theorem 14.5.3. *Let $f(t,s)$ be a continuous function on $\mathbb{R} \times \mathbb{R}$, and assume that*

$$\int_{\mathbb{R}} \left(\int_{\mathbb{R}} |f(t,s)| ds \right) dt < \infty. \tag{14.5.2}$$

Then,

$$\int_{\mathbb{R}} \left(\int_{\mathbb{R}} |f(t,s)| dt \right) ds < \infty,$$

and the double integral $\iint_{\mathbb{R}^2} f(t,s) dt ds$ converges and can be computed as

$$\iint_{\mathbb{R}^2} f(t,s) dt ds = \int_{\mathbb{R}} \left(\int_{\mathbb{R}} f(t,s) ds \right) dt = \int_{\mathbb{R}} \left(\int_{\mathbb{R}} f(t,s) dt \right) ds.$$

14.6   Interchanging integration and derivation

Interchanging integration and derivation occurs in particular when studying functions defined by integrals. We mention two useful results.

Theorem 14.6.1. *Let $f(t,s)$ be a continuous function on $[a,b] \times (c,d)$ and assume that $\frac{\partial f}{\partial s}$ exists and is continuous on $[a,b] \times (c,d)$. Then, the function $g(s) = \int_a^b f(t,s) dt$ is differentiable with respect to s and*

$$g'(s) = \int_a^b \frac{\partial f}{\partial s}(t,s) dt.$$

See for instance [63, (8.22.2) p. 179].

The case where $[a,b]$ in the previous Theorem is replaced by $[0,\infty)$ (or more generally, by a non compact interval) is more involved. A proof of the following theorem uses the dominated convergence theorem (see Theorem 17.5.2 for the latter).

Theorem 14.6.2. *Let $f(t,s)$ be a continuous function on $[a,\infty) \times (c,d)$ and assume that $\frac{\partial f}{\partial s}$ exists and is continuous on $[0,\infty) \times (c,d)$. Let $s_0 \in (c,d)$ and let $(h_n)_{n \in \mathbb{N}}$ be a sequence of numbers with limit 0 and such that $s_0 + h_n \in (c,d)$ for all $n \in \mathbb{N}$. Assume that there is a function $g(t)$ such that*

$$\left| \frac{f(t,s_0 + h_n) - f(t,s_0)}{h_n} \right| \le g(t), \quad n = 1,2,\ldots, \quad and \quad \int_0^\infty g(t) dt < \infty.$$

Then, the function $g(s) = \int_0^\infty f(t,s) dt$ is differentiable with respect to s at the point s_0 and

$$g'(s_0) = \int_0^\infty \frac{\partial f}{\partial s}(t,s_0) dt.$$

14.7 Interchanging sum or products and limit

The following result, due to Tannery (see [120, Appendix] for a discussion and applications), is useful for instance to prove that

$$\lim_{N\to\infty} \left(1 + \frac{z}{N}\right)^N = e^z, \quad z \in \mathbb{C}. \tag{14.7.1}$$

It is also used in Exercise 3.7.16.

Theorem 14.7.1. *Let $(c_{nN})_{n,N=1,\ldots}$ be a doubly indexed sequence of complex numbers with the following properties:*

(1) *for every $N \in \mathbb{N}$, the limit*

$$\lim_{N\to\infty} c_{nN} \stackrel{\text{def.}}{=} c_n$$

     exists, and

(2) *there is a sequence $(d_n)_{n=1,\ldots}$ of positive numbers such that*

$$|c_{nN}| \leq d_n \quad \text{and} \quad \sum_{n=1}^{\infty} d_n < \infty.$$

Then, it holds that

$$\lim_{N\to\infty} \sum_{n=1}^{\infty} c_{nN} = \sum_{n=1}^{\infty} c_n.$$

A direct proof of this fact is done as follows. Fix $\epsilon > 0$. We first note that the series $\sum_{n=1}^{\infty} c_n$ is absolutely convergent since

$$\lim_{N\to\infty} |c_{nN}| = |\lim_{N\to\infty} c_{nN}| = |c_n| \leq d_n, \quad n = 1, 2, \ldots.$$

Since $|c_{nN}| \leq d_n$, there exists n_0 such that, for all $N \in \mathbb{N}$,

$$n \geq n_0 \implies \sum_{n=n_0}^{\infty} |c_{nN}| \leq \epsilon.$$

The important point is that n_0 does not depend on N. Since $\sum_{n=1}^{\infty} |c_n|$ is absolutely convergent, we can suppose that n_0 is also chosen such that

$$\sum_{n=n_0+1}^{\infty} |c_n| \leq \epsilon.$$

We now write

$$\sum_{n=1}^{\infty} c_{nN} - \sum_{n=1}^{\infty} c_n = \sum_{n=1}^{n_0}(c_{nN} - c_n) + \sum_{n=n_0+1}^{\infty} c_{nN} - \sum_{n=n_0+1}^{\infty} c_n,$$

and choose N_0 such that, for all $n \in \{1, \ldots, n_0\}$,

$$N \geq N_0 \Longrightarrow |c_{nN} - c_n| \leq \epsilon.$$

Thus for $N \geq N_0$, we have

$$\left| \sum_{n=1}^{\infty} c_{nN} - \sum_{n=1}^{\infty} c_n \right| \leq (n_0 + 2)\epsilon,$$

and this concludes the proof. For a related discussion, see for instance [112, pp. 198–199] and [53, pp. 207–209]. We can also view this theorem as a consequence of the dominated convergence theorem (see Theorem 17.5.2), with the measure

$$\mu\{n\} = 1, \quad n = 1, 2, \ldots.$$

In the case of (14.7.1) we have (see [112, p. 200])

$$c_{0N} = 1 \quad \text{and} \quad c_{nN} = z^n \frac{\prod_{\ell=1}^{n-1}(1 - \frac{\ell}{N})}{n!}, \quad n = 1, 2, \ldots,$$

(note that $c_{nN} = 0$ for $n \geq N$), and the conditions of the theorem are readily seen to hold, with

$$c_n = \frac{z^n}{n!} \quad \text{and} \quad d_n = \frac{(2|z|)^n}{n!}.$$

We denote by $\ln(1 + z)$ the function analytic in the open unit disk which takes value 1 at the origin and such that

$$\exp(\ln(1 + z)) = 1 + z, \quad z \in \mathbb{D}.$$

See Exercises 4.4.13 and 6.3.1 for the latter. Using the bounds

$$\frac{|z|}{2} \leq |\ln(1 + z)| \leq \frac{3|z|}{2}, \quad \text{for} \quad |z| \leq 1/2,$$

(see [5, p. 192] and Exercise 4.4.13), we deduce easily the following result from Theorem 14.7.1.

Theorem 14.7.2. *Let $(a_{nN})_{n, N=1, \ldots}$ be a doubly indexed sequence of complex numbers, with the following properties: For every $N \in \mathbb{N}$ it holds that:*

(1)
$$\sum_{n=1}^{\infty} |a_{nN}| < +\infty,$$

and

(2) *the limit*

$$\lim_{N \to \infty} a_{nN} \stackrel{\text{def.}}{=} a_n$$

exists, and

(3) *there exists a sequence of positive numbers d_n such that*

$$|a_{nN}| \leq d_n \quad \text{and} \quad \sum_{n=1}^{\infty} d_n < \infty.$$

Then, the product $\prod_{n=1}^{\infty}(1+a_n)$ converges absolutely, and we have

$$\lim_{N\to\infty} \prod_{n=1}^{\infty}(1+a_{nN}) = \prod_{n=1}^{\infty}(1+a_n).$$

Proof. Using the estimates (4.4.18), we see that one of the series

$$\sum_{n=1}^{\infty} \ln(1+a_{nN}) \quad \text{and} \quad \sum_{n=1}^{\infty} a_{nN},$$

converges absolutely if and only if the other converges absolutely, and similarly for the series

$$\sum_{n=1}^{\infty} \ln(1+a_n) \quad \text{and} \quad \sum_{n=1}^{\infty} a_n.$$

Using Theorem 14.7.1 we see that

$$\lim_{N\to\infty} \sum_{n=1}^{\infty} \ln(1+a_{nN}) = \sum_{n=1}^{\infty} \ln(1+a_n).$$

The result follows since

$$\frac{\prod_{n=1}^{\infty}(1+a_{nN})}{\prod_{n=1}^{\infty}(1+a_n)} = \exp\left\{ \sum_{n=1}^{\infty} \ln(1+a_{nN}) - \sum_{n=1}^{\infty} \ln(1+a_n) \right\}. \qquad \square$$

Chapter 15

Some Topology

Topology intervenes in complex variables at various levels. First of all analytic functions are defined in open sets. Connectedness plays a key role in the proof of the uniqueness theorem for analytic functions. The space of functions analytic in an open set is endowed with the topology of uniform convergence on compact sets. This makes this set a metrizable space, and its underlying structure stresses the role of compactness, and plays a key role in the proof of Riemann's theorem on conformal equivalence of open simply-connected sets (different from $\mathbb{C}$ itself) with the open unit disk

15.1 Point topology

Definition 15.1.1. Let E be a non-empty set. A family $\mathcal{O} \subset \mathcal{P}(E)$ is called a *topology* if:

1. $\emptyset$ and E belong to $\mathcal{O}$.
2. $\mathcal{O}$ is closed under finite intersection.
3. $\mathcal{O}$ is closed under arbitrary union.

A space E endowed with a topology is called a *topological space*, and the elements of $\mathcal{O}$ are called *open*. A set $F \subset E$ is called *closed* if $E \setminus F$ is open. A subset $A \subset E$ is a topological space when endowed with its induced topology

$$\mathcal{O}_A = \{A \cap O \, ; \, O \in \mathcal{O}\}.$$

Every subfamily $\mathcal{M} \subset \mathcal{P}(E)$ generates a topology $\mathcal{O}(\mathcal{M})$ with the property that it is contained in any other topology containing $\mathcal{M}$. More precisely,

$$\mathcal{O}(\mathcal{M}) = \bigcap_{\substack{\mathcal{M} \subset \mathcal{O}, \\ \mathcal{O} \text{ topology}}} \mathcal{O}.$$

Of special importance are the notions of Hausdorff, compact, connected, and of arc-connected spaces and sets. A topological space $(E, \mathcal{O})$ is called Hausdorff if for every pair of different points $x_1, x_2 \in E$ there exist non-intersecting open sets O_1 and O_2 such that $x_1 \in O_1$ and $x_2 \in O_2$. It is called *compact* if it is Hausdorff and if every open covering of E admits a finite sub-covering. It is called *sequentially compact* if every infinite sequence of points admits a convergent subsequence. In general, sequential compactness is not related to compactness (for a counterexample, see [205, pp. 125–126]), but the two notions are equivalent in the case of metric spaces; see Definition 15.1.4 below. The topological space $(E, \mathcal{O})$ is called *connected* if it cannot be written in a non-trivial way as a union of two open sets. A subset A of a topological space is said to be connected if it is connected for its induced topology $\mathcal{O}_A$.

A function f from a topological space $(E_1, \mathcal{O}_1)$ into a topological space $(E_2, \mathcal{O}_2)$ is *continuous* if

$$\forall O_2 \subset \mathcal{O}_2, \qquad f^{-1}(O_2) \in \mathcal{O}_1.$$

An important fact used in Section 15.6 is:

Theorem 15.1.2. *The continuous image of a connected set is connected.*

The following application of the previous theorem is taken, together with its solution, from [38, Exercise 18.9, p. 127].

Exercise 15.1.3. *Describe the connected subsets of $\mathbb{R}$ and show that there is no continuous function f such that*

$$f(\mathbb{Q}) \subset \mathbb{R} \setminus \mathbb{Q},$$
$$f(\mathbb{R} \setminus \mathbb{Q}) \subset \mathbb{Q}. \tag{15.1.1}$$

Hint. The result is striking for $\mathbb{Q}$, but the rational numbers could be replaced by any countable subset of the real numbers, and the proof would be the same.

We also recall:

Definition 15.1.4. A *metric* on a space E is a map

$$d : E \times E \longrightarrow [0, \infty)$$

with the following properties: For all $x, y, z \in E$ it holds that:

(a) $d(x, y) = 0 \iff x = y$.

(b) $d(x, y) = d(y, x)$.

(c) $d(x, y) \leq d(x, z) + d(z, y)$.

A pair (E, d) where d is a metric is called a metric space.

Inequality (c) is called the triangle inequality.

Exercise 15.1.5. *Let* $\| \cdot \|_{\mathbb{R}^3}$ *denote the Euclidean norm in* $\mathbb{R}^3$,

$$\|(u, v, w)\|_{\mathbb{R}^3} = \sqrt{u^2 + v^2 + w^2}, \quad (u, v, w) \in \mathbb{R}^3.$$

Show that the function

$$d(z, w) = \|\varphi^{-1}(z) - \varphi^{-1}(w)\|_{\mathbb{R}^3} \tag{15.1.2}$$

$$= \begin{cases} \dfrac{2|z-w|}{\sqrt{1+|z|^2}\sqrt{1+|w|^2}}, & z, w \in \mathbb{C}, \\[2mm] \dfrac{2|w|}{\sqrt{1+|w|^2}}, & z = \infty \text{ and } w \in \mathbb{C}, \\[2mm] 0, & z = w = \infty, \end{cases} \tag{15.1.3}$$

where φ *is defined by (2.1.2) and (2.1.3), is a metric on the Riemann sphere.*

Hilbert and Banach spaces are important cases of metric spaces. The space of functions analytic in an open set Ω with the topology of uniform convergence on compact sets is a metric space which is not a Banach space; see Section 10.1.

A metric defines a topology in a natural way, namely the topology generated by the sets

$$B(x, \rho) = \{y \in E, \ d(x, y) < \rho\}.$$

Exercise 15.1.6. *Let* $\mathbb{R}$ *with the (usual) topology defined by the absolute value. Show that open sets are countable unions of disjoint intervals.*

Exercise 15.1.7 (see [190, p. 7]). *Let* $[-\infty, +\infty]$ *denote the real line to which have been added two points denoted by* $\pm\infty$ *(and which do not belong to* $\mathbb{R}$). *Define a set* O *to be open if it is empty or if it is a (not necessarily disjoint) union of sets of the following forms:*

(i) O *open in* $\mathbb{R}$.

(ii) $\{-\infty\} \cup (-\infty, a)$ *where* $a \in \mathbb{R}$.

(iii) $\{+\infty\} \cup (b, \infty)$.

Show that this defines a topology. Is $[-\infty, \infty]$ *Hausdorff with this topology?*

The absolute value defines a metric on $\mathbb{R}$, and hence a topology. The interval $[0, 1]$ is endowed with the induced topology. The space E is called *arc-connected* if for every two points a and b in E there is a continuous map

$$\gamma : [0, 1] \longrightarrow E$$

such that $\gamma(0) = a$ and $\gamma(1) = b$.

Exercise 15.1.8. *The set* $\mathbb{C} \setminus \{0\}$ *is arc-connected.*

15.2   Compact spaces

We first recall the following fact:

Theorem 15.2.1. *Let f be a continuous map from a compact space E into a Hausdorff space F. Then, $f(E)$ is compact in the induced topology.*

Exercise 15.2.2. *Show that the space $[-\infty, \infty]$ defined in Exercise 15.1.7 is compact with the topology defined there.*

The following exercise deals with the Cantor set. See [23, p. 79].

Exercise 15.2.3. *Let*

$$U_0 = \left(\frac{1}{3}, \frac{2}{3} \right),$$

and for $n \geq 1$,

$$U_n = \bigcup_{(\epsilon_1,\ldots,\epsilon_n)\in\{0,2\}^n} \left(3^{-n-1} + \sum_{k=1}^{n} \frac{\epsilon_k}{3^k}, 2 \cdot 3^{-n-1} + \sum_{k=1}^{n} \frac{\epsilon_k}{3^k} \right).$$

Let

$$C = [0,1] \setminus \bigcup_{n \in \mathbb{N}_0} U_n.$$

Show that C is compact, not countable, but that the total length of the U_n is equal to 1.

We have recalled in Section 10.1 that the space $A(\Omega)$ of functions analytic in an open connected set Ω, endowed with uniform convergence on compact subsets, is a metric space. A family $\mathcal{F}$ in $A(\Omega)$ is called *normal* if from every infinite sequence one can extract a convergence subsequence. One does not require that the limit belongs to $\mathcal{F}$. The limit always belongs to $\mathcal{F}$ if and only if $\mathcal{F}$ is sequentially compact.

15.3   Compactification

We have seen that $[-\infty, \infty]$ endowed with the topology defined in Exercise 15.1.7 (which contains the natural topology of $\mathbb{R}$) is compact. When we consider the Riemann sphere, we get a compactification of the complex plane by adding *one* point. More generally, given a locally compact space $(X, \mathcal{T})$, add an element ω not belonging to X, and let $\widehat{X} = X \cup \{\omega\}$. In the application in this book, $X = \mathbb{C}$ and $\mathcal{T}$ is the usual topology of the plane, and ω, the point at infinity, is denoted by ∞.

Let $\widehat{\mathcal{T}} \subset \mathcal{P}(\widehat{X})$ be defined as follows:

$$A \in \widehat{\mathcal{T}} \iff \begin{cases} A \in \mathcal{T}, & \text{or,} \\ A = \widehat{X} \setminus K, & \text{where } K \subset X \text{ is compact, or,} \\ A = \widehat{X}. \end{cases}$$

Theorem 15.3.1. *The family of sets $\widehat{\mathcal{T}}$ defines a Hausdorff topology on $\widehat{X}$, and $\widehat{X}$ endowed with this topology is compact. Furthermore, the identity is bi-continuous from $(X, \mathcal{T})$ onto $X \subset \widehat{X}$.*

The reader can find for instance in [39, Corollaire, p. IX.21] a necessary and sufficient condition for $\widehat{X}$ to be metrizable: $\widehat{X}$ is metrizable if and only if X is metrizable and countable at infinity.

For this section, see also [2, Exercice 17, p. 51].

15.4 Plane topology

We now specialize some of the previous definitions in the setting of the plane.

Definition 15.4.1. Let $z_0 \in \mathbb{C}$ and $r > 0$. The set

$$B(z_0, r) = \{z \in \mathbb{C} \ : \ |z - z_0| < r\} \tag{15.4.1}$$

is called the *open disk* with *center* z_0 and *radius* r.

The set

$$\overline{B(z_0, r)} = \{z \in \mathbb{C} \ : \ |z - z_0| \le r\} \tag{15.4.2}$$

is called the closed disk with center z_0 and radius r.

Definition 15.4.2. A subset Ω of the complex plane is said to be *open* if the following condition holds:

$$\forall z \in \Omega, \quad \exists r > 0 \text{ such that } B(z, r) \subset \Omega.$$

In particular, every open disk is open in the sense of Definition 15.4.2.

The set

$$\mathbb{C} \setminus (-\infty, 0]$$

is open, but the set

$$\mathbb{C} \setminus (-\infty, 0) \tag{15.4.3}$$

is not. A set will be closed if its complement is open. Another and more direct characterization can be given in terms of limits. Of course there are sets which are neither open nor closed, for instance $\mathbb{C} \setminus [-1, 0)$.

Definition 15.4.3. A subset V of the complex plane is called a *neighborhood* of the point $z_0 \in \mathbb{C}$ if there is $r > 0$ such that

$$B(z_0, r) \subset V.$$

Neighborhoods of infinity are defined as follows:

Definition 15.4.4. A subset of the complex plane is called a *neighborhood of infinity* if it contains a set of the form

$$\{z;\ |z| > R\}$$

for some $R > 0$.

A set $\Omega \in \mathbb{C}$ will be said to be *bounded* if

$$\Omega \subset B(0, R)$$

for some $R > 0$. Neighborhoods of infinity are example of sets which are *not* bounded.

The distance of a point z to a set A is defined to be

$$d(z, A) = \inf_{a \in A} |z - a|,$$

and the closure of the set A is

$$\overline{A} = \{x \in \mathbb{C},\ \text{such that } d(x, A) = 0\}.$$

The boundary of the set A is

$$\partial A = \overline{A} \cap (\overline{\mathbb{C} \setminus A}).$$

In $\mathbb{C}$, sets which are both closed and bounded have a special property: They are compact.

The notion of a simply-connected set plays an important role in solving global problems in complex variables. Star-shaped sets and convex sets form two very important examples of simply-connected sets, and will be sufficient for most applications.

Definition 15.4.5. A subset Ω of the complex plane is called *star-shaped* if there is a point $z_0 \in \Omega$ with the property

$$\forall z \in \Omega, \quad [z_0, z] \subset \Omega.$$

For z_1 and z_2 in $\mathbb{R}^2$ we denote by $[z_1, z_2]$ the interval with endpoints z_1 and z_2:

$$[z_1, z_2] = \{z_1 + t(z_2 - z_1)\,;\, t \in [0, 1]\}.$$

Definition 15.4.6. A subset Ω of the complex plane is called *convex* if

$$\forall z, w \in \Omega, \quad [z, w] \subset \Omega.$$

A convex set is in particular star-shaped. One can take any point $z_0 \in \Omega$ in Definition 15.4.5. This example also shows that z_0 is in general not unique. The set $\Omega = \mathbb{C} \setminus (-\infty, 0]$ is an important example of a star-shaped set which is not convex. In this case one can take $z_0 = x_0$ for any $x_0 > 0$.

Another important result is Jordan's curve theorem. Take a circle in the plane. It is geometrically clear that it divides the plane into two open parts, one bounded (the unit disk), and one unbounded. Jordan's curve theorem asserts that this result is true for any closed simple curve. See for instance [95, p. 68], [118, Chapter 3].

Lemma 15.4.7. *An arc-connected subset of $\mathbb{R}^2$ is connected. For open sets of $\mathbb{R}^2$ the two notions are equivalent.*

For a counterexample when the set is not open, take for instance the graph of the function $\sin(1/x)$ together with the closed interval with end points $(0, -1)$ and $(0, 1)$.

15.5 Some points of algebraic topology

Algebraic topology pops up in the theory of one complex variable as soon as one wants to make precise how two contours are close to each other. Two fundamental notions are used, homotopy and homology.

To characterize simply-connected domains in terms of homology or homotopy, one needs to be able to compute homology or homotopy groups of an open subset of the complex plane.

Let Ω be a domain which is not simply-connected. Then, not every analytic function will have a primitive in Ω, and not every non-vanishing analytic function will have an analytic logarithm. It is important to know a minimal (in a sense to be made precise) set of closed curves for which the conditions

$$\int_C f(z)dz = 0 \quad (resp.) \quad \int_C \frac{f'(z)}{f(z)}dz = 0$$

ensure the existence of a primitive in Ω, or respectively, of an analytic logarithm for a *given* function f.

A case of importance is the disk with p holes, defined as follows (see for instance [102, § 5.6, p. 45]): Consider $p+1$ Jordan curves $C_0, \ldots, C_p$, and assume that the interiors of $C_1, \ldots, C_p$ do not intersect, and are all in the interior of C_0. The homology group of this set is generated by Jordan curves, say $D_1, \ldots, D_p$ such that $C_i \subset D_i$ for $i = 1, \ldots, p$, and $D_i \cap D_j = \emptyset$ for $i \neq j$, while the homotopy group is the free group with p generators (see, e.g., [102, p. 96]).

15.6   A proof of the fundamental theorem of algebra

The following proof of the fundamental theorem of algebra was given to the author
as an exam problem in the winter of 1974–1975 in the *classe de mathématiques
spéciales* by Professor Maurice Crestey at the Lycée Louis-le-Grand, in Paris. The
topological facts used in the solution have been reviewed in Section 15.1. In view
of the identification between the complex numbers and $\mathbb{R}^2$ we denote the points of
$\mathbb{R}^2$ by the letters $z, z_1, z_2, \ldots$. We also note that here the derivative of a polynomial
is defined algebraically via $(z^n)' = nz^{n-1}$ for $n \geq 1$ and not using the notion of
$\mathbb{C}$-differentiability.

(1) Let $z_0, \ldots, z_N$ be a finite set of points in $\mathbb{R}^2$. Prove that $\mathbb{R}^2 \setminus \{z_0, \ldots, z_N\}$ is
    arc-connected.

(2) Let f be a continuous function from $\mathbb{R}^2$ into itself such that, for some $R_0 \geq 0$,

$$\forall R > R_0, \quad \exists r > 0 \quad \text{such that:} \quad |z| > r \Longrightarrow |f(z)| > R. \qquad (15.6.1)$$

Show that the image of a closed set under f is a closed set. Show that every
non-constant polynomial satisfies (15.6.1).

(3) Let $Q \in \mathbb{C}[X]$ be a non-constant polynomial with $Q'(0) = 0$:

$$Q(X) = q_0 + q_2 X^2 + \cdots + q_N X^N,$$

and define

$$\rho = \frac{1}{1 + \sum_{k=2}^{N} k|q_k|}. \qquad (15.6.2)$$

Define a sequence of complex numbers by:

$$u_0 = 0, \; u_1 = Q(u_0), \ldots, u_{\ell+1} = Q(u_\ell), \ldots. \qquad (15.6.3)$$

Show that

$$|q_0| \leq \rho^2 \Longrightarrow |u_\ell| \leq \rho, \quad \forall \ell \geq 0. \qquad (15.6.4)$$

Assuming $|q_0| \leq \rho^2$, show that $(u_\ell)_{\ell \in \mathbb{N}_0}$ is a Cauchy sequence and that its
limit u satisfies

$$u = Q(u).$$

(4) Let $P \in \mathbb{C}[X]$ be a non-constant polynomial. Let $z_0 \in \mathbb{C}$ be such that
    $P'(z_0) \neq 0$. Show that for w close enough to $P(z_0)$ the equation $P(z_0+z) = w$
    admits at least one solution.

(5) Let
$$A = \text{Ran } P, \quad \text{and} \quad B = \{P(z), z \in \mathbb{C}, P'(z) \neq 0\}.$$

Using the topological properties of A and B, show that $A = \mathbb{C}$. In particular
$0 \in A$, and the equation $P(z) = 0$ has at least one solution. It follows
then from the factor theorem (see Exercise 1.5.2) that P has n roots, where
$n = \deg P$.

Solution. (1) Take two points w_1 and w_2 in $\mathbb{R}^2 \setminus \{z_0, \ldots, z_N\}$, and consider the interval $[w_1, w_2]$. If none of the points z_j belongs to this interval, we take this interval to join w_1 and w_2. Let now j_0 be such that z_{j_0} belongs to $[w_1, w_2]$. There exists $\epsilon > 0$ such that the closed disk with center z_{j_0} and radius ϵ contains no other points z_k, $k \neq j_0$. We replace the intersection

$$[w_1, w_2] \cap \{|z - z_{j_0}| \leq \epsilon\} \tag{15.6.5}$$

by one of the half-circles of the circle $|z - z_{j_0}| = \epsilon$ which join the endpoints of (15.6.5). Repeating this construction for all the points $z_j \in [w_1, w_2]$ we construct a continuous curve which lies in $\mathbb{R}^2 \setminus \{z_0, \ldots, z_N\}$, and connects w_1 and w_2.

(2) We first recall the following: $\mathbb{R}^2$ is a complete metric space, and in a complete metric space, a set is closed if and only if it is complete; see [47, Corollaire, p. 83]. Let E be a closed subset of $\mathbb{R}^2$, and let $(z_n)_{n \in \mathbb{N}}$ be a sequence of elements in E such that $(f(z_n))_{n \in \mathbb{N}}$ is a Cauchy sequence in $f(E)$. In particular $(f(z_n))_{n \in \mathbb{N}}$ converges to a point in $\mathbb{R}^2$ and thus is bounded:

$$M = \sup_{n \in \mathbb{N}} |f(z_n)| < \infty.$$

Apply (15.6.1) to $R = \max(R_0, M + 1)$. There exists $r > 0$ such that

$$|z| > r \implies |f(z)| > M + 1,$$

and in particular all $|z_n| \leq r$ since $\sup_{n \in \mathbb{N}} |f(z_n)| = M$. Therefore the sequence $(z_n)_{n \in \mathbb{N}}$ has a converging subsequence, say $(z_{n_k})_{k \in \mathbb{N}}$, which converges to a point, say w. Since E is closed, $w \in E$. Since f is continuous,

$$\lim_{k \to \infty} f(z_{n_k}) = f(w) \in f(E).$$

Since the sequence $(f(z_n))_{n \in \mathbb{N}}$ converges, its limit is equal to the limit of any of its subsequence, and so

$$\lim_{n \to \infty} f(z_n) = f(w) \in f(E),$$

and so $f(E)$ is closed. Let now

$$p(z) = p_0 + \cdots + p_N z^N$$

be a non-constant polynomial, with $p_N \neq 0$. For $z \neq 0$ we have

$$p(z) = p_N z^N (1 + r(z))$$

where

$$r(z) = \frac{1}{p_N} \left(\frac{p_0}{z^N} + \frac{p_1}{z^{N-1}} + \cdots + \frac{p_{N-1}}{z} \right).$$

Let r_0 be such that

$$|z| > r_0 \implies |r(z)| < 1/2,$$

and let R_0 be defined by

$$r_0 = \left(\frac{2R_0}{|p_N|} \right)^{\frac{1}{N}}. \tag{15.6.6}$$

For $|z| > r_0$ we have

$$|p(z)| > \frac{|p_N| \cdot |z|^N}{2},$$

from which (15.6.1) follows with R_0 as in (15.6.6) and

$$r = \left(\frac{2R}{|p_N|} \right)^{\frac{1}{N}}.$$

(3) We prove (15.6.4) by induction. For $\ell = 0$ the result is clear. Assume that the assumption holds at rank ℓ. Then

$$|u_{\ell+1}| \leq |q_0| + \sum_{k=2}^{N} |u_\ell|^k |q_k|$$

$$\leq \rho^2 + \sum_{k=2}^{N} \rho^k |q_k| \qquad \text{(since } |q_0| \leq \rho^2 \text{ and } |u_\ell| \leq \rho\text{)}$$

$$= \rho^2 \left\{ 1 + \sum_{k=2}^{N} \rho^{k-2} |q_k| \right\}$$

$$\leq \rho^2 \left\{ 1 + \sum_{k=2}^{N} |q_k| \right\} \qquad \text{(since } |\rho| < 1\text{)}$$

$$\leq \rho^2 \left\{ 1 + \sum_{k=2}^{N} k |q_k| \right\} = \rho.$$

Hence

$$|u_{\ell+1}| \leq \rho.$$

Assuming that $|q_0| \leq \rho^2$, we now show that $(u_\ell)_{\ell \in \mathbb{N}_0}$ is a Cauchy sequence. We first remark that, for $k \geq 2$,

$$\left| \sum_{j=0}^{k-1} u_\ell^j u_{\ell-1}^{k-1-j} \right| \leq k \rho^{k-1} \leq k \rho. \tag{15.6.7}$$

We have

$$|u_{\ell+1} - u_\ell| = |Q(u_\ell) - Q(u_{\ell-1})| = \left| \sum_{k=2}^{N} q_k (u_\ell^k - u_{\ell-1}^k) \right|.$$

But we have

$$|u_\ell^k - u_{\ell-1}^k| = |u_\ell - u_{\ell-1}| \cdot \left| \sum_{j=0}^{k-1} u_\ell^j u_{\ell-1}^{k-1-j} \right| \leq |u_\ell - u_{\ell-1}| \cdot k\rho,$$

where we have used (15.6.7). Thus, with

$$K = \rho \left\{ \sum_{k=2}^N k|q_k| \right\} = \frac{\sum_{k=2}^N k|q_k|}{1 + \sum_{k=2}^N k|q_k|} < 1,$$

we have

$$|u_{\ell+1} - u_\ell| = \left| \sum_{k=2}^N q_k(u_\ell^k - u_{\ell-1}^k) \right| \leq K|u_\ell - u_{\ell-1}|. \qquad (15.6.8)$$

Since Q is not a constant, we have $K \neq 0$ and $(u_\ell)_{\ell \in \mathbb{N}_0}$ is a Cauchy sequence. Indeed, it follows from (15.6.8) that

$$|u_{\ell+1} - u_\ell| \leq K^{\ell+1}|u_1 - u_0|, \quad \ell = 0, 1, \dots$$

and, for $\ell, m \in \mathbb{N}_0$,

$$|u_{\ell+m+1} - u_\ell| \leq \sum_{k=\ell}^{m+\ell} |u_{k+1} - u_k|$$

$$\leq (K^{\ell+m+1} + \cdots + K^{\ell+1})|u_1 - u_0|$$

$$\leq \frac{K^\ell |u_1 - u_0|}{1 - K}.$$

Let $u = \lim_{\ell \to \infty} u_\ell$. Since Q is a continuous function we have

$$\lim_{\ell \to \infty} Q(u_\ell) = Q(u),$$

and hence $u = Q(u)$ in view of (15.6.3).

(4) The equation $P(z_0 + z) = w$ can be rewritten as

$$Q(z) = z$$

where the polynomial Q defined by

$$Q(z) = \frac{w + zP'(z_0) - P(z_0 + z)}{P'(z_0)}$$

is such that $Q'(0) = 0$. Let

$$Q(z) = q_0(w) + z^2 q_2 + \cdots + z^N q_N,$$

where N is the degree of P. Note that

$$q_0(w) = \frac{w - P(z_0)}{P'(z_0)},$$

and that the coefficients $q_2, \ldots, q_n$ do not depend on w. Let ρ be defined as in (15.6.2). For w close enough to $P(z_0)$ we have

$$|q_0(w)| \le \rho^2,$$

and hence, by (3), the equation $Q(z) = z$ has a solution. Thus, for such w, the equation $P(z_0 + z) = w$ has also a solution.

(5) We first assume that the equation $P'(z) = 0$ has solutions. These form a finite set, say $z_0, \ldots, z_M$. The set

$$\mathbb{C} \setminus \{P(z_0), \ldots, P(z_M)\}$$

is open, as the complement of a finite set. By (1), it is arc-connected, and therefore connected. In view of (2), the set $A = \operatorname{Ran} P$ is closed, and hence $\mathbb{C} \setminus A$ is open. In view of (4), the set

$$B = A \setminus \{P(z_0), \ldots, P(z_M)\}$$

is open. We have

$$\mathbb{C} \setminus \{P(z_0), \ldots, P(z_M)\} = (\mathbb{C} \setminus A) \cup (A \setminus \{P(z_0), \ldots, P(z_M)\}).$$

Since $\mathbb{C} \setminus \{P(z_0), \ldots, P(z_M)\}$ is connected, we have that

$$A = \mathbb{C} \quad \text{or} \quad A = \{P(z_0), \ldots, P(z_M)\}.$$

This last possibility is excluded since A is connected (recall that the continuous image of a connected set is connected). Thus $A = \mathbb{C}$.

Assume now that the equation $P'(z) = 0$ has no solution. The set A is then closed thanks to (2) and open thanks to (4). Since $A \ne \emptyset$ and since $\mathbb{C}$ is connected, we have that $A = \mathbb{C}$.   □

Remark 15.6.1. Rather than assuming in the last stage of the proof that $P'(z) = 0$ has no solution we can proceed by induction. Since every polynomial of degree 2^9 has a root and is the derivative of a polynomial of degree 3, we get that every polynomial of degree 3 has roots. By induction we get that since every polynomial of degree n has a root and is the derivative of a polynomial of degree $n+1$, every polynomial of degree $n+1$ has a root. So every polynomial has a root.

We remark that the first claim in the problem is a special case of the following result (see [42, Theorem 1.24, p. 27]):

---

[9]We could also begin with polynomials of degree 3 or 4.

Theorem 15.6.2. *Let Ω be an open connected subset of $\mathbb{C}$ and let $A \subset \Omega$ without limit points in Ω. Then, $\Omega \setminus A$ is connected.*

Remark 15.6.3. It is interesting to compare the analytic tools used in the above proof and the ones used in Question 1.5.1.

15.7 Solutions

Solution of Exercise 15.1.3. We follow [38, p. 228], and proceed by contradiction. In view of (15.1.1) the image $f(\mathbb{R})$ is not reduced to a point. It is connected. Since connected subsets of $\mathbb{R}$ are intervals, and $f(\mathbb{R})$ has more than one point, $f(\mathbb{R})$ is an interval not reduced to a point, and hence is not countable. By (15.1.1), $f(\mathbb{R})$ is countable, and hence we obtain a contradiction. $\square$

Solution of Exercise 15.1.5. The map φ is one-to-one and onto between $\mathbb{C} \cup \{\infty\}$ and $\mathbb{R}^3$, and therefore

$$d(z, w) = 0 \iff \varphi^{-1}(z) = \varphi^{-1}(w) \iff z = w.$$

Furthermore,

$$d(z, w) = \|\varphi^{-1}(z) - \varphi^{-1}(w)\|_{\mathbb{R}^3} = \|\varphi^{-1}(w) - \varphi^{-1}(z)\|_{\mathbb{R}^3} = d(w, z)$$

and, for $z_1, z_2, z_2 \in \mathbb{S}_2$, we have

$$\begin{aligned} d(z_1, z_3) &= \|\varphi^{-1}(z_1) - \varphi^{-1}(z_3)\|_{\mathbb{R}^3} \\ &\leq \|\varphi^{-1}(z_1) - \varphi^{-1}(z_2)\|_{\mathbb{R}^3} + \|\varphi^{-1}(z_2) - \varphi^{-1}(z_3)\|_{\mathbb{R}^3} \\ &= d(z_1, z_2) + d(z_2, z_3). \end{aligned}$$

Therefore d defines a metric. We show that it is given by (15.1.3). Let $z = a + ib$ and $w = c + id$ be the cartesian forms of z and w. We have for $z, w \in \mathbb{C}$:

$$\begin{aligned} d(z, w)^2 &= \left(\frac{2a}{1 + |z|^2} - \frac{2c}{1 + |w|^2} \right)^2 + \left(\frac{2b}{1 + |z|^2} - \frac{2d}{1 + |w|^2} \right)^2 \\ &\quad + \left(-\frac{2}{1 + |z|^2} + \frac{2}{1 + |w|^2} \right)^2 \qquad \left(\text{since } \frac{|z|^2 - 1}{|z|^2 + 1} = 1 - \frac{2}{1 + |z|^2} \right) \\ &= 4 \left\{ \frac{a^2 + b^2 + 1}{(1 + |z|^2)^2} + \frac{c^2 + d^2 + 1}{(1 + |w|^2)^2} - \frac{2ac + 2bd}{(1 + |z|^2)(1 + |w|^2)} \right. \\ &\qquad \left. - \frac{2}{(1 + |z|^2)(1 + |w|^2)} \right\} \\ &= 4 \left\{ \frac{1}{1 + |z|^2} + \frac{1}{1 + |w|^2} - \frac{2ac + 2bd}{(1 + |z|^2)(1 + |w|^2)} - \frac{2}{(1 + |z|^2)(1 + |w|^2)} \right\} \\ &= \frac{4|z - w|^2}{(1 + |z|^2)(1 + |w|^2)}, \end{aligned}$$

which is formula (15.1.2) for complex numbers z and w. The case where z or w is the point at infinity is treated in the same way. □

Solution of Exercise 15.1.6. Let O be an open subset of the real line. Every $x \in O$ is contained in an open interval contained in O:

$$\forall x \in O, \; \exists \epsilon_x > 0, \;\; (x - \epsilon_x, x + \epsilon_x) \subset O.$$

We define an equivalence class in O as follows: x and y will be said to be equivalent if they are in a common open interval which is included in O. We have indeed an equivalence relation, and each equivalence class is an open interval. O is the union of the equivalent classes. Since two equivalent classes do not intersect, we can associate to each of them a different rational number, and so the union is at most countable. □

Solution of Exercise 15.1.7. We show that the topology is Hausdorff. Take two points a and b in $[-\infty, \infty]$ with $a \neq b$. If both a and b are in $\mathbb{R}$, they are included in two disjoint intervals of $\mathbb{R}$, open in the usual topology of $\mathbb{R}$. These intervals are also open in the topology of $[-\infty, \infty]$. Assume now that $a = -\infty$ and $b \in \mathbb{R}$. Let c, d, e be real numbers such that $c < d < b < e$. Then

$$-\infty \in \{-\infty\} \cup (-\infty, c) \quad \text{and} \quad b \in (d, e),$$

and therefore $-\infty$ and b are in disjoint open subsets of $[-\infty, \infty]$. The case of a real point and ∞ is treated in the same way. Assume now that $a = -\infty$ and $b = \infty$ and let $c \in \mathbb{R}$. Then

$$-\infty \in \{-\infty\} \cup (-\infty, c) \quad \text{and} \quad \infty \in \{\infty\} \cup (c, \infty),$$

and so a and b are in disjoint open subsets of $[-\infty, \infty]$. □

In Exercise 15.2.2 we show that $[-\infty, \infty]$ endowed with the topology $\mathcal{O}$ is a compact space.

Solution of Exercise 15.1.8. Take two given different points z_1 and z_2 in $\mathbb{C} \setminus \{0\}$. If the origin does not belong to the interval $[z_1, z_2]$, the two points are connected by $[z_1, z_2]$. Suppose now that $0 \in [z_1, z_2]$. Take any triangle with one side equal to $[z_1, z_2]$, and let z_3 be its third vertex. The path built from the intervals $[z_1, z_3]$ and $[z_3, z_2]$ lies in $\mathbb{C} \setminus \{0\}$, and so the set is arc-connected. □

Solution of Exercise 15.2.2. Let $(O_j)_{j \in J}$ be an open covering of $[-\infty, \infty]$, indexed by the set J. There are indices j_1 and j_2 such that $-\infty \in O_{j_1}$ and $\infty \in O_{j_2}$. If $[-\infty, \infty] = O_{j_1} \cup O_{j_2}$, we have already a finite sub-covering of $[-\infty, \infty]$. Assume now that $O_{j_1} \cup O_{j_2} \subsetneq [-\infty, \infty]$. The set O_{j_1} contains a set of the form $\{-\infty\} \cup (-\infty, a_1)$, while the set O_{j_2} contains a set of the form $\{\infty\} \cup (a_2, \infty)$. We have $a_1 \leq a_2$ since $O_{j_1} \cup O_{j_2} \subsetneq [-\infty, \infty]$. Let c and d be real numbers such that $c < a_1 \leq a_2 < d$. The set $[c, d]$ is compact in $\mathbb{R}$, and $(O_j \setminus \{-\infty, \infty\})_{j \in J}$ is a covering of $[c, d]$ made of open sets of $\mathbb{R}$. There is therefore a finite sub-cover of $[c, d]$,

$$[c, d] \subset \bigcup_{n=3}^{N} (O_{j_n} \setminus \{-\infty, \infty\}).$$

and therefore, as a subset of $[-\infty, \infty]$,

$$[c, d] \subset \bigcup_{n=3}^{N} O_{j_n}.$$

Thus $\bigcup_{n=3}^{N} O_{j_n}$ together with O_{j_1} and O_{j_2} gives a finite sub-cover of $[-\infty, \infty]$. Therefore, $[-\infty, \infty]$ is compact. □

Solution of Exercise 15.2.3. Each of the sets U_n is open and so is the union $\bigcup_{n \in \mathbb{N}_0} U_n$.

Therefore $\mathbb{R} \setminus \bigcup_{n \in \mathbb{N}_0} U_n$ is closed in $\mathbb{R}$ and so is

$$C = [0, 1] \cap \bigcup_{n \in \mathbb{N}_0} U_n.$$

Since $[0, 1]$ is compact it follows that C is also compact (since it is closed and bounded in $\mathbb{R}$, or, if you prefer, since it is a closed subset of a compact).

Each U_n is itself the disjoint union of 2^n open intervals, each of Lebesgue measure $2/3^{n+1}$. It follows that

$$\sum_{n=0}^{\infty} \lambda(U_n) = \sum_{n=0}^{\infty} \frac{2^n}{3^{n+1}} = 1.$$

Hence, C has Lebesgue measure 0. We will leave it to the student to check that C is the set of points in $[0, 1]$ which have *a* triadic expansion containing only 0's and 2's. For instance,

$$\frac{1}{3} = \sum_{n=2}^{\infty} \frac{2}{3^{n+1}} \in C.$$ □

We refer to [38, Ch. 23] for a discussion of the Cantor set.

Remark 15.7.1. The exercise shows in fact that the Cantor set C has measure 0. It is also of the first category, meaning that it is a countable union of nowhere dense sets. In general the two notions, being of measure zero and being of first category, are quite different. See Remark 17.1.2.

Chapter 16

Some Functional Analysis Essentials

In the previous chapters we have tried to illustrate via various exercises some connections between the theory of functions of a complex variable and functional analysis. In the present chapter we review some of the notions which have been used.

16.1 Hilbert and Banach spaces

Let V be a vector space on $\mathbb{C}$. We recall that a *norm* is a map from V into $[0, \infty)$, often denoted by a symbol such as $\| \cdot \|_V$, and with the following properties:

(a) For every $u \in V$ and $\lambda \in \mathbb{C}$,

$$\|\lambda u\|_V = |\lambda| \cdot \|u\|_V.$$

(b) For every $u, v \in V$,

$$\|u + v\|_V \le \|u\|_V + \|v\|_V \quad \text{(triangle inequality)}.$$

(c) Let $u \in V$. Then
$$\|u\|_V = 0 \quad \Longleftrightarrow \quad u = 0.$$

A norm defines a metric via the formula

$$d(u, v) = \|u - v\|_V. \qquad (16.1.1)$$

The space V is called a Banach space if it is complete in the corresponding topology. In the above definition, one can replace the complex numbers by the real numbers and consider a vector space over the real numbers. The simplest Banach space over $\mathbb{R}$ is certainly $\mathbb{R}$ itself, endowed with norm the absolute value. Similarly, the simplest Banach space over the complex numbers is $\mathbb{C}$ itself, still with norm the absolute value. The following example is a classical exercise in calculus classes, and we will skip its proof.

Question 16.1.1. *The space of complex-valued continuous functions defined on $[0,1]$ with the supremum norm*

$$\|x\|_\infty = \max_{t \in [0,1]} |x(t)|$$

is a Banach space.

The space in the next exercise is denoted by H_∞. It corresponds to the limit case $p = \infty$ in (5.6.9).

Exercise 16.1.2. *The space of functions analytic in the open unit disk with the supremum norm*

$$\|f\|_\infty = \sup_{z \in \mathbb{D}} |f(z)|$$

is a Banach space.

Two norms $\| \cdot \|_1$ and $\| \cdot \|_2$ on a vector space V are called equivalent if there exist two strictly positive numbers m and M such that

$$m\|v\|_1 \le \|v\|_2 \le M\|v\|_1, \quad \forall v \in V. \tag{16.1.2}$$

We use in a number of places in this book the fact that all norms are equivalent in a finite-dimensional vector space. See for instance the proof of Exercise 11.5.6.

An *inner product* on V is a map $[\cdot, \cdot]$ from $V \times V$ into $\mathbb{C}$ with the following properties:

(a) For every $u, v, w \in V$ and $\lambda, \mu \in \mathbb{C}$,

$$[\lambda u + \mu v, w] = \lambda[u, w] + \mu[v, w].$$

(b) For every $u, v \in V$,

$$[u, v] = \overline{[v, u]}.$$

(c) For every $u \in V$,

$$[u, u] \ge 0, \tag{16.1.3}$$

and

(d) There is equality in (16.1.3) if and only if $u = 0$.

In other words, $[\cdot,\cdot]$ is a non-degenerate positive sesquilinear form. If only conditions (a), (b) and (c) are in force, the sesquilinear form is called a degenerate inner product. For instance, for a non-negative matrix $M \in \mathbb{C}^{p \times p}$, the map

$$[c, d] = d^* M c \tag{16.1.4}$$

defines an inner product on $\mathbb{C}^p$ if and only if M is strictly positive. Otherwise, (16.1.4) defines a degenerate inner product. Recall the Cauchy–Schwarz inequality

$$|[u, v]| \le \sqrt{[u, u]} \cdot \sqrt{[v, v]}, \quad \forall u, v \in V, \tag{16.1.5}$$

which holds for any possibly degenerate inner product. See Remark 16.3.3 for a quick proof. See Exercise 16.3.4 for an example.

Given a (non-degenerate) inner product $[\cdot, \cdot]$, the map

$$\|u\| = \sqrt{[u, u]}$$

defines a norm on V, and hence a metric via the formula (16.1.1). This metric induces in turn a topology on V.

Definition 16.1.3. The vector space V endowed with the inner product $[\cdot, \cdot]$ is called a *pre-Hilbert space*. It is called a *Hilbert space* if it is complete when endowed with the topology induced by the associated norm.

Exercise 16.1.4. *Show that the set of complex-valued continuous functions continuous on $[0, 1]$ endowed with the inner product*

$$[x, y] = \int_0^1 x(t)\overline{y(t)}dt \tag{16.1.6}$$

is a pre-Hilbert space. Show that it is not complete.

One application of integration theory will be to have an explicit description of the completion of the pre-Hilbert space appearing in Exercise 16.1.4.

Question 16.1.5. *Show that the space ℓ_2 of square summable sequences $\mathbf{z} = (z_n)_{n \in \mathbb{N}_0}$ endowed with the inner product*

$$\langle \mathbf{z}, \mathbf{w} \rangle_{\ell_2} = \sum_{n=0}^{\infty} z_n \overline{w_n}$$

is a Hilbert space.

We have already encountered in this work important examples of Hilbert spaces, whose elements are functions analytic in some open subset of the complex plane. For instance, the Hardy space (see Definition 5.6.11), the Fock space (see Definition 5.6.13), and the Bergman space (see Definition 9.5.2).

Two elements u and v in a pre-Hilbert space are called orthogonal if

$$[u, v] = 0,$$

and two sets in a pre-Hilbert space are called orthogonal if every element of the first set is orthogonal to every element of the second set.

Let X be a subset (and in particular, X is not necessarily a linear space) of the Hilbert space $\mathcal{H}$. We denote by $X^\perp$ the set of elements orthogonal to all the elements of $\mathcal{H}$,

$$X^\perp = \{h \in \mathcal{H} \; ; [x, h] = 0 \; \forall x \in X\}.$$

We note that $X^\perp$ is a closed subspace of $\mathcal{H}$. A key fact in the geometry of Hilbert spaces is:

Theorem 16.1.6. *Let $M \subset \mathcal{H}$ be a closed subspace of the Hilbert space $\mathcal{H}$. Then, every element $h \in \mathcal{H}$ can be written in a unique way as*

$$h = m + n,$$

where $m \in M$ and $n \in M^\perp$.

A linear map (the term *linear operator* is also used in this context) T from the Hilbert space $\mathcal{H}_1$ into the Hilbert space $\mathcal{H}_2$ is said to be continuous if it is continuous with respect to the topologies induced by the respective norms of $\mathcal{H}_1$ and $\mathcal{H}_2$. Continuity is equivalent to the existence of a $K > 0$ such that

$$\|Th_1\|_{\mathcal{H}_2} \le K\|h_1\|_{\mathcal{H}_1}, \quad \forall h_1 \in \mathcal{H}_1. \tag{16.1.7}$$

This last condition means that the operator T is bounded. The smallest K in (16.1.7) is called the norm of the operator. The Fourier transform, which is an important example of a bounded operator (between appropriate spaces), has been presented in Section 13.5.

We will denote by $\mathbf{L}(\mathcal{H}_1, \mathcal{H}_2)$ the space of continuous linear operators from the Hilbert space $\mathcal{H}_1$ into the Hilbert space $\mathcal{H}_2$, and set $\mathbf{L}(\mathcal{H}) = \mathbf{L}(\mathcal{H}, \mathcal{H})$.

Theorem 16.1.7.

(a) *Let $T \in \mathbf{L}(\mathcal{H}_2, \mathcal{H}_3)$ and $U \in \mathbf{L}(\mathcal{H}_1, \mathcal{H}_2)$. Then*

$$TU \in \mathbf{L}(\mathcal{H}_1, \mathcal{H}_3) \text{ and } \|TU\| \le \|T\| \cdot \|U\|.$$

(b) *Let $T \in \mathbf{L}(\mathcal{H}, \mathcal{H})$. Then, for every $n \in \mathbb{N}$,*

$$\|T^n\| \le \|T\|^n.$$

Question 16.1.8. *Show that the backward-shift operator*

$$T(z_0, z_1, z_2, \ldots) = (z_1, z_2, \ldots)$$

is bounded from ℓ_2 into itself.

In the next question, see [189, Exercice 14, p. 70] for a hint on how to show that the Cesàro operator is bounded. For a proof that its norm is equal to 2, see [201, Theorem 1, p. 154].

Question 16.1.9. *Show that the Cesàro operator defined in (7.2.10)*

$$\left(\frac{\sum_{j=0}^{n} a_j}{n+1}\right)_{n\in\mathbb{N}_0}$$

is bounded from ℓ_2 into itself, and that its norm is equal to 2.

When the Hilbert space is finite-dimensional, say of dimension N, an operator T is really a matrix and the eigenvalues are exactly the numbers z for which $T - zI$ is not invertible. For arbitrary Hilbert spaces one has the following definition:

Definition 16.1.10. Let $\mathcal{H}$ be a Hilbert space, and let T be a (not necessarily continuous) linear operator from $\mathcal{H}$ into itself. The *spectrum* of T is the set of numbers for which $T - zI_{\mathcal{H}}$ is not boundedly invertible. The *resolvent set* of T is the complement of the spectrum.

The resolvent set is denoted by $\rho(T)$ and the spectrum by $\sigma(T)$. We have for continuous operators:

Exercise 16.1.11. *Let $(\mathcal{H}, [\cdot,\cdot]_{\mathcal{H}})$ be a Hilbert space, and let T be a linear continuous operator from $\mathcal{H}$ into itself. Then $\sigma(T) \neq \emptyset$.*

Bounded linear operator with values in $\mathbb{C}$ have an important characterization, described in the following theorem.

Theorem 16.1.12 (Riesz' representation theorem). *Let $(\mathcal{H}, [\cdot,\cdot]_{\mathcal{H}})$ be a Hilbert space. For every $g \in \mathcal{H}$, the map*

$$h \mapsto [h, g]_{\mathcal{H}} \tag{16.1.8}$$

is continuous, and conversely every linear map φ from the Hilbert space $\mathcal{H}$ into $\mathbb{C}$ is of the form (16.1.8) for a uniquely defined $h_\varphi \in \mathcal{H}$. Furthermore

$$\|\varphi\| = \|h_\varphi\|_{\mathcal{H}}.$$

In this expression, $\|\varphi\|$ denotes the operator norm of φ, that is

$$\|\varphi\| = \sup_{\substack{h\in\mathcal{H}\\h\neq0}} \frac{|\varphi(h)|}{\|h\|_{\mathcal{H}}}.$$

We conclude with the definition of differentiable functions.

Definition 16.1.13. Let $\mathcal{B}_1$ and $\mathcal{B}_2$ be two Banach spaces, with respective norms $\|\cdot\|_1$ and $\|\cdot\|_2$, and let Ω be an open subset of $\mathcal{B}_1$. The function f from Ω into $\mathcal{B}_2$ is differentiable at the point $b_1 \in \mathcal{B}_1$ if there exists a bounded linear operator df_{b_1} from $\mathcal{B}_1$ into $\mathcal{B}_2$ (called the differential of f at b_1), such that

$$\lim_{h \to 0} \frac{\|f(b_1 + h) - f(b_1) - df_{b_1}(h)\|_2}{\|h\|_1} = 0. \tag{16.1.9}$$

Remark 16.1.14. When $\mathcal{B}_1 = \mathcal{B}_2 = \mathbb{C}$ the differential is a complex number, while it is (identified with a) matrix in $\mathbb{R}^{2 \times 2}$ when $\mathcal{B}_1 = \mathcal{B}_2 = \mathbb{R}^2$. See (4.2.4) for the first case and (14.1.1) applied to the real and imaginary parts of the given function for the second case. In that last case, the function is $\mathbb{C}$-differentiable if and only if this matrix is of the form (1.1.2), that is, if and only if it corresponds to a complex number.

16.2 Countably normed spaces

The set of functions analytic in an (say connected) open set Ω is a countably normed space, when endowed with the family of norms

$$\|f\|_n = \max_{z \in K_n} |f(z)|,$$

where $(K_n)_{n \in \mathbb{N}}$ is an increasing family of compact sets, such that $K_n \subset \overset{\circ}{K}_{n+1}$ (the interior of K_{n+1}) and whose union is Ω.

Exercise 16.2.1. *Let $\mathcal{V}$ be a vector space on $\mathbb{R}$ or $\mathbb{C}$, endowed with a countable number of norms $\|\cdot\|_n$, $n = 1, 2, \ldots$, such that*

$$\|x\|_1 \le \|x\|_2 \le \cdots \le \|x\|_n \le \|x\|_{n+1} \le \cdots, \quad \forall x \in \mathcal{V}.$$

(a) *Describe the smallest topology for which all these norms are continuous.*

(b) *Let $\mathcal{T}$ be this topology. Show that $(\mathcal{V}, \mathcal{T})$ is metrizable, with the metric*

$$d(x, y) = \sum_{n=1}^{\infty} \frac{1}{2^n} \frac{\|x - y\|_n}{1 + \|x - y\|_n}, \quad x, y \in \mathcal{V}.$$

16.3 Reproducing kernel Hilbert spaces

We begin with a definition:

Definition 16.3.1. A matrix $A \in \mathbb{C}^{n \times n}$ is called *non-negative* (or *positive*) if

$$c^* A c \ge 0, \quad \forall c \in \mathbb{C}^n. \tag{16.3.1}$$

In other words, if $A = (a_{\ell,j})_{\ell,j=1,\ldots,n}$ it is required that

$$\sum_{\ell,j=1}^{n} \overline{c_\ell} c_j a_{\ell,j} \geq 0. \tag{16.3.2}$$

In particular, $c^* A c = c^* A^* c$, and the polarization identity

$$d^* B c = \frac{1}{4} \sum_{k=0}^{3} i^k (c + i^k d)^* B(c + i^k d), \quad B \in \mathbb{C}^{n \times n}, \tag{16.3.3}$$

with $B = A - A^*$ implies that a positive matrix is Hermitian: $A = A^*$. The polarization identity is proved by a direct and simple computation as follows:

$$\frac{1}{4} \sum_{k=0}^{3} i^k (c + i^k d)^* B(c + i^k d) = \frac{1}{4} \sum_{k=0}^{3} (i^k c^* + d^*) B(c + i^k d)$$

$$= d^* B c + \frac{\sum_{k=0}^{3} i^k}{4} c^* B c + \frac{\sum_{k=0}^{3} i^{2k}}{4} c^* B d$$

$$+ \frac{\sum_{k=0}^{3} i^k}{4} d^* B d$$

$$= d^* B c.$$

We will use the notation

$$A \geq 0$$

to say that a matrix is positive.

The following result is well known:

Proposition 16.3.2. *Let $A \in \mathbb{C}^{n \times n}$. The following are equivalent:*

(1) $A \geq 0$.
(2) $A = A^*$ *and all its eigenvalues (which are real since A is assumed Hermitian) are positive or equal to 0.*
(3) *A admits a positive square root*

$$A = B^2, \quad \text{where} \quad B \in \mathbb{C}^{n \times n} \quad \text{and} \quad B \geq 0.$$

(4) *A admits a factorization*

$$A = C^* C \tag{16.3.4}$$

for some matrix $C \in \mathbb{C}^{p \times n}$.

Remark 16.3.3. The determinant of a positive matrix is greater or equal to 0. This allows to give a quick proof of the Cauchy–Schwarz inequality. Indeed, in the notation of (16.1.5), let $\lambda, \mu \in \mathbb{C}$. The inequality

$$[\lambda u + \mu v, \lambda u + \mu v] \geq 0$$

can be rewritten as

$$\begin{pmatrix} \bar{\lambda} & \bar{\mu} \end{pmatrix} \begin{pmatrix} [u,u] & [v,u] \\ [u,v] & [v,v] \end{pmatrix} \begin{pmatrix} \lambda \\ \mu \end{pmatrix} \geq 0.$$

Since this inequality holds for all $\lambda, \mu \in \mathbb{C}$, the matrix

$$\begin{pmatrix} [u,u] & [v,u] \\ [u,v] & [v,v] \end{pmatrix} \geq 0.$$

Computing its determinant leads to the Cauchy–Schwarz inequality.

Exercise 16.3.4. Let $A \in \mathbb{C}^{n \times n}$ be a positive matrix. Show that (see [185, Exercice 8.4.11, 1°fc, p. 303]; the author took the result from an exercises sheet from his student times at the Lycée Louis-le-Grand)

$$(c^* A^2 c)^2 \leq (c^* A c)(c^* A^3 c), \quad c \in \mathbb{C}^n. \tag{16.3.5}$$

Hint. Consider $\mathbb{C}^n$ endowed with the inner product

$$[c,d]_A = d^* A c, \quad c, d \in \mathbb{C}^n.$$

Remark 16.3.5. Inequality (16.3.5) is trivial for diagonal matrices, and can also be proved using the diagonalization of a positive matrix (that is, the spectral theorem for such matrices). The spectral theorem for Hermitian operators in Hilbert space allows to extend (16.3.5) to the Hilbert space setting. The reader might want to prove the following inequality, which is due to Heinz (in a slightly weaker form; see [116, p. 421]) and Kato (see [134]), and contains (16.3.5) as a (very) special case. The results of Heinz and Kato are in the setting of unbounded operators.

Question 16.3.6. Let C, D, Q be $\mathbb{C}^{n \times n}$ and such that C and D are positive and

$$\|Qc\| \leq \|Dc\|, \quad and \quad \|Q^*c\| \leq \|Cc\|, \quad \forall c \in \mathbb{C}^n.$$

Then,

$$|c^* Q d|^2 \leq (c^* D^{2t} c)(d^* C^{2-2t} d), \quad c, d \in \mathbb{C}^n \quad and \quad t \in [0,1]. \tag{16.3.6}$$

See also [54, p. 655] for a discussion. Inequality (16.3.5) corresponds to the choices $c = d$, $C = D = Q = A^2$ and $t = 1/4$ (see the previous proposition for the notion of positive square root of a positive matrix).

Given two positive matrices A and B in $\mathbb{C}^{n \times n}$, the notation $A \leq B$ means that the matrix $B - A$ is positive.

Exercise 16.3.7. Let A be a positive matrix. Show that $A^2 \leq A$ if and only if A is a contraction.

The product of two positive matrices need not be Hermitian, let alone positive. But there are two important operations which preserve positivity.

Exercise 16.3.8.

(1) Let $A = (a_{\ell,j})_{\ell,j=1,\ldots,n}$ and $B = (b_{\ell,j})_{\ell,j=1,\ldots,n}$ be two positive matrices in $\mathbb{C}^{n \times n}$. Then the matrix defined by

$$(A \cdot B)_{\ell,j} = a_{\ell,j} b_{\ell,j}, \quad \ell, j = 1, \ldots, n,$$

is positive.

(2) Let $A \in \mathbb{C}^{p \times p}$ and $B \in \mathbb{C}^{q \times q}$ be two positive matrices of possibly different sizes. Then, the matrix $A \otimes B \in \mathbb{C}^{pq \times pq}$ defined by

$$A \otimes B = \begin{pmatrix} a_{11}B & a_{12}B & \cdots & a_{1p}B \\ a_{21}B & a_{22}B & \cdots & a_{2p}B \\ \vdots & & & \vdots \\ a_{p1}B & a_{p2}B & \cdots & a_{pp}B \end{pmatrix}$$

is positive.

The matrix $A \cdot B$ is called the Schur (or Hadamard) product of A and B, while $A \otimes B$ is their tensor product.

The following exercise is inspired by the notion of Markov product of two positive definite functions appearing in Marek Bozejko's paper [41].

Exercise 16.3.9. Let $A, B \in \mathbb{C}^{n \times n}$ be positive contractions. Then the matrix

$$\begin{pmatrix} A & AB \\ BA & B \end{pmatrix} \tag{16.3.7}$$

is positive.

A Hilbert space $\mathcal{H}$ of *functions* defined on a set Ω is called a reproducing kernel Hilbert space if, for every $w \in \Omega$, the linear functional

$$f \mapsto f(w)$$

is continuous. By the Riesz representation theorem (see Theorem 16.1.12), there exists a uniquely determined element $k_w \in \mathcal{H}$ such that

$$[f, k_w] = f(w).$$

The function k_w is called the reproducing kernel of $\mathcal{H}$. It is a function defined on Ω and we will use the notation

$$k_w(z) = k(z, w).$$

Exercise 16.3.10. *Show that the reproducing kernel has the following properties:*

(a) *For every $z, w \in \Omega$,*

$$k(z, w) = \overline{k(w, z)}.$$

(b) *For every $N \in \mathbb{N}$, every $c_1, \ldots, c_N \in \mathbb{C}$ and every $w_1, \ldots, w_N \in \Omega$,*

$$\sum_{\ell, j=1}^{N} c_j \overline{c_\ell} k(w_\ell, w_j) \geq 0. \qquad (16.3.8)$$

Equivalent to (16.3.8) is to say that, for every $N \in \mathbb{N}$, all the $N \times N$ matrices with (ℓ, j) entry $k(w_\ell, w_j)$ are positive.

Definition 16.3.11. A function defined on a set Ω and for which (16.3.8) holds for all possible choices of $N, c_1, \ldots, c_N, w_1, \ldots, w_N$ is called a *positive definite kernel*. We will also say *positive definite function*. When Ω is a subset of the complex numbers, the function $f(z)$ is called *positive definite* if the associated kernel $f(z - \overline{w})$ is positive definite.

It is easy to check that the sum of two positive definite kernels is still positive definite. It is a bit more difficult to show that the product of two positive definite kernels is still positive definite. One uses then the result presented in Exercise 16.3.8.

Exercise 16.3.12.

(a) *Show that the function $\cos z$ is positive definite on $\mathbb{C}$.*
(b) *Show that the function (3.7.11)*

$$\prod_{n=1}^{\infty} \cos \left(\frac{t}{\rho^n} \right)$$

is positive definite on $\mathbb{R}$.

Question 16.3.13. *In a reproducing kernel Hilbert space, convergence in norm implies pointwise convergence.*

Bochner's theorem characterizes positive definite functions on the real line. It has far-reaching generalizations due to L. Schwartz and R. Minlos. We refer the student to [110] for these. See also Remark 17.9.2.

Theorem 16.3.14 (Bochner's theorem). *A function f defined on the real line and continuous at the origin is positive definite if and only if it is of the form*

$$f(t) = \int_{\mathbb{R}} e^{itx} d\mu(x),$$

where $d\mu$ is a positive measure on the real line such that $\mu(\mathbb{R}) < \infty$.

See Exercise 8.2.1 for an illustration.

There is a one-to-one correspondence between positive definite kernels on the set Ω and reproducing kernel Hilbert spaces of functions in Ω. This fundamental result originates with the works of Moore and Aronszajn. See [24].

Elements of a reproducing kernel Hilbert space can be characterized as follows:

Theorem 16.3.15. *Let $\mathcal{H}(K)$ be a reproducing kernel Hilbert space of functions defined in a set Ω, with reproducing kernel $K(z, w)$. The function f belongs to $\mathcal{H}(K)$ if and only if there exists a number $M > 0$ such that the kernel*

$$K(z, w) - \frac{f(z)\overline{f(w)}}{M^2}$$

is positive definite in Ω. The smallest such M is equal to the norm of f in $\mathcal{H}(K)$.

With this result, we propose the following exercise:

Exercise 16.3.16. *Let $|v| < 1$ and let $b_v(z) = \frac{z-v}{1-\overline{v}z}$, the associated Blaschke factor (1.1.44). Define, for f analytic in the open unit disk,*

$$(T_v(f))(z) = \frac{\sqrt{1 - |v|^2}}{1 - \overline{v}z} f(b_v(z)). \tag{16.3.9}$$

Prove that $T_v(f)$ belongs to the Hardy space $\mathbf{H}_2(\mathbb{D})$ when f does.

We refer to Exercise 2.3.9 for a characterization of the Moebius transforms which map $\mathbb{D}$ into itself. We send the reader to [14] for more on the above exercise and its relations to multiscale systems.

Finally we mention the connections between the analyticity of the kernel and of the elements of the associated reproducing kernel Hilbert space.

Theorem 16.3.17. *Let Ω be an open subset of $\mathbb{C}$ and let $K(z, w)$ be positive definite in Ω. Assume that for every $w \in \Omega$, the function $z \mapsto K(z, w)$ is analytic in Ω. Then, the elements of $\mathcal{H}(K)$ are analytic in Ω.*

We mention that any reproducing kernel Hilbert space of functions analytic in $\Omega \subset \mathbb{C}$ can be given a new inner product for which point evaluations are not bounded for w in a dense subset of Ω. See [65], and [15] for an illustration in the case of the Hardy space.

16.4 Solutions

Solution of Exercise 16.1.2. We will not prove that $\| \cdot \|_\infty$ indeed defines a norm, and focus on the completeness of the space. Let $(f_n)_{n \in \mathbb{N}}$ be a Cauchy sequence with respect to the norm $\| \cdot \|_\infty$:

$$\forall \epsilon > 0, \ \exists N \in \mathbb{N}, \ n, m \geq N \Longrightarrow \sup_{z \in \mathbb{D}} |f_n(z) - f_m(z)| < \epsilon. \tag{16.4.1}$$

In particular, for every $r \in (0, 1)$ we have that

$$\forall \epsilon > 0, \ \exists N \in \mathbb{N}, \ n, m \geq N \implies \max_{|z| \leq r} |f_n(z) - f_m(z)| < \epsilon.$$

It follows that for every $z \in \mathbb{D}$, the pointwise limit $f(z) = \lim_{n \to \infty} f_n(z)$ exists and that the limit function is analytic in $\mathbb{D}$ (recall that a series of functions which converges uniformly on compact sets is analytic). Take $\epsilon = 1$ and let $m \to \infty$ in (16.4.1). We have that

$$\sup_{z \in \mathbb{D}} |f_n(z) - f(z)| \leq 1,$$

and so

$$\sup_{z \in \mathbb{D}} |f(z)| \leq 1 + \|f\|_\infty,$$

and the limit $f \in H_\infty$. With ϵ arbitrary, letting $m \to \infty$ in (16.4.1) we finally obtain that

$$\lim_{n \to \infty} \|f - f_n\|_\infty = 0,$$

and so H_∞ is complete.   □

Solution of Exercise 16.1.4. We leave the proof that (16.1.6) is an inner product to the reader. Consider the function

$$x(t) = \begin{cases} 1, & t \in [0, 1/2), \\ 0, & t \in [1/2, 1]. \end{cases}$$

It is easily approximated by continuous functions in the inner product induced by (16.1.6). Take for instance the functions

$$x_n(t) = \begin{cases} 1, & t \in [0, 1/2), \\ -nt + 1 + \frac{n}{2}, & t \in [1/2, 1/2 + 1/n], \\ 0, & t \in [1/2 + 1/n, 1], \end{cases}$$

for $n = 2, 3, \ldots$. Then

$$\int_0^1 (x(t) - x_n(t))^2 dt = \frac{1}{3n} \to 0$$

as $n \to \infty$, but x is not continuous.   □

The real question is the following: Consider the set of, say real-valued, functions $\mathcal{R}[0, 1]$ which are Riemann integrable on $[0, 1]$ and say that $x, y \in \mathcal{R}[0, 1]$ are equivalent, $x \sim y$, if

$$\int_0^1 (x(t) - y(t))^2 dt = 0.$$

That $\mathcal{R}[0,1]/\sim$ endowed with (16.1.4) is a pre-Hilbert space is quite clear, but the question is to show that is not complete. This can be seen, in a somewhat indirect way, as follows (we refer the reader to Chapter 17 for some of the facts used in the discussion). We know that a Riemann integrable function (say, on $[0,1]$) is Lebesgue integrable with respect to the Lebesgue measure on $[0,1]$. So we have a natural isometric inclusion from $\mathcal{R}[0,1]/\sim$ into $\mathbf{L}_2[0,1]$. Since $\mathcal{R}[0,1]/\sim$ contains continuous functions and since these are dense in $\mathbf{L}_2[0,1]$ (see Theorem 17.8.1), we obtain that the closure of $\mathcal{R}[0,1]/\sim$ is $\mathbf{L}_2[0,1]$. The inclusion is strict since there are functions which are Lebesgue integrable, but not Riemann integrable.

Solution of Exercise 16.1.11. We assume that T is such that $\sigma(T) = \emptyset$ and proceed in a number of steps to obtain a contradiction.

Step 1: *Let $z \in \mathbb{C}$ and let $u \in \mathbb{C}$ be such that*

$$|u - z| \cdot \|(T - uI_{\mathcal{H}})^{-1}\|_{\mathcal{H}} \leq \frac{1}{2}.$$

Then,

$$\|(T - uI_{\mathcal{H}})^{-1} - (T - zI_{\mathcal{H}})^{-1}\|_{\mathcal{H}} \leq 2|u - z| \cdot \|(T - uI_{\mathcal{H}})^{-2}\|_{\mathcal{H}}.$$

This comes from the resolvent identity

$$(T - zI_{\mathcal{H}})^{-1} - (T - wI_{\mathcal{H}})^{-1} = (z - w)(T - zI_{\mathcal{H}})^{-1}(T - wI_{\mathcal{H}})^{-1},$$

since

$$
\begin{aligned}
(T - uI_{\mathcal{H}})^{-1} - (T - zI_{\mathcal{H}})^{-1} &= (u - z)(T - uI_{\mathcal{H}})^{-1}(T - zI_{\mathcal{H}})^{-1}\\
&= (u - z)(T - uI_{\mathcal{H}})^{-2}(I_{\mathcal{H}} + (u - z)(T - uI_{\mathcal{H}})^{-1})^{-1}.
\end{aligned}
$$

Step 2: *For every $h \in \mathcal{H}$ the function*

$$z \mapsto x_h(z) = [(T - zI_{\mathcal{H}})^{-1}h, h]_{\mathcal{H}}$$

is entire, and its derivative is given by the formula

$$x_h'(z) = [(T - zI_{\mathcal{H}})^{-2}h, h]_{\mathcal{H}}.$$

Indeed, we have

$$
\begin{aligned}
\frac{x_h(z) - x_h(w)}{z - w} &= \frac{[(T - zI_{\mathcal{H}})^{-1}h, h]_{\mathcal{H}} - [(T - wI_{\mathcal{H}})^{-1}h, h]_{\mathcal{H}}}{z - w}\\
&= \frac{[((T - zI_{\mathcal{H}})^{-1} - (T - wI_{\mathcal{H}})^{-1})h, h]_{\mathcal{H}}}{z - w}\\
&= [(T - zI_{\mathcal{H}})^{-1}(T - wI_{\mathcal{H}})^{-1}h, h]_{\mathcal{H}},
\end{aligned}
$$

where we have used the resolvent identity. Therefore

$$\frac{x_h(z) - x_h(w)}{z - w} - [(T - zI_\mathcal{H})^{-2}h, h]_\mathcal{H}$$
$$= [(T - zI_\mathcal{H})^{-1}(T - wI_\mathcal{H})^{-1} - (T - zI_\mathcal{H})^{-2})h, h]_\mathcal{H},$$

and we conclude by using the inequality proved in Step 1.

Step 3: *The function x_h is bounded.*

We first note that, for $|z| > \|T\|$ we have

$$(zI_\mathcal{H} - T)^{-1} = \frac{1}{z}(I_\mathcal{H} - \frac{T}{z})^{-1}$$
$$= \frac{1}{z}\sum_{n=0}^{\infty}\frac{T^n}{z^n}$$

so that, for $|z| \geq 2\|T\|$, we have

$$\|(zI_\mathcal{H} - T)^{-1}\|_\mathcal{H} \leq \frac{1}{|z|}\sum_{n=0}^{\infty}\frac{\|T\|^n}{|z|^n}$$
$$= \frac{1}{|z| - \|T\|}_\mathcal{H}$$
$$\leq \frac{1}{\|T\|_\mathcal{H}}.$$

The function x_h is also bounded by continuity in $|z| \leq 2\|T\|$. So x_h is constant by Liouville's theorem. Since it goes to 0 at infinity, it would be identically equal to 0, and we obtain a contradiction with the premise of the proof. □

Solution of Exercise 16.2.1. We follow [109].

(a) A basis of open neighborhoods of the origin is given by the sets of the form

$$\bigcap_{i=1}^{m}\{x\,;\,\|x\|_{n_i} < \epsilon_i\},\tag{16.4.2}$$

where m varies in $\mathbb{N}$, $n_1, \ldots, n_m \in \mathbb{N}$, and $\epsilon_1, \ldots, \epsilon_m$ are all strictly positive numbers. We remark that any such neighborhood contains the set

$$\{x\,;\,\|x\|_{n_0} < \epsilon_0\},$$

where $M = \max\{n_1, \ldots, n_m\}$ and $\epsilon = \min\{\epsilon_1, \ldots, \epsilon_m\}$.

(b) We set, for $r > 0$,

$$B_d(0, r) = \{x \in V\,;\,d(0, x) < r\}.$$

In view of the preceding remark, it is enough to show that

(i) for every $M \in \mathbb{N}$ and $\epsilon > 0$ there exists $r > 0$ such that

$$B_d(0, r) \subset \{x \, ; \, \|x\|_M < \epsilon\},$$

and

(ii) that, conversely, for every $r > 0$ there exists M and ϵ as above such that

$$\{x \, ; \, \|x\|_M < \epsilon\} \subset B_d(0, r). \tag{16.4.3}$$

We first prove (i): Let us take r such that $r2^M < 1$. The condition

$$\sum_{n=1}^{\infty} \frac{1}{2^n} \frac{\|x\|_n}{1 + \|x\|_n} < r$$

implies in particular that

$$\frac{1}{2^M} \frac{\|x\|_M}{1 + \|x\|_M} < r,$$

or equivalently

$$\|x\|_M < \frac{r2^M}{1 - r2^M}.$$

This last expression will be less than ϵ as soon as

$$r2^M < \frac{\epsilon}{1 + \epsilon}.$$

We now prove (ii). Consider a ball $B_d(0, r)$. Since the distance is bounded by 1, it is enough to consider the case $r < 1$. Let $M \in \mathbb{N}$ be such that

$$\sum_{n=M+1}^{\infty} \frac{1}{2^n} < \frac{r}{2}.$$

Since the norms are increasing and since the function $u \mapsto \frac{u}{1+u}$ is increasing, we moreover have that

$$\sum_{n=1}^{M} \frac{1}{2^n} \frac{\|x\|_n}{1 + \|x\|_n} \leq \frac{\|x\|_M}{1 + \|x\|_M} \times \left(\frac{1}{2} + \cdots + \frac{1}{2^M} \right).$$

Choose $\epsilon > 0$ such that

$$\frac{\epsilon}{1 + \epsilon} \times \left(\frac{1}{2} + \cdots + \frac{1}{2^M} \right) < \frac{r}{2}.$$

Then, for $\|x\|_M < \epsilon$ we have that

$$\frac{\|x\|_M}{1 + \|x\|_M} \times \left(\frac{1}{2} + \cdots + \frac{1}{2^M} \right) < \frac{r}{2}.$$

For such ϵ, (16.4.3) is in force. $\square$

Solution of Exercise 16.3.4. The sesquilinear form $[c, d]_A$ defines a (possibly degenerate) inner product, and the Cauchy–Schwarz inequality holds there:

$$|[c, d]_A|^2 \le [c, c]_A \cdot [d, d]_A, \quad c, d \in \mathbb{C}^n.$$

The result is obtained by setting $d = Ac$ in this inequality   □

Solution of Exercise 16.3.7. Write now $A = U^*DU$, where U is unitary and D is a diagonal matrix with entries the eigenvalues of A. Then $A^2 \le A$ holds if and only if $D^2 \le D$ holds, that is if and only if every eigenvalue, say λ, of A satisfies $\lambda^2 \le \lambda$. This last condition holds if and only if $\lambda \in [0, 1]$. The result follows.   □

Solution of Exercise 16.3.8. By linearity, it is enough to prove the claims for rank 1 positive matrices since every non-negative matrix $A \in \mathbb{C}^{n \times n}$ can be written as a sum of such matrices:

$$A = \sum_{j=1}^{N} u_j u_j^*, \quad u_j \in \mathbb{C}^n.$$

We begin with (1). Let thus $A = aa^*$ and $B = bb^*$ where $a, b \in \mathbb{C}^n$. We write

$$a = \begin{pmatrix} a_1 \\ a_2 \\ \vdots \\ a_n \end{pmatrix} \quad \text{and} \quad b = \begin{pmatrix} b_1 \\ b_2 \\ \vdots \\ b_n \end{pmatrix}.$$

Then

$$(aa^*)_{\ell,k} = a_\ell \overline{a_k} \quad \text{and} \quad (bb^*)_{\ell,k} = b_\ell \overline{b_k}, \quad \ell, k = 1, \ldots, n,$$

and so

$$((aa^*) \cdot (bb^*))_{\ell,k} = a_\ell b_\ell \overline{a_k b_k}, \quad \ell, k = 1, \ldots, n.$$

It follows that $(aa^*) \cdot (bb^*) = cc^*$, with

$$c = \begin{pmatrix} a_1 b_1 \\ a_2 b_2 \\ \vdots \\ a_n b_n \end{pmatrix}.$$

We now turn to (2). We now take $a \in \mathbb{C}^p$ and $b \in \mathbb{C}^q$. We show that

$$aa^* \otimes bb^* = (a \otimes b)(a \otimes b)^* \tag{16.4.4}$$

from which the positivity of $aa^* \otimes bb^*$ will follow. To prove (16.4.4) we note

$$(aa^*) \otimes (bb^*) = \begin{pmatrix} a_1\overline{a_1}bb^* & a_1\overline{a_2}bb^* & \cdots & a_1\overline{a_p}bb^* \\ a_2\overline{a_1}bb^* & a_2\overline{a_2}bb^* & \cdots & a_2\overline{a_p}bb^* \\ \vdots & \vdots & \vdots & \vdots \\ a_p\overline{a_1}bb^* & a_p\overline{a_2}bb^* & \cdots & a_p\overline{a_p}bb^* \end{pmatrix}$$

$$= \left((a \otimes b)(\overline{a_1}b^*) \quad (a \otimes b)(\overline{a_2}b^*) \quad \cdots \quad (a \otimes b)(\overline{a_p}b^*) \right)$$

$$= (a \otimes b)(a \otimes b)^*. \qquad \square$$

Solution of Exercise 16.3.9. Writing

$$\begin{pmatrix} A & AB \\ BA & B \end{pmatrix} = \begin{pmatrix} I_n & A \\ 0 & I_n \end{pmatrix} \begin{pmatrix} A - ABA & 0 \\ 0 & B \end{pmatrix} \begin{pmatrix} I_n & 0 \\ A & I_n \end{pmatrix},$$

and since $B \geq 0$, we see that (16.3.7) is positive if and only if $A - ABA \geq 0$. But $B \leq I_n$ and so

$$ABA \leq A^2 \leq A,$$

where we have used the fact that A is also a contraction and Exercise 16.3.7 to obtain the second inequality. Hence $A - ABA \geq 0$. $\qquad \square$

Solution of Exercise 16.3.10. (a) For every $z, w \in \Omega$ we have

$$[k_w, k_z] = k_w(z) = k(z, w) \quad \text{and} \quad [k_z, k_w] = k_z(w) = k(w, z)$$

and

$$[k_w, k_z] = \overline{[k_z, k_w]}.$$

The result follows.

(b) Let $f = \sum_{j=1}^N c_j k_{w_j}$. We have $[f, f] \geq 0$. Therefore,

$$0 \leq [f, f]$$

$$= \sum_{\ell, j=1}^N [c_j k_{w_j}, c_\ell k_{w_\ell}]$$

$$= \sum_{\ell, j=1}^N c_j \overline{c_\ell} k(w_\ell, w_j). \qquad \square$$

Solution of Exercise 16.3.12. It suffices to write

$$\cos(z - \overline{w}) = (\cos z)(\overline{\cos w}) + (\sin z)(\overline{\sin w})$$

and to notice that both the kernels

$$(\cos z)(\overline{\cos w}) \quad \text{and} \quad (\sin z)(\overline{\sin w})$$

are positive definite in $\mathbb{C}$. $\qquad \square$

Solution of Exercise 16.3.16. We follow [14], and divide the proof into a number of steps:

Step 1: *The formula*

$$\frac{1 - b_v(z)\overline{b_v(w)}}{1 - z\overline{w}} = \frac{1 - |v|^2}{(1 - z\overline{v})(1 - \overline{w}v)} \tag{16.4.5}$$

holds, where z, w are in the domain of definition of b_v.

This is just formula (1.1.51).

Step 2: *A function f defined in $\mathbb{D}$ is analytic there and belongs to $\mathbf{H}_2(\mathbb{D})$, with $\|f\|_{\mathbf{H}_2(\mathbb{D})} \leq 1$, if and only if the kernel*

$$\frac{1}{1 - z\overline{w}} - f(z)\overline{f(w)} \tag{16.4.6}$$

is positive definite in $\mathbb{D}$.

This follows from Theorem 16.3.15 since $\mathbf{H}_2(\mathbb{D})$ is the reproducing kernel Hilbert space with reproducing kernel $\frac{1}{1-z\overline{w}}$.

Step 3: *Compute*

$$\Delta(z, w) = \frac{1}{1 - z\overline{w}} - (T_v f)(z)\overline{(T_v f(w))}$$

for $z, w \in \mathbb{D}$.

Using (16.4.5) we can write:

$$\Delta(z, w) = \frac{1 - |v|^2}{(1 - b_v(z)\overline{b_v(w)})(1 - z\overline{v})(1 - \overline{w}v)}$$

$$- \frac{1 - |v|^2}{(1 - z\overline{v})(1 - \overline{w}v)} f(b_v(z))\overline{f(b_v(w))}$$

$$= \frac{1 - |v|^2}{(1 - z\overline{v})(1 - \overline{w}v)}$$

$$\times \left\{ \frac{1}{1 - b_v(z)\overline{b_v(w)}} - f(b_v(z))\overline{f(b_v(w))} \right\}.$$

The kernel

$$\frac{1}{1 - b_v(z)\overline{b_v(w)}} - f(b_v(z))\overline{f(b_v(w))}$$

is positive definite in $\mathbb{D}$ since the kernel (16.4.6) is positive definite there. It follows that $\Delta(z, w)$ is positive definite in the open unit disk. By Step 2 the function $T_v(f)$ belongs to $\mathbf{H}_2(\mathbb{D})$ and has norm less than or equal to 1. Hence the operator T_v is a contraction from $\mathbf{H}_2(\mathbb{D})$ into itself.   □

Chapter 17

A Brief Survey of Integration

In this chapter we briefly review some notions and results related to integration. We in particular discuss the following topics:

1. Algebras and σ-algebras.
2. Measurable functions.
3. Measures.
4. The main convergence theorems.
5. Complete measures.

Although we use in this book integration theory only on the real line, we have chosen to present the essentials of the general theory. We recommend to the interested student the books of Rudin [190] and of Folland [85] for a complete, but relatively short, discussion of integration. The industrious might want to look at the series of books of Bourbaki on integration.

17.1 Introduction

Students who begin to learn complex analysis usually have only a knowledge of the Riemann integral, and not of the general theory of integration. This is certainly good enough to define large families of functions via integrals (see for instance Exercises 4.4.19 and 4.4.20). But Riemann integrable functions do not have nice properties with respect to limits, as is illustrated in the following exercise.

Exercise 17.1.1. *Give an example of a uniformly bounded sequence of functions* $(f_n)_{n \in \mathbb{N}}$ *which are Riemann integrable on the interval* $[0, 1]$, *which converge pointwise, but whose pointwise limit is not Riemann integrable.*

Remark 17.1.2. In fact, a pointwise limit of continuous functions of a real variable (that is, a Baire function of class 1) need not be Riemann integrable either. We refer to [105, p. 40], [136, Theorem 23.18, p. 185], [174, Theorem 7.3, p. 32 and Theorem 7.4, p. 33] for a characterization of pointwise limits of continuous functions. In the setting of a Polish space, the result, due to Hahn, states that the set of pointwise convergence is a $F_{\sigma\delta}$ set. A proof, in the setting of metric spaces, can be found in [142, pp. 63–67]. In the case of domain of definition equal to $\mathbb{R}$, and when the functions are continuous on a dense set of points, these are the functions whose set of points of discontinuity is of first category. Recall that this means that the set is a countable union of nowhere dense sets (that is, of sets whose closure has an empty interior). On the other hand, a function, say defined and bounded on a compact interval, is Riemann integrable if and only if its set of points of discontinuity has Lebesgue measure 0 (see, e.g., [174, Theorem 7.5, p. 33]).

It is therefore difficult to have a general theorem which allows interchanging limit and integration for Riemann integrable functions. As explained in the papers [151], [169], maybe the first example of such a result is Arzelà's bounded (or dominated) convergence theorem, appearing in [25], and which reads as follows (see [106, p. 144], [151, Theorem A, p. 970]):

Theorem 17.1.3. *Let* $f_1, f_2, \ldots$ *be a uniformly bounded sequence of Riemann integrable functions on the interval* $[a, b]$, *converging pointwise to the function* f, *and* assume *that* f *is also Riemann integrable. Then,*

$$\lim_{n\to\infty} \int_a^b |f(t) - f_n(t)| dt = 0. \tag{17.1.1}$$

We refer to [151] for a proof, and also mention the papers [106], [210] for related discussions and results.

Another, and easier, such result is Weierstrass' theorem (see Theorem 14.4.1), but the hypothesis of uniform convergence in that theorem is very strong (one can also consider only Riemann integrable functions; see [106, Theorem 1]). Furthermore, Weierstrass' theorem concerns only continuous functions with compact support. A number of interesting examples, such as the Gamma function, are defined in terms of integrals of continuous functions on an infinite interval.

The above discussion gives a first motivation to go beyond the Riemann integral. Another, and related, motivation is as follows. The space of, say continuous, complex-valued functions defined on a compact interval $[a, b]$, and endowed with the metric

$$d(f, g) = \left(\int_a^b |f(t) - g(t)|^2 dt \right)^{1/2} \tag{17.1.2}$$

is not complete. See Exercise 16.1.4, and the discussion after the proof of this exercise. From the general theory of metric spaces we know that it is isometrically included in a complete metric space, unique up to an isometry of metric spaces. For the problems at hand in signal processing and in the theory of linear systems, this abstract completion is not too useful. It is more appropriate to consider the Lebesgue spaces $\mathbf{L}_2(a,b)$. The construction of these spaces is one of the keystones of the integration theory which we review briefly in this chapter. For engineers, the Lebesgue space $\mathbf{L}_2(\mathbb{R}, dx)$ plays a fundamental role, and models the space of signals with finite energy. These are of course not the only motivation and advantages of modern integration theory. A third, and very important, motivation to introduce measure theory in the study of analytic functions, is the study of boundary behaviour of a function. For instance, and this is quite beyond the scope of the present book, a function (say f) which is analytic and bounded in the open unit disk admits almost everywhere radial (and in fact non-tangential) boundary values. For radial limit, this means that, at the possible exception of a subset of $[0, 2\pi]$ of Lebesgue measure zero, the limit

$$\lim_{r \uparrow 1} f(re^{i\theta})$$

exists.

Another very important fact, not touched upon here, is Riesz' theorem on the dual of the space of continuous functions on a locally compact Hausdorff space. The space $C_0[0,1]$, endowed with the maximum norm

$$\|x\|_\infty = \max_{t \in [0,1]} |x(t)|$$

is a Banach space, and it is a natural question to ask what is its dual, that is, to describe the set of its linear continuous functionals. For instance

$$\varphi(x) = x(0), \qquad \text{and} \qquad \varphi(x) = \int_0^1 x(t)x_\varphi(t)dt$$

where $x_\varphi \in C_0[0,1]$ are such functionals. But the description of all functionals requires measure theory. See for instance [189, Théorème 6.19, p. 126]. See also [85, pp. 57–58] for a discussion of the advantages of Lebesgue's theory of integration.

17.2 σ-algebras and measures

Definition 17.2.1. Let E be a (non-empty) set. A family $\mathcal{A} \subset \mathcal{P}(E)$ is called an *algebra* if it satisfies the following:

1. The empty set $\emptyset$ belongs to $\mathcal{A}$.
2. $\mathcal{A}$ is closed under complementation: If $A \in \mathcal{A}$ then $E \setminus A \in \mathcal{A}$.
3. $\mathcal{A}$ is closed under finite union: If $(A_n)_{n=1}^N$ is a sequence of elements of $\mathcal{A}$, then

$$\bigcup_{n=1}^{N} A_n \in \mathcal{A}.$$

It is called a *σ-algebra* if it is closed under countable union: If $(A_n)_{n \in \mathbb{N}}$ is a sequence of sets belonging to $\mathcal{A}$, then

$$\bigcup_{n \in \mathbb{N}} A_n \in \mathcal{A}.$$

A key feature in the definition of a σ-algebra is that one considers *countable* unions, and not arbitrary unions. Recall that, in the definition of a topology, *arbitrary unions* come into play. See Definition 15.1.1. A couple $(E, \mathcal{A})$ where $\mathcal{A}$ is a σ-algebra of subsets of E is called a measurable space.

Question 17.2.2.

(a) *The intersection of any family of σ-algebras is a σ-algebra.*

(b) *Any subset $\mathcal{F} \subset \mathcal{P}(E)$ generates a σ-algebra $\sigma(\mathcal{F})$ which contains $\mathcal{F}$ and is minimal in the following sense: If $\mathcal{A}$ is another σ-algebra which contains $\mathcal{F}$, then, $\sigma(\mathcal{F}) \subset \mathcal{A}$.*

When in the previous exercise E is a topological space and one takes for $\mathcal{F}$ the family $\mathcal{O}$ of open sets, the σ-algebra generated by $\mathcal{O}$ is called the σ-algebra of Borel sets of E.

Exercise 17.2.3. *Let $\mathbb{R}$ be endowed with its usual topology (defined by the absolute value). Show that $\mathbb{Q}$ is a Borel set.*

Definition 17.2.4. Let $(E_1, \mathcal{A}_1)$ and $(E_2, \mathcal{A}_2)$ be two measurable spaces. A function f from E_1 into E_2 is called *measurable* if

$$\forall A \in \mathcal{A}_2, \ f^{-1}(A) \in \mathcal{A}_1.$$

Question 17.2.5. *Assume in Definition 17.2.4 that E_2 is a topological space with topology $\mathcal{O}_2$, and that $\mathcal{A}_2$ is the associated σ-algebra of Borel sets. Show that a function f from E_1 into E_2 is measurable if and only if*

$$\forall O \in \mathcal{O}_2, \ f^{-1}(O) \in \mathcal{A}_1.$$

In real analysis, a pointwise limit does not keep any reasonable property of functions (and in particular, and as already noted, does not keep Riemann integrability). A key fact which links the definition of a σ-algebra and sequences of measurable functions is presented in the following exercise.

Question 17.2.6. *Let $[-\infty, \infty]$ and its topology be as in Exercise 15.1.7. Let $(f_n)_{n \in \mathbb{N}}$ be a sequence of measurable functions from $(E_1, \mathcal{A}_1)$ with values in $[-\infty, \infty]$, and assume that the pointwise limit*

$$\lim_{n \to \infty} f_n(x)$$

exists in $[-\infty, \infty]$ for every x. Then, the function $\lim_{n \to \infty} f_n$ is measurable.

Similar results hold for sup, inf, lim inf and lim sup.

Question 17.2.7. *Let $(E, \mathcal{A})$ be a measurable set, and let $A_1, \ldots, A_N$ be measurable sets which are pairwise disjoint and such that*

$$E = \bigcup_{j=1}^{N} A_j.$$

(a) *Let $f_1, \ldots, f_N \in \mathbb{C}$. Show that the function*

$$f(x) = \sum_{j=1}^{N} f_j 1_{A_j} \qquad (17.2.1)$$

is measurable from E into the complex numbers.

(b) *Assume now that the numbers f_j are real or equal to $\pm\infty$. Show that the function (17.2.1) is measurable from E into the topological space $[-\infty, \infty]$*

Functions of the form (17.2.1) are called *simple functions*. A key result that allows us to proceed is the following (see [189, Théorème 1.17, p. 15]): *A measurable function with values in $[0, \infty]$ is the increasing limit of measurable functions of the form* (17.2.1). See the discussion after the definition (17.3.2) of the Lebesgue integral.

Exercise 17.2.8. *Show that the Dirichlet function*

$$\varphi(x) = \begin{cases} 1, & \text{if } x \in \mathbb{Q}, \\ 0, & \text{if } x \in \mathbb{R} \setminus \mathbb{Q}, \end{cases} \qquad (17.2.2)$$

is a measurable simple function.

17.3 Positive measures and integrals

Given an algebra $\mathcal{A}$, a pre-measure (see [85, p. 30]) is a function from $\mathcal{A}$ into $[0, \infty]$ with the following properties:

1. There exists an element $A \in \mathcal{A}$ such that $\mu(A) < \infty$.
2. For any countable union $A_1, A_2, \ldots$ of pairwise disjoint elements of $\mathcal{A}$ such that

$$\bigcup_{n=1}^{\infty} A_n \in \mathcal{A},$$

we have

$$\mu\left(\bigcup_{n=1}^{\infty} A_n\right) = \sum_{n=1}^{\infty} \mu(A_n).$$

A positive measure on a σ-algebra $\mathcal{A}$ is a map

$$\mu : \mathcal{A} \longrightarrow [0, \infty]$$

with the following properties:

1. There exists an element $A \in \mathcal{A}$ such that $\mu(A) < \infty$.
2. For any countable union $A_1, A_2, \ldots$ of pairwise disjoint elements of $\mathcal{A}$ we have

$$\mu\left(\bigcup_{n=1}^{\infty} A_n\right) = \sum_{n=1}^{\infty} \mu(A_n).$$

The measure is called a Borel measure if $\mathcal{A}$ is the sigma-algebra generated by a topology. A triple $(E, \mathcal{A}, \mu)$ where $(E, \mathcal{A})$ is a measurable space and μ is a positive measure on $\mathcal{A}$ is called a measured space. All measures considered in this chapter are positive, and we will use the term measure for positive measure in the remaining of the chapter.

Question 17.3.1. *Let $(E, \mathcal{A}, \mu)$ be a measured space, and let $(A_n)_{n \in \mathbb{N}}$ be a decreasing sequence of measurable sets. Assume that $\mu(A_1) < \infty$. Show that*

$$\lim_{n \to \infty} \mu(A_n) = \mu\left(\bigcap_{n=1}^{\infty} A_n\right).$$

Is the claim still true if we do not assume that $\mu(A_1) < \infty$?

The integral of a function of the form (17.2.1) is defined by

$$\int_E f d\mu \stackrel{\text{def.}}{=} \sum_{j=1}^{N} f_j \mu(A_j). \tag{17.3.1}$$

The representation (17.2.1) is not unique, but it is easy to see that the right side of (17.3.1) does not depend on the given representation of f.

Exercise 17.3.2. *Consider the Dirichlet function (17.2.2) restricted to $[0, 1]$. Show that it is a measurable simple function and compute its integral.*

The integral $\int_E f d\mu$ is defined to be

$$\int_E f d\mu = \sup_{\substack{s \leq f \\ s \text{ simple}}} \int_E s d\mu. \tag{17.3.2}$$

An important fact is that, given any measurable function from $(E, \mathcal{A})$ into $[0, \infty]$, there exists an increasing sequence of simple functions $(f_n)_{n \in \mathbb{N}}$ which converges pointwise to f. It is in particular used to show that the integral is additive, and to prove the version of the monotone convergence theorem for series of functions (see Theorem 17.5.3 below for the latter). See for instance [190, p. 22].

Exercise 17.3.3. *Let $(E, \mathcal{A}, d\mu)$ be a measured space, and let f be a measurable function with values in $[0, \infty]$, such that $\int_E f d\mu = 0$. Show that the set of points where f is strictly positive is measurable and has measure equal to 0.*

17.4 Functions with values in $[-\infty, \infty]$

Let f be a measurable function from the measured space $(E, \mathcal{A}, d\mu)$ into $[-\infty, \infty]$. Then, $|f|$ is also measurable, and, by the previous section, one can define

$$\int_E |f| d\mu \in [0, \infty].$$

In the case of functions with values in $[-\infty, \infty]$, one restricts oneself to functions for which

$$\int_E |f| d\mu < \infty,$$

and one defines

$$\int_E f d\mu \stackrel{\text{def.}}{=} \frac{1}{2} \left\{ \int_E (|f| + f) d\mu - \int_E (|f| - f) d\mu \right\}.$$

17.5 The main theorems

Theorem 17.5.1 (The monotone convergence theorem). *Let $(E, \mathcal{A}, d\mu)$ be a measured space, and let $(F_n)_{n \in \mathbb{N}}$ be an increasing family of measurable functions with values in $[0, \infty]$. Then*

$$\lim_{n \to \infty} \int_E F_n(x) d\mu(x) = \int_E (\lim_{n \to \infty} F_n(x)) d\mu(x). \tag{17.5.1}$$

Theorem 17.5.2 (The dominated convergence theorem). *Let $(E, \mathcal{A}, d\mu)$ be a measured space, and let $(F_n)_{n \in \mathbb{N}}$ be a family of measurable functions with values in $[-\infty, \infty]$ such that:*

(a) *The limit*

$$F(x) = \lim_{n \to \infty} F_n(x)$$

exists for all $x \in E$.

(b) *There exists a measurable function G such that*

$$|F_n(x)| \leq G(x), \quad \forall x \in E,$$

and

$$\int_E G(x) d\mu(x) < \infty.$$

Then, F is integrable and

$$\lim_{n\to\infty} \int_E F_n(x)d\mu(x) = \int_E F(x)d\mu(x). \qquad (17.5.2)$$

These two theorems have been used in the text for a number of exercises. See for instance Exercises 3.5.7, 4.2.16 and 5.6.14. They allow us to simplify proofs even in the case of continuous functions. See for instance Exercise 17.6.3 below. It is often convenient to present them in terms of series of functions rather than sequences of functions. One then has:

Theorem 17.5.3 (The monotone convergence theorem for series). *Let $(E, \mathcal{A}, d\mu)$ be a measured space, and let $(f_n)_{n\in\mathbb{N}}$ be a series of measurable functions with values in $[0, \infty]$. Then*

$$\sum_{n=1}^{\infty} \int_E f_n(x)d\mu(x) = \int_E \left(\sum_{n=1}^{\infty} f_n(x) \right) d\mu(x). \qquad (17.5.3)$$

Theorem 17.5.4 (The dominated convergence theorem for series). *Let $(E, \mathcal{A}, d\mu)$ be a measured space, and let $(f_n)_{n\in\mathbb{N}}$ be a family of measurable functions with values in $[-\infty, \infty]$ such that there exists a measurable function g such that*

$$\sum_{n=1}^{\infty} |f_n(x)| \le g(x), \quad \forall x \in E,$$

and

$$\int_E g(x)d\mu(x) < \infty.$$

Then, $f = \sum_{n=1}^{\infty} f_n$ is integrable and

$$\int_E f(x)d\mu(x) = \int_E \left(\sum_{n=1}^{\infty} f_n(x) \right) d\mu(x). \qquad (17.5.4)$$

We conclude this section with the following exercise, pertaining to infinite-dimensional analysis. See also the remark after the solution. We note that, in the notation of the exercise, the function $f(a) = e^{-\frac{\|a\|^2}{2}}$ is positive definite since

$$f(a - b) = e^{-\frac{\|a\|^2}{2}} e^{\langle a,b \rangle} e^{-\frac{\|b\|^2}{2}}, \quad a, b \in \ell_{2,\mathbb{R}}.$$

Exercise 17.5.5. *Consider the space $\ell_{2,\mathbb{R}}$ of sequences $a = (a_n)_{n\in\mathbb{N}}$ of real numbers such that $\|a\|^2 = \sum_{n=1}^{\infty} a_n^2 < \infty$. Show that there is no positive Borel measure P on $\ell_{2,\mathbb{R}}$ such that*

$$e^{-\frac{\|a\|^2}{2}} = \int_{\ell_{2,\mathbb{R}}} e^{i\langle a,b \rangle} dP(b), \qquad (17.5.5)$$

where $\langle a, b \rangle$ denotes the inner product in $\ell_{2,\mathbb{R}}$.

17.6   Carathéodory's theorem and the Lebesgue measure

We now discuss the following problem: Let E be a set, and suppose that we are given a family of elements $\mathcal{F}$ of $\mathcal{P}(E)$, and an additive function μ on $\mathcal{F}$. Can we extend μ to a measure on the σ-algebra generated by $\mathcal{F}$. For instance, is there a Borel measure on $\mathbb{R}$ such that

$$\mu(a, b) = b - a \qquad\qquad (17.6.1)$$

for all finite open intervals (a, b)? Cases of importance are when $\mathcal{F}$ is an algebra, or when E is an infinite product of spaces and $\mathcal{F}$ is the algebra of cylinders. In case of an algebra, the main theorem is as follows (see for instance [212, Théorème fondamental, p. 27], and for a more general statement, [85, Theorem 1.14, p. 31]).

Theorem 17.6.1. *Assume that the measure μ on the algebra $\mathcal{A}$ is σ-finite. Then, there is a unique extension of μ to the σ-algebra generated by $\mathcal{A}$.*

The existence of a unique measure, called the Lebesgue measure, satisfying (17.6.1) follows from the previous result.

Theorem 17.6.2. *Let f be Riemann-integrable on the interval $[a, b]$. Then, f is Lebesgue integrable and the two integrals coincide.*

We will use the notation $\mathbf{L}_2[a, b]$, or $\mathbf{L}_2(a, b)$, for the Lebesgue space of square integrable functions with respect to the Lebesgue measure.

As we have already pointed out, measure and integration theory allow us to simplify proofs of results already in the case of a Riemann integrable function. As an example, we have:

Exercise 17.6.3. *Let f be a continuous function on $\mathbb{R}$ such that the (generalized) Riemann integral*

$$\int_{\mathbb{R}} |f(t)| dt < \infty.$$

Show that the function

$$F(x) = \int_{\mathbb{R}} e^{-ixt} f(t) dt$$

is continuous on $\mathbb{R}$.

The function F is the Fourier transform of f. See Section 13.5.

17.7   Completion of measures

Let $(E, \mathcal{A}, d\mu)$ be a measured space, and let f be a measurable function defined on E and with values in $[0, \infty]$. Assume that

$$\int_E f(x) d\mu(x) = 0.$$

This equality is equivalent to the fact that the set of points where $f(x) > 0$ or is equal to $+\infty$ has measure 0 (see Exercise 17.3.3), but it does not imply that f is equal to the zero function. One says that the set of points where f does not vanish is negligible. More generally, a (possibly non-measurable) subset X of E is called negligible if there exists a measurable set A such that

$$X \subset A \quad \text{and} \quad \mu(A) = 0.$$

The measure is said to be complete if all negligible sets are measurable. Clearly, every countable set is negligible for the Lebesgue measure on the real line, but there are also negligible sets which are not countable, as is illustrated by the next question.

Question 17.7.1. *The Cantor set (see Exercise 15.2.3) is negligible for the Lebesgue measure.*

Theorem 17.7.2. *Let $(E, \mathcal{A}, d\mu)$ be a measured space. The family $\mathcal{A}^*$ of subsets U of E for which there exist $A, B \in \mathcal{A}$ such that*

$$A \subset U \subset B, \quad \mu(B - A) = 0$$

is a σ-algebra. The formula

$$\mu^*(U) = \mu(A)$$

defines a complete measure which extends μ.

Let $(E, \mathcal{A}, d\mu)$ be a measured space. Two measurable functions f and g are said to be equivalent if they differ on a set of measure 0. We note that the monotone convergence theorem and the dominated convergence theorem have also versions in which instead of functions one considers equivalence classes of functions. We leave to the reader the formulation of these results.

Given two equivalent measurable functions f and g we have

$$\int_E |f(x)|^2 d\mu(x) = \int_E |g(x)|^2 d\mu(x)$$

where the integral is possibly equal to $+\infty$. We denote by $\widetilde{f}$ the equivalence class of the function f.

Definition 17.7.3. The space $\mathbf{L}_2(E, \mathcal{A}, d\mu)$ is the space of equivalence classes of measurable functions such that

$$\|\widetilde{f}\|_2^2 \stackrel{\text{def.}}{=} \int_E |f(x)|^2 d\mu(x) < \infty, \quad f \in \widetilde{f}. \tag{17.7.1}$$

Theorem 17.7.4. *The space $\mathbf{L}_2(E, \mathcal{A}, d\mu)$ endowed with the sesquilinear form*

$$\langle \widetilde{f}, \widetilde{g} \rangle = \int_E f(x)\overline{g(x)} d\mu(x)$$

is a Hilbert space.

Question 17.7.5. *Show that the functions*

$$f_n(t) = e^{\frac{-i\pi t n}{F}}, \quad n \in \mathbb{Z}, \tag{17.7.2}$$

form an orthonormal basis of $\mathbf{L}_2(-F, F)$, *with the normalized inner product*

$$\langle f, g \rangle = \frac{1}{2F} \int_{(-F,F)} f(t)\overline{g(t)}dt. \tag{17.7.3}$$

17.8   Density results

The following theorem implies that the closure of the continuous functions on a compact interval $[a, b]$ of the real line, endowed with the norm (17.1.2), is indeed the space $\mathbf{L}_2[a, b]$. See also the discussion after the proof of Exercise 16.1.4 in the preceding chapter.

Theorem 17.8.1 (see [189, Théorème 3.14, p. 66]). *Let E be a locally compact Hausdorff space, let $\mathcal{A}$ be a σ-algebra of E which contains the Borel sets of E, and let μ be a positive Borel measure on $\mathcal{A}$. Then, the continuous functions with compact support are dense in* $\mathbf{L}_2(E, \mathcal{A}, d\mu)$.

Quite often it is important to consider rational functions rather than continuous functions. This is illustrated in the following exercise (recall that $\mathbb{T}$ denotes the unit circle).

Exercise 17.8.2. *Let $d\mu$ be a positive and finite Borel measure on $[0, 2\pi]$. Assume that $f \in \mathbf{L}_2([0, 2\pi], d\mu)$ is such that*

$$\int_0^{2\pi} \frac{\overline{f(t)}d\mu(t)}{e^{it} - z} = 0, \quad \forall z \in \mathbb{C} \setminus \mathbb{T}.$$

Show that $f = 0$.

Similarly, but we will not present the proof here, if $d\mu$ is a positive Borel measure on the real line such that (5.5.20) holds:

$$\int_{\mathbb{R}} \frac{d\mu(t)}{t^2 + 1} < \infty,$$

then the linear span of the functions of the form $\frac{1}{z-w}$, where w spans $\mathbb{C} \setminus \mathbb{R}$, is dense in $\mathbf{L}_2(d\mu)$.

17.9   Solutions

For the convenience of the reader we recall the definition of a Riemann integrable function. Let $[a, b]$ be a compact interval. The function f bounded on $[a, b]$ is said to be Riemann integrable with integral $I \overset{\text{def.}}{=} \int_a^b f(t)dt$ if for every $\epsilon > 0$ there exists $\delta > 0$ with the following property: For every subdivision

$$a = a_0 < a_1 < a_2 < a_3 < \cdots < a_{n-1} < a_n = b \quad \text{with} \quad \max_{k=1,\ldots,n} |a_k - a_{k-1}| < \delta,$$

and any choice of numbers $\xi_k \in [a_{k-1}, a_k]$ (with $k = 1, \ldots, n$),

$$\left| I - \sum_{k=1}^{n} f(\xi_k)(a_k - a_{k-1}) \right| < \epsilon.$$

The expression

$$\sum_{k=1}^{n} f(\xi_k)(a_k - a_{k-1})$$

is called a Riemann sum.

Solution of Exercise 17.1.1. We consider the Dirichlet function (see (17.2.2)) restricted to $[0,1]$. It is well known, and easily seen by computing Riemann sums, that this function is not integrable in the sense of Riemann. Indeed, for any subdivision $0 = a_0 < a_1 < a_2 < \cdots < a_{n-1} < a_n = 1$ the choice of rational ξ_k leads to

$$\sum_{k=1}^{n} f(\xi_k)(a_k - a_{k-1}) = 1,$$

while the choice of irrational ξ_k leads to

$$\sum_{k=1}^{n} f(\xi_k)(a_k - a_{k-1}) = 0.$$

On the other hand, let $(q_n)_{n \in \mathbb{N}}$ be an enumeration of the rational numbers in $[0,1]$, and let

$$\varphi_N(x) = \begin{cases} 1, & \text{if } x \in \{q_1, \ldots, q_N\}, \\ 0, & \text{otherwise.} \end{cases}$$

Each φ_N is Riemann integrable and we have

$$\lim_{N \to \infty} \varphi_N(x) = \varphi(x), \quad x \in [0,1]. \qquad \square$$

Remark 17.9.1. The above example already appears in [106], where the reader can also find an example of a sequence of Riemann integrable functions, with a pointwise limit which is Riemann integrable, but for which limit and integral cannot be interchanged.

Solution of Exercise 17.2.3. Remark first that any singleton is closed, and hence measurable. Let now $\{q_n, n \in \mathbb{N}\}$ be a bijection between $\mathbb{N}$ and $\mathbb{Q}$. Since

$$\mathbb{Q} = \bigcup_{n=1}^{\infty} \{q_n\},$$

it follows that $\mathbb{Q}$ is measurable. $\qquad \square$

We remark that the set $\mathbb{Q}$ is neither closed nor open in $\mathbb{R}$.

Solution of Exercise 17.2.8. We have

$$\varphi(x) = 1_{\mathbb{R}\setminus\mathbb{Q}}(x) = 1 \cdot 1_{\mathbb{R}\setminus\mathbb{Q}}(x) + 0 \cdot 1_{\mathbb{R}\setminus\mathbb{Q}}(x),$$

and hence φ is a measurable step function since $\mathbb{Q}$ (and hence $\mathbb{R}\setminus\mathbb{Q}$) is measurable in view of Exercise 17.2.3. □

Solution of Exercise 17.3.2. We first note that every singleton has Lebesgue measure zero since, for $x \in \mathbb{R}$,

$$\{x\} = \bigcap_{n=1}^{\infty} \left(x - \frac{1}{n}, x + \frac{1}{n} \right).$$

In particular every countable set has Lebesgue measure zero, and so

$$\mu([0,1] \cap \mathbb{Q}) = 0.$$

Therefore we have

$$\mu([0,1] \cap (\mathbb{R}\setminus\mathbb{Q})) = 1,$$

and the integral is equal to 1. □

Solution of Exercise 17.3.3. For $n \in \mathbb{N}$ let

$$E_n = \left\{ x \in E \, ; \, f(x) \in \left[\frac{1}{n}, \infty\right] \right\}.$$

The set E_n is measurable. Furthermore, we claim that $\mu(E_n) = 0$. Indeed, assume that $\mu(E_n) \neq 0$ (and in particular the case $\mu(E_n) = \infty$ is not excluded). There exists $M > 0$ such that $\mu(E_n) \geq M$. Thus

$$0 = \int_E f(x)d\mu(x) \geq \frac{1}{n} \int_E 1_{E_n} d\mu(x) \geq \frac{\mu(E_n)}{n} \geq \frac{M}{n} > 0,$$

which is impossible.

The set of points where $f(x) \neq 0$ is $\bigcup_{n \in \mathbb{N}} E_n$ and hence has measure 0 since

$$\mu\left(\bigcup_{n \in \mathbb{N}} E_n\right) \leq \sum_{n=1}^{\infty} \mu(E_n) = 0.$$ □

Solution of Exercise 17.5.5. We first note the following. The space $\ell_{2,\mathbb{R}}$ is a Hilbert space, and for every $a \in \ell_{2,\mathbb{R}}$ the function $b \mapsto e^{i\langle a,b\rangle}$ is continuous, and hence is measurable with respect to the Borel sigma-algebra. In particular, the integral on the right side of (17.5.5) will make sense as soon as P is a finite Borel measure. We proceed by contradiction to prove that no such measure exists. Assume that

P exists such that (17.5.5) holds. Setting $a = 0$ in (17.5.5) we see that P is a probability measure. Let, for $n \in \mathbb{N}$, $e^{(n)}$ denote the element of $\ell_{2,\mathbb{R}}$ with all entries equal to 0, besides the nth one, equal to 1, and set $a = e^{(n)}$ in (17.5.5). We have

$$e^{-\frac{1}{2}} = \int_{\ell_{2,\mathbb{R}}} e^{i\langle e^{(n)},b\rangle}\,dP(b) = \int_{\ell_{2,\mathbb{R}}} e^{ib_n}\,dP(b).$$

For every $b = (b_n)_{n\in\mathbb{N}} \in \ell_{2,\mathbb{R}}$,

$$\lim_{n\to\infty} b_n = 0.$$

Since $|e^{ib_n}| \le 1$, the dominated convergence theorem with $f_n(b) = e^{ib_n}$ and $f(b) = g(b) \equiv 1$ leads to

$$e^{-\frac{1}{2}} = \int_{\ell_{2,\mathbb{R}}} e^{i\langle e^{(n)},b\rangle}\,dP(b)$$

$$= \lim_{n\to\infty} \int_{\ell_{2,\mathbb{R}}} e^{i\langle e^{(n)},b\rangle}\,dP(b)$$

$$= \int_{\ell_{2,\mathbb{R}}} \lim_{n\to\infty} e^{i\langle e^{(n)},b\rangle}\,dP(b)$$

$$= \int_{\ell_{2,\mathbb{R}}} dP(b) = 1,$$

since such a P would be a probability measure. We thus obtain a contradiction. $\square$

Remark 17.9.2. Let $\mathscr{S}$ denote the set of sequences of real numbers $(a_n)_{n\in\mathbb{N}}$ such that

$$\sum_{n=1}^{\infty} n^{2p} a_n^{2p} < \infty, \quad p = 0, 1, 2, \ldots.$$

Then, $\mathscr{S}$ is a nuclear Fréchet space, with dual the space $\mathscr{S}'$ of real sequences such that

$$\sum_{n=1}^{\infty} n^{2p} a_n^{-2p} < \infty$$

for some $p \in \mathbb{N}_0$. The Bochner–Minlos theorem asserts that there is a Borel measure on $\mathscr{S}'$ such that

$$e^{-\frac{\|a\|^2}{2}} = \int_{\mathscr{S}'} e^{i\langle b,a\rangle}\,dP(b),$$

where the brackets denote the duality between $\mathscr{S}$ and its dual. The argument in the proof of the preceding exercise cannot be applied since

$$\lim_{n\to\infty} \langle b, e^{(n)}\rangle \not\to 0$$

in general for $b \in \mathscr{S}'$. See for instance [122, Appendix A].

Solution of Exercise 17.6.3. Let $x \in \mathbb{R}$ and let $(x_n)_{n\in\mathbb{N}}$ be a sequence of real numbers such that $\lim_{n\to\infty} x_n = x$. Set

$$f_n(t) = e^{-ix_n t} f(t), \quad n \in \mathbb{N},$$

and

$$g(t) = |f(t)|.$$

The functions f_n and g belong to $\mathbf{L}_1(\mathbb{R}, dx)$ and we have

$$\lim_{n\to\infty} f_n(t) = e^{-ixt} f(t), \quad \forall t \in \mathbb{R}.$$

Furthermore

$$|f_n(t)| \le g(t), \quad \forall t \in \mathbb{R}.$$

The dominated convergence theorem allows us to conclude that

$$\lim_{n\to\infty} F(x_n) = \lim_{n\to\infty} \int_{\mathbb{R}} f_n(t)dt = \int_{\mathbb{R}} (\lim_{n\to\infty} f_n(t))dt = F(x),$$

and F is continuous at the point x. □

Solution of Exercise 17.8.2. We first remark that for every $z \in \mathbb{C} \setminus \mathbb{T}$ the function

$$t \mapsto \frac{1}{e^{it} - z}$$

is bounded,

$$\left| \frac{1}{e^{it} - z} \right| \le \begin{cases} \frac{1}{1-|z|}, & \text{if } |z| < 1, \\ \frac{1}{|z|-1}, & \text{if } |z| > 1, \end{cases}$$

and hence belongs to $\mathbf{L}_2([0, 2\pi], d\mu)$. Let z be in the open unit disk. Using the dominated convergence theorem for series with

$$f_n(t) = z^n \overline{f(t)} e^{-i(n+1)t}, \quad n = 0, 1, \dots,$$

and

$$g(t) = \frac{|f(t)|}{1 - |z|},$$

we have

$$\int_0^{2\pi} \frac{\overline{f(t)}d\mu(t)}{e^{it} - z} = \int_0^{2\pi} \frac{\overline{f(t)}e^{-it}d\mu(t)}{1 - e^{-it}z}$$

$$= \int_0^{2\pi} \overline{f(t)}e^{-it} \left(\sum_{n=0}^{\infty} (e^{-it}z)^n \right) d\mu(t)$$

$$= \sum_{n=0}^{\infty} z^n \int_0^{2\pi} \overline{f(t)}e^{-i(n+1)t}d\mu(t).$$

This power series is identically equal to 0 in $\mathbb{D}$, and therefore

$$\int_0^{2\pi} \overline{f(t)}e^{-i(n+1)t}d\mu(t), \quad n = 0, 1, \ldots.$$

Take now $|z| > 1$. Another application of the dominated convergence theorem for series of functions gives:

$$\int_0^{2\pi} \frac{\overline{f(t)}d\mu(t)}{e^{it} - z} = \frac{1}{z}\int_0^{2\pi} \frac{\overline{f(t)}d\mu(t)}{z^{-1}e^{it} - 1}$$

$$= -\frac{1}{z}\int_0^{2\pi} \overline{f(t)}(\sum_{n=0}^{\infty}(e^{it}z^{-1})^n)d\mu(t)$$

$$= -\sum_{n=0}^{\infty} z^{-n-1}\int_0^{2\pi} \overline{f(t)}e^{int}d\mu(t).$$

This power series is in fact a Laurent expansion in $|z| > 1$ and identically equal to 0 there, and therefore

$$\int_0^{2\pi} \overline{f(t)}e^{int}d\mu(t), \quad n = 0, 1, \ldots.$$

It follows that f is orthogonal to all trigonometric polynomials. Since every continuous function on $[0, 2\pi]$ (with same values at 0 and 2π) can be approximated in the supremum norm by trigonometric polynomials (see for instance [189, Théorème 4.25, p. 87]), it follows that f is orthogonal to all continuous functions. By Theorem 17.8.1, we have $f = 0$, μ-a.e.   □

Bibliography

[1] M. Ablowitz and A.S. Fokas. *Complex variables. Introduction and Applications.* Cambridge texts in mathematics. Cambridge University Press, 1997.

[2] S. Abou-Jaoudé and J. Chevalier. *Cahiers de mathématiques. Analyse I. Topologie.* O.C.D.L., 65 rue Claude-Bernard, Paris 5, 1971.

[3] W.W. Adams and E.G. Straus. Non-archimedian analytic functions taking the same values at the same points. *Illinois J. Math.*, 15:418–424, 1971.

[4] L. Ahlfors. Open Riemann surfaces and extremal problems on compact subregions. *Comment. Math. Helv.*, 24:100–123, 1950.

[5] L. Ahlfors. *Complex analysis.* McGraw-Hill Book Co., third edition, 1978.

[6] D. Alpay. *The Schur algorithm, reproducing kernel spaces and system theory.* American Mathematical Society, Providence, RI, 2001. Translated from the 1998 French original by Stephen S. Wilson, Panoramas et Synthèses.

[7] D. Alpay. *An advanced complex analysis problem book. Topological Vector Spaces, Functional Analysis, and Hilbert spaces of analytic functions.* Birkhäuser/Springer Basel AG, Basel, 2015.

[8] D. Alpay, V. Bolotnikov, and L. Rodman. Tangential interpolation with symmetries and two-point interpolation problem for matrix-valued H_2-functions. *Integral Equations Operator Theory*, 32(1):1–28, 1998.

[9] D. Alpay, A. Dijksma, J. Rovnyak, and H. de Snoo. *Schur functions, operator colligations, and reproducing kernel Pontryagin spaces*, volume 96 of *Operator theory: Advances and Applications*. Birkhäuser Verlag, Basel, 1997.

[10] D. Alpay and H. Dym. On a new class of realization formulas and their applications. *Linear Algebra Appl.*, 241/243:3–84, 1996.

[11] D. Alpay and I. Gohberg. Unitary rational matrix functions. In I. Gohberg, editor, *Topics in interpolation theory of rational matrix-valued functions*, volume 33 of *Operator Theory: Advances and Applications*, pages 175–222. Birkhäuser Verlag, Basel, 1988.

[12] D. Alpay, P. Jorgensen, I. Lewkowicz, and D. Volok. A new realization of rational functions, with applications to linear combination interpolation. *Complex Variables and Elliptic Equations*, 00:00–00, 2016.

[13] D. Alpay, P. Jorgensen, R. Seager, and D. Volok. On discrete analytic functions: Products, rational functions and reproducing kernels. *Journal of Applied Mathematics and Computing*, 41:393–426, 2013.

[14] D. Alpay and M. Mboup. Discrete-time multi-scale systems. *Integral Equations and Operator Theory*, 68:163–191, 2010.

[15] D. Alpay and T.M. Mills. A family of Hilbert spaces which are not reproducing kernels Hilbert spaces. *J. Anal. Appl.*, 1(2):107–111, 2003.

[16] D. Alpay and M. Porat. Iterated Fock spaces. Preprint.

[17] D. Alpay and V. Vinnikov. Analogues d'espaces de de Branges sur des surfaces de Riemann. *C.R. Acad. Sci. Paris Sér. I Math.*, 318:1077–1082, 1994.

[18] É. Amar and É. Matheron. *Analyse complexe*. Cassini, Paris, 2004.

[19] M. Andersson. *Topics in complex analysis*. Universitext: Tracts in Mathematics. Springer, 1996.

[20] T. Andreescu and D. Andrica. *Complex numbers from A to... Z*. Birkhäuser Boston Inc., Boston, MA, 2006. Translated and revised from the 2001 Romanian original.

[21] T. Andreescu and B. Enescu. *Mathematical Olympiad Treasures*. Birkhäuser, second edition, 2011.

[22] T. Apostol. *Calculus. Volume II*. Xerox College Publishing, Waltham, Massachusetts, second edition, 1969.

[23] J.-M. Arnaudies. *L'intégrale de Lebesgue sur la droite*. Vuibert, 1997.

[24] N. Aronszajn. Theory of reproducing kernels. *Trans. Amer. Math. Soc.*, 68:337–404, 1950.

[25] C. Arzelà. Sulla integrazione per serie. *Atti Acc. Linecei Rend., Rome*, 1:532–537, 596–599, 1885.

[26] R.B. Ash. *Information theory*. Dover Publications Inc., New York, 1990. Corrected reprint of the 1965 original.

[27] M. Audin. Un cours sur les fonctions spéciales. `http://www-irma.u-strasbg.fr/~maudin/fonctionsspe1109.pdf`.

[28] J. Bak and D.J. Newman. *Complex analysis*. Undergraduate Texts in Mathematics. Springer-Verlag, New York, 1982.

[29] J. Barros-Neto. *An introduction to the theory of distributions*. Marcel Dekker, 1973.

[30] H. Bart, I. Gohberg, M.A. Kaashoek, and A.C.M. Ran. *Factorization of matrix and operator functions: the state space method*, volume 178 of *Operator Theory: Advances and Applications*. Birkhäuser Verlag, Basel, 2008. Linear Operators and Linear Systems.

[31] C. Berenstein and R. Gay. *Complex variables*, volume 125 of *Graduate Texts in Mathematics*. Springer-Verlag, New York, 1991. An introduction.

[32] C. Berg, J. Christensen, and P. Ressel. *Harmonic analysis on semigroups*, volume 100 of *Graduate Texts in Mathematics*. Springer-Verlag, New York, 1984. Theory of positive definite and related functions.

[33] S. Bergman and M. Schiffer. *Kernel functions and elliptic differential equations in mathematical physics*. Academic Press, 1953.

[34] B. Berndt and B. Yeap. Explicit evaluations and reciprocity theorems for finite trigonometric sums. *Adv. in Appl. Math.*, 29(3):358–385, 2002.

[35] A.I. Bobenko, Ch. Mercat, and Y.B. Suris. Linear and nonlinear theories of discrete analytic functions. Integrable structure and isomonodromic Green's function. *J. Reine Angew. Math.*, 583:117–161, 2005.

[36] N. Boccara. *Fonctions analytiques*. Collection Mathématiques pour l'ingénieur. Ellipses, 32 rue Bargue, Paris 75015, 1996.

[37] A. Bogomolny. Coaxal circles theorem. http://www.cut-the-knot.org/Curriculum/Geometry/CoaxalCircles.shtml.

[38] H. Boualem and R. Brouzet. *La Planète ℝ. Voyage au pays des nombres réels*. Dunod, Paris, 2002.

[39] N. Bourbaki. *Topologie générale. Chapitres 5 à 10*. Diffusion C.C.L.S., Paris, 1974.

[40] J. Boyd. Large-degree asymptotics and exponential asymptotics for Fourier, Chebyshev and Hermite coefficients and Fourier transforms. *J. Eng. Math.*, pages 355–399, 2009.

[41] M. Bożejko. Positive-definite kernels, length functions on groups and a noncommutative von Neumann inequality. *Studia Math.*, 95(2):107–118, 1989.

[42] R.B. Burckel. *An introduction to classical complex analysis, Vol. 1*. Birkhäuser, 1979.

[43] D. Burns and S. Krantz. Rigidity of holomorphic mappings and a new Schwarz lemma at the boundary. *J. Amer. Math. Soc.*, 7(3):661–676, 1994.

[44] B. Calvo, J. Doyen, A. Calvo, and F. Boschet. *Exercices d'algèbre*. Armand Colin, 1970.

[45] H. Cartan. *Théorie élémentaire des fonctions analytiques d'une ou plusieurs variables complexes*. Hermann, Paris, 1975.

[46] G. Choquet. *Topology*, volume XIX of *Pure and Applied Mathematics*. Academic Press, 1966.

[47] G. Choquet. *Cours d'analyse, Tome II: Topologie*. Masson, 120 bd Saint-Germain, Paris VI, 1973.

[48] W. Chu and A. Marini. Partial fractions and trigonometric identities. *Adv. in Appl. Math.*, 23(2):115–175, 1999.

[49] R.V. Churchill and J.W. Brown. *Complex variables and applications. Fifth edition*. McGraw-Hill, 1990.

[50] S. Colombo. *Les transformations de Mellin et de Hankel: Applications à la physique mathématique*. Monographies du Centre d'Études Mathématiques en vue des Applications: B. – Méthodes de Calcul. Centre National de la Recherche Scientifique, Paris, 1959.

[51] P. Colwell. *Blaschke products, bounded analytic functions*. Ann Arbor, the University of Michigan press, 1969.

[52] E.H. Connell and P. Porcelli. An algorithm of J. Schur and the Taylor series. *Proc. Amer. Math. Soc.*, 13:232–235, 1962.

[53] E.T. Copson. *An introduction to the theory of functions of a complex variable*. Oxford University Press, 1972. First edition 1935.

[54] H. Cordes, A. Jensen, S.T. Kuroda, G. Ponce, B. Simon, and M. Taylor. Tosio Kato (1917–1999). *Notices of the AMS*, 47(6):650–657, 2000.

[55] E. Corominas and F. Sunyer Balaguer. Conditions for an infinitely differentiable function to be a polynomial. *Revista Mat. Hisp.-Amer.* (4), 14:26–43, 1954.

[56] P. Giesy and D. Still. Another elementary proof that $\sum 1/k^2 = \pi^2/6$. *Math. Mag.*, 45:148–149, 1972.

[57] D.E. Daykin, C.E. Linderholm, A. Wilansky, H.E. Debrunner, and H.S. Witsenhausen. Problems and Solutions: Solutions of Advanced Problems: 5914. *Amer. Math. Monthly*, 81(7):787–788, 1974.

[58] D.E. Daykin and A. Wilansky. Sets of complex numbers. *Math. Mag.*, 47:228–229, 1974.

[59] G. de Barra. *Measure theory and integration*. Horwood Publishing Series. Mathematics and Its Applications. Horwood Publishing Limited, Chichester, 2003. Revised edition of the 1981 original.

[60] L. de Branges. *Hilbert spaces of entire functions*. Prentice Hall Inc., Englewood Cliffs, N.J., 1968.

[61] P.N. de Souza and J.N. Silva. *Berkeley problems in mathematics*. Problem Books in Mathematics. Springer-Verlag, New York, third edition, 2004.

[62] P.N. de Souza and J.N. Silva. *Berkeley problems in mathematics*. Problem Books in Mathematics. Springer-Verlag, New York, 1998.

[63] J. Dieudonné. *Éléments d'analyse, Volume 1: fondements de l'analyse moderne*. Gauthier-Villars, Paris, 1969.

[64] J.D. Dixon. A brief proof of Cauchy's integral theorem. *Proc. Amer. Math. Soc.*, 29:625–626, 1971.

[65] W. Donoghue and P. Masani. A class of invalid assertions concerning function Hilbert spaces. *Bol. Soc. Mat. Mexicana* (2), 28(2):77–80, 1983.

[66] W.F. Donoghue. *Distributions and Fourier transforms*, volume 32 of *Pure and Applied Mathematics*. Academic Press, New York, 1969.

[67] W.F. Donoghue. *Monotone matrix functions and analytic continuation*, volume 207 of *Die Grundlehren der mathematischen Wissenschaften*. Springer-Verlag, 1974.

[68] J. Doyle, B. Francis, and A. Tannenbaum. *Feedback control theory*. Macmillan Publishing Company, New York, 1992.

[69] P.L. Duren. *Theory of H^p spaces*. Academic press, New York, 1970.

[70] D. Dutkay and P.E.T. Jorgensen. Wavelets on fractals. *Rev. Mat. Iberoam.*, 22(1):131–180, 2006.

[71] H. Dym and H.P. McKean. *Fourier series and integrals*. Academic Press, 1972.

[72] H. Dym and H.P. McKean. *Gaussian processes, function theory and the inverse spectral problem*. Academic Press, 1976.

[73] S. Elaydi. *An introduction to difference equations*. Undergraduate Texts in Mathematics. Springer, New York, third edition, 2005.

[74] M.A. Evgrafov. *Analytic functions*. Dover Publications Inc., New York, 1978. Translated from the Russian, reprint of the 1966 original English translation, edited and with a foreword by Bernard R. Gelbaum.

[75] M.A. Evgravof, K. Béjanov, Y. Sidorov, M. Fédoruk, and M. Chabounine. *Recueil de problèmes sur la théorie des fonctions analytiques*. Éditions Mir, Moscou, 1974.

[76] P. Faurre, M. Clerget, and F. Germain. *Opérateurs rationnels positifs*, volume 8 of *Méthodes Mathématiques de l'Informatique [Mathematical Methods of Information Science]*. Dunod, Paris, 1979. Application à l'hyperstabilité et aux processus aléatoires.

[77] J. Favard. *Cours d'analyse de l'École Polytechnique. Tome 1. Introduction-Operations*. Gauthier-Villars, 1968.

[78] J. Fay. *Theta functions on Riemann surfaces*. Springer-Verlag, Berlin, 1973. Lecture Notes in Mathematics, Vol. 352.

[79] C. Fefferman. An easy proof of the fundamental theorem of algebra. *Am. Math. Mon.*, 74:854–855, 1967.

[80] S.D. Fisher. *Complex Variable. Second Edition*. Dover, 1999.

[81] W. Fisher and I. Lieb. *Funkionentheorie*. Vieweg Verlag, 2005. 9th edition.

[82] D. Flament. *Histoire des nombres complexes. Entre algèbre et géométrie*. CNRS Editions, 15 rue Malebranche. 75005. Paris, 2003.

[83] H. Flanders. On the Fresnel integrals. *The American Mathematical Monthly*, 89:264–266, 1982.

[84] F.J Flanigan. *Complex variables. Harmonic and analytic functions*. Dover books on advanced mathematics. Dover, 1983.

[85] G.B. Folland. *Real analysis*. Pure and Applied Mathematics (New York). John Wiley & Sons Inc., New York, second edition, 1999. Modern techniques and their applications, a Wiley-Interscience Publication.

[86] L.R. Ford. *An introduction to the theory of automorphic functions.* Number 6 in Edinburgh Mathematical Tracts. Bell and Sons, 1915.

[87] M.K. Fort, Jr. Some properties of continuous functions. *Amer. Math. Monthly*, 59:372–375, 1952.

[88] S. Francinou, H. Gianella, and S. Nicolas. *Exercices de mathématiques. Oraux X-ENS. Analyse 1.* Cassini, 2003.

[89] A.E. Frazho and W. Bhosri. *An operator perspective on signals and systems*, volume 204 of *Operator Theory: Advances and Applications*. Birkhäuser Verlag, Basel, 2010. Linear Operators and Linear Systems.

[90] M.W. Frazier. *An introduction to wavelets through linear algebra.* Undergraduate Texts in Mathematics. Springer-Verlag, New York, 1999.

[91] E. Freitag and R. Busam. *Complex analysis.* Springer, 2005.

[92] E. Freitag and R. Busam. *Funktionentheorie 1.* Springer, 2006. 4. korrigierte und erweiterte Auflage.

[93] O. Frostman. Sur les produits de Blaschke. *Kungl. Fysiografiska Sällskapets i Lund Förhandlingar [Proc. Roy. Physiog. Soc. Lund]*, 12(15):169–182, 1942.

[94] P.A. Fuhrmann. *Linear systems and operators in Hilbert space.* McGraw-Hill international book company, 1981.

[95] W. Fulton. *Algebraic topology*, volume 153 of *Graduate Texts in Mathematics*. Springer-Verlag, New York, 1995. A first course.

[96] F. G.-M. *Cours d'algèbre élémentaire.* Librairie générale, 77 rue de Vaugirard, Paris VI-ème, 1927. (Frère Gabriel-Marie et Réunion de Professeurs.)

[97] Th. Gamelin. *Complex analysis.* Undergraduate Texts in Mathematics. Springer-Verlag, New York, 2001.

[98] P.R. Garabedian. The classes L_p and conformal mapping. *Trans. Amer. Math. Soc.*, 69:392–415, 1950.

[99] P.R. Garabedian. a simple proof of a simple version of the Riemann mapping theorem by simple functional analysis. *Amer. Math. Monthly*, 98(9):824–826, 1991.

[100] Y. Genin, P. van Dooren, T. Kailath, J.M. Delosme, and M. Morf. On Σ-lossless transfer functions and related questions. *Linear Algebra Appl.*, 50:251–275, 1983.

[101] J.P. Gilman, I. Kra, and R.E. Rodríguez. *Complex analysis*, volume 245 of *Graduate Texts in Mathematics*. Springer, New York, 2007. In the spirit of Lipman Bers.

[102] C. Godbillon. *Éléments de topologie algébrique.* Hermann, Paris, 1971.

[103] R. Godement. *Cours d'algèbre.* Hermann, 1987.

[104] I. Gohberg, S. Goldberg, and M.A. Kaashoek. *Classes of linear operators. Vol. II*, volume 63 of *Operator Theory: Advances and Applications*. Birkhäuser Verlag, Basel, 1993.

[105] S. Gonnord and N. Tosel. *Thèmes d'analyse pour l'agrégation. Topologie et analyse fonctionnelle.* Ellipses. Éditions Marketing S.A. 32 rue Bargues, Paris 75015, 1996.

[106] R.A. Gordon. A convergence theorem for the Riemann integral. *Math. Mag.*, 73(2):141–147, 2000.

[107] R. Goulfier. *Mathématiques. Exercices d'oral* 1974–1975 *avec corrigés.* Bréal, 310–320 Bd de la Boissière, 93100, Montreuil, 1975.

[108] J.D. Gray and M.J. Morris. When is a function that satisfies the Cauchy–Riemann equations analytic. *The American Mathematical Monthly*, pages 246–256, 1978.

[109] I.M. Guelfand and G.E. Shilov. *Les distributions. Tome* 2. Collection Universitaire de Mathématiques, no. 15. Dunod, Paris, 1964.

[110] I.M. Guelfand and N.Y. Vilenkin. *Les distributions. Tome 4: Applications de l'analyse harmonique.* Collection Universitaire de Mathématiques, No. 23. Dunod, Paris, 1967.

[111] R. Gunning. *Lectures on Riemann surfaces*, volume 2 of *Mathematical notes, Princeton University press.* Springer-Verlag, Berlin, Heidelberg, New York, 1966.

[112] E. Hairer and G. Wanner. *Analysis by its history.* Undergraduate Texts in Mathematics. Readings in Mathematics. Springer, New York, 2008. Corrected reprint of the 1996 original [MR1410751].

[113] G. Hamel. Eine charakteristische Eigenschaft beschränkter analytischer Funktionen. *Math. Ann.*, 78(1):257–269, 1917.

[114] G.H. Hardy. *Divergent Series.* Oxford, at the Clarendon Press, 1949.

[115] B. Hauchecorne. *Les contre-exemples en mathématiques.* Edition Marketing. 32 rue Bargue, 75740, Paris cedex 15, 2007. Seconde édition, revue et augmentée.

[116] E. Heinz. Beiträge zur Störungstheorie der Spektralzerlegung. *Math. Ann.*, 123:415–438, 1951.

[117] J.W. Helton. *Operator theory, analytic functions, matrices and electrical engineering*, volume 68 of *CBMS Lecture Notes.* Amer. Math. Soc., Rhodes Island, 1987.

[118] M. Henle. *A combinatorial introduction to topology.* W.H. Freeman and Co., San Francisco, Calif., 1979. A Series of Books in Mathematical Sciences.

[119] F. Hiai and D. Petz. *The semicircle law, free random variables and entropy*, volume 77 of *Mathematical Surveys and Monographs.* American Mathematical Society, Providence, RI, 2000.

[120] J. Hofbauer. A simple proof of $1 + \frac{1}{2^2} + \frac{1}{3^2} + \cdots = \frac{\pi^2}{6}$ and related identities. *Amer. Math. Monthly*, 109(2):196–200, 2002.

[121] K. Hoffman. *Banach spaces of analytic functions.* Dover Publications Inc., New York, 1988. Reprint of the 1962 original.

[122] H. Holden, B. Øksendal, J. Ubøe, and T. Zhang. *Stochastic partial differential equations*. Probability and its Applications. Birkhäuser Boston Inc., Boston, MA, 1996.

[123] F. Holme. A simple calculation of $1 + \frac{1}{2^2} + \frac{1}{3^2} + \cdots = \frac{\pi^2}{6}$. *Nordisk Matematisk Tidskrift*, 18:91–92, 120, 1970.

[124] R. Honsberger. *Mathematical gems. III*, volume 9 of *The Dolciani Mathematical Expositions*. Mathematical Association of America, Washington, DC, 1985.

[125] R. Honsberger. *From Erdős to Kiev*, volume 17 of *The Dolciani Mathematical Expositions*. Mathematical Association of America, Washington, DC, 1996. Problems of Olympiad caliber.

[126] N. Jacobson. *Lectures in abstract algebra. Volume I – Basic concepts*. D. Van Nostrand Company, Princeton, New Jersey, 1966.

[127] F. Jacobzon, S. Reich, and D. Shoikhet. Linear fractional mappings: invariant sets, semigroups and commutativity. *J. Fixed Point Theory Appl.*, 5(1):63–91, 2009.

[128] W.B. Jones, O. Njåstad, and W. J. Thron. Schur fractions, Perron–Carathéodory fractions and Szegő polynomials, a survey. In *Analytic theory of continued fractions, II (Pitlochry/Aviemore, 1985)*, volume 1199 of *Lecture Notes in Math.*, pages 127–158. Springer, Berlin, 1986.

[129] P.E.T. Jorgensen and S. Pedersen. Spectral theory for Borel sets in $\mathbf{R}^n$ of finite measure. *J. Funct. Anal.*, 107(1):72–104, 1992.

[130] M. Kaashoek. State space theory of rational matrix functions and applications. In P. Lancaster, editor, *Lectures on operator theory and its applications*, volume 3 of *Fields Institute Monographs*, pages 233–333. American Mathematical Society, 1996.

[131] T. Kailath. A theorem of I. Schur and its impact on modern signal processing. In I. Gohberg, editor, *I. Schur methods in operator theory and signal processing*, volume 18 of *Oper. Theory Adv. Appl.*, pages 9–30. Birkhäuser, Basel, 1986.

[132] R.E. Kalman, P.L. Falb, and M.A. Arbib. *Topics in mathematical system theory*. McGraw-Hill Book Co., New York, 1969.

[133] Th. Kaluza. Über die Koeffizienten reziproker Potenzreihen. *Math. Z.*, 28(1):161–170, 1928.

[134] T. Kato. Notes on some inequalities for linear operators. *Math. Ann.*, 125:208–212, 1952.

[135] R.P. Kaufman and N.W. Rickert. An inequality concerning measures. *Bull. Amer. Math. Soc.*, 72:672–676, 1966.

[136] A.S. Kechris. *Classical descriptive set theory*, volume 156 of *Graduate Texts in Mathematics*. Springer-Verlag, New York, 1995.

[137] H. Kneser. *Funktionentheorie*. Studia Mathematica, Bd. 13. Vandenhoeck & Ruprecht, Göttingen, 1958.

[138] K. Knopp. *Problem book in the theory of functions. Volume II*. Dover, New York, 1952.

[139] H. Koch. *Einführung in die klassische Mathematik I*. Springer-Verlag, 1986.

[140] A.N. Kolmogorov. *Selected works of A.N. Kolmogorov. Vol. III*, volume 27 of *Mathematics and its Applications (Soviet Series)*. Kluwer Academic Publishers Group, Dordrecht, 1993. Information theory and the theory of algorithms, with a biography of Kolmogorov by N.N. Bogolyubov, B.V. Gnedenko and S.L. Sobolev, with commentaries by R.L. Dobrushin, A. Kh. Shen′, V.M. Tikhomirov, Ya.M. Barzdin [Jānis Bārzdiņš], Ya.G. Sinaĭ, V.A. Uspenski [V.A. Uspenskiĭ] and A.L. Semyonov, edited by A.N. Shiryayev [A.N. Shiryaev], translated from the 1987 Russian original by A.B. Sossinsky [A.B. Sosinskiĭ].

[141] B. Korenblum. An extension of the Nevanlinna theory. *Acta Math.*, 135(3-4):187–219, 1975.

[142] T.W. Körner. The behavior of power series on their circle of convergence. In *Banach spaces, harmonic analysis, and probability theory (Storrs, Conn., 1980/1981)*, volume 995 of *Lecture Notes in Math.*, pages 56–94. Springer, Berlin-New York, 1983.

[143] S. Lang. *Algebra (third edition)*. Addison-Wesley, 1993.

[144] S. Lang. *Complex analysis*, volume 103 of *Graduate Texts in Mathematics*. Springer-Verlag, New York, fourth edition, 1999.

[145] P.D. Lax. On the existence of Green's function. *Proc. Amer. Math. Soc.*, 3:526–531, 1952.

[146] N.N. Lebedev. *Special functions and their applications*. Dover Publications Inc., New York, 1972. Revised edition, translated from the Russian and edited by Richard A. Silverman, unabridged and corrected republication.

[147] I.E. Leonard. More on the Fresnel integrals. *The American Mathematical Monthly*, 95:431–433, 1988.

[148] Ta-Tsien Li. *Problems and solutions in mathematics*. World Scientific, 1998.

[149] M.A. Lifshits. *Gaussian random functions*, volume 322 of *Mathematics and its Applications*. Kluwer Academic Publisher, 1995.

[150] D.H. Luecking and L.A. Rubel. *Complex analysis*. Universitext. Springer-Verlag, New York, 1984. A functional analysis approach.

[151] W.A.J. Luxemburg. Arzelà's dominated convergence theorem for the Riemann integral. *Amer. Math. Monthly*, 78:970–979, 1971.

[152] M.A. Maingueneau. 30 *semaines de khôlles en MATH, première partie*. Ellipses. Edition Marketing, 32 rue Bargue, 75015 Paris, 1994.

[153] A.I. Markushevich. *Theory of analytic functions*. Moscow, 1950.

[154] D. Martin. *Complex numbers*, volume 4 of *Solving problems in mathematics*. Oliver and Boyd, Edinburgh and London, 1967.

[155] M.J. Martín. Composition operators with linear fractional symbols and their adjoints. In *Proceedings of the First Advanced Course in Operator Theory and Complex Analysis*, pages 105–112. Univ. Sevilla Secr. Publ., Seville, 2006.

[156] J.H. Mathews and R.W. Howell. *Complex analysis for mathematics and engineering*. Jones and Bartlett Publishers, Sudbury, Massachusetts, 1997.

[157] J.E. Maxfield and M.W. Maxfield. *Abstract algebra and solution by radicals*. Dover Publications Inc., New York, 1992. Corrected reprint of the 1971 original.

[158] M.L. Mehta. *Random matrices*, volume 142 of *Pure and Applied Mathematics (Amsterdam)*. Elsevier/Academic Press, Amsterdam, third edition, 2004.

[159] E. Meijering. A chronology of interpolation. From ancient astronomy to modern signal and image processing. *Proceedings of the IEEE*, pages 319–342, 2002.

[160] J.W. Milnor. *Topology from the differentiable viewpoint*. Princeton Landmarks in Mathematics. Princeton University Press, Princeton, NJ, 1997. Based on notes by David W. Weaver, revised reprint of the 1965 original.

[161] D.S. Mitrinović, J.E. Pečarić, and A.M. Fink. *Classical and new inequalities in analysis*. Dordrecht: Kluwer Academic Publishers, 1993.

[162] D. Mumford. *Tata lectures on Theta, I*, volume 28 of *Progress in mathematics*. Birkhäuser Verlag, Basel, 1983.

[163] R. Narasimhan. *Compact Riemann surfaces*. Lectures in mathematics, ETH Zürich. Birkhäuser Verlag, Basel, 1992.

[164] R. Narasimhan and Y. Nievergelt. *Complex analysis in one variable*. Birkhäuser Boston Inc., Boston, MA, second edition, 2001.

[165] I.P. Natanson. *Theory of functions of a real variable. Volume I.* Frederick Ungar Publishing Co., New York, 1955. Translated by Leo F. Boron with the collaboration of Edwin Hewitt.

[166] T. Needham. *Visual complex analysis*. The Clarendon Press Oxford University Press, New York, 1997.

[167] Z. Nehari. *Conformal mapping*. McGraw-Hill Book Co., Inc., New York, Toronto, London, 1952.

[168] Z. Nehari. On bounded bilinear forms. *Ann. Math.*, 65:153–162, 1957.

[169] C.P. Niculescu and Fl. Popovici. The monotone convergence theorem for the Riemann integral. *An. Univ. Craiova Ser. Mat. Inform.*, 38(2):55–58, 2011.

[170] N.K. Nikolski. *Operators, functions, and systems: an easy reading. Vol. 1*, volume 92 of *Mathematical Surveys and Monographs*. American Mathematical Society, Providence, RI, 2002. Hardy, Hankel, and Toeplitz, translated from the French by Andreas Hartmann.

[171] N.K. Nikolski. *Operators, functions, and systems: an easy reading. Vol. 2*, volume 93 of *Mathematical Surveys and Monographs*. American Mathematical Society, Providence, RI, 2002. Model operators and systems, translated from the French by Andreas Hartmann and revised by the author.

[172] C.D. Olds. The Fresnel integrals. *The American Mathematical Monthly*, 75:285–286, 1968.

[173] R. Osserman. From Schwarz to Pick to Ahlfors and beyond. *Notices of the AMS*, 46(8):868–873, 1999.

[174] J.C. Oxtoby. *Measure and category*, volume 2 of *Graduate Texts in Mathematics*. Springer-Verlag, New York, second edition, 1987.

[175] J.F. Pabion. *Éléments d'analyse complexe. Licence de Mathématiques*. Ellipses, 1995.

[176] E. Pap. *Complex analysis through examples and exercises*. Kluwer, 1999.

[177] I. Papadimitriou. A simple proof of the formula $\sum_{k=1}^{\infty} k^{-2} = \pi^2/6$. *Amer. Math. Monthly*, 80:424–425, 1973.

[178] J.R. Partington. *Interpolation, identification and sampling*. Oxford University Press, 1997.

[179] D. Pedoe. *Circles*. MAA Spectrum. Mathematical Association of America, Washington, DC, 1995. A mathematical view, revised reprint of the 1979 edition, with a biographical appendix on Karl Feuerbach by Laura Guggenbuhl.

[180] C. Pisot and M. Zamansky. *Mathématiques générales. Tome 5*. Dunod, Paris, 1972.

[181] A.K. Pizer. Research Problems: A Problem on Rational Functions. *Amer. Math. Monthly*, 80(5):552–553, 1973.

[182] G. Polya and G. Szegö. *Problems and Theorems in Analysis I. Series. Integral Calculus. Theory of functions*. Springer, 1998. First published in 1924 as vol. 193 of the *Grundlehren der mathematischen Wissenschaften*.

[183] G. Polya and G. Szegö. *Problems and Theorems in Analysis II. Theory of functions, zeros, polynomials, determinants, number theory, geometry*. Springer, 1998. First published in 1924 as vol. 193 of the *Grundlehren der mathematischen Wissenschaften*.

[184] H.A. Priestley. *Introduction to complex analysis*. Oxford University Press, second (revised) edition, 2003.

[185] E. Ramis. *Exercices d'algèbre avec solutions développées*. Masson, Paris, 1970.

[186] E. Ramis. *Exercices d'analyse avec solutions développées*. Masson, Paris, 1972.

[187] R. Remmert. *Theory of complex functions*, volume 122 of *Graduate Texts in Mathematics*. Springer-Verlag, New York, 1991. Translated from the second German edition by Robert B. Burckel, Readings in Mathematics.

[188] R. Remmert. *Classical topics in complex function theory*, volume 172 of *Graduate Texts in Mathematics*. Springer-Verlag, New York, 1998. Translated from the German by Leslie Kay.

[189] W. Rudin. *Analyse réelle et complexe*. Masson, Paris, 1980.

[190] W. Rudin. *Real and complex analysis*. McGraw-Hill, 1982.

[191] W. Rudin. Elementary Problems E 3325. *Amer. Math. Monthly*, 96(5):445, 1989.

[192] S. Saks and A. Zygmund. *Analytic functions*. Monografie Matematyczne, Tom XXVIII. Polskie Towarzystwo Matematyczne, Warszawa, 1952. Translated by E.J. Scott.

[193] G. Sansone and J. Gerretsen. *Lectures on the theory of functions of a complex variable. I. Holomorphic functions*. P. Noordhoff, Groningen, 1960.

[194] I. Schur. Über die Potenzreihen, die im Innern des Einheitskreises beschränkt sind, I. *Journal für die Reine und Angewandte Mathematik*, 147:205–232, 1917. English translation in: I. Schur methods in operator theory and signal processing. (Operator theory: Advances and Applications OT 18 (1986), Birkhäuser Verlag), Basel.

[195] S.L. Segal. *Nine introductions in complex analysis*, volume 53 of *North-Holland Mathematics Studies*. North-Holland Publishing Co., Amsterdam, 1981. Notas de Matemática [Mathematical Notes], 80.

[196] C.E. Shannon. A mathematical theory of communication. *Bell System Tech. J.*, 27:379–423, 623–656, 1948.

[197] C.E. Shannon. Communication in the presence of noise. *Proc. I.R.E.*, 37:10–21, 1949.

[198] D.O. Shklarsky, N.N. Chentzov, and I.M. Yaglom. *The USSR Olympiad problem book*. Dover Publications, Inc., New York, 1993. Selected problems and theorems of elementary mathematics, translated from the third Russian edition by John Maykovich, revised and edited by Irving Sussman, reprint of the 1962 translation.

[199] C.L. Siegel. *Topics in complex function theory, Volume II*. Wiley Classics Library. Wiley-Interscience, 1988.

[200] R.A. Silverman. *Complex analysis with applications*. Dover Publications Inc., New York, 1984. Reprint of the 1974 original.

[201] A.G. Siskakis. Composition semigroups and the Cesàro operator on H^p. *J. London Math. Soc.* (2), 36(1):153–164, 1987.

[202] V.I. Smirnov. Sur les valeurs limites des fonctions régulières à l'intérieur d'un cercle. *Journal de la Société Physique-Mathématique de Léningrad*, 2:22–37, 1928. English Translation in: Topics in interpolation theory, volume 95 in the series Operator Theory: Advances and Applications, (1997), pp. 481–494.

[203] K.T. Smith. *Power series from a computational point of view*. Springer-Verlag, 1987.

[204] M.R. Speigel and J. Liu. *Mathematical handbook of formulas and tables.* McGraw-Hill, second edition, 1999.

[205] L. Steen and J.A. Seebach. *Counterexamples in topology.* Dover Publications Inc., Mineola, NY, 1995. Reprint of the second (1978) edition.

[206] E.M. Stein and R. Shakarchi. *Complex analysis.* Princeton Lectures in Analysis, II. Princeton University Press, Princeton, NJ, 2003.

[207] P. Tauvel. *Analyse complexe. Exercices corrigés.* Dunod, 1999.

[208] A. Tétrel. *Solutions de questions de Mathématiques Spéciales posées aux examens oraux des Écoles Polytechnique, Normale, Centrale, etc., première partie, algèbre, analyse, mécanique.* Librairie Croville, 20 rue de la Sorbonne, Paris, onzième edition, 1963.

[209] R. Thibault. *Solutions de questions de mathématiques spéciales.* Librairie Croville, 20 rue de la Sorbonne, Paris, 1963.

[210] B.S. Thomson. Monotone convergence theorem for the Riemann integral. *Amer. Math. Monthly,* 117(6):547–550, 2010.

[211] A. Tissier. *Mathématiques générales. Agrégation interne de mathématiques. Exercices avec solutions.* Bréal, 310–320 Bd de la Boissière 93100, Montreuil, 1991.

[212] A. Tortrat. *Calcul des probabilités.* Masson et Cie, éditeurs, Paris, 1963.

[213] J.F. Tourpines, C. Lanos, P.Y. Cantet, A. Passy, and F. Darles. *L'oral de mathématiques aux concours des grandes écoles scientifiques. Exercices corrigés. Concours communs Mines Ponts. 2° édition.* Marketing. 32 rue Bargue. 75015 Paris, 1974.

[214] F. Treves. *Topological vector spaces, distributions and kernels.* Academic Press, 1967.

[215] Yu.Yu. Trokhimchuk. *Continuous mappings and conditions of monogeneity.* Translated from Russian. I. Program for Scientific Translations, Jerusalem, 1964.

[216] H. van Leeuwen and H. Maassen. A q deformation of the Gauss distribution. *J. Math. Phys.,* 36(9):4743–4756, 1995.

[217] J. Vauthier. *Algèbre et analyse. Exercices corrigés. Grand Oral de Polytechnique.* Éditions Eska, 30 rue de Domrémy, 75013, Paris, 1985.

[218] J. Vauthier and J.-J. Prat. *Exercices de mathématiques Oral du C.A.P.E.S avec rappels de cours.* Masson, 1978.

[219] C. Wagschal. *Fonctions holomorphes. Équations différentielles. Exercices corrigés.* Hermann, 293 rue Lecourbe, 75015, Paris, 2003.

[220] E.T. Whittaker. On the functions which are represented by the expansions of interpolation- theory. *Proceedings of the Royal Society of Edinburgh,* 35:181–194, 1915.

[221] J.M. Whittaker. *Interpolatory Function Theory*, volume 33 of *Cambridge Tracts in Mathematics and Mathematical Physics*. Cambridge Univ. Press, Cambridge, U.K., 1935.

[222] K.S. Williams and K. Hardy. *The Red Book of mathematical problems*. Dover Publications, Inc., Mineola, NY, 1996. Corrected reprint of the *The Red Book: 100 practice problems for undergraduate mathematics competitions* [Integer Press, Ottawa, ON, 1988].

[223] M. Wirth. Continuité des racines d'un polynôme comme fonctions des coefficients. *Revue de Mathématiques Spéciales*, pages 100–107, 1987/1988.

[224] A.M. Yaglom and I.M. Yaglom. *Challenging mathematical problems with elementary solutions. Vol. II.* Dover Publications, Inc., New York, 1987. Problems from various branches of mathematics, translated from the Russian by James McCawley, Jr., reprint of the 1967 edition.

[225] Marguerite Yourcenar. *Sous bénéfice d'inventaire*, volume 110 of *Folio Essais*. Gallimard, 1988. Édition Gallimard, 1962, première édition; 1978 pour l'édition définitive.

[226] S. Zaremba. L'équation biharmonique et une classe remarquable de fonctions fondamentales harmoniques. *Bulletin international de l'Académie des Sciences de Cracovie*, pages 147–197, 1907.

[227] A.H. Zemanian. *Realizability theory for continuous linear systems*. Dover Publications, Inc., New York, 1995.

Index

analytic continuation, 48
analytic function, 151, 166
analytic square root, 238, 239, 291, 299
Apollonius circle, 69
arc, 209
arc length line integral, 211
area computation, 231

backward-shift operator, 221, 548
Bernoulli numbers, 344
Bernoulli's lemma, 117
Bessel function, 168
biharmonic functions, 424
Blaschke factor, 19, 71
Blaschke product, 120, 303
Bohr's inequality, 225
Borel sets, 566
Borel's theorem, 233
bounds
    for the function $1 - \cos z$, 121
    for the function $\ln(1 + z)$, 526
    for Weierstrass factors, 120
Brouwer's theorem, 156, 351

Cantor set, 532, 543, 572
Cartan's theorem, 303
cartesian form, 79
Catalan constant, 308
Catalan numbers, 96
Cauchy multiplication theorem, 515
Cauchy product, 169, 515
Cauchy–Goursat theorem, 217
Cauchy–Riemann equations, 16, 158
    in polar coordinates, 421
Cauchy–Schwarz inequality, 95, 547
Cesàro operator, 345

characteristic polynomial, 475
closed contour, 211
closed Jordan curve, 211
completing the square, 17, 63, 86
complex number
    absolute value, 13
    conjugate, 12
    formula for the argument, 16
    modulus, 13
    polar representation, 13
    purely imaginary, 12
    roots of order n, 14, 15
concatenation, 210
confinement lemma, 30
conformal, 159
conjugate harmonic, 419
continuous arc, 209
continuous logarithm, 156, 178
continuous square root, 155, 156, 177
contour (closed), 211
convolution, 5, 169, 462, 515
    and power series, 105
    example, 197
    operator, 461
cross-ratio, 72
curve, 209

decimation operator, 300
derivative
    formula, 158
    logarithmic derivative, 157
Dirichlet integral, 218, 256
Dirichlet problem, 424
downsampling operator, 300

9 783319 421797